SI Units Used in Mechanics

Quantity	Unit	SI Symbol
(Base Units)		
Length	meter*	m
Mass	kilogram	kg
Time	second	s
(Derived Units)		
Acceleration, linear	meter/second^2	m/s^2
Acceleration, angular	radian/second^2	rad/s^2
Area	meter^2	m^2
Density	kilogram/meter^3	kg/m^3
Force	newton	N $(= \text{kg} \cdot \text{m/s}^2)$
Frequency	hertz	Hz $(= 1/\text{s})$
Impulse, linear	newton-second	N $\cdot$ s
Impulse, angular	newton-meter-second	N $\cdot$ m $\cdot$ s
Moment of force	newton-meter	N $\cdot$ m
Moment of inertia, area	meter^4	m^4
Moment of inertia, mass	kilogram-meter^2	kg $\cdot$ m^2
Momentum, linear	kilogram-meter/second	kg $\cdot$ m/s $(= \text{N} \cdot \text{s})$
Momentum, angular	kilogram-meter^2/second	kg $\cdot$ m^2/s $(= \text{N} \cdot \text{m} \cdot \text{s})$
Power	watt	W $(= \text{J/s} = \text{N} \cdot \text{m/s})$
Pressure, stress	pascal	Pa $(= \text{N/m}^2)$
Product of inertia, area	meter^4	m^4
Product of inertia, mass	kilogram-meter^2	kg $\cdot$ m^2
Spring constant	newton/meter	N/m
Velocity, linear	meter/second	m/s
Velocity, angular	radian/second	rad/s
Volume	meter^3	m^3
Work, energy	joule	J $(= \text{N} \cdot \text{m})$
(Supplementary and Other Acceptable Units)		
Distance (navigation)	nautical mile	$(= 1{,}852 \text{ km})$
Mass	ton (metric)	t $(= 1000 \text{ kg})$
Plane angle	degrees (decimal)	°
Plane angle	radian	—
Speed	knot	(1.852 km/h)
Time	day	d
Time	hour	h
Time	minute	min

*Also spelled *metre*.

SI Unit Prefixes

Multiplication Factor	Prefix	Symbol
$1\ 000\ 000\ 000\ 000 = 10^{12}$	tera	T
$1\ 000\ 000\ 000 = 10^{9}$	giga	G
$1\ 000\ 000 = 10^{6}$	mega	M
$1\ 000 = 10^{3}$	kilo	k
$100 = 10^{2}$	hecto	h
$10 = 10$	deka	da
$0.1 = 10^{-1}$	deci	d
$0.01 = 10^{-2}$	centi	c
$0.001 = 10^{-3}$	milli	m
$0.000\ 001 = 10^{-6}$	micro	μ
$0.000\ 000\ 001 = 10^{-9}$	nano	n
$0.000\ 000\ 000\ 001 = 10^{-12}$	pico	p

Selected Rules for Writing Metric Quantities

1. (a) Use prefixes to keep numerical values generally between 0.1 and 1000.
   (b) Use of the prefixes hecto, deka, deci, and centi should generally be avoided except for certain areas or volumes where the numbers would be awkward otherwise.
   (c) Use prefixes only in the numerator of unit combinations. The one exception is the base unit kilogram. (*Example:* write kN/m not N/mm; J/kg not mJ/g)
   (d) Avoid double prefixes. (*Example:* write GN not kMN)
2. Unit designations
   (a) Use a dot for multiplication of units. (*Example:* write N $\cdot$ m not Nm)
   (b) Avoid ambiguous double solidus. (*Example:* write N/m^2 not N/m/m)
   (c) Exponents refer to entire unit. (*Example:* mm^2 means (mm)^2)
3. Number grouping
   Use a space rather than a comma to separate numbers in groups of three, counting from the decimal point in both directions. *Example:* 4 607 321.048 72)
   Space may be omitted for numbers of four digits. (*Example:* 4296 or 0.0476)

FOREWORD

This series of textbooks was begun in 1951 by the late Dr. James L. Meriam. At that time, the books represented a revolutionary transformation in undergraduate mechanics education. They became the definitive textbooks for the decades that followed as well as models for other engineering mechanics texts that have subsequently appeared. Published under slightly different titles prior to the 1978 First Editions, this textbook series has always been characterized by logical organization, clear and rigorous presentation of the theory, instructive sample problems, and a rich collection of real-life problems, all with a high standard of illustration. In addition to the U.S. versions, the books have appeared in SI versions and have been translated into many foreign languages. These texts collectively represent an international standard for undergraduate texts in mechanics.

The innovations and contributions of Dr. Meriam (1917–2000) to the field of engineering mechanics cannot be overstated. He was one of the premier engineering educators of the second half of the twentieth century. Dr. Meriam earned his B.E., M. Eng., and Ph.D. degrees from Yale University. He had early industrial experience with Pratt and Whitney Aircraft and the General Electric Company. During the Second World War he served in the U.S. Coast Guard. He was a member of the faculty of the University of California–Berkeley, Dean of Engineering at Duke University, a faculty member at the California Polytechnic State University–San Luis Obispo, and visiting professor at the University of California–Santa Barbara, finally retiring in 1990. Professor Meriam always placed great emphasis on teaching, and this trait was recognized by his students wherever he taught. At Berkeley in 1963, he was the first recipient of the Outstanding Faculty Award of Tau Beta Pi, given primarily for excellence in teaching. In 1978, he received the Distinguished Educator Award for Outstanding Service to Engineering Mechanics Education from the American Society for Engineering Education, and in 1992 was the Society's recipient of the Benjamin Garver Lamme Award, which is ASEE's highest annual national award.

Dr. L. Glenn Kraige, coauthor of the *Engineering Mechanics* series since the early 1980s, has also made significant contributions to mechanics education. Dr. Kraige earned his B.S., M.S., and Ph.D. degrees at the University of Virginia, principally in aerospace engineering, and he currently serves as the W. S. "Pete" White Professor of Engineering Science and Mechanics at Virginia Polytechnic Institute and State University. During the mid 1970s, I had the singular pleasure of chairing Professor Kraige's graduate committee and take particular pride in the fact that he was the first of my three dozen Ph.D. graduates. Professor Kraige was invited by Professor Meriam to team with him and thereby ensure that the Meriam legacy of textbook authorship excellence was carried forward to future generations. For the past two and a half decades, this highly successful team of authors has made an enormous and global impact on the education of several generations of engineers.

In addition to his widely recognized research and publications in the field of spacecraft dynamics, Professor Kraige has devoted his attention to the teaching of mechanics at both introductory and advanced levels. His outstanding teaching has been widely recognized and has earned him teaching awards at the departmental, college, university, state, regional, and national levels. These include the Francis J. Maher Award for excellence in education in the Department of Engineering Science and Mechanics, the Wine Award for excellence in university teaching, and the Outstanding Educator Award from the State Council of Higher Education for the Commonwealth of Virginia. In 1996, the Mechanics Division of ASEE bestowed upon him the Archie Higdon Distinguished Educator Award. The Carnegie Foundation for the Advancement of Teaching and the Council for Advancement and Support of Education awarded him the distinction of Virginia Professor of the Year for 1997. In his teaching, Professor Kraige stresses the development of analytical capabilities along with the strengthening of physical insight and engineering judgment. Since the early 1980s, he has worked on personal-computer software designed to enhance the teaching/learning process in statics, dynamics, strength of materials, and higher-level areas of dynamics and vibrations.

The Sixth Edition of *Engineering Mechanics* continues the same high standards set by previous editions and adds new features of help and interest to students. It contains a vast collection of interesting and instructive problems. The faculty and students privileged to teach or study from Professors Meriam and Kraige's *Engineering Mechanics* will benefit from the several decades of investment by two highly accomplished educators. Following the pattern of the previous editions, this textbook stresses the application of theory to actual engineering situations, and at this important task it remains the best.

John L. Junkins
Distinguished Professor of Aerospace Engineering
Holder of the George J. Eppright Chair Professorship in Engineering
Texas A&M University
College Station, Texas

PREFACE

Engineering mechanics is both a foundation and a framework for most of the branches of engineering. Many of the topics in such areas as civil, mechanical, aerospace, and agricultural engineering, and of course engineering mechanics itself, are based upon the subjects of statics and dynamics. Even in a discipline such as electrical engineering, practitioners, in the course of considering the electrical components of a robotic device or a manufacturing process, may find themselves first having to deal with the mechanics involved.

Thus, the engineering mechanics sequence is critical to the engineering curriculum. Not only is this sequence needed in itself, but courses in engineering mechanics also serve to solidify the student's understanding of other important subjects, including applied mathematics, physics, and graphics. In addition, these courses serve as excellent settings in which to strengthen problem-solving abilities.

PHILOSOPHY

The primary purpose of the study of engineering mechanics is to develop the capacity to predict the effects of force and motion while carrying out the creative design functions of engineering. This capacity requires more than a mere knowledge of the physical and mathematical principles of mechanics; also required is the ability to visualize physical configurations in terms of real materials, actual constraints, and the practical limitations which govern the behavior of machines and structures. One of the primary objectives in a mechanics course is to help the student develop this ability to visualize, which is so vital to problem formulation. Indeed, the construction of a meaningful mathematical model is often a more important experience than its solution. Maximum progress is made when the principles and their limitations are learned together within the context of engineering application.

There is a frequent tendency in the presentation of mechanics to use problems mainly as a vehicle to illustrate theory rather than to develop theory for the purpose of solving problems. When the first view is allowed to predominate, problems tend to become overly idealized and unrelated to engineering with the result that the exercise becomes dull, academic, and uninteresting. This approach deprives the student of valuable experience in formulating problems and thus of discovering the need for and meaning of theory. The second view provides by far the stronger motive for learning theory and leads to a better balance between theory and application. The crucial role played by interest and purpose in providing the strongest possible motive for learning cannot be overemphasized.

Furthermore, as mechanics educators, we should stress the understanding that, at best, theory can only approximate the real world of mechanics rather than the view that the real world approximates the theory. This difference in philosophy is indeed basic and distinguishes the *engineering* of mechanics from the *science* of mechanics.

Over the past several decades, several unfortunate tendencies have occurred in engineering education. First, emphasis on the geometric and physical meanings of prerequisite mathematics appears to have diminished. Second, there has been a significant reduction and even elimination of instruction in graphics, which in the past enhanced the visualization and representation of mechanics problems. Third, in advancing the mathematical level of our treatment of mechanics, there has been a tendency to allow the notational manipulation of vector operations to mask or replace geometric visualization. Mechanics is inherently a subject which depends on geometric and physical perception, and we should increase our efforts to develop this ability.

A special note on the use of computers is in order. The experience of formulating problems, where reason and judgment are developed, is vastly more important for the student than is the manipulative exercise in carrying out the solution. For this reason, computer usage must be carefully controlled. At present, constructing free-body diagrams and formulating governing equations are best done with pencil and paper. On the other hand, there are instances in which the *solution* to the governing equations can best be carried out and displayed using the computer. Computer-oriented problems should be genuine in the sense that there is a condition of design or criticality to be found, rather than "makework" problems in which some parameter is varied for no apparent reason other than to force artificial use of the computer. These thoughts have been kept in mind during the design of the computer-oriented problems in the Sixth Edition. To conserve adequate time for problem formulation, it is suggested that the student be assigned only a limited number of the computer-oriented problems.

As with previous editions, this Sixth Edition of *Engineering Mechanics* is written with the foregoing philosophy in mind. It is intended primarily for the first engineering course in mechanics, generally taught in the second year of study. *Engineering Mechanics* is written in a style which is both concise and friendly. The major emphasis is on basic principles and methods rather than on a multitude of special cases. Strong effort has been made to show both the cohesiveness of the relatively few fundamental ideas and the great variety of problems which these few ideas will solve.

PEDAGOGICAL FEATURES

The basic structure of this textbook consists of an article which rigorously treats the particular subject matter at hand, followed by one or more Sample Problems, followed by a group of Problems. There is a Chapter Review at the end of each chapter which summarizes the main points in that chapter, followed by a Review Problem set.

Problems

The 86 Sample Problems appear on specially colored pages by themselves. The solutions to typical statics problems are presented in detail. In addition, explanatory and cautionary notes (Helpful Hints) in blue type are number-keyed to the main presentation.

There are 1020 homework exercises, of which approximately 50 percent are new to the Sixth Edition. The problem sets are divided into *Introductory Problems* and *Representative Problems*. The first section consists of simple, uncomplicated problems designed to help students gain confidence with the new topic, while most of the problems in the second section are of average difficulty and length. The problems are generally arranged in order of increasing difficulty. More difficult exercises appear near the end of the *Representative Problems* and are marked with the symbol ▶. *Computer-Oriented Problems*, marked with an asterisk, appear in a special section at the conclusion of the *Review Problems* at the end of each chapter. The answers to all odd-numbered problems and to all difficult problems have been provided.

In recognition of the need for emphasis on SI units, there are approximately two problems in SI units for every one in U.S. customary units. This apportionment between the two sets of units permits anywhere from a 50–50 emphasis to a 100-percent SI treatment.

A notable feature of the Sixth Edition, as with all previous editions, is the wealth of interesting and important problems which apply to engineering design. Whether directly identified as such or not, virtually all of the problems deal with principles and procedures inherent in the design and analysis of engineering structures and mechanical systems.

Illustrations

In order to bring the greatest possible degree of realism and clarity to the illustrations, this textbook series continues to be produced in full color. It is important to note that color is used consistently for the identification of certain quantities:

- *red* for forces and moments,
- *green* for velocity and acceleration arrows,
- *orange dashes* for selected trajectories of moving points.

Subdued colors are used for those parts of an illustration which are not central to the problem at hand. Whenever possible, mechanisms or objects which commonly have a certain color will be portrayed in that color. All of the fundamental elements of technical illustration which have been an essential part of this *Engineering Mechanics* series of textbooks have been retained. The author wishes to restate the conviction that a high standard of illustration is critical to any written work in the field of mechanics.

Features New to this Edition

While retaining the hallmark features of all previous editions, we have incorporated these improvements:

- All theory portions have been reexamined in order to maximize rigor, clarity, readability, and level of friendliness.
- Key Concepts areas within the theory presentation have been specially marked and highlighted.
- The Chapter Reviews are highlighted and feature itemized summaries.
- Approximately 50 percent of the homework problems are new to this Sixth Edition. All new problems have been independently solved in order to ensure a high degree of accuracy.
- New Sample Problems have been added, including ones with computer-oriented solutions.

- All Sample Problems are printed on specially colored pages for quick identification.
- Within-the-chapter photographs have been added in order to provide additional connection to actual situations in which statics has played a major role.

ORGANIZATION

In Chapter 1, the fundamental concepts necessary for the study of mechanics are established.

In Chapter 2, the properties of forces, moments, couples, and resultants are developed so that the student may proceed directly to the equilibrium of nonconcurrent force systems in Chapter 3 without unnecessarily belaboring the relatively trivial problem of the equilibrium of concurrent forces acting on a particle.

In both Chapters 2 and 3, analysis of two-dimensional problems is presented in Section A before three-dimensional problems are treated in Section B. With this arrangement, the instructor may cover all of Chapter 2 before beginning Chapter 3 on equilibrium, or the instructor may cover the two chapters in the order 2A, 3A, 2B, 3B. The latter order treats force systems and equilibrium in two dimensions and then treats these topics in three dimensions.

Application of equilibrium principles to simple trusses and to frames and machines is presented in Chapter 4 with primary attention given to two-dimensional systems. A sufficient number of three-dimensional examples are included, however, to enable students to exercise more general vector tools of analysis.

The concepts and categories of distributed forces are introduced at the beginning of Chapter 5, with the balance of the chapter divided into two main sections. Section A treats centroids and mass centers; detailed examples are presented to help students master early applications of calculus to physical and geometrical problems. Section B includes the special topics of beams, flexible cables, and fluid forces, which may be omitted without loss of continuity of basic concepts.

Chapter 6 on friction is divided into Section A on the phenomenon of dry friction and Section B on selected machine applications. Although Section B may be omitted if time is limited, this material does provide a valuable experience for the student in dealing with both concentrated and distributed friction forces.

Chapter 7 presents a consolidated introduction to virtual work with applications limited to single-degree-of-freedom systems. Special emphasis is placed on the advantage of the virtual-work and energy method for interconnected systems and stability determination. Virtual work provides an excellent opportunity to convince the student of the power of mathematical analysis in mechanics.

Moments and products of inertia of areas are presented in Appendix A. This topic helps to bridge the subjects of statics and solid mechanics. Appendix C contains a summary review of selected topics of elementary mathematics as well as several numerical techniques which the student should be prepared to use in computer-solved problems. Useful tables of physical constants, centroids, and moments of inertia are contained in Appendix D.

SUPPLEMENTS

The following items have been prepared to complement this textbook:

Instructor's Manual

Prepared by the authors and independently checked fully worked solutions to all problems in the text are available to faculty by contacting their local Wiley representative.

Instructor Lecture Resources

(Available on the text website at www.wiley.com/college/meriam):

WileyPlus: A complete online learning system to help prepare and present lectures, assign and manage homework, keep track of student progress, and customize your course content and delivery. See the description in front of the book for more information, and the website for a demonstration. Talk to your Wiley representative for details on setting up your Wiley-Plus course.

Lecture software specifically designed to aid the lecturer, especially in larger classrooms. Written by the author and incorporating figures from the textbooks, this software is based on the Macromedia Flash platform. Major use of animation, concise review of the theory, and numerous sample problems make this tool extremely useful for student self-review of the material.

Web-based simulations of representative applications in statics that allow for "what-if" analysis. Developed by Richard Stanley at Kettering University, these simulations allow an instructor to explore a problem from the text by changing variables and seeing the new results develop both visually and numerically. Available from the book website and as part of the WileyPlus package.

All *figures* in the text are available in electronic format for use in creating lecture presentations.

All *Sample Problems* are available as electronic files for display and discussion in the classroom.

Transparencies for over 40 solved problems, similar to those in the text, available in .pdf format for use in lecture or for self-study by students.

Extension sample problems build on sample problems from the text and show how computational tools can be used to investigate a variety of "what if" scenarios. Available to both faculty and students, these were developed by Brian Harper at Ohio State University.

Solving Mechanics Problems with . . . .

A series of booklets introduces the use of computational software in the solution of mechanics problems. Developed by Brian Harper at Ohio State University, the booklets are available for Matlab, MathCAD, and Maple. Please contact your local Wiley representative for more information, or visit the book website at www.wiley.com/college/meriam.

ACKNOWLEDGMENTS

Special recognition is due Dr. A. L. Hale, formerly of Bell Telephone Laboratories, for his continuing contribution in the form of invaluable suggestions and accurate checking of the manuscript. Dr. Hale has rendered similar service for all previous versions of this entire series of mechanics books, dating back to the 1950s. He reviews all aspects of the books, including all old and new text and figures. Dr. Hale carries out an independent solution to each new homework exercise and provides the author with suggestions and needed corrections to the solutions which appear in the *Instructor's Manual*. Dr. Hale is well known for being extremely accurate in his work, and his fine knowledge of the English language is a great asset which aids every use of this textbook. In addition to his normal contributions, Dr. Hale has been the prime proofreader for the Sixth Edition.

I would like to thank the faculty members of the Department of Engineering Science and Mechanics at VPI&SU who regularly offer constructive suggestions. These include Scott L. Hendricks, Saad A. Ragab, Norman E. Dowling, Michael W. Hyer, and J. Wallace Grant. In addition, my assistant of thirty years, Vanessa McCoy, is recognized for her long-term contribution to this textbook project.

The following individuals (listed in alphabetical order) provided feedback on the Fifth Edition, reviewed samples of the Sixth Edition, or otherwise contributed to the Sixth Edition:

Michael Ales, *U.S. Merchant Marine Academy*
Joseph Arumala, *University of Maryland Eastern Shore*
Eric Austin, *Clemson University*
Stephen Bechtel, *Ohio State University*
Peter Birkemoe, *University of Toronto*
Achala Chatterjee, *San Bernardino Valley College*
Yi-chao Chen, *University of Houston*
Mary Cooper, *Cal Poly San Luis Obispo*
Mukaddes Darwish, *Texas Tech University*
Kurt DeGoede, *Elizabethtown College*
John DesJardins, *Clemson University*
Larry DeVries, *University of Utah*
Craig Downing, *Southeast Missouri State University*
William Drake, *Missouri State University*
Raghu Echempati, *Kettering University*
Amelito Enriquez, *Canada College*
Sven Esche, *Stevens Institute of Technology*
Wallace Franklin, *U.S. Merchant Marine Academy*
Barry Goodno, *Georgia Institute of Technology*
Robert Harder, *George Fox University*
Javier Hasbun, *University of West Georgia*
Javad Hashemi, *Texas Tech University*
Scott Hendricks, *Virginia Tech*
Robert Hyers, *University of Massachusetts, Amherst*
Matthew Ikle, *Adams State College*
Duane Jardine, *University of New Orleans*
Qing Jiang, *University of California, Riverside*
Jennifer Kadlowec, *Rowan University*
Robert Kern, *Milwaukee School of Engineering*
John Krohn, *Arkansas Tech University*
Keith Lindler, *United States Naval Academy*
Francisco Manzo-Robledo, *Washington State University*
Geraldine Milano, *New Jersey Institute of Technology*
Saeed Niku, *Cal Poly San Luis Obispo*
Wilfrid Nixon, *University of Iowa*
Karim Nohra, *University of South Florida*
Vassilis Panoskaltsis, *Case Western Reserve University*
Chandra Putcha, *California State University, Fullerton*
Blayne Roeder, *Purdue University*
Eileen Rossman, *Cal Poly San Luis Obispo*
Nestor Sanchez, *University of Texas, San Antonio*
Scott Schiff, *Clemson University*
Sergey Smirnov, *Texas Tech University*

Sally Steadman, *University of South Alabama*
Ertugrul Taciroglu, *UCLA*
Constantine Tarawneh, *University of Texas*
John Turner, *University of Wyoming*
Mohammed Zikry, *North Carolina State University*

The contributions by the staff of John Wiley & Sons, Inc., including Editor Joe Hayton, Senior Production Editor Lisa Wojcik, Senior Designer Kevin Murphy, Senior Illustration Editor Sigmund Malinowski, and Senior Photograph Editor Lisa Gee, reflect a high degree of professional competence and are duly recognized. I wish to especially acknowledge the critical production efforts of Christine Cervoni of Camelot Editorial Services, LLC. The talented illustrators of Precision Graphics continue to maintain a high standard of illustration excellence.

Finally, I wish to state the extremely significant contribution of my family. In addition to providing patience and support for this project, my wife Dale has managed the preparation of the manuscript for the Sixth Edition and has been a key individual in checking all stages of the proof. In addition, my son David has contributed problem ideas, illustrations, and solutions to a number of the problems.

I am extremely pleased to participate in extending the time duration of this textbook series well past the fifty-year mark. In the interest of providing you with the best possible educational materials over future years, I encourage and welcome all comments and suggestions. Please address your comments to kraige@vt.edu.

L. Glenn Kraige

Blacksburg, Virginia

CONTENTS

CHAPTER **6**

FRICTION   339

Structures which support large forces must be designed with the principles of mechanics foremost in mind. In this view of Pittsburgh, one can see a variety of such structures.

1

INTRODUCTION TO STATICS

CHAPTER OUTLINE

1/1   MECHANICS

Mechanics is the physical science which deals with the effects of forces on objects. No other subject plays a greater role in engineering analysis than mechanics. Although the principles of mechanics are few, they have wide application in engineering. The principles of mechanics are central to research and development in the fields of vibrations, stability and strength of structures and machines, robotics, rocket and spacecraft design, automatic control, engine performance, fluid flow, electrical machines and apparatus, and molecular, atomic, and subatomic behavior. A thorough understanding of this subject is an essential prerequisite for work in these and many other fields.

Mechanics is the oldest of the physical sciences. The early history of this subject is synonymous with the very beginnings of engineering. The earliest recorded writings in mechanics are those of Archimedes (287–212 **B.C.**) on the principle of the lever and the principle of buoyancy. Substantial progress came later with the formulation of the laws of vector combination of forces by Stevinus (1548–1620), who also formulated most of the principles of statics. The first investigation of a dynamics problem is credited to Galileo (1564–1642) for his experiments with falling stones. The accurate formulation of the laws of motion, as well as the law of gravitation, was made by Newton (1642–1727), who also conceived the idea of the infinitesimal in mathematical analysis. Substantial contributions to the development of mechanics were also made by da Vinci, Varignon, Euler, D'Alembert, Lagrange, Laplace, and others.

In this book we will be concerned with both the development of the principles of mechanics and their application. The principles of mechanics as a science are rigorously expressed by mathematics, and thus

Sir Isaac Newton

© S. Terry/Photo Researchers, Inc.

mathematics plays an important role in the application of these principles to the solution of practical problems.

The subject of mechanics is logically divided into two parts: *statics*, which concerns the equilibrium of bodies under action of forces, and *dynamics*, which concerns the motion of bodies. *Engineering Mechanics* is divided into these two parts, *Vol. 1 Statics* and *Vol. 2 Dynamics*.

1/2   BASIC CONCEPTS

The following concepts and definitions are basic to the study of mechanics, and they should be understood at the outset.

Space is the geometric region occupied by bodies whose positions are described by linear and angular measurements relative to a coordinate system. For three-dimensional problems, three independent coordinates are needed. For two-dimensional problems, only two coordinates are required.

Time is the measure of the succession of events and is a basic quantity in dynamics. Time is not directly involved in the analysis of statics problems.

Mass is a measure of the inertia of a body, which is its resistance to a change of velocity. Mass can also be thought of as the quantity of matter in a body. The mass of a body affects the gravitational attraction force between it and other bodies. This force appears in many applications in statics.

Force is the action of one body on another. A force tends to move a body in the direction of its action. The action of a force is characterized by its *magnitude*, by the *direction* of its action, and by its *point of application*. Thus force is a vector quantity, and its properties are discussed in detail in Chapter 2.

A *particle* is a body of negligible dimensions. In the mathematical sense, a particle is a body whose dimensions are considered to be near zero so that we may analyze it as a mass concentrated at a point. We often choose a particle as a differential element of a body. We may treat a body as a particle when its dimensions are irrelevant to the description of its position or the action of forces applied to it.

Rigid body. A body is considered rigid when the change in distance between any two of its points is negligible for the purpose at hand. For instance, the calculation of the tension in the cable which supports the boom of a mobile crane under load is essentially unaffected by the small internal deformations in the structural members of the boom. For the purpose, then, of determining the external forces which act on the boom, we may treat it as a rigid body. Statics deals primarily with the calculation of external forces which act on rigid bodies in equilibrium. Determination of the internal deformations belongs to the study of the mechanics of deformable bodies, which normally follows statics in the curriculum.

1/3   SCALARS AND VECTORS

We use two kinds of quantities in mechanics—scalars and vectors. *Scalar quantities* are those with which only a magnitude is associated. Examples of scalar quantities are time, volume, density, speed, energy, and mass. *Vector quantities*, on the other hand, possess direction as well as magnitude, and must obey the parallelogram law of addition as described later in this article. Examples of vector quantities are displacement, velocity, acceleration, force, moment, and momentum. Speed is a scalar. It is the magnitude of velocity, which is a vector. Thus velocity is specified by a direction as well as a speed.

Vectors representing physical quantities can be classified as free, sliding, or fixed.

A *free vector* is one whose action is not confined to or associated with a unique line in space. For example, if a body moves without rotation, then the movement or displacement of any point in the body may be taken as a vector. This vector describes equally well the direction and magnitude of the displacement of every point in the body. Thus, we may represent the displacement of such a body by a free vector.

A *sliding vector* has a unique line of action in space but not a unique point of application. For example, when an external force acts on a rigid body, the force can be applied at any point along its line of action without changing its effect on the body as a whole,* and thus it is a sliding vector.

A *fixed vector* is one for which a unique point of application is specified. The action of a force on a deformable or nonrigid body must be specified by a fixed vector at the point of application of the force. In this instance the forces and deformations within the body depend on the point of application of the force, as well as on its magnitude and line of action.

Conventions for Equations and Diagrams

A vector quantity $\mathbf{V}$ is represented by a line segment, Fig. 1/1, having the direction of the vector and having an arrowhead to indicate the sense. The length of the directed line segment represents to some convenient scale the magnitude $|\mathbf{V}|$ of the vector, which is printed with lightface italic type V. For example, we may choose a scale such that an arrow one inch long represents a force of twenty pounds.

In scalar equations, and frequently on diagrams where only the magnitude of a vector is labeled, the symbol will appear in lightface italic type. Boldface type is used for vector quantities whenever the directional aspect of the vector is a part of its mathematical representation. When writing vector equations, *always* be certain to preserve the mathematical distinction between vectors and scalars. In handwritten work, use a distinguishing mark for each vector quantity, such as an underline, $\underline{V}$, or an arrow over the symbol, $\vec{V}$, to take the place of boldface type in print.

Working with Vectors

The direction of the vector $\mathbf{V}$ may be measured by an angle θ from some known reference direction as shown in Fig. 1/1. The negative of $\mathbf{V}$ is a vector $-\mathbf{V}$ having the same magnitude as $\mathbf{V}$ but directed in the sense opposite to $\mathbf{V}$, as shown in Fig. 1/1.

Figure 1/1

---

*This is the *principle of transmissibility*, which is discussed in Art. 2/2.

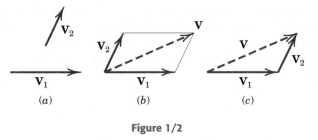

Figure 1/2

Vectors must obey the parallelogram law of combination. This law states that two vectors $\mathbf{V}_1$ and $\mathbf{V}_2$, treated as free vectors, Fig. 1/2a, may be replaced by their equivalent vector $\mathbf{V}$, which is the diagonal of the parallelogram formed by $\mathbf{V}_1$ and $\mathbf{V}_2$ as its two sides, as shown in Fig. 1/2b. This combination is called the *vector sum*, and is represented by the vector equation

$$\mathbf{V} = \mathbf{V}_1 + \mathbf{V}_2$$

where the plus sign, when used with the vector quantities (in boldface type), means *vector* and not *scalar* addition. The scalar sum of the magnitudes of the two vectors is written in the usual way as $V_1 + V_2$. The geometry of the parallelogram shows that $V \neq V_1 + V_2$.

The two vectors $\mathbf{V}_1$ and $\mathbf{V}_2$, again treated as free vectors, may also be added head-to-tail by the triangle law, as shown in Fig. 1/2c, to obtain the identical vector sum $\mathbf{V}$. We see from the diagram that the order of addition of the vectors does not affect their sum, so that $\mathbf{V}_1 + \mathbf{V}_2 = \mathbf{V}_2 + \mathbf{V}_1$.

The difference $\mathbf{V}_1 - \mathbf{V}_2$ between the two vectors is easily obtained by adding $-\mathbf{V}_2$ to $\mathbf{V}_1$ as shown in Fig. 1/3, where either the triangle or parallelogram procedure may be used. The difference $\mathbf{V}'$ between the two vectors is expressed by the vector equation

$$\mathbf{V}' = \mathbf{V}_1 - \mathbf{V}_2$$

where the minus sign denotes *vector subtraction*.

Any two or more vectors whose sum equals a certain vector $\mathbf{V}$ are said to be the *components* of that vector. Thus, the vectors $\mathbf{V}_1$ and $\mathbf{V}_2$ in Fig. 1/4a are the components of $\mathbf{V}$ in the directions 1 and 2, respectively. It is usually most convenient to deal with vector components which are mutually perpendicular; these are called *rectangular components*. The

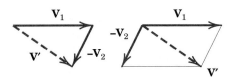

Figure 1/3

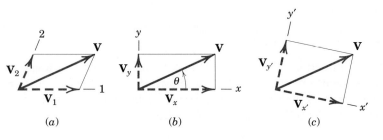

Figure 1/4

vectors $\mathbf{V}_x$ and $\mathbf{V}_y$ in Fig. 1/4*b* are the *x*- and *y*-components, respectively, of $\mathbf{V}$. Likewise, in Fig. 1/4*c*, $\mathbf{V}_{x'}$ and $\mathbf{V}_{y'}$ are the *x'*- and *y'*-components of $\mathbf{V}$. When expressed in rectangular components, the direction of the vector with respect to, say, the *x*-axis is clearly specified by the angle θ, where

$$\theta = \tan^{-1}\frac{V_y}{V_x}$$

A vector $\mathbf{V}$ may be expressed mathematically by multiplying its magnitude V by a vector $\mathbf{n}$ whose magnitude is one and whose direction coincides with that of $\mathbf{V}$. The vector $\mathbf{n}$ is called a *unit vector*. Thus,

$$\mathbf{V} = V\mathbf{n}$$

In this way both the magnitude and direction of the vector are conveniently contained in one mathematical expression. In many problems, particularly three-dimensional ones, it is convenient to express the rectangular components of $\mathbf{V}$, Fig. 1/5, in terms of unit vectors $\mathbf{i}$, $\mathbf{j}$, and $\mathbf{k}$, which are vectors in the *x*-, *y*-, and *z*-directions, respectively, with unit magnitudes. Because the vector $\mathbf{V}$ is the vector sum of the components in the *x*-, *y*-, and *z*-directions, we can express $\mathbf{V}$ as follows:

$$\boxed{\mathbf{V} = V_x\mathbf{i} + V_y\mathbf{j} + V_z\mathbf{k}}$$

We now make use of the *direction cosines l, m*, and *n* of $\mathbf{V}$, which are defined by

$$l = \cos\theta_x \qquad m = \cos\theta_y \qquad n = \cos\theta_z$$

Thus, we may write the magnitudes of the components of $\mathbf{V}$ as

$$\boxed{V_x = lV \qquad V_y = mV \qquad V_z = nV}$$

where, from the Pythagorean theorem,

$$\boxed{V^2 = V_x{}^2 + V_y{}^2 + V_z{}^2}$$

Note that this relation implies that $l^2 + m^2 + n^2 = 1$.

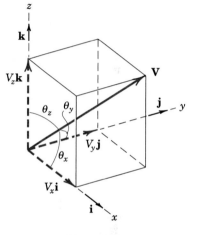

Figure 1/5

1/4   NEWTON'S LAWS

Sir Isaac Newton was the first to state correctly the basic laws governing the motion of a particle and to demonstrate their validity.* Slightly reworded with modern terminology, these laws are:

Law 1. A particle remains at rest or continues to move with *uniform velocity* (in a straight line with a constant speed) if there is no unbalanced force acting on it.

---

*Newton's original formulations may be found in the translation of his *Principia* (1687) revised by F. Cajori, University of California Press, 1934.

Law II. The acceleration of a particle is proportional to the vector sum of forces acting on it, and is in the direction of this vector sum.

Law III. The forces of action and reaction between interacting bodies are equal in magnitude, opposite in direction, and *collinear* (they lie on the same line).

The correctness of these laws has been verified by innumerable accurate physical measurements. Newton's second law forms the basis for most of the analysis in dynamics. As applied to a particle of mass m, it may be stated as

$$\boxed{\mathbf{F} = m\mathbf{a}} \tag{1/1}$$

where $\mathbf{F}$ is the vector sum of forces acting on the particle and $\mathbf{a}$ is the resulting acceleration. This equation is a *vector* equation because the direction of $\mathbf{F}$ must agree with the direction of $\mathbf{a}$, and the magnitudes of $\mathbf{F}$ and $m\mathbf{a}$ must be equal.

Newton's first law contains the principle of the equilibrium of forces, which is the main topic of concern in statics. This law is actually a consequence of the second law, since there is no acceleration when the force is zero, and the particle either is at rest or is moving with a uniform velocity. The first law adds nothing new to the description of motion but is included here because it was part of Newton's classical statements.

The third law is basic to our understanding of force. It states that forces always occur in pairs of equal and opposite forces. Thus, the downward force exerted on the desk by the pencil is accompanied by an upward force of equal magnitude exerted on the pencil by the desk. This principle holds for all forces, variable or constant, regardless of their source, and holds at every instant of time during which the forces are applied. Lack of careful attention to this basic law is the cause of frequent error by the beginner.

In the analysis of bodies under the action of forces, it is absolutely necessary to be clear about which force of each action-reaction pair is being considered. It is necessary first of all to *isolate* the body under consideration and then to consider only the one force of the pair which acts *on* the body in question.

1/5 UNITS

In mechanics we use four fundamental quantities called *dimensions*. These are length, mass, force, and time. The units used to measure these quantities cannot all be chosen independently because they must be consistent with Newton's second law, Eq. 1/1. Although there are a number of different systems of units, only the two systems most commonly used in science and technology will be used in this text. The four fundamental dimensions and their units and symbols in the two systems are summarized in the following table.

QUANTITY	DIMENSIONAL SYMBOL	SI UNITS		U.S. CUSTOMARY UNITS	
		UNIT	SYMBOL	UNIT	SYMBOL
Mass	M	Base units { kilogram	kg	slug	—
Length	L	meter	m	Base units { foot	ft
Time	T	second	s	second	sec
Force	F	newton	N	pound	lb

SI Units

The International System of Units, abbreviated SI (from the French, Système International d'Unités), is accepted in the United States and throughout the world, and is a modern version of the metric system. By international agreement, SI units will in time replace other systems. As shown in the table, in SI, the units kilogram (kg) for mass, meter (m) for length, and second (s) for time are selected as the base units, and the newton (N) for force is derived from the preceding three by Eq. 1/1. Thus, force (N) = mass (kg) $\times$ acceleration (m/s^2) or

$$N = kg \cdot m/s^2$$

Thus, 1 newton is the force required to give a mass of 1 kg an acceleration of 1 m/s^2.

Consider a body of mass m which is allowed to fall freely near the surface of the earth. With only the force of gravitation acting on the body, it falls with an acceleration g toward the center of the earth. This gravitational force is the *weight W* of the body, and is found from Eq. 1/1:

$$W\,(N) = m\,(kg) \times g\,(m/s^2)$$

U.S. Customary Units

The U.S. customary, or British system of units, also called the foot-pound-second (FPS) system, has been the common system in business and industry in English-speaking countries. Although this system will in time be replaced by SI units, for many more years engineers must be able to work with both SI units and FPS units, and both systems are used freely in *Engineering Mechanics*.

As shown in the table, in the U.S. or FPS system, the units of feet (ft) for length, seconds (sec) for time, and pounds (lb) for force are selected as base units, and the slug for mass is derived from Eq. 1/1. Thus, force (lb) = mass (slugs) $\times$ acceleration (ft/sec^2), or

$$slug = \frac{lb \cdot sec^2}{ft}$$

Therefore, 1 slug is the mass which is given an acceleration of 1 ft/sec^2 when acted on by a force of 1 lb. If W is the gravitational force or weight and g is the acceleration due to gravity, Eq. 1/1 gives

$$m\,(slugs) = \frac{W\,(lb)}{g\,(ft/sec^2)}$$

Note that seconds is abbreviated as *s* in SI units, and as *sec* in FPS units.

In U.S. units the pound is also used on occasion as a unit of mass, especially to specify thermal properties of liquids and gases. When distinction between the two units is necessary, the force unit is frequently written as lbf and the mass unit as lbm. In this book we use almost exclusively the force unit, which is written simply as lb. Other common units of force in the U.S. system are the *kilopound* (kip), which equals 1000 lb, and the *ton*, which equals 2000 lb.

The International System of Units (SI) is termed an *absolute* system because the measurement of the base quantity mass is independent of its environment. On the other hand, the U.S. system (FPS) is termed a *gravitational* system because its base quantity force is defined as the gravitational attraction (weight) acting on a standard mass under specified conditions (sea level and 45° latitude). A standard pound is also the force required to give a one-pound mass an acceleration of 32.1740 ft/sec^2.

In SI units the kilogram is used *exclusively* as a unit of mass—*never* force. In the MKS (meter, kilogram, second) gravitational system, which has been used for many years in non-English-speaking countries, the kilogram, like the pound, has been used both as a unit of force and as a unit of mass.

Primary Standards

Primary standards for the measurements of mass, length, and time have been established by international agreement and are as follows:

Mass. The kilogram is defined as the mass of a specific platinum–iridium cylinder which is kept at the International Bureau of Weights and Measures near Paris, France. An accurate copy of this cylinder is kept in the United States at the National Institute of Standards and Technology (NIST), formerly the National Bureau of Standards, and serves as the standard of mass for the United States.

Length. The meter, originally defined as one ten-millionth of the distance from the pole to the equator along the meridian through Paris, was later defined as the length of a specific platinum–iridium bar kept at the International Bureau of Weights and Measures. The difficulty of accessing the bar and reproducing accurate measurements prompted the adoption of a more accurate and reproducible standard of length for the meter, which is now defined as 1 650 763.73 wavelengths of a specific radiation of the krypton-86 atom.

Time. The second was originally defined as the fraction 1/(86 400) of the mean solar day. However, irregularities in the earth's rotation led to difficulties with this definition, and a more accurate and reproducible standard has been adopted. The second is now defined as the duration of 9 192 631 770 periods of the radiation of a specific state of the cesium-133 atom.

For most engineering work, and for our purpose in studying mechanics, the accuracy of these standards is considerably beyond our

The standard kilogram

needs. The standard value for gravitational acceleration g is its value at sea level and at a 45° latitude. In the two systems these values are

SI units $\quad\quad g = 9.806\ 65\ \text{m/s}^2$

U.S. units $\quad\quad g = 32.1740\ \text{ft/sec}^2$

The approximate values of 9.81 m/s^2 and 32.2 ft/sec^2, respectively, are sufficiently accurate for the vast majority of engineering calculations.

Unit Conversions

The characteristics of SI units are shown inside the front cover of this book, along with the numerical conversions between U.S. customary and SI units. In addition, charts giving the approximate conversions between selected quantities in the two systems appear inside the back cover for convenient reference. Although these charts are useful for obtaining a feel for the relative size of SI and U.S. units, in time engineers will find it essential to think directly in terms of SI units without converting from U.S. units. In statics we are primarily concerned with the units of length and force, with mass needed only when we compute gravitational force, as explained previously.

Figure 1/6 depicts examples of force, mass, and length in the two systems of units, to aid in visualizing their relative magnitudes.

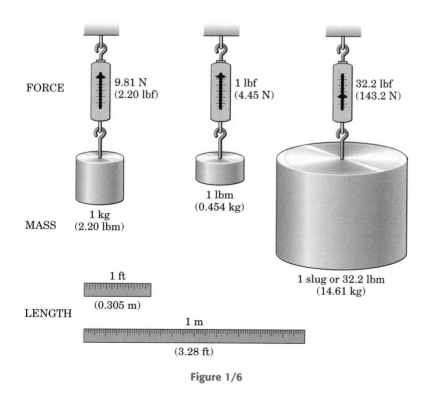

Figure 1/6

1/6   LAW OF GRAVITATION

In statics as well as dynamics we often need to compute the weight of a body, which is the gravitational force acting on it. This computation depends on the *law of gravitation*, which was also formulated by Newton. The law of gravitation is expressed by the equation

$$F = G\frac{m_1 m_2}{r^2} \tag{1/2}$$

where F = the mutual force of attraction between two particles

G = a universal constant known as the *constant of gravitation*

m_1, m_2 = the masses of the two particles

r = the distance between the centers of the particles

The mutual forces F obey the law of action and reaction, since they are equal and opposite and are directed along the line joining the centers of the particles, as shown in Fig. 1/7. By experiment the gravitational constant is found to be $G = 6.673(10^{-11})$ m³/(kg·s²).

Gravitational Attraction of the Earth

Gravitational forces exist between every pair of bodies. On the surface of the earth the only gravitational force of appreciable magnitude is the force due to the attraction of the earth. For example, each of two iron spheres 100 mm in diameter is attracted to the earth with a gravitational force of 37.1 N, which is its weight. On the other hand, the force of mutual attraction between the spheres if they are just touching is 0.000 000 095 1 N. This force is clearly negligible compared with the earth's attraction of 37.1 N. Consequently the gravitational attraction of the earth is the only gravitational force we need to consider for most engineering applications on the earth's surface.

The gravitational attraction of the earth on a body (its weight) exists whether the body is at rest or in motion. Because this attraction is a force, the weight of a body should be expressed in newtons (N) in SI units and in pounds (lb) in U.S. customary units. Unfortunately in common practice the mass unit kilogram (kg) has been frequently used as a measure of weight. This usage should disappear in time as SI units become more widely used, because in SI units the kilogram is used exclusively for mass and the newton is used for force, including weight.

For a body of mass m near the surface of the earth, the gravitational attraction F on the body is specified by Eq. 1/2. We usually denote the

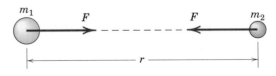

Figure 1/7

magnitude of this gravitational force or weight with the symbol W. Because the body falls with an acceleration g, Eq. 1/1 gives

$$W = mg \qquad \text{(1/3)}$$

The weight W will be in newtons (N) when the mass m is in kilograms (kg) and the acceleration of gravity g is in meters per second squared (m/s^2). In U.S. customary units, the weight W will be in pounds (lb) when m is in slugs and g is in feet per second squared. The standard values for g of 9.81 m/s^2 and 32.2 ft/sec^2 will be sufficiently accurate for our calculations in statics.

The true weight (gravitational attraction) and the apparent weight (as measured by a spring scale) are slightly different. The difference, which is due to the rotation of the earth, is quite small and will be neglected. This effect will be discussed in *Vol. 2 Dynamics*.

1/7   ACCURACY, LIMITS, AND APPROXIMATIONS

The number of significant figures in an answer should be no greater than the number of figures justified by the accuracy of the given data. For example, suppose the 24-mm side of a square bar was measured to the nearest millimeter, so we know the side length to two significant figures. Squaring the side length gives an area of 576 mm^2. However, according to our rule, we should write the area as 580 mm^2, using only two significant figures.

When calculations involve small differences in large quantities, greater accuracy in the data is required to achieve a given accuracy in the results. Thus, for example, it is necessary to know the numbers 4.2503 and 4.2391 to an accuracy of five significant figures to express their difference 0.0112 to three-figure accuracy. It is often difficult in lengthy computations to know at the outset how many significant figures are needed in the original data to ensure a certain accuracy in the answer. Accuracy to three significant figures is considered satisfactory for most engineering calculations.

In this text, answers will generally be shown to three significant figures unless the answer begins with the digit 1, in which case the answer will be shown to four significant figures. For purposes of calculation, consider all data given in this book to be exact.

Differentials

The *order* of differential quantities frequently causes misunderstanding in the derivation of equations. Higher-order differentials may always be neglected compared with lower-order differentials when the mathematical limit is approached. For example, the element of volume ΔV of a right circular cone of altitude h and base radius r may be taken to be a circular slice a distance x from the vertex and of thickness Δx. The expression for the volume of the element is

$$\Delta V = \frac{\pi r^2}{h^2} \left[x^2 \, \Delta x + x(\Delta x)^2 + \tfrac{1}{3}(\Delta x)^3 \right]$$

Note that, when passing to the limit in going from ΔV to dV and from Δx to dx, the terms containing $(\Delta x)^2$ and $(\Delta x)^3$ drop out, leaving merely

$$dV = \frac{\pi r^2}{h^2} x^2 \, dx$$

which gives an exact expression when integrated.

Small-Angle Approximations

When dealing with small angles, we can usually make use of simplifying approximations. Consider the right triangle of Fig. 1/8 where the angle θ, expressed in radians, is relatively small. If the hypotenuse is unity, we see from the geometry of the figure that the arc length $1 \times \theta$ and $\sin \theta$ are very nearly the same. Also $\cos \theta$ is close to unity. Furthermore, $\sin \theta$ and $\tan \theta$ have almost the same values. Thus, for small angles we may write

$$\sin \theta \cong \tan \theta \cong \theta \qquad \cos \theta \cong 1$$

provided that the angles are expressed in radians. These approximations may be obtained by retaining only the first terms in the series expansions for these three functions. As an example of these approximations, for an angle of 1°

$$1° = 0.017\,453 \text{ rad} \qquad \tan 1° = 0.017\,455$$

$$\sin 1° = 0.017\,452 \qquad \cos 1° = 0.999\,848$$

If a more accurate approximation is desired, the first two terms may be retained, and they are

$$\sin \theta \cong \theta - \theta^3/6 \qquad \tan \theta \cong \theta + \theta^3/3 \qquad \cos \theta \cong 1 - \theta^2/2$$

where the angles must be expressed in radians. (To convert degrees to radians, multiply the angle in degrees by $\pi/180°$.) The error in replacing the sine by the angle for 1° (0.0175 rad) is only 0.005 percent. For 5° (0.0873 rad) the error is 0.13 percent, and for 10° (0.1745 rad), the error is still only 0.51 percent. As the angle θ approaches zero, the following relations are true in the mathematical limit:

$$\sin d\theta = \tan d\theta = d\theta \qquad \cos d\theta = 1$$

where the differential angle $d\theta$ must be expressed in radians.

1/8   PROBLEM SOLVING IN STATICS

We study statics to obtain a quantitative description of forces which act on engineering structures in equilibrium. Mathematics establishes the relations between the various quantities involved and enables us to predict effects from these relations. We use a dual thought process in

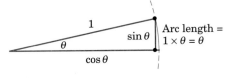

Figure 1/8

solving statics problems: We think about both the physical situation and the corresponding mathematical description. In the analysis of every problem, we make a transition between the physical and the mathematical. One of the most important goals for the student is to develop the ability to make this transition freely.

Making Appropriate Assumptions

We should recognize that the mathematical formulation of a physical problem represents an ideal description, or *model*, which approximates but never quite matches the actual physical situation. When we construct an idealized mathematical model for a given engineering problem, certain approximations will always be involved. Some of these approximations may be mathematical, whereas others will be physical.

For instance, it is often necessary to neglect small distances, angles, or forces compared with large distances, angles, or forces. Suppose a force is distributed over a small area of the body on which it acts. We may consider it to be a concentrated force if the dimensions of the area involved are small compared with other pertinent dimensions.

We may neglect the weight of a steel cable if the tension in the cable is many times greater than its total weight. However, if we must calculate the deflection or sag of a suspended cable under the action of its weight, we may not ignore the cable weight.

Thus, what we may assume depends on what information is desired and on the accuracy required. We must be constantly alert to the various assumptions called for in the formulation of real problems. The ability to understand and make use of the appropriate assumptions in the formulation and solution of engineering problems is certainly one of the most important characteristics of a successful engineer. One of the major aims of this book is to provide many opportunities to develop this ability through the formulation and analysis of many practical problems involving the principles of statics.

Using Graphics

Graphics is an important analytical tool for three reasons:

1. We use graphics to represent a physical system on paper with a sketch or diagram. Representing a problem geometrically helps us with its physical interpretation, especially when we must visualize three-dimensional problems.

2. We can often obtain a graphical solution to problems more easily than with a direct mathematical solution. Graphical solutions are both a practical way to obtain results, and an aid in our thought processes. Because graphics represents the physical situation and its mathematical expression simultaneously, graphics helps us make the transition between the two.

3. Charts or graphs are valuable aids for representing results in a form which is easy to understand.

Formulating Problems and Obtaining Solutions

In statics, as in all engineering problems, we need to use a precise and logical method for formulating problems and obtaining their solutions. We formulate each problem and develop its solution through the following sequence of steps.

1. Formulate the problem:
   (*a*) State the given data.
   (*b*) State the desired result.
   (*c*) State your assumptions and approximations.
2. Develop the solution:
   (*a*) Draw any diagrams you need to understand the relationships.
   (*b*) State the governing principles to be applied to your solution.
   (*c*) Make your calculations.
   (*d*) Ensure that your calculations are consistent with the accuracy justified by the data.
   (*e*) Be sure that you have used consistent units throughout your calculations.
   (*f*) Ensure that your answers are reasonable in terms of magnitudes, directions, common sense, etc.
   (*g*) Draw conclusions.

Keeping your work neat and orderly will help your thought process and enable others to understand your work. The discipline of doing orderly work will help you develop skill in formulation and analysis. Problems which seem complicated at first often become clear when you approach them with logic and discipline.

The Free-Body Diagram

The subject of statics is based on surprisingly few fundamental concepts and involves mainly the application of these basic relations to a variety of situations. In this application the *method* of analysis is all important. In solving a problem, it is essential that the laws which apply be carefully fixed in mind and that we apply these principles literally and exactly. In applying the principles of mechanics to analyze forces acting on a body, it is essential that we *isolate* the body in question from all other bodies so that a complete and accurate account of all forces acting on this body can be taken. This *isolation* should exist mentally and should be represented on paper. The diagram of such an isolated body with the representation of *all* external forces acting *on* it is called a *free-body diagram*.

The free-body-diagram method is the key to the understanding of mechanics. This is so because the *isolation* of a body is the tool by which *cause* and *effect* are clearly separated, and by which our attention is clearly focused on the literal application of a principle of mechanics. The technique of drawing free-body diagrams is covered in Chapter 3, where they are first used.

Numerical Values versus Symbols

In applying the laws of statics, we may use numerical values to represent quantities, or we may use algebraic symbols, and leave the answer as a formula. When numerical values are used, the magnitude of each quantity expressed in its particular units is evident at each stage of the calculation. This is useful when we need to know the magnitude of each term.

The symbolic solution, however, has several advantages over the numerical solution. First, the use of symbols helps to focus our attention on the connection between the physical situation and its related mathematical description. Second, we can use a symbolic solution repeatedly for obtaining answers to the same type of problem, but having different units or numerical values. Third, a symbolic solution enables us to make a dimensional check at every step, which is more difficult to do when numerical values are used. In any equation representing a physical situation, the dimensions of every term on both sides of the equation must be the same. This property is called *dimensional homogeneity*.

Thus, facility with both numerical and symbolic forms of solution is essential.

Solution Methods

Solutions to the problems of statics may be obtained in one or more of the following ways.

1. Obtain mathematical solutions by hand, using either algebraic symbols or numerical values. We can solve most problems this way.

2. Obtain graphical solutions for certain problems.

3. Solve problems by computer. This is useful when a large number of equations must be solved, when a parameter variation must be studied, or when an intractable equation must be solved.

Many problems can be solved with two or more of these methods. The method utilized depends partly on the engineer's preference and partly on the type of problem to be solved. The choice of the most expedient method of solution is an important aspect of the experience to be gained from the problem work. There are a number of problems in *Vol. 1 Statics* which are designated as *Computer-Oriented Problems*. These problems appear at the end of the Review Problem sets and are selected to illustrate the type of problem for which solution by computer offers a distinct advantage.

1/9   CHAPTER REVIEW

This chapter has introduced the concepts, definitions, and units used in statics, and has given an overview of the procedure used to formulate and solve problems in statics. Now that you have finished this chapter, you should be able to do the following:

1. Express vectors in terms of unit vectors and perpendicular components, and perform vector addition and subtraction.
2. State Newton's laws of motion.
3. Perform calculations using SI and U.S. units, using appropriate accuracy.
4. Express the law of gravitation and calculate the weight of an object.
5. Apply simplifications based on differential and small-angle approximations.
6. Describe the methodology used to formulate and solve statics problems.

Sample Problem 1/1

Determine the weight in newtons of a car whose mass is 1400 kg. Convert the mass of the car to slugs and then determine its weight in pounds.

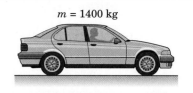

$m = 1400$ kg

Solution. From relationship 1/3, we have

① $$W = mg = 1400(9.81) = 13\ 730\ \text{N} \qquad\qquad Ans.$$

From the table of conversion factors inside the front cover of the textbook, we see that 1 slug is equal to 14.594 kg. Thus, the mass of the car in slugs is

② $$m = 1400\ \text{kg}\left[\frac{1\ \text{slug}}{14.594\ \text{kg}}\right] = 95.9\ \text{slugs} \qquad\qquad Ans.$$

Finally, its weight in pounds is

③ $$W = mg = (95.9)(32.2) = 3090\ \text{lb} \qquad\qquad Ans.$$

As another route to the last result, we can convert from kg to lbm. Again using the table inside the front cover, we have

$$m = 1400\ \text{kg}\left[\frac{1\ \text{lbm}}{0.45359\ \text{kg}}\right] = 3090\ \text{lbm}$$

The weight in pounds associated with the mass of 3090 lbm is 3090 lb, as calculated above. We recall that 1 lbm is the amount of mass which under standard conditions has a weight of 1 lb of force. We rarely refer to the U.S. mass unit lbm in this textbook series, but rather use the slug for mass. The sole use of slug, rather than the unnecessary use of two units for mass, will prove to be powerful and simple—especially in dynamics.

Helpful Hints

① Our calculator indicates a result of 13 734 N. Using the rules of significant-figure display used in this textbook, we round the written result to four significant figures, or 13 730 N. Had the number begun with any digit other than 1, we would have rounded to three significant figures.

② A good practice with unit conversion is to multiply by a factor such as $\left[\dfrac{1\ \text{slug}}{14.594\ \text{kg}}\right]$, which has a value of 1, because the numerator and the denominator are equivalent. Make sure that cancellation of the units leaves the units desired; here the units of kg cancel, leaving the desired units of slug.

③ Note that we are using a previously calculated result (95.9 slugs). We must be sure that when a calculated number is needed in subsequent calculations, it is retained in the calculator to its full accuracy, (95.929834 · · ·) until it is needed. This may require storing it in a register upon its initial calculation and recalling it later. We must not merely punch 95.9 into our calculator and proceed to multiply by 32.2—this practice will result in loss of numerical accuracy. Some individuals like to place a small indication of the storage register used in the right margin of the work paper, directly beside the number stored.

Sample Problem 1/2

Use Newton's law of universal gravitation to calculate the weight of a 70-kg person standing on the surface of the earth. Then repeat the calculation by using $W = mg$ and compare your two results. Use Table D/2 as needed.

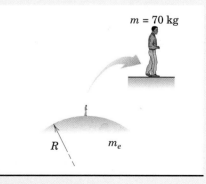

$m = 70$ kg

R m_e

Solution. The two results are

① $$W = \frac{Gm_e m}{R^2} = \frac{(6.673\cdot 10^{-11})(5.976\cdot 10^{24})(70)}{[6371\cdot 10^3]^2} = 688\ \text{N} \qquad\qquad Ans.$$

$$W = mg = 70(9.81) = 687\ \text{N} \qquad\qquad Ans.$$

The discrepancy is due to the fact that Newton's universal gravitational law does not take into account the rotation of the earth. On the other hand, the value $g = 9.81$ m/s² used in the second equation does account for the earth's rotation. Note that had we used the more accurate value $g = 9.80665$ m/s² (which likewise accounts for the earth's rotation) in the second equation, the discrepancy would have been larger (686 N would have been the result).

Helpful Hint

① The effective distance between the mass centers of the two bodies involved is the radius of the earth.

Sample Problem 1/3

For the vectors $\mathbf{V}_1$ and $\mathbf{V}_2$ shown in the figure,

(a) determine the magnitude S of their vector sum $\mathbf{S} = \mathbf{V}_1 + \mathbf{V}_2$

(b) determine the angle α between $\mathbf{S}$ and the positive x-axis

(c) write $\mathbf{S}$ as a vector in terms of the unit vectors $\mathbf{i}$ and $\mathbf{j}$ and then write a unit vector $\mathbf{n}$ along the vector sum $\mathbf{S}$

(d) determine the vector difference $\mathbf{D} = \mathbf{V}_1 - \mathbf{V}_2$

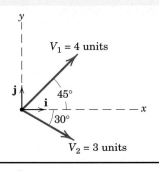

Solution   (a) We construct to scale the parallelogram shown in Fig. a for adding $\mathbf{V}_1$ and $\mathbf{V}_2$. Using the law of cosines, we have

$$S^2 = 3^2 + 4^2 - 2(3)(4) \cos 105°$$

$$S = 5.59 \text{ units} \qquad Ans.$$

① (b) Using the law of sines for the lower triangle, we have

$$\frac{\sin 105°}{5.59} = \frac{\sin(\alpha + 30°)}{4}$$

$$\sin(\alpha + 30°) = 0.692$$

$$(\alpha + 30°) = 43.8° \qquad \alpha = 13.76° \qquad Ans.$$

(c) With knowledge of both S and α, we can write the vector $\mathbf{S}$ as

$$\mathbf{S} = S[\mathbf{i} \cos \alpha + \mathbf{j} \sin \alpha]$$

$$= S[\mathbf{i} \cos 13.76° + \mathbf{j} \sin 13.76°] = 5.43\mathbf{i} + 1.328\mathbf{j} \text{ units} \qquad Ans.$$

② Then $\mathbf{n} = \dfrac{\mathbf{S}}{S} = \dfrac{5.43\mathbf{i} + 1.328\mathbf{j}}{5.59} = 0.971\mathbf{i} + 0.238\mathbf{j} \qquad Ans.$

(d) The vector difference $\mathbf{D}$ is

$$\mathbf{D} = \mathbf{V}_1 - \mathbf{V}_2 = 4(\mathbf{i} \cos 45° + \mathbf{j} \sin 45°) - 3(\mathbf{i} \cos 30° - \mathbf{j} \sin 30°)$$

$$= 0.230\mathbf{i} + 4.33\mathbf{j} \text{ units} \qquad Ans.$$

The vector $\mathbf{D}$ is shown in Fig. b as $\mathbf{D} = \mathbf{V}_1 + (-\mathbf{V}_2)$.

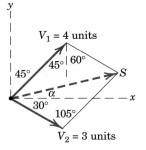

(a)

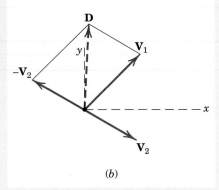

(b)

Helpful Hints

① You will frequently use the laws of cosines and sines in mechanics. See Art. C/6 of Appendix C for a review of these important geometric principles.

② A unit vector may always be formed by dividing a vector by its magnitude. Note that a unit vector is dimensionless.

PROBLEMS

1/1 Determine the angles made by the vector $\mathbf{V} = -36\mathbf{i} + 15\mathbf{j}$ with the positive x- and y-axes. Write the unit vector $\mathbf{n}$ in the direction of $\mathbf{V}$.

Ans. $\theta_x = 157.4°$, $\theta_y = 67.4°$
$\mathbf{n} = -0.923\mathbf{i} + 0.385\mathbf{j}$

1/2 Determine the magnitude of the vector sum $\mathbf{V} = \mathbf{V}_1 + \mathbf{V}_2$ and the angle θ_x which $\mathbf{V}$ makes with the positive x-axis. Complete both graphical and algebraic solutions.

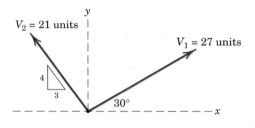

Problem 1/2

1/3 For the given vectors $\mathbf{V}_1$ and $\mathbf{V}_2$ of Prob. 1/2, determine the magnitude of the vector difference $\mathbf{V}' = \mathbf{V}_2 - \mathbf{V}_1$ and the angle θ_x which $\mathbf{V}'$ makes with the positive x-axis. Complete both graphical and algebraic solutions.

Ans. $V' = 36.1$ units, $\theta_x = 174.8°$

1/4 A force is specified by the vector $\mathbf{F} = 160\mathbf{i} + 80\mathbf{j} - 120\mathbf{k}$ N. Calculate the angles made by $\mathbf{F}$ with the positive x-, y-, and z-axes.

1/5 What is the mass in both slugs and kilograms of a 1000-lb beam?

Ans. $m = 31.1$ slugs, $m = 453$ kg

1/6 From the gravitational law calculate the weight W (gravitational force with respect to the earth) of an 80-kg man in a spacecraft traveling in a circular orbit 250 km above the earth's surface. Express W in both newtons and pounds.

1/7 Determine the weight in newtons of a woman whose weight in pounds is 125. Also, find her mass in slugs and in kilograms. Determine your own weight in newtons.

Ans. $W = 556$ N, $m = 3.88$ slugs, $m = 56.7$ kg

1/8 Suppose that two nondimensional quantities are exactly $A = 8.67$ and $B = 1.429$. Using the rules for significant figures as stated in this chapter, express the four quantities $(A + B)$, $(A - B)$, (AB), and (A/B).

1/9 Compute the magnitude F of the force which the sun exerts on the earth. Perform the calculation first in pounds and then convert your result to newtons. Refer to Table D/2 for necessary physical quantities.

Ans. $F = 7.97(10^{21})$ lb, $F = 3.55(10^{22})$ N

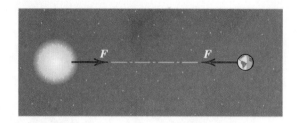

Problem 1/9

1/10 The element of volume ΔV of the right circular cone of altitude h and base radius r is formed by slicing the cone at a distance x from the vertex. If the slice is of finite thickness Δx, show that its volume ΔV is $[\pi r^2/h^2]$ $[x^2\,\Delta x + x(\Delta x)^2 + \frac{1}{3}(\Delta x)^3]$. (Recall the formula for the volume of a cone.) Explain what happens to the second and third terms when Δx becomes the infinitesimal dx.

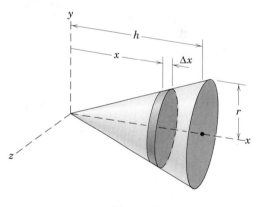

Problem 1/10

1/11 What is the percent error n in replacing the sine of 20° by the value of the angle in radians? Repeat for the tangent of 20°, and explain the qualitative difference in the two error percentages.

Ans. $n = 2.06\%$, $n = 4.09\%$

The properties of force systems must be thoroughly understood by the engineers who design devices such as this crane and its lifting harness. Try to visualize the forces present in the various parts of this system.

FORCE SYSTEMS

2/1 INTRODUCTION

In this and the following chapters, we study the effects of forces which act on engineering structures and mechanisms. The experience gained here will help you in the study of mechanics and in other subjects such as stress analysis, design of structures and machines, and fluid flow. This chapter lays the foundation for a basic understanding not only of statics but also of the entire subject of mechanics, and you should master this material thoroughly.

2/2 FORCE

Before dealing with a group or *system* of forces, it is necessary to examine the properties of a single force in some detail. A force has been defined in Chapter 1 as an action of one body on another. In dynamics we will see that a force is defined as an action which tends to cause acceleration of a body. A force is a *vector quantity*, because its effect depends on the direction as well as on the magnitude of the action. Thus, forces may be combined according to the parallelogram law of vector addition.

The action of the cable tension on the bracket in Fig. 2/1a is represented in the side view, Fig. 2/1b, by the force vector $\mathbf{P}$ of magnitude P. The effect of this action on the bracket depends on P, the angle θ, and the location of the point of application A. Changing any one of these three specifications will alter the effect on the bracket, such as the force

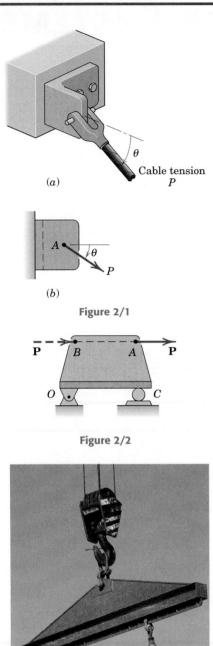

(a)

(b)

Figure 2/1

Figure 2/2

The forces associated with this lifting rig must be carefully identified, classified, and analyzed in order to provide a safe and effective working environment.

in one of the bolts which secure the bracket to the base, or the internal force and deformation in the material of the bracket at any point. Thus, the complete specification of the action of a force must include its *magnitude*, *direction*, and *point of application*, and therefore we must treat it as a fixed vector.

External and Internal Effects

We can separate the action of a force on a body into two effects, *external* and *internal*. For the bracket of Fig. 2/1 the effects of **P** external to the bracket are the reactive forces (not shown) exerted on the bracket by the foundation and bolts because of the action of **P**. Forces external to a body can be either *applied* forces or *reactive* forces. The effects of **P** internal to the bracket are the resulting internal forces and deformations distributed throughout the material of the bracket. The relation between internal forces and internal deformations depends on the material properties of the body and is studied in strength of materials, elasticity, and plasticity.

Principle of Transmissibility

When dealing with the mechanics of a rigid body, we ignore deformations in the body and concern ourselves with only the net external effects of external forces. In such cases, experience shows us that it is not necessary to restrict the action of an applied force to a given point. For example, the force **P** acting on the rigid plate in Fig. 2/2 may be applied at *A* or at *B* or at any other point on its line of action, and the net external effects of **P** on the bracket will not change. The external effects are the force exerted on the plate by the bearing support at *O* and the force exerted on the plate by the roller support at *C*.

This conclusion is summarized by the *principle of transmissibility*, which states that a force may be applied at any point on its given line of action without altering the resultant effects of the force *external* to the *rigid* body on which it acts. Thus, whenever we are interested in only the resultant external effects of a force, the force may be treated as a *sliding* vector, and we need specify only the *magnitude*, *direction*, and *line of action* of the force, and not its *point of application*. Because this book deals essentially with the mechanics of rigid bodies, we will treat almost all forces as sliding vectors for the rigid body on which they act.

Force Classification

Forces are classified as either *contact* or *body* forces. A contact force is produced by direct physical contact; an example is the force exerted on a body by a supporting surface. On the other hand, a body force is generated by virtue of the position of a body within a force field such as a gravitational, electric, or magnetic field. An example of a body force is your weight.

Forces may be further classified as either *concentrated* or *distributed*. Every contact force is actually applied over a finite area and is therefore really a distributed force. However, when the dimensions of the area are very small compared with the other dimensions of the body, we may consider the force to be concentrated at a point with negligible loss of accuracy. Force can be distributed over an *area*, as in the case of mechanical contact, over a *volume* when a body force such as weight is acting, or over a *line*, as in the case of the weight of a suspended cable.

The *weight* of a body is the force of gravitational attraction distributed over its volume and may be taken as a concentrated force acting through the center of gravity. The position of the center of gravity is frequently obvious if the body is symmetric. If the position is not obvious, then a separate calculation, explained in Chapter 5, will be necessary to locate the center of gravity.

We can measure a force either by comparison with other known forces, using a mechanical balance, or by the calibrated movement of an elastic element. All such comparisons or calibrations have as their basis a primary standard. The standard unit of force in SI units is the newton (N) and in the U.S. customary system is the pound (lb), as defined in Art. 1/5.

Action and Reaction

According to Newton's third law, the *action* of a force is always accompanied by an *equal* and *opposite reaction*. It is essential to distinguish between the action and the reaction in a pair of forces. To do so, we first *isolate* the body in question and then identify the force exerted *on* that body (not the force exerted *by* the body). It is very easy to mistakenly use the wrong force of the pair unless we distinguish carefully between action and reaction.

Concurrent Forces

Two or more forces are said to be *concurrent at a point* if their lines of action intersect at that point. The forces $\mathbf{F}_1$ and $\mathbf{F}_2$ shown in Fig. 2/3a have a common point of application and are concurrent at the point A. Thus, they can be added using the parallelogram law in their common plane to obtain their sum or *resultant* $\mathbf{R}$, as shown in Fig. 2/3a. The resultant lies in the same plane as $\mathbf{F}_1$ and $\mathbf{F}_2$.

Suppose the two concurrent forces lie in the same plane but are applied at two different points as in Fig. 2/3b. By the principle of transmissibility, we may move them along their lines of action and complete their vector sum $\mathbf{R}$ at the point of concurrency A, as shown in Fig. 2/3b. We can replace $\mathbf{F}_1$ and $\mathbf{F}_2$ with the resultant $\mathbf{R}$ without altering the external effects on the body upon which they act.

We can also use the triangle law to obtain $\mathbf{R}$, but we need to move the line of action of one of the forces, as shown in Fig. 2/3c. If we add the same two forces as shown in Fig. 2/3d, we correctly preserve the magnitude and direction of $\mathbf{R}$, but we lose the correct line of action, because $\mathbf{R}$ obtained in this way does not pass through A. Therefore this type of combination should be avoided.

We can express the sum of the two forces mathematically by the vector equation

$$\mathbf{R} = \mathbf{F}_1 + \mathbf{F}_2$$

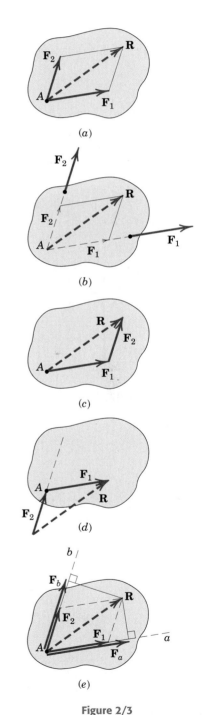

(a)

(b)

(c)

(d)

(e)

Figure 2/3

Vector Components

In addition to combining forces to obtain their resultant, we often need to replace a force by its *vector components* in directions which are convenient for a given application. The vector sum of the components must equal the original vector. Thus, the force **R** in Fig. 2/3*a* may be replaced by, or *resolved* into, two vector components **F**$_1$ and **F**$_2$ with the specified directions by completing the parallelogram as shown to obtain the magnitudes of **F**$_1$ and **F**$_2$.

The relationship between a force and its vector components along given axes must not be confused with the relationship between a force and its perpendicular* projections onto the same axes. Figure 2/3*e* shows the perpendicular projections **F**$_a$ and **F**$_b$ of the given force **R** onto axes *a* and *b*, which are parallel to the vector components **F**$_1$ and **F**$_2$ of Fig. 2/3*a*. Figure 2/3*e* shows that the components of a vector are not necessarily equal to the projections of the vector onto the same axes. Furthermore, the vector sum of the projections **F**$_a$ and **F**$_b$ is not the vector **R**, because the parallelogram law of vector addition must be used to form the sum. The components and projections of **R** are equal only when the axes *a* and *b* are perpendicular.

A Special Case of Vector Addition

To obtain the resultant when the two forces **F**$_1$ and **F**$_2$ are parallel as in Fig. 2/4, we use a special case of addition. The two vectors are combined by first adding two equal, opposite, and collinear forces **F** and −**F** of convenient magnitude, which taken together produce no external effect on the body. Adding **F**$_1$ and **F** to produce **R**$_1$, and combining with the sum **R**$_2$ of **F**$_2$ and −**F** yield the resultant **R**, which is correct in magnitude, direction, and line of action. This procedure is also useful for graphically combining two forces which have a remote and inconvenient point of concurrency because they are almost parallel.

It is usually helpful to master the analysis of force systems in two dimensions before undertaking three-dimensional analysis. Thus the remainder of Chapter 2 is subdivided into these two categories.

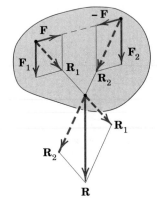

Figure 2/4

SECTION A   TWO-DIMENSIONAL FORCE SYSTEMS

2/3   RECTANGULAR COMPONENTS

The most common two-dimensional resolution of a force vector is into rectangular components. It follows from the parallelogram rule that the vector **F** of Fig. 2/5 may be written as

$$\mathbf{F} = \mathbf{F}_x + \mathbf{F}_y \tag{2/1}$$

where **F**$_x$ and **F**$_y$ are *vector components* of **F** in the *x*- and *y*-directions. Each of the two vector components may be written as a scalar times the

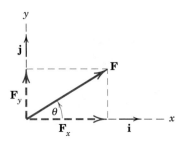

Figure 2/5

*Perpendicular projections are also called *orthogonal* projections.

appropriate unit vector. In terms of the unit vectors **i** and **j** of Fig. 2/5, $\mathbf{F}_x = F_x\mathbf{i}$ and $\mathbf{F}_y = F_y\mathbf{j}$, and thus we may write

$$\mathbf{F} = F_x\mathbf{i} + F_y\mathbf{j} \qquad (2/2)$$

where the scalars F_x and F_y are the x and y *scalar components* of the vector **F**.

The scalar components can be positive or negative, depending on the quadrant into which **F** points. For the force vector of Fig. 2/5, the x and y scalar components are both positive and are related to the magnitude and direction of **F** by

$$\boxed{\begin{array}{ll} F_x = F \cos \theta & F = \sqrt{F_x^2 + F_y^2} \\[2mm] F_y = F \sin \theta & \theta = \tan^{-1} \dfrac{F_y}{F_x} \end{array}} \qquad (2/3)$$

Conventions for Describing Vector Components

We express the magnitude of a vector with lightface italic type in print; that is, $|\mathbf{F}|$ is indicated by F, a quantity which is always *nonnegative*. However, the scalar components, also denoted by lightface italic type, will include sign information. See Sample Problems 2/1 and 2/3 for numerical examples which involve both positive and negative scalar components.

When both a force and its vector components appear in a diagram, it is desirable to show the vector components of the force with dashed lines, as in Fig. 2/5, and show the force with a solid line, or vice versa. With either of these conventions it will always be clear that a force and its components are being represented, and not three separate forces, as would be implied by three solid-line vectors.

Actual problems do not come with reference axes, so their assignment is a matter of arbitrary convenience, and the choice is frequently up to the student. The logical choice is usually indicated by the way in which the geometry of the problem is specified. When the principal dimensions of a body are given in the horizontal and vertical directions, for example, you would typically assign reference axes in these directions.

Determining the Components of a Force

Dimensions are not always given in horizontal and vertical directions, angles need not be measured counterclockwise from the x-axis, and the origin of coordinates need not be on the line of action of a force. Therefore, it is essential that we be able to determine the correct components of a force no matter how the axes are oriented or how the angles are measured. Figure 2/6 suggests a few typical examples of vector resolution in two dimensions.

Memorization of Eqs. 2/3 is not a substitute for understanding the parallelogram law and for correctly projecting a vector onto a reference axis. A neatly drawn sketch always helps to clarify the geometry and avoid error.

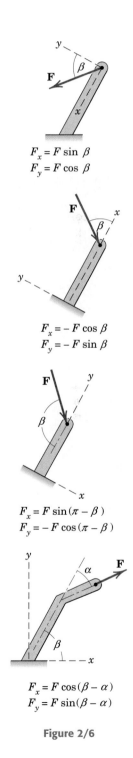

$F_x = F \sin \beta$
$F_y = F \cos \beta$

$F_x = -F \cos \beta$
$F_y = -F \sin \beta$

$F_x = F \sin (\pi - \beta)$
$F_y = -F \cos (\pi - \beta)$

$F_x = F \cos (\beta - \alpha)$
$F_y = F \sin(\beta - \alpha)$

Figure 2/6

Rectangular components are convenient for finding the sum or resultant $\mathbf{R}$ of two forces which are concurrent. Consider two forces $\mathbf{F}_1$ and $\mathbf{F}_2$ which are originally concurrent at a point O. Figure 2/7 shows the line of action of $\mathbf{F}_2$ shifted from O to the tip of $\mathbf{F}_1$ according to the triangle rule of Fig. 2/3. In adding the force vectors $\mathbf{F}_1$ and $\mathbf{F}_2$, we may write

$$\mathbf{R} = \mathbf{F}_1 + \mathbf{F}_2 = (F_{1_x}\mathbf{i} + F_{1_y}\mathbf{j}) + (F_{2_x}\mathbf{i} + F_{2_y}\mathbf{j})$$

or

$$R_x\mathbf{i} + R_y\mathbf{j} = (F_{1_x} + F_{2_x})\mathbf{i} + (F_{1_y} + F_{2_y})\,\mathbf{j}$$

from which we conclude that

$$R_x = F_{1_x} + F_{2_x} = \Sigma F_x$$
$$R_y = F_{1_y} + F_{2_y} = \Sigma F_y$$

(2/4)

The term ΣF_x means "the algebraic sum of the x scalar components". For the example shown in Fig. 2/7, note that the scalar component F_{2_y} would be negative.

The structural elements in the foreground transmit concentrated forces to the brackets at both ends.

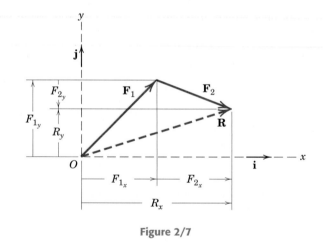

Figure 2/7

Sample Problem 2/1

The forces $\mathbf{F}_1$, $\mathbf{F}_2$, and $\mathbf{F}_3$, all of which act on point A of the bracket, are specified in three different ways. Determine the x and y scalar components of each of the three forces.

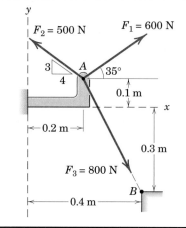

Solution.   The scalar components of $\mathbf{F}_1$, from Fig. a, are

$$F_{1_x} = 600 \cos 35° = 491 \text{ N} \qquad \textit{Ans.}$$

$$F_{1_y} = 600 \sin 35° = 344 \text{ N} \qquad \textit{Ans.}$$

The scalar components of $\mathbf{F}_2$, from Fig. b, are

$$F_{2_x} = -500(\tfrac{4}{5}) = -400 \text{ N} \qquad \textit{Ans.}$$

$$F_{2_y} = 500(\tfrac{3}{5}) = 300 \text{ N} \qquad \textit{Ans.}$$

Note that the angle which orients $\mathbf{F}_2$ to the x-axis is never calculated. The cosine and sine of the angle are available by inspection of the 3-4-5 triangle. Also note that the x scalar component of $\mathbf{F}_2$ is negative by inspection.

The scalar components of $\mathbf{F}_3$ can be obtained by first computing the angle α of Fig. c.

$$\alpha = \tan^{-1}\left[\frac{0.2}{0.4}\right] = 26.6°$$

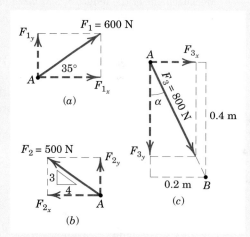

① Then $F_{3_x} = F_3 \sin \alpha = 800 \sin 26.6° = 358 \text{ N} \qquad \textit{Ans.}$

$$F_{3_y} = -F_3 \cos \alpha = -800 \cos 26.6° = -716 \text{ N} \qquad \textit{Ans.}$$

Alternatively, the scalar components of $\mathbf{F}_3$ can be obtained by writing $\mathbf{F}_3$ as a magnitude times a unit vector $\mathbf{n}_{AB}$ in the direction of the line segment AB. Thus,

② $$\mathbf{F}_3 = F_3\mathbf{n}_{AB} = F_3\frac{\overrightarrow{AB}}{\overline{AB}} = 800\left[\frac{0.2\mathbf{i} - 0.4\mathbf{j}}{\sqrt{(0.2)^2 + (-0.4)^2}}\right]$$

$$= 800\,[0.447\mathbf{i} - 0.894\mathbf{j}]$$

$$= 358\mathbf{i} - 716\mathbf{j} \text{ N}$$

The required scalar components are then

$$F_{3_x} = 358 \text{ N} \qquad \textit{Ans.}$$

$$F_{3_y} = -716 \text{ N} \qquad \textit{Ans.}$$

which agree with our previous results.

Helpful Hints

① You should carefully examine the geometry of each component determination problem and not rely on the blind use of such formulas as $F_x = F \cos \theta$ and $F_y = F \sin \theta$.

② A unit vector can be formed by dividing *any* vector, such as the geometric position vector $\overrightarrow{AB}$, by its length or magnitude. Here we use the overarrow to denote the vector which runs from A to B and the overbar to determine the distance between A and B.

Sample Problem 2/2

Combine the two forces **P** and **T**, which act on the fixed structure at B, into a single equivalent force **R**.

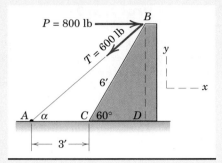

Graphical solution. The parallelogram for the vector addition of forces **T** and ① **P** is constructed as shown in Fig. a. The scale used here is 1 in. = 800 lb; a scale of 1 in. = 200 lb would be more suitable for regular-size paper and would give greater accuracy. Note that the angle a must be determined prior to construction of the parallelogram. From the given figure

$$\tan \alpha = \frac{\overline{BD}}{\overline{AD}} = \frac{6 \sin 60°}{3 + 6 \cos 60°} = 0.866 \qquad \alpha = 40.9°$$

Measurement of the length R and direction θ of the resultant force **R** yields the approximate results

$$R = 525 \text{ lb} \qquad \theta = 49° \qquad \textit{Ans.}$$

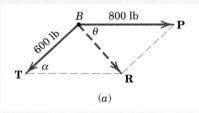

(a)

Geometric solution. The triangle for the vector addition of **T** and **P** is shown ② in Fig. b. The angle α is calculated as above. The law of cosines gives

$$R^2 = (600)^2 + (800)^2 - 2(600)(800) \cos 40.9° = 274{,}300$$

$$R = 524 \text{ lb} \qquad \textit{Ans.}$$

From the law of sines, we may determine the angle θ which orients **R**. Thus,

$$\frac{600}{\sin \theta} = \frac{524}{\sin 40.9°} \qquad \sin \theta = 0.750 \qquad \theta = 48.6° \qquad \textit{Ans.}$$

Algebraic solution. By using the x-y coordinate system on the given figure, we may write

$$R_x = \Sigma F_x = 800 - 600 \cos 40.9° = 346 \text{ lb}$$

$$R_y = \Sigma F_y = -600 \sin 40.9° = -393 \text{ lb}$$

The magnitude and direction of the resultant force **R** as shown in Fig. c are then

$$R = \sqrt{R_x^2 + R_y^2} = \sqrt{(346)^2 + (-393)^2} = 524 \text{ lb} \qquad \textit{Ans.}$$

$$\theta = \tan^{-1} \frac{|R_y|}{|R_x|} = \tan^{-1} \frac{393}{346} = 48.6° \qquad \textit{Ans.}$$

The resultant **R** may also be written in vector notation as

$$\mathbf{R} = R_x \mathbf{i} + R_y \mathbf{j} = 346\mathbf{i} - 393\mathbf{j} \text{ lb} \qquad \textit{Ans.}$$

Helpful Hints

① Note the repositioning of **P** to permit parallelogram addition at B.

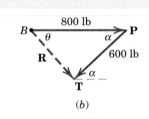

(b)

② Note the repositioning of **F** so as to preserve the correct line of action of the resultant **R**.

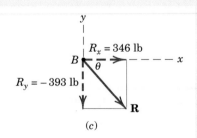

(c)

Sample Problem 2/3

The 500-N force **F** is applied to the vertical pole as shown. (1) Write **F** in terms of the unit vectors **i** and **j** and identify both its vector and scalar components. (2) Determine the scalar components of the force vector **F** along the x'- and y'-axes. (3) Determine the scalar components of **F** along the x- and y'-axes.

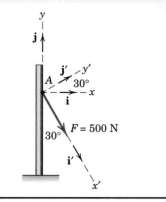

Solution. **Part (1).** From Fig. a we may write **F** as

$$\mathbf{F} = (F \cos \theta)\mathbf{i} - (F \sin \theta)\mathbf{j}$$

$$= (500 \cos 60°)\mathbf{i} - (500 \sin 60°)\mathbf{j}$$

$$= (250\mathbf{i} - 433\mathbf{j}) \text{ N} \qquad\qquad Ans.$$

The scalar components are $F_x = 250$ N and $F_y = -433$ N. The vector components are $\mathbf{F}_x = 250\mathbf{i}$ N and $\mathbf{F}_y = -433\mathbf{j}$ N.

Part (2). From Fig. b we may write **F** as $\mathbf{F} = 500\mathbf{i}'$ N, so that the required scalar components are

$$F_{x'} = 500 \text{ N} \qquad F_{y'} = 0 \qquad\qquad Ans.$$

Part (3). The components of **F** in the x- and y'-directions are nonrectangular and are obtained by completing the parallelogram as shown in Fig. c. The magnitudes of the components may be calculated by the law of sines. Thus,

①

$$\frac{|F_x|}{\sin 90°} = \frac{500}{\sin 30°} \qquad |F_x| = 1000 \text{ N}$$

$$\frac{|F_{y'}|}{\sin 60°} = \frac{500}{\sin 30°} \qquad |F_{y'}| = 866 \text{ N}$$

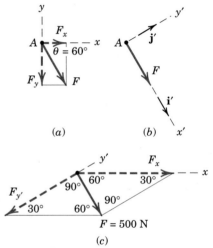

The required scalar components are then

$$F_x = 1000 \text{ N} \qquad F_{y'} = -866 \text{ N} \qquad\qquad Ans.$$

Helpful Hint

① Obtain F_x and $F_{y'}$ graphically and compare your results with the calculated values.

Sample Problem 2/4

Forces $\mathbf{F}_1$ and $\mathbf{F}_2$ act on the bracket as shown. Determine the projection F_b of their resultant **R** onto the b-axis.

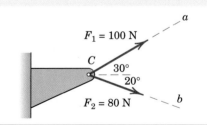

Solution. The parallelogram addition of $\mathbf{F}_1$ and $\mathbf{F}_2$ is shown in the figure. Using the law of cosines gives us

$$R^2 = (80)^2 + (100)^2 - 2(80)(100) \cos 130° \qquad R = 163.4 \text{ N}$$

The figure also shows the orthogonal projection $\mathbf{F}_b$ of **R** onto the b-axis. Its length is

$$F_b = 80 + 100 \cos 50° = 144.3 \text{ N} \qquad\qquad Ans.$$

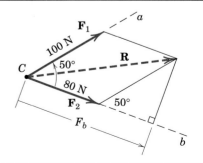

Note that the components of a vector are in general not equal to the projections of the vector onto the same axes. If the a-axis had been perpendicular to the b-axis, then the projections and components of **R** would have been equal.

PROBLEMS

Introductory Problems

2/1 The force **F** has a magnitude of 800 N. Express **F** as a vector in terms of the unit vectors **i** and **j**. Identify the x and y scalar components of **F**.

> Ans. $\mathbf{F} = -459\mathbf{i} + 655\mathbf{j}$ N
> $F_x = -459$ N, $F_y = 655$ N

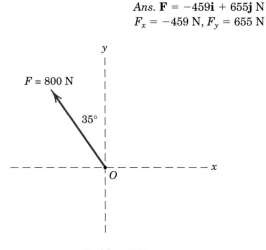

Problem 2/1

2/2 The magnitude of the force **F** is 600 lb. Express **F** as a vector in terms of the unit vectors **i** and **j**. Identify both the scalar and vector components of **F**.

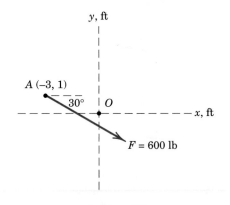

Problem 2/2

2/3 The slope of the 4.8-kN force **F** is specified as shown in the figure. Express **F** as a vector in terms of the unit vectors **i** and **j**.

> Ans. $\mathbf{F} = -2.88\mathbf{i} - 3.84\mathbf{j}$ kN

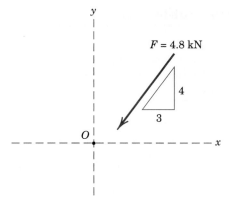

Problem 2/3

2/4 The line of action of the 4800-lb force **F** runs through the points A and B as shown in the figure. Determine the x and y scalar components of **F**.

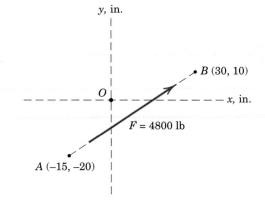

Problem 2/4

2/5 A cable stretched between the fixed supports A and B is under a tension T of 900 lb. Express the tension as a vector using the unit vectors **i** and **j**, first, as a force $\mathbf{T}_A$ acting on A and second, as a force $\mathbf{T}_B$ acting on B.

> Ans. $\mathbf{T}_A = 749\mathbf{i} - 499\mathbf{j}$ lb, $\mathbf{T}_B = -749\mathbf{i} + 499\mathbf{j}$ lb

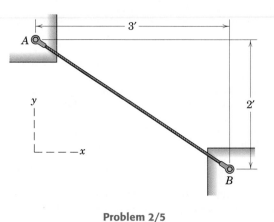

Problem 2/5

2/6 The 1800-N force **F** is applied to the end of the I-beam. Express **F** as a vector using the unit vectors **i** and **j**.

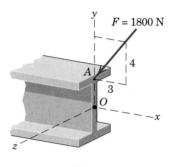

Problem 2/6

2/7 The two structural members, one of which is in tension and the other in compression, exert the indicated forces on joint O. Determine the magnitude of the resultant **R** of the two forces and the angle θ which **R** makes with the positive x-axis.

Ans. $R = 3.61$ kN, $\theta = 206°$

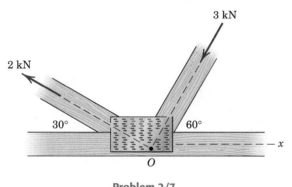

Problem 2/7

2/8 Two forces are applied to the construction bracket as shown. Determine the angle θ which makes the resultant of the two forces vertical. Determine the magnitude R of the resultant.

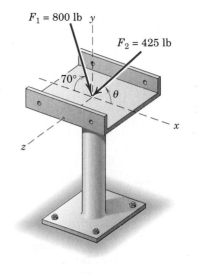

Problem 2/8

Representative Problems

2/9 In the design of a control mechanism, it is determined that rod AB transmits a 260-N force **P** to the crank BC. Determine the x and y scalar components of **P**.

Ans. $P_x = -240$ N
$P_y = -100$ N

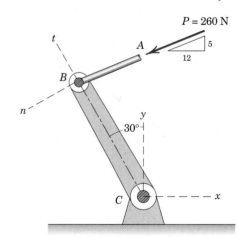

Problem 2/9

2/10 For the mechanism of Prob. 2/9, determine the scalar components P_t and P_n of **P** which are tangent and normal, respectively, to crank BC.

2/11 The *t*-component of the force **F** is known to be 75 N. Determine the *n*-component and the magnitude of **F**.

$$\text{Ans. } F_n = -62.9 \text{ N}, F = 97.9 \text{ N}$$

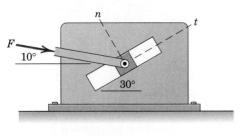

Problem 2/11

2/12 A force **F** of magnitude 800 lb is applied to point *C* of the bar *AB* as shown. Determine both the *x-y* and the *n-t* components of **F**.

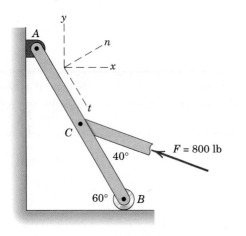

Problem 2/12

2/13 The two forces shown act at point *A* of the bent bar. Determine the resultant **R** of the two forces.

$$\text{Ans. } \mathbf{R} = 2.35\mathbf{i} - 3.45\mathbf{j} \text{ kips}$$

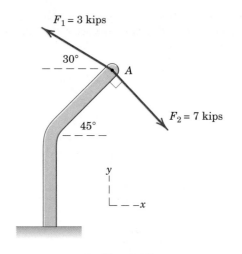

Problem 2/13

2/14 To satisfy design limitations it is necessary to determine the effect of the 2-kN tension in the cable on the shear, tension, and bending of the fixed I-beam. For this purpose replace this force by its equivalent of two forces at *A*, F_t parallel and F_n perpendicular to the beam. Determine F_t and F_n.

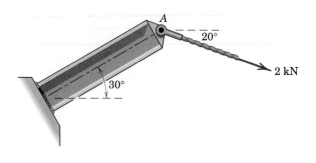

Problem 2/14

2/15 Determine the magnitude F_s of the tensile spring force in order that the resultant of $\mathbf{F}_s$ and $\mathbf{F}$ is a vertical force. Determine the magnitude R of this vertical resultant force.

Ans. $F_s = 60$ lb, $R = 103.9$ lb

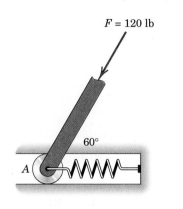

$F = 120$ lb

$60°$

A

Problem 2/15

2/16 The ratio of the lift force L to the drag force D for the simple airfoil is $L/D = 10$. If the lift force on a short section of the airfoil is 50 lb, compute the magnitude of the resultant force $\mathbf{R}$ and the angle θ which it makes with the horizontal.

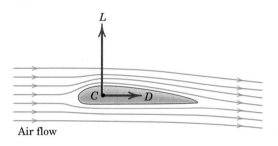

L

C D

Air flow

Problem 2/16

2/17 Determine the components of the 2-kN force along the oblique axes a and b. Determine the projections of $\mathbf{F}$ onto the a- and b-axes.

Ans. $F_a = 0.598$ kN, $F_b = 1.633$ kN
$P_a = 1.414$ kN, $P_b = 1.932$ kN

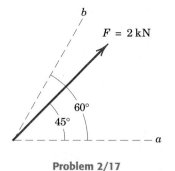

b

$F = 2$ kN

$60°$

$45°$

a

Problem 2/17

2/18 Determine the scalar components R_a and R_b of the force $\mathbf{R}$ along the nonrectangular axes a and b. Also determine the orthogonal projection P_a of $\mathbf{R}$ onto axis a.

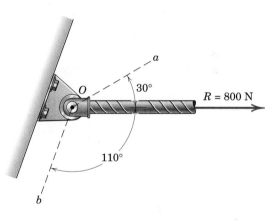

a

O $30°$ $R = 800$ N

$110°$

b

Problem 2/18

2/19 Determine the resultant $\mathbf{R}$ of the two forces shown by (a) applying the parallelogram rule for vector addition and (b) summing scalar components.

Ans. $\mathbf{R} = 520\mathbf{i} - 700\mathbf{j}$ N

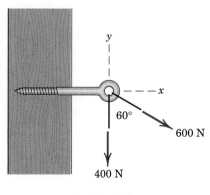

y

x

$60°$

600 N

400 N

Problem 2/19

2/20 It is desired to remove the spike from the timber by applying force along its horizontal axis. An obstruction A prevents direct access, so that two forces, one 400 lb and the other **P**, are applied by cables as shown. Compute the magnitude of **P** necessary to ensure a resultant **T** directed along the spike. Also find T.

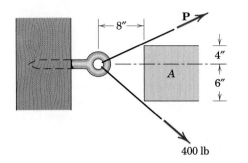

Problem 2/20

2/21 At what angle θ must the 800-lb force be applied in order that the resultant **R** of the two forces has a magnitude of 2000 lb? For this condition, determine the angle β between **R** and the vertical.

Ans. $\theta = 51.3°$, $\beta = 18.19°$

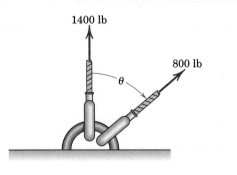

Problem 2/21

2/22 The unstretched length of the spring of modulus $k = 1.2$ kN/m is 100 mm. When pin P is in the position $\theta = 30°$, determine the x- and y-components of the force which the spring exerts on the pin. (The force in a spring is given by $F = kx$, where x is the deflection from the unstretched length.)

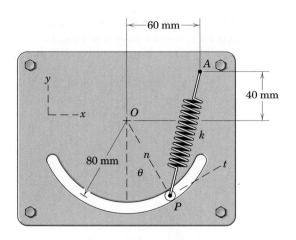

Problem 2/22

2/23 Refer to the statement and figure of Prob. 2/22. When pin P is in the position $\theta = 20°$, determine the n- and t-components of the force F which the spring of modulus $k = 1.2$ kN/m exerts on the pin.

Ans. $F_n = 19.18$ N, $F_t = 13.84$ N

2/24 The cable AB prevents bar OA from rotating clockwise about the pivot O. If the cable tension is 750 N, determine the n- and t-components of this force acting on point A of the bar.

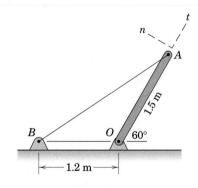

Problem 2/24

2/25 At what angle θ must the 400-lb force be applied in order that the resultant **R** of the two forces have a magnitude of 1000 lb? For this condition what will be the angle β between **R** and the horizontal?

Ans. $\theta = 51.3°$, $\beta = 18.19°$

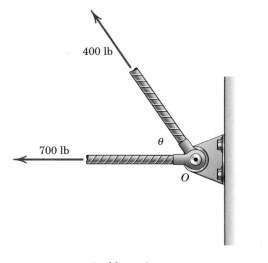

Problem 2/25

2/26 In the design of the robot to insert the small cylindrical part into a close-fitting circular hole, the robot arm must exert a 90-N force P on the part parallel to the axis of the hole as shown. Determine the components of the force which the part exerts *on* the robot along axes (*a*) parallel and perpendicular to the arm AB, and (*b*) parallel and perpendicular to the arm BC.

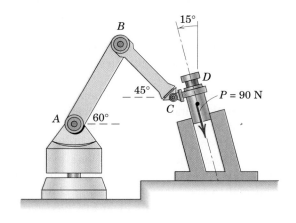

Problem 2/26

2/27 The guy cables AB and AC are attached to the top of the transmission tower. The tension in cable AC is 8 kN. Determine the required tension T in cable AB such that the net effect of the two cable tensions is a downward force at point A. Determine the magnitude R of this downward force.

Ans. $T = 5.68$ kN, $R = 10.21$ kN

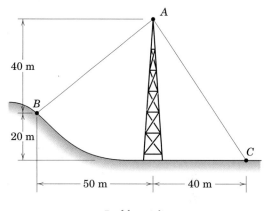

Problem 2/27

2/28 The gusset plate is subjected to the two forces shown. Replace them by two equivalent forces, F_x in the x-direction and F_a in the a-direction. Determine the magnitudes of F_x and F_a. Solve geometrically or graphically.

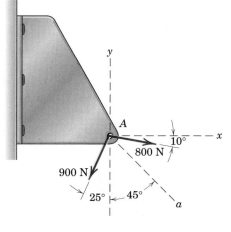

Problem 2/28

2/4 MOMENT

In addition to the tendency to move a body in the direction of its application, a force can also tend to rotate a body about an axis. The axis may be any line which neither intersects nor is parallel to the line of action of the force. This rotational tendency is known as the *moment* **M** of the force. Moment is also referred to as *torque*.

As a familiar example of the concept of moment, consider the pipe wrench of Fig. 2/8*a*. One effect of the force applied perpendicular to the handle of the wrench is the tendency to rotate the pipe about its vertical axis. The magnitude of this tendency depends on both the magnitude *F* of the force and the effective length *d* of the wrench handle. Common experience shows that a pull which is not perpendicular to the wrench handle is less effective than the right-angle pull shown.

Moment about a Point

Figure 2/8*b* shows a two-dimensional body acted on by a force **F** in its plane. The magnitude of the moment or tendency of the force to rotate the body about the axis *O-O* perpendicular to the plane of the body is proportional both to the magnitude of the force and to the *moment arm d*, which is the perpendicular distance from the axis to the line of action of the force. Therefore, the magnitude of the moment is defined as

$$M = Fd \qquad (2/5)$$

The moment is a vector **M** perpendicular to the plane of the body. The sense of **M** depends on the direction in which **F** tends to rotate the body. The right-hand rule, Fig. 2/8*c*, is used to identify this sense. We represent the moment of **F** about *O-O* as a vector pointing in the direction of the thumb, with the fingers curled in the direction of the rotational tendency.

The moment **M** obeys all the rules of vector combination and may be considered a sliding vector with a line of action coinciding with the moment axis. The basic units of moment in SI units are newton-meters (N·m), and in the U.S. customary system are pound-feet (lb-ft).

When dealing with forces which all act in a given plane, we customarily speak of the moment *about a point*. By this we mean the moment with respect to an axis normal to the plane and passing through the point. Thus, the moment of force **F** about point *A* in Fig. 2/8*d* has the magnitude *M = Fd* and is counterclockwise.

Moment directions may be accounted for by using a stated sign convention, such as a plus sign (+) for counterclockwise moments and a minus sign (−) for clockwise moments, or vice versa. Sign consistency within a given problem is essential. For the sign convention of Fig. 2/8*d*, the moment of **F** about point *A* (or about the *z*-axis passing through point *A*) is positive. The curved arrow of the figure is a convenient way to represent moments in two-dimensional analysis.

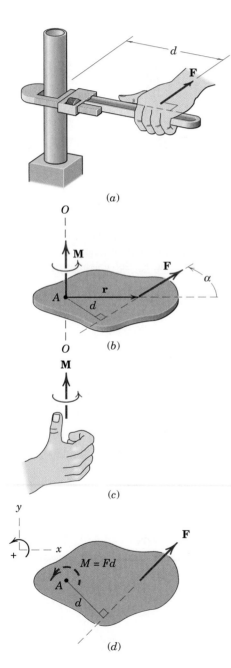

(a)

(b)

(c)

(d)

Figure 2/8

The Cross Product

In some two-dimensional and many of the three-dimensional problems to follow, it is convenient to use a vector approach for moment calculations. The moment of **F** about point A of Fig. 2/8b may be represented by the cross-product expression

$$\boxed{\mathbf{M} = \mathbf{r} \times \mathbf{F}} \tag{2/6}$$

where **r** is a position vector which runs from the moment reference point A to *any* point on the line of action of **F**. The magnitude of this expression is given by*

$$M = Fr \sin \alpha = Fd \tag{2/7}$$

which agrees with the moment magnitude as given by Eq. 2/5. Note that the moment arm $d = r \sin \alpha$ does not depend on the particular point on the line of action of **F** to which the vector **r** is directed. We establish the direction and sense of **M** by applying the right-hand rule to the sequence **r** × **F**. If the filgers of the right hand are curled in the direction of rotation from the positive sense of **r** to the positive sense of **F**, then the thumb points in the positive sense of **M**.

We must maintain the sequence **r** × **F**, because the sequence **F** × **r** would produce a vector with a sense opposite to that of the correct moment. As was the case with the scalar approach, the moment **M** may be thought of as the moment about point A or as the moment about the line O-O which passes through point A and is perpendicular to the plane containing the vectors **r** and **F**. When we evaluate the moment of a force about a given point, the choice between using the vector cross product or the scalar expression depends on how the geometry of the problem is specified. If we know or can easily determine the perpendicular distance between the line of action of the force and the moment center, then the scalar approach is generally simpler. If, however, **F** and **r** are not perpendicular and are easily expressible in vector notation, then the cross-product expression is often preferable.

In Section B of this chapter, we will see how the vector formulation of the moment of a force is especially useful for determining the moment of a force about a point in three-dimensional situations.

Varignon's Theorem

One of the most useful principles of mechanics is *Varignon's theorem*, which states that the moment of a force about any point is equal to the sum of the moments of the components of the force about the same point.

---

*See item 7 in Art. C/7 of Appendix C for additional information concerning the cross product.

To prove this theorem, consider the force **R** acting in the plane of the body shown in Fig. 2/9a. The forces **P** and **Q** represent any two non-rectangular components of **R**. The moment of **R** about point O is

$$\mathbf{M}_O = \mathbf{r} \times \mathbf{R}$$

Because **R** = **P** + **Q**, we may write

$$\mathbf{r} \times \mathbf{R} = \mathbf{r} \times (\mathbf{P} + \mathbf{Q})$$

Using the distributive law for cross products, we have

$$\mathbf{M}_O = \mathbf{r} \times \mathbf{R} = \mathbf{r} \times \mathbf{P} + \mathbf{r} \times \mathbf{Q} \tag{2/8}$$

which says that the moment of **R** about O equals the sum of the moments about O of its components **P** and **Q**. This proves the theorem.

Varignon's theorem need not be restricted to the case of two components, but it applies equally well to three or more. Thus we could have used any number of concurrent components of **R** in the foregoing proof.*

Figure 2/9b illustrates the usefulness of Varignon's theorem. The moment of **R** about point O is Rd. However, if d is more difficult to determine than p and q, we can resolve **R** into the components **P** and **Q**, and compute the moment as

$$M_O = Rd = -pP + qQ$$

where we take the clockwise moment sense to be positive.

Sample Problem 2/5 shows how Varignon's theorem can help us to calculate moments.

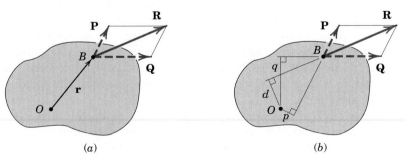

(a)   (b)

Figure 2/9

---

*As originally stated, Varignon's theorem was limited to the case of two concurrent components of a given force. See *The Science of Mechanics*, by Ernst Mach, originally published in 1883.

Sample Problem 2/5

Calculate the magnitude of the moment about the base point O of the 600-N force in five different ways.

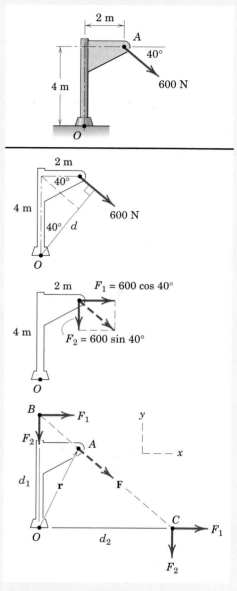

Solution.   **(I)**   The moment arm to the 600-N force is

$$d = 4 \cos 40° + 2 \sin 40° = 4.35 \text{ m}$$

① By $M = Fd$ the moment is clockwise and has the magnitude

$$M_O = 600(4.35) = 2610 \text{ N·m} \qquad Ans.$$

(II)   Replace the force by its rectangular components at A

$$F_1 = 600 \cos 40° = 460 \text{ N}, \qquad F_2 = 600 \sin 40° = 386 \text{ N}$$

By Varignon's theorem, the moment becomes

② $$M_O = 460(4) + 386(2) = 2610 \text{ N·m} \qquad Ans.$$

(III)   By the principle of transmissibility, move the 600-N force along its line of action to point B, which eliminates the moment of the component F_2. The moment arm of F_1 becomes

$$d_1 = 4 + 2 \tan 40° = 5.68 \text{ m}$$

and the moment is

$$M_O = 460(5.68) = 2610 \text{ N·m} \qquad Ans.$$

③ **(IV)**   Moving the force to point C eliminates the moment of the component F_1. The moment arm of F_2 becomes

$$d_2 = 2 + 4 \cot 40° = 6.77 \text{ m}$$

and the moment is

$$M_O = 386(6.77) = 2610 \text{ N·m} \qquad Ans.$$

(V)   By the vector expression for a moment, and by using the coordinate system indicated on the figure together with the procedures for evaluating cross products, we have

④ $$\mathbf{M}_O = \mathbf{r} \times \mathbf{F} = (2\mathbf{i} + 4\mathbf{j}) \times 600(\mathbf{i} \cos 40° - \mathbf{j} \sin 40°)$$

$$= -2610\mathbf{k} \text{ N·m}$$

The minus sign indicates that the vector is in the negative z-direction. The magnitude of the vector expression is

$$M_O = 2610 \text{ N·m} \qquad Ans.$$

Helpful Hints

① The required geometry here and in similar problems should not cause difficulty if the sketch is carefully drawn.

② This procedure is frequently the shortest approach.

③ The fact that points B and C are not on the body proper should not cause concern, as the mathematical calculation of the moment of a force does not require that the force be on the body.

④ Alternative choices for the position vector $\mathbf{r}$ are $\mathbf{r} = d_1\mathbf{j} = 5.68\mathbf{j}$ m and $\mathbf{r} = d_2\mathbf{i} = 6.77\mathbf{i}$ m.

Sample Problem 2/6

The trap door OA is raised by the cable AB, which passes over the small frictionless guide pulleys at B. The tension everywhere in the cable is T, and this tension applied at A causes a moment M_O about the hinge at O. Plot the quantity M_O/T as a function of the door elevation angle θ over the range $0 \leq \theta \leq 90°$ and note minimum and maximum values. What is the physical significance of this ratio?

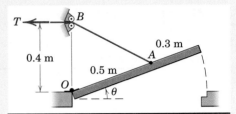

Solution. We begin by constructing a figure which shows the tension force $\mathbf{T}$ acting directly on the door, which is shown in an arbitrary angular position θ. It should be clear that the direction of $\mathbf{T}$ will vary as θ varies. In order to deal with this variation, we write a unit vector $\mathbf{n}_{AB}$ which "aims" $\mathbf{T}$:

①
$$\mathbf{n}_{AB} = \frac{\mathbf{r}_{AB}}{r_{AB}} = \frac{\mathbf{r}_{OB} - \mathbf{r}_{OA}}{r_{AB}}$$

Using the x-y coordinates of our figure, we can write

②
$$\mathbf{r}_{OB} = 0.4\mathbf{j} \text{ m and } \mathbf{r}_{OA} = 0.5(\cos\theta\mathbf{i} + \sin\theta\mathbf{j}) \text{ m}$$

So

$$\mathbf{r}_{AB} = \mathbf{r}_{OB} - \mathbf{r}_{OA} = 0.4\mathbf{j} - (0.5)(\cos\theta\mathbf{i} + \sin\theta\mathbf{j})$$
$$= -0.5\cos\theta\mathbf{i} + (0.4 - 0.5\sin\theta)\mathbf{j} \text{ m}$$

and

$$r_{AB} = \sqrt{(0.5\cos\theta)^2 + (0.4 - 0.5\sin\theta)^2}$$
$$= \sqrt{0.41 - 0.4\sin\theta} \text{ m}$$

The desired unit vector is

$$\mathbf{n}_{AB} = \frac{\mathbf{r}_{AB}}{r_{AB}} = \frac{-0.5\cos\theta\mathbf{i} + (0.4 - 0.5\sin\theta)\mathbf{j}}{\sqrt{0.41 - 0.4\sin\theta}}$$

Our tension vector can now be written as

$$\mathbf{T} = T\mathbf{n}_{AB} = T\left[\frac{-0.5\cos\theta\mathbf{i} + (0.4 - 0.5\sin\theta)\mathbf{j}}{\sqrt{0.41 - 0.4\sin\theta}}\right]$$

③ The moment of $\mathbf{T}$ about point O, as a vector, is $\mathbf{M}_O = \mathbf{r}_{OB} \times \mathbf{T}$, where $\mathbf{r}_{OB} = 0.4\mathbf{j}$ m, or

$$\mathbf{M}_O = 0.4\mathbf{j} \times T\left[\frac{-0.5\cos\theta\mathbf{i} + (0.4 - 0.5\sin\theta)\mathbf{j}}{\sqrt{0.41 - 0.4\sin\theta}}\right]$$
$$= \frac{0.2T\cos\theta}{\sqrt{0.41 - 0.4\sin\theta}}\mathbf{k}$$

The magnitude of $\mathbf{M}_O$ is

$$M_O = \frac{0.2T\cos\theta}{\sqrt{0.41 - 0.4\sin\theta}}$$

and the requested ratio is

$$\frac{M_O}{T} = \frac{0.2\cos\theta}{\sqrt{0.41 - 0.4\sin\theta}} \qquad\qquad Ans.$$

which is plotted in the accompanying graph. The expression M_O/T is the moment arm d (in meters) which runs from O to the line of action of $\mathbf{T}$. It has a maximum value of 0.4 m at $\theta = 53.1°$ (at which point $\mathbf{T}$ is horizontal) and a minimum value of 0 at $\theta = 90°$ (at which point $\mathbf{T}$ is vertical). The expression is valid even if T varies.

This sample problem treats moments in two-dimensional force systems, and it also points out the advantages of carrying out a solution for an arbitrary position, so that behavior over a range of positions can be examined.

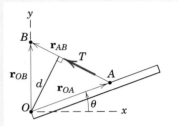

Helpful Hints

① Recall that any unit vector can be written as a vector divided by its magnitude. In this case the vector in the numerator is a position vector.

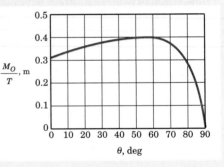

② Recall that any vector may be written as a magnitude times an "aiming" unit vector.

③ In the expression $\mathbf{M} = \mathbf{r} \times \mathbf{F}$, the position vector $\mathbf{r}$ runs *from* the moment center *to* any point on the line of action of $\mathbf{F}$. Here, $\mathbf{r}_{OB}$ is more convenient than $\mathbf{r}_{OA}$.

PROBLEMS

Introductory Problems

2/29 The 10-kN force is applied at point A. Determine the moment of **F** about point O. Determine the points on the x- and y-axes about which the moment of **F** is zero.

Ans. $M_O = 16$ kN·m CW
$(x, y) = (2.67, 0)$ m and $(0, 2)$ m

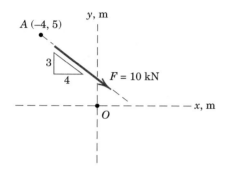

Problem 2/29

2/30 Determine the moment of the 200-lb force about point A and about point O.

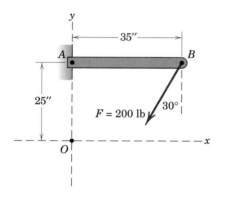

Problem 2/30

2/31 Determine the moment of the 50-N force (a) about point O by Varignon's theorem and (b) about point C by a vector approach.

Ans. $M_O = 179.6$ N·mm CCW, $\mathbf{M}_C = 1616\mathbf{k}$ N·mm

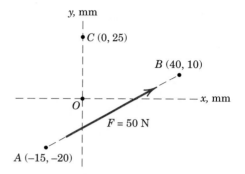

Problem 2/31

2/32 The force of magnitude F acts along the edge of the triangular plate. Determine the moment of **F** about point O.

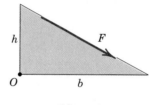

Problem 2/32

2/33 In steadily turning the water pump, a person exerts the 120-N force on the handle as shown. Determine the moment of this force about point O.

Ans. $M_O = 14.74$ N·m CW

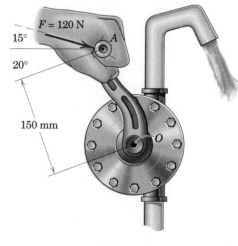

Problem 2/33

2/34 The throttle-control sector pivots freely at O. If an internal torsional spring exerts a return moment M = 1.8 N·m on the sector when in the position shown, for design purposes determine the necessary throttle-cable tension T so that the net moment about O is zero. Note that when T is zero, the sector rests against the idle-control adjustment screw at R.

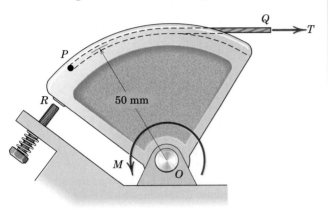

Problem 2/34

2/35 A force **F** of magnitude 60 N is applied to the gear. Determine the moment of **F** about point O.
Ans. M_O = 5.64 N·m CW

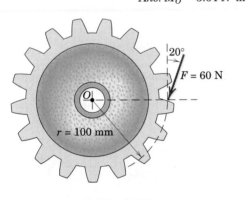

Problem 2/35

2/36 Calculate the moment of the 250-N force on the handle of the monkey wrench about the center of the bolt.

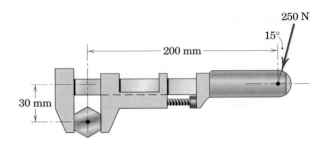

Problem 2/36

2/37 A mechanic pulls on the 13-mm combination wrench with the 140-N force shown. Determine the moment of this force about the bolt center O.
Ans. M_O = 13.10 N·m CCW

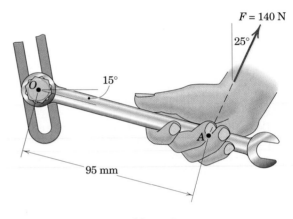

Problem 2/37

2/38 As a trailer is towed in the forward direction, the force F = 120 lb is applied as shown to the ball of the trailer hitch. Determine the moment of this force about point O.

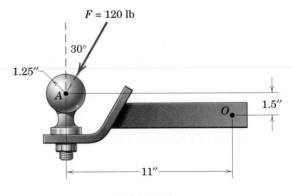

Problem 2/38

Representative Problems

2/39 A portion of a mechanical coin sorter works as follows: Pennies and dimes roll down the 20° incline, the last triangular portion of which pivots freely about a horizontal axis through O. Dimes are light enough (2.28 grams each) so that the triangular portion remains stationary, and the dimes roll into the right collection column. Pennies, on the other hand, are heavy enough (3.06 grams each) so that the triangular portion pivots clockwise, and the pennies roll into the left collection column. Determine the moment about O of the weight of the penny in terms of the slant distance s in millimeters.

$$Ans. \ M_O = 0.1335 + 0.0282s \ \text{N·mm} \ (s \ \text{in mm})$$

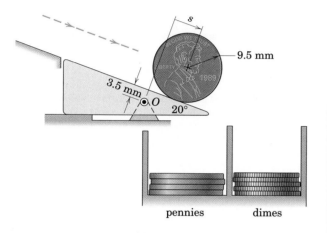

pennies   dimes

Problem 2/39

2/40 Elements of the lower arm are shown in the figure. The weight of the forearm is 5 lb with mass center at G. Determine the combined moment about the elbow pivot O of the weights of the forearm and the sphere. What must the biceps tension force be so that the overall moment about O is zero?

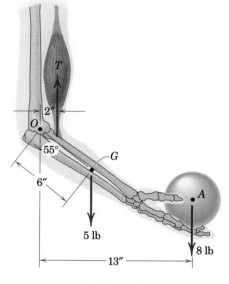

Problem 2/40

2/41 A 32-lb pull T is applied to a cord, which is wound securely around the inner hub of the drum. Determine the moment of T about the drum center C. At what angle θ should T be applied so that the moment about the contact point P is zero?

$$Ans. \ M_C = 160 \ \text{lb-in. CW}, \ \theta = 51.3°$$

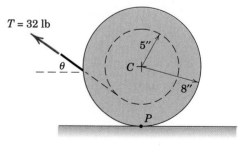

Problem 2/41

2/42 A force of 200 N is applied to the end of the wrench to tighten a flange bolt which holds the wheel to the axle. Determine the moment M produced by this force about the center O of the wheel for the position of the wrench shown.

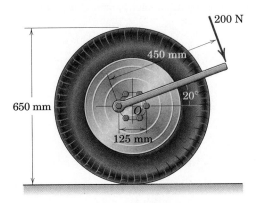

Problem 2/42

2/43 In order to raise the flagpole OC, a light frame OAB is attached to the pole and a tension of 780 lb is developed in the hoisting cable by the power winch D. Calculate the moment M_O of this tension about the hinge point O.

Ans. M_O = 5010 lb-ft CCW

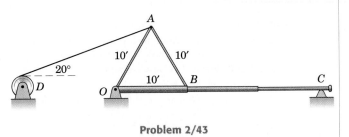

Problem 2/43

2/44 The uniform work platform, which has a mass per unit length of 28 kg/m, is simply supported by cross rods A and B. The 90-kg construction worker starts from point B and walks to the right. At what location s will the combined moment of the weights of the man and platform about point B be zero?

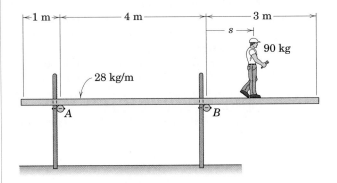

Problem 2/44

2/45 In raising the pole from the position shown, the tension T in the cable must supply a moment about O of 72 kN·m. Determine T.

Ans. T = 8.65 kN

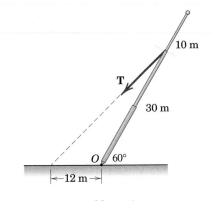

Problem 2/45

2/46 The force exerted by the plunger of cylinder *AB* on the door is 40 N directed along the line *AB*, and this force tends to keep the door closed. Compute the moment of this force about the hinge *O*. What force F_C normal to the plane of the door must the door stop at *C* exert on the door so that the combined moment about *O* of the two forces is zero?

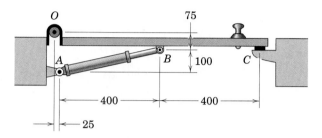

Dimensions in millimeters

Problem 2/46

2/47 The 2-lb force is applied to the handle of the hydraulic control valve as shown. Calculate the moment of this force about point *O*.

Ans. $M_O = 22.5$ lb-in. CW

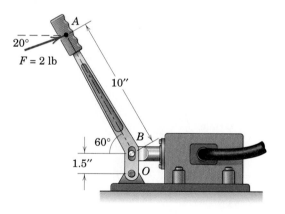

Problem 2/47

2/48 Calculate the moment M_A of the 200-N force about point *A* by using three scalar methods and one vector method.

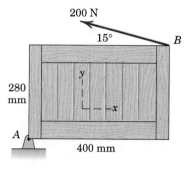

Problem 2/48

2/49 An exerciser begins with his arm in the relaxed vertical position *OA*, at which the elastic band is unstretched. He then rotates his arm to the horizontal position *OB*. The elastic modulus of the band is $k = 60$ N/m—that is, 60 N of force is required to stretch the band each additional meter of elongation. Determine the moment about *O* of the force which the band exerts on the hand *B*.

Ans. $M_O = 26.8$ N·m CCW

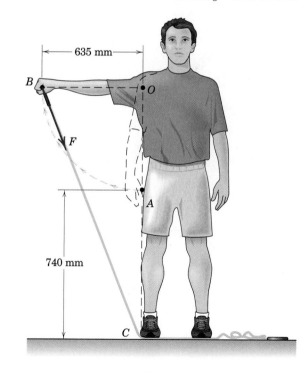

Problem 2/49

2/50 (*a*) Calculate the moment of the 90-N force about point *O* for the condition $\theta = 15°$. Also, determine the value of θ for which the moment about *O* is (*b*) zero and (*c*) a maximum.

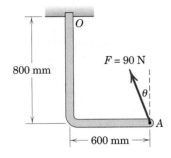

Problem 2/50

2/51 The small crane is mounted along the side of a pickup bed and facilitates the handling of heavy loads. When the boom elevation angle is $\theta = 40°$, the force in the hydraulic cylinder *BC* is 4.5 kN, and this force applied at point *C* is in the direction from *B* to *C* (the cylinder is in compression). Determine the moment of this 4.5-kN force about the boom pivot point *O*.

Ans. $M_O = 0.902$ kN·m CW

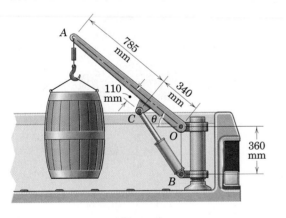

Problem 2/51

2/52 Design criteria require that the robot exert the 90-N force on the part as shown while inserting a cylindrical part into the circular hole. Determine the moment about points *A*, *B*, and *C* of the force which the part exerts on the robot.

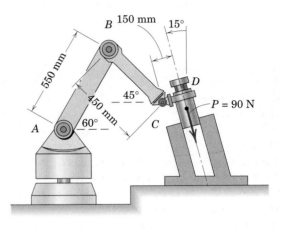

Problem 2/52

2/53 The masthead fitting supports the two forces shown. Determine the magnitude of **T** which will cause no bending of the mast (zero moment) at point *O*.

Ans. T = 4.04 kN

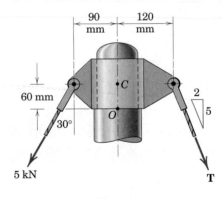

Problem 2/53

2/54 The piston, connecting rod, and crankshaft of a diesel engine are shown in the figure. The crank throw *OA* is half the stroke of 8 in., and the length *AB* of the rod is 14 in. For the position indicated, the rod is under a compression along *AB* of 3550 lb. Determine the moment *M* of this force about the crankshaft axis *O*.

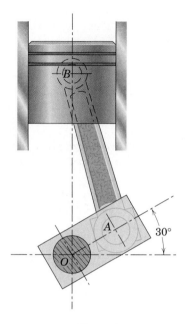

Problem 2/54

2/55 The 120-N force is applied as shown to one end of the curved wrench. If $\alpha = 30°$, calculate the moment of *F* about the center *O* of the bolt. Determine the value of α which would maximize the moment about *O*; state the value of this maximum moment.

Ans. $M_O = 41.5$ N·m CW
$\alpha = 33.2°, (M_O)_{max} = 41.6$ N·m CW

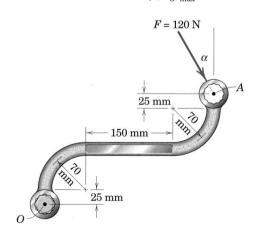

Problem 2/55

▶**2/56** If the combined moment of the two forces about point *C* is zero, determine
  (a) the magnitude of the force **P**
  (b) the magnitude *R* of the resultant of the two forces
  (c) the coordinates *x* and *y* of the point *A* on the rim of the wheel about which the combined moment of the two forces is a maximum
  (d) the combined moment M_A of the two forces about *A*.

Ans. (a) $P = 61.6$ lb
   (b) $R = 141.3$ lb
   (c) $x = 4.90$ in., $y = 6.32$ in.
   (d) $M_A = 1577$ lb-in. CW

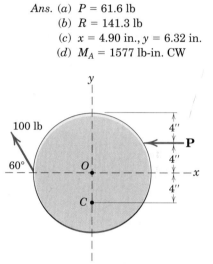

Problem 2/56

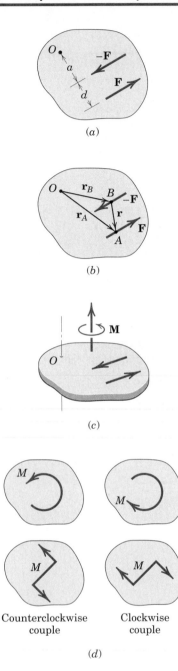

(a)

(b)

(c)

Counterclockwise
couple

Clockwise
couple

(d)

Figure 2/10

2/5   COUPLE

The moment produced by two equal, opposite, and noncollinear forces is called a *couple*. Couples have certain unique properties and have important applications in mechanics.

Consider the action of two equal and opposite forces **F** and −**F** a distance *d* apart, as shown in Fig. 2/10*a*. These two forces cannot be combined into a single force because their sum in every direction is zero. Their only effect is to produce a tendency of rotation. The combined moment of the two forces about an axis normal to their plane and passing through any point such as *O* in their plane is the couple **M**. This couple has a magnitude

$$M = F(a + d) - Fa$$

or

$$M = Fd$$

Its direction is counterclockwise when viewed from above for the case illustrated. Note especially that the magnitude of the couple is independent of the distance *a* which locates the forces with respect to the moment center *O*. It follows that the moment of a couple has the same value for all moment centers.

Vector Algebra Method

We may also express the moment of a couple by using vector algebra. With the cross-product notation of Eq. 2/6, the combined moment about point *O* of the forces forming the couple of Fig. 2/10*b* is

$$\mathbf{M} = \mathbf{r}_A \times \mathbf{F} + \mathbf{r}_B \times (-\mathbf{F}) = (\mathbf{r}_A - \mathbf{r}_B) \times \mathbf{F}$$

where $\mathbf{r}_A$ and $\mathbf{r}_B$ are position vectors which run from point *O* to arbitrary points *A* and *B* on the lines of action of **F** and −**F**, respectively. Because $\mathbf{r}_A - \mathbf{r}_B = \mathbf{r}$, we can express **M** as

$$\mathbf{M} = \mathbf{r} \times \mathbf{F}$$

Here again, the moment expression contains no reference to the moment center *O* and, therefore, is the same for all moment centers. Thus, we may represent **M** by a free vector, as show in Fig. 2/10*c*, where the direction of **M** is normal to the plane of the couple and the sense of **M** is established by the right-hand rule.

Because the couple vector **M** is always perpendicular to the plane of the forces which constitute the couple, in two-dimensional analysis we can represent the sense of a couple vector as clockwise or counterclockwise by one of the conventions shown in Fig. 2/10*d*. Later, when we deal with couple vectors in three-dimensional problems, we will make full use of vector notation to represent them, and the mathematics will automatically account for their sense.

Equivalent Couples

Changing the values of *F* and *d* does not change a given couple as long as the product *Fd* remains the same. Likewise, a couple is not affected if the forces act in a different but parallel plane. Figure 2/11

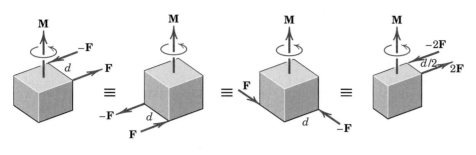

Figure 2/11

shows four different configurations of the same couple **M**. In each of the four cases, the couples are equivalent and are described by the same free vector which represents the identical tendencies to rotate the bodies.

Force–Couple Systems

The effect of a force acting on a body is the tendency to push or pull the body in the direction of the force, and to rotate the body about any fixed axis which does not intersect the line of the force. We can represent this dual effect more easily by replacing the given force by an equal parallel force and a couple to compensate for the change in the moment of the force.

The replacement of a force by a force and a couple is illustrated in Fig. 2/12, where the given force **F** acting at point A is replaced by an equal force **F** at some point B and the counterclockwise couple $M = Fd$. The transfer is seen in the middle figure, where the equal and opposite forces **F** and $-$**F** are added at point B without introducing any net external effects on the body. We now see that the original force at A and the equal and opposite one at B constitute the couple $M = Fd$, which is counterclockwise for the sample chosen, as shown in the right-hand part of the figure. Thus, we have replaced the original force at A by the same force acting at a different point B and a couple, without altering the external effects of the original force on the body. The combination of the force and couple in the right-hand part of Fig. 2/12 is referred to as a *force–couple system*.

By reversing this process, we can combine a given couple and a force which lies in the plane of the couple (normal to the couple vector) to produce a single, equivalent force. Replacement of a force by an equivalent force–couple system, and the reverse procedure, have many applications in mechanics and should be mastered.

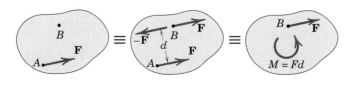

Figure 2/12

Sample Problem 2/7

The rigid structural member is subjected to a couple consisting of the two 100-N forces. Replace this couple by an equivalent couple consisting of the two forces **P** and −**P**, each of which has a magnitude of 400 N. Determine the proper angle θ.

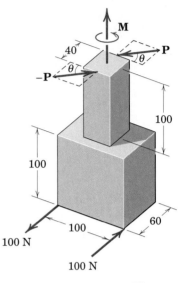

100

100

100

60

100 N

100 N

Dimensions in millimeters

Solution. The original couple is counterclockwise when the plane of the forces is viewed from above, and its magnitude is

$$[M = Fd] \qquad\qquad M = 100(0.1) = 10 \text{ N·m}$$

The forces **P** and −**P** produce a counterclockwise couple

$$M = 400(0.040) \cos \theta$$

① Equating the two expressions gives

$$10 = (400)(0.040) \cos \theta$$

$$\theta = \cos^{-1} \frac{10}{16} = 51.3° \qquad\qquad Ans.$$

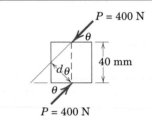

P = 400 N

40 mm

P = 400 N

Helpful Hint

① Since the two equal couples are parallel free vectors, the only dimensions which are relevant are those which give the perpendicular distances between the forces of the couples.

Sample Problem 2/8

Replace the horizontal 80-lb force acting on the lever by an equivalent system consisting of a force at O and a couple.

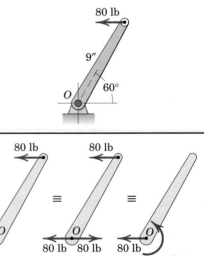

80 lb

9″

60°

O

80 lb   80 lb

≡   ≡

O   O   O

80 lb   80 lb   80 lb

624 lb-in.

Solution. We apply two equal and opposite 80-lb forces at O and identify the counterclockwise couple

$$[M = Fd] \qquad\qquad M = 80(9 \sin 60°) = 624 \text{ lb-in.} \qquad\qquad Ans.$$

① Thus, the original force is equivalent to the 80-lb force at O and the 624-lb-in. couple as shown in the third of the three equivalent figures.

Helpful Hint

① The reverse of this problem is often encountered, namely, the replacement of a force and a couple by a single force. Proceeding in reverse is the same as replacing the couple by two forces, one of which is equal and opposite to the 80-lb force at O. The moment arm to the second force would be $M/F = 624/80 = 7.79$ in., which is 9 sin 60°, thus determining the line of action of the single resultant force of 80 lb.

PROBLEMS

Introductory Problems

2/57 The caster unit is subjected to the pair of 80-lb forces shown. Determine the moment associated with these forces.

Ans. $M = 112$ lb-in. CW

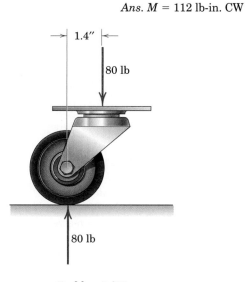

Problem 2/57

2/58 A force $F = 60$ N acts along the line AB. Determine the equivalent force–couple system at point C.

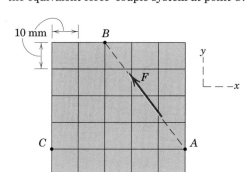

Problem 2/58

2/59 The top view of a revolving entrance door is shown. Two persons simultaneously approach the door and exert force of equal magnitudes as shown. If the resulting moment about the door pivot axis at O is 25 N·m, determine the force magnitude F.

Ans. $F = 16.18$ N

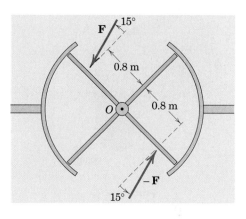

Problem 2/59

2/60 The indicated force–couple system is applied to a small shaft at the center of the plate. Replace this system by a single force and specify the coordinate of the point on the x-axis through which the line of action of this resultalt force passes.

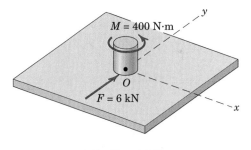

Problem 2/60

2/61 The bracket is spot welded to the end of the shaft at point O. To show the effect of the 200-lb force on the weld, replace the force by its equivalent of a force and couple $\mathbf{M}$ at O. Express $\mathbf{M}$ in vector notation.

Ans. $\mathbf{M} = -800\mathbf{i}$ lb-in.

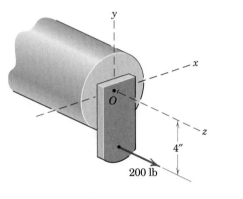

Problem 2/61

2/62 As part of a test, the two aircraft engines are revved up and the propeller pitches are adjusted so as to result in the fore and aft thrusts shown. What force F must be exerted by the ground on each of the main braked wheels at A and B to counteract the turning effect of the two propeller thrusts? Neglect any effects of the nose wheel C, which is turned 90° and unbraked.

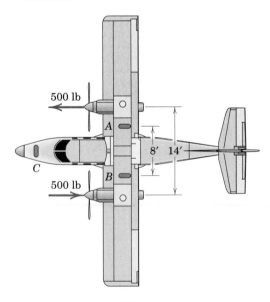

Problem 2/62

2/63 Replace the 10-kN force acting on the steel column by an equivalent force–couple system at point O. This replacement is frequently done in the design of structures.

Ans. $R = 10$ kN, $M_O = 0.75$ kN·m CCW

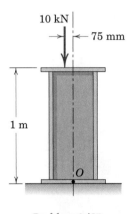

Problem 2/63

2/64 Each propeller of the twin-screw ship develops a full-speed thrust of 300 kN. In maneuvering the ship, one propeller is turning full speed ahead and the other full speed in reverse. What thrust P must each tug exert on the ship to counteract the effect of the ship's propellers?

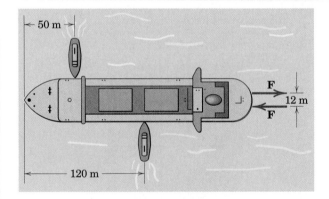

Problem 2/64

Representative Problems

2/65 A lug wrench is used to tighten a square-head bolt. If 50-lb forces are applied to the wrench as shown, determine the magnitude F of the equal forces exerted on the four contact points on the 1-in. bolt head so that their external effect on the bolt is equivalent to that of the two 50-lb forces. Assume that the forces are perpendicular to the flats of the bolt head.

Ans. F = 700 lb

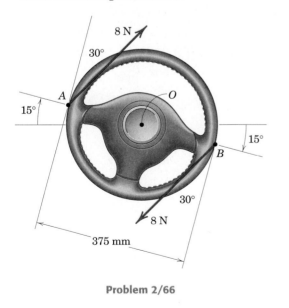

Problem 2/65

2/66 During a steady right turn, a person exerts the forces shown on the steering wheel. Note that each force consists of a tangential component and a radially-inward component. Determine the moment exerted about the steering column at O.

Problem 2/66

2/67 The 180-N force is applied to the end of body OAB. If $\theta = 50°$, determine the equivalent force–couple system at the shaft axis O.

Ans. $\mathbf{F} = -169.1\mathbf{i} - 61.6\mathbf{j}$ N, $M_O = 41.9$ N·m CCW

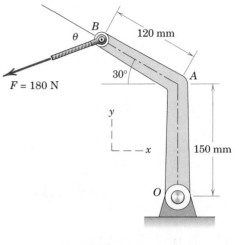

Problem 2/67

2/68 A force $\mathbf{F}$ of magnitude 50 N is exerted on the automobile parking-brake lever at the position $x = 250$ mm. Replace the force by an equivalent force–couple system at the pivot point O.

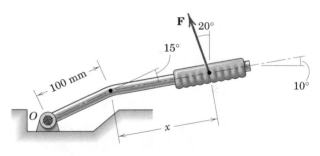

Problem 2/68

2/69 The tie-rod AB exerts the 250-N force on the steering knuckle AO as shown. Replace this force by an equivalent force–couple system at O.

Ans. $\mathbf{F} = 43.4\mathbf{i} + 246\mathbf{j}$ N, $M_O = 60.0$ N·m CW

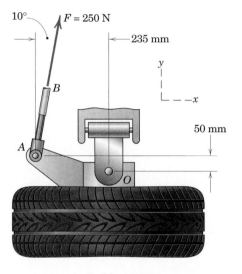

10° $F = 250$ N

235 mm

y

x

B

50 mm

A

O

Problem 2/69

2/70 The combined drive wheels of a front-wheel-drive automobile are acted on by a 7000-N normal reaction force and a friction force $\mathbf{F}$, both of which are exerted by the road surface. If it is known that the resultant of these two forces makes a 15° angle with the vertical, determine the equivalent force–couple system at the car mass center G. Treat this as a two-dimensional problem.

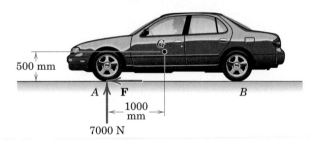

G

500 mm

A $\mathbf{F}$ B

1000 mm

7000 N

Problem 2/70

2/71 The system consisting of the bar OA, two identical pulleys, and a section of thin tape is subjected to the two 180-N tensile forces shown in the figure. Determine the equivalent force–couple system at point O.

Ans. $M = 21.7$ N·m CCW

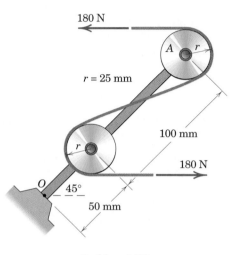

180 N

A r

$r = 25$ mm

100 mm

180 N

O 45°

50 mm

Problem 2/71

2/72 Calculate the moment M_B of the 900-N force about the bolt at B. Simplify your work by first replacing the force by its equivalent force–couple system at A.

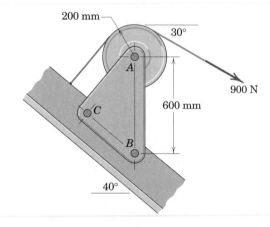

200 mm

30°

A

900 N

C

600 mm

B

40°

Problem 2/72

2/73 The bracket is fastened to the girder by means of the two rivets A and B and supports the 520-lb force. Replace this force by a force acting along the centerline between the rivets and a couple. Then redistribute this force and couple by replacing it by two forces, one at A and the other at B, and ascertain the forces supported by the rivets.

Ans. $F_A = 231$ lb
$F_B = 751$ lb

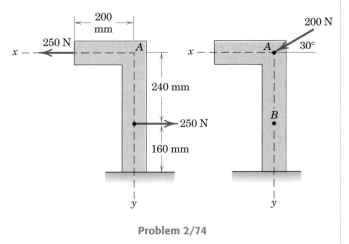

Problem 2/73

2/74 The angle plate is subjected to the two 250-N forces shown. It is desired to replace these forces by an equivalent set consisting of the 200-N force applied at A and a second force applied at B. Determine the y-coordinate of B.

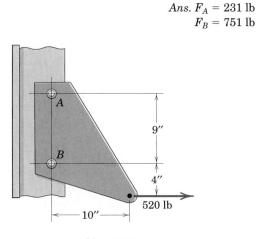

Problem 2/74

2/75 The weld at O can support a maximum of 500 lb of force along each of the n- and t-directions and a maximum of 1000 lb-ft of moment. Determine the allowable range for the direction θ of the 600-lb force applied at A. The angle θ is restricted to $0 \le \theta \le 90°$.

Ans. $23.6° \le \theta \le 56.4°$

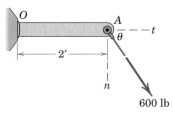

Problem 2/75

2/76 The device shown is a part of an automobile seat-back-release mechanism. The part is subjected to the 4-N force exerted at A and a 300-N·mm restoring moment exerted by a hidden torsional spring. Determine the y-intercept of the line of action of the single equivalent force.

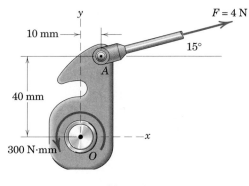

Problem 2/76

2/6   Resultants

The properties of force, moment, and couple were developed in the previous four articles. Now we are ready to describe the resultant action of a group or *system* of forces. Most problems in mechanics deal with a system of forces, and it is usually necessary to reduce the system to its simplest form to describe its action. The *resultant* of a system of forces is the simplest force combination which can replace the original forces without altering the external effect on the rigid body to which the forces are applied.

Equilibrium of a body is the condition in which the resultant of all forces acting on the body is zero. This condition is studied in statics. When the resultant of all forces on a body is not zero, the acceleration of the body is obtained by equating the force resultant to the product of the mass and acceleration of the body. This condition is studied in dynamics. Thus, the determination of resultants is basic to both statics and dynamics.

The most common type of force system occurs when the forces all act in a single plane, say, the *x-y* plane, as illustrated by the system of three forces $\mathbf{F}_1$, $\mathbf{F}_2$, and $\mathbf{F}_3$ in Fig. 2/13*a*. We obtain the magnitude and direction of the resultant force $\mathbf{R}$ by forming the *force polygon* shown in part *b* of the figure, where the forces are added head-to-tail in any sequence. Thus, for any system of coplanar forces we may write

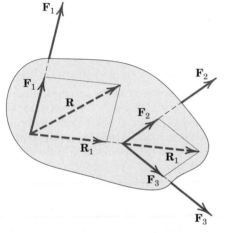

(*a*)

$$\mathbf{R} = \mathbf{F}_1 + \mathbf{F}_2 + \mathbf{F}_3 + \cdots = \Sigma\mathbf{F}$$
$$R_x = \Sigma F_x \qquad R_y = \Sigma F_y \qquad R = \sqrt{(\Sigma F_x)^2 + (\Sigma F_y)^2}$$
$$\theta = \tan^{-1}\frac{R_y}{R_x} = \tan^{-1}\frac{\Sigma F_y}{\Sigma F_x}$$

(2/9)

Graphically, the correct line of action of $\mathbf{R}$ may be obtained by preserving the correct lines of action of the forces and adding them by the parallelogram law. We see this in part *a* of the figure for the case of three forces where the sum $\mathbf{R}_1$ of $\mathbf{F}_2$ and $\mathbf{F}_3$ is added to $\mathbf{F}_1$ to obtain $\mathbf{R}$. The principle of transmissibility has been used in this process.

Algebraic Method

We can use algebra to obtain the resultant force and its line of action as follows:

1. Choose a convenient reference point and move all forces to that point. This process is depicted for a three-force system in Figs. 2/14*a* and *b*, where M_1, M_2, and M_3 are the couples resulting from the transfer of forces $\mathbf{F}_1$, $\mathbf{F}_2$, and $\mathbf{F}_3$ from their respective original lines of action to lines of action through point O.

2. Add all forces at O to form the resultant force $\mathbf{R}$, and add all couples to form the resultant couple M_O. We now have the single force–couple system, as shown in Fig. 2/14*c*.

3. In Fig. 2/14*d*, find the line of action of $\mathbf{R}$ by requiring $\mathbf{R}$ to have a moment of M_O about point O. Note that the force systems of Figs. 2/14*a* and 2/14*d* are equivalent, and that $\Sigma(Fd)$ in Fig. 2/14*a* is equal to Rd in Fig. 2/14*d*.

(*b*)

Figure 2/13

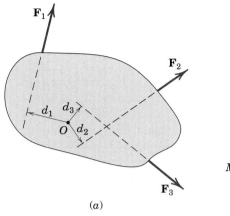

(a)

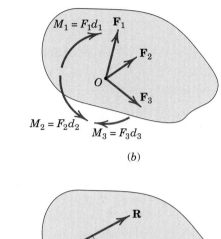

(b)

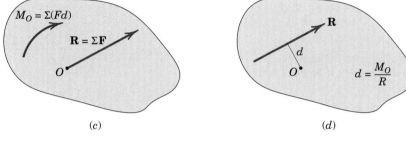

(c)   (d)

Figure 2/14

Principle of Moments

This process is summarized in equation form by

$$\mathbf{R} = \Sigma\mathbf{F}$$
$$M_O = \Sigma M = \Sigma(Fd) \qquad \textbf{(2/10)}$$
$$Rd = M_O$$

The first two of Eqs. 2/10 reduce a given system of forces to a force–couple system at an arbitrarily chosen but convenient point O. The last equation specifies the distance d from point O to the line of action of $\mathbf{R}$, and states that the moment of the resultant force about any point O equals the sum of the moments of the original forces of the system about the same point. This extends Varignon's theorem to the case of *nonconcurrent* force systems; we call this extension the *principle of moments*.

For a concurrent system of forces where the lines of action of all forces pass through a common point O, the moment sum ΣM_O about that point is zero. Thus, the line of action of the resultant $\mathbf{R} = \Sigma\mathbf{F}$, determined by the first of Eqs. 2/10, passes through point O. For a parallel force system, select a coordinate axis in the direction of the forces. If the resultant force $\mathbf{R}$ for a given force system is zero, the resultant of the system need not be zepo because the resultant may be a couple. The three forces in Fig. 2/15, for instance, have a zero resultant force but have a resultant clockwise couple $M = F_3 d$.

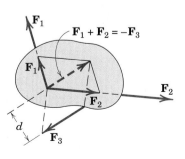

Figure 2/15

Sample Problem 2/9

Determine the resultant of the four forces and one couple which act on the plate shown.

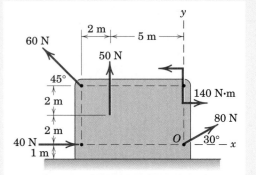

Solution.   Point O is selected as a convenient reference point for the force–couple system which is to represent the given system.

$[R_x = \Sigma F_x]$ $\qquad$ $R_x = 40 + 80 \cos 30° - 60 \cos 45° = 66.9$ N

$[R_y = \Sigma F_y]$ $\qquad$ $R_y = 50 + 80 \sin 30° + 60 \cos 45° = 132.4$ N

$[R = \sqrt{R_x{}^2 + R_y{}^2}]$ $\qquad$ $R = \sqrt{(66.9)^2 + (132.4)^2} = 148.3$ N $\qquad$ *Ans.*

$\left[\theta = \tan^{-1}\dfrac{R_y}{R_x}\right]$ $\qquad$ $\theta = \tan^{-1}\dfrac{132.4}{66.9} = 63.2°$ $\qquad$ *Ans.*

① $[M_O = \Sigma(Fd)]$ $\qquad$ $M_O = 140 - 50(5) + 60 \cos 45°(4) - 60 \sin 45°(7)$

$\qquad\qquad\qquad\qquad = -237$ N·m

The force–couple system consisting of $\mathbf{R}$ and M_O is shown in Fig. *a*.

We now determine the final line of action of $\mathbf{R}$ such that $\mathbf{R}$ alone represents the original system.

$[Rd = |M_O|]$ $\qquad$ $148.3d = 237$ $\qquad$ $d = 1.600$ m $\qquad$ *Ans.*

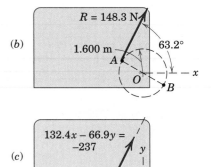

Hence, the resultant $\mathbf{R}$ may be applied at any point on the line which makes a $63.2°$ angle with the x-axis and is tangent at point A to a circle of 1.600-m radius with center O, as shown in part b of the figure. We apply the equation $Rd = M_O$ in an absolute-value sense (ignoring any sign of M_O) and let the physics of the situation, as depicted in Fig. *a*, dictate the final placement of $\mathbf{R}$. Had M_O been counterclockwise, the correct line of action of $\mathbf{R}$ would have been the tangent at point B.

The resultant $\mathbf{R}$ may also be located by determining its intercept distance b to point C on the x-axis, Fig. *c*. With R_x and R_y acting through point C, only R_y exerts a moment about O so that

$$R_y b = |M_O| \qquad \text{and} \qquad b = \frac{237}{132.4} = 1.792 \text{ m}$$

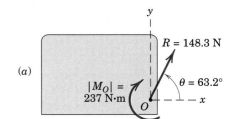

Alternatively, the y-intercept could have been obtained by noting that the moment about O would be due to R_x only.

A more formal approach in determining the final line of action of $\mathbf{R}$ is to use the vector expression

$$\mathbf{r} \times \mathbf{R} = \mathbf{M}_O$$

where $\mathbf{r} = x\mathbf{i} + y\mathbf{j}$ is a position vector running from point O to any point on the line of action of $\mathbf{R}$. Substituting the vector expressions for $\mathbf{r}$, $\mathbf{R}$, and $\mathbf{M}_O$ and carrying out the cross product result in

$$(x\mathbf{i} + y\mathbf{j}) \times (66.9\mathbf{i} + 132.4\mathbf{j}) = -237\mathbf{k}$$

$$(132.4x - 66.9y)\mathbf{k} = -237\mathbf{k}$$

Thus, the desired line of action, Fig. *c*, is given by

$$132.4x - 66.9y = -237$$

② By setting $y = 0$, we obtain $x = -1.792$ m, which agrees with our earlier calculation of the distance b.

Helpful Hints

① We note that the choice of point O as a moment center eliminates any moments due to the two forces which pass through O. Had the clockwise sign convention been adopted, M_O would have been $+237$ N·m, with the plus sign indicating a sense which agrees with the sign convention. Either sign convention, of course, leads to the conclusion of a clockwise moment M_O.

② Note that the vector approach yields sign information automatically, whereas the scalar approach is more physically oriented. You should master both methods.

PROBLEMS

Introductory Problems

2/77 Calculate the magnitude of the tension **T** and the angle θ for which the eye bolt will be under a resultant downward force of 15 kN.

Ans. $T = 12.85$ kN, $\theta = 38.9°$

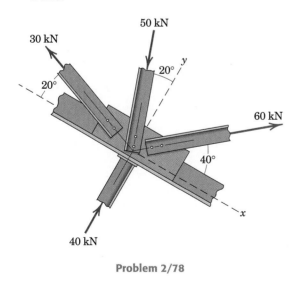

Problem 2/77

2/78 Determine the resultant **R** of the four forces acting on the gusset plate. Also find the magnitude of **R** and the angle θ_x which the resultant makes with the x-axis.

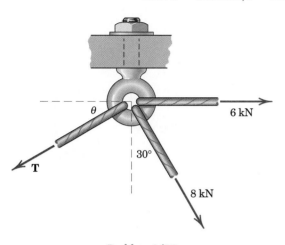

Problem 2/78

2/79 Determine the equivalent force–couple system at the origin O for each of the three cases of forces being applied to the edge of a circular disk. If the resultant can be so expressed, replace this force–couple system with a stand-alone force.

Ans. (a) **R** $= -2F$**j** along $x = -r$
(b) **R** $= -F$**i** along $y = 3r$
(c) **R** $= -F$**i** along $y = -r$

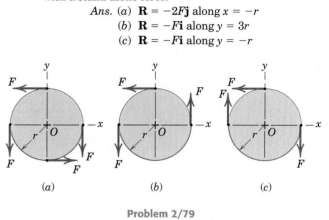

Problem 2/79

2/80 Determine the height h above the base B at which the resultant of the three forces acts.

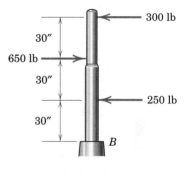

Problem 2/80

2/81 Where does the resultant of the two forces act?

Ans. 10.70 m to the left of A

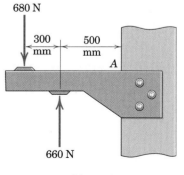

Problem 2/81

2/82 Under nonuniform and slippery road conditions, the two forces shown are exerted on the two rear-drive wheels of the pickup truck, which has a limited-slip rear differential. Determine the y-intercept of the resultant of this force system.

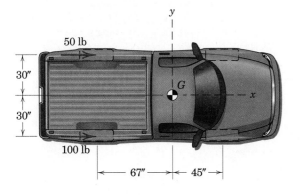

Problem 2/82

2/83 If the resultant of the two forces and couple M passes through point O, determine M.

Ans. $M = 148.0$ N·m CCW

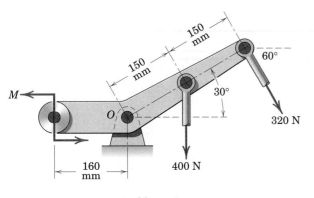

Problem 2/83

2/84 Determine the magnitude of the force **F** applied to the handle which will make the resultant of the three forces pass through O.

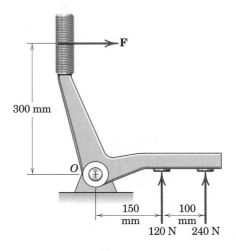

Problem 2/84

2/85 Determine and locate the resultant **R** of the two forces and one couple acting on the I-beam.

Ans. $R = 4$ kN down at $x = 5$ m

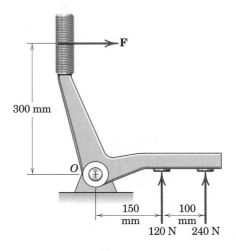

Problem 2/85

2/86 A commercial airliner with four jet engines, each producing 90 kN of forward thrust, is in a steady, level cruise when engine number 3 suddenly fails. Determine and locate the resultant of the three remaining engine thrust vectors. Treat this as a two-dimensional problem.

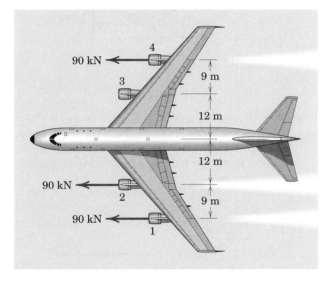

Problem 2/86

2/87 Replace the three forces acting on the bent pipe by a single equivalent force **R**. Specify the distance x from point O to the point on the x-axis through which the line of action of **R** passes.
Ans. $\mathbf{R} = -50\mathbf{i} + 20\mathbf{j}$ lb, $x = 65$ in. (off pipe)

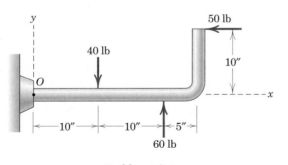

Problem 2/87

Representative Problems

2/88 The directions of the two thrust vectors of an experimental aircraft can be independently changed from the conventional forward direction within limits. For the thrust configuration shown, determine the equivalent force–couple system at point O. Then replace this force–couple system by a single force and specify the point on the x-axis through which the line of action of this resultant passes. These results are vital to assessing design performance.

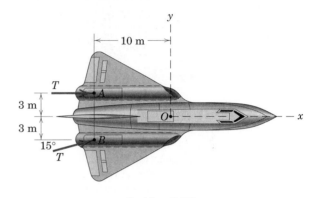

Problem 2/88

2/89 Determine the resultant **R** of the three forces acting on the simple truss. Specify the points on the x- and y-axes through which **R** must pass.
Ans. $\mathbf{R} = -15\mathbf{i} - 47.3\mathbf{j}$ kN
$x = 7.42$ m, $y = -23.4$ m

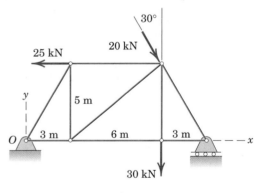

Problem 2/89

2/90 The gear and attached V-belt pulley are turning counterclockwise and are subjected to the tooth load of 1600 N and the 800-N and 450-N tensions in the V-belt. Represent the action of these three forces by a resultant force **R** at O and a couple of magnitude M. Is the unit slowing down or speeding up?

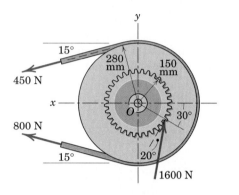

Problem 2/90

2/91 The design specifications for the attachment at A for this beam depend on the magnitude and location of the applied loads. Represent the resultant of the three forces and couple by a single force **R** at A and a couple M. Specify the magnitude of **R**.

$$Ans. \ \mathbf{R} = 1.879\mathbf{i} - 1.684\mathbf{j} \text{ kN}, \ R = 2.52 \text{ kN}$$
$$M = 14.85 \text{ kN·m CW}$$

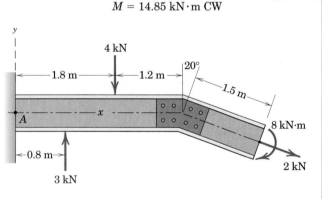

Problem 2/91

2/92 In the equilibrium position shown, the resultant of the three forces acting on the bell crank passes through the bearing O. Determine the vertical force **P**. Does the result depend on θ?

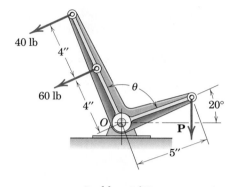

Problem 2/92

2/93 Two integral pulleys are subjected to the belt tensions shown. If the resultant **R** of these forces passes through the center O, determine T and the magnitude of **R** and the counterclockwise angle θ it makes with the x-axis.

$$Ans. \ T = 60 \text{ N}, \ R = 193.7 \text{ N}, \ \theta = 34.6°$$

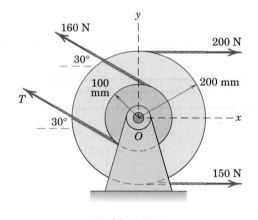

Problem 2/93

2/94 While sliding a desk toward the doorway, three students exert the forces shown in the overhead view. Determine the equivalent force–couple system at point A. Then determine the equation of the line of action of the resultant force.

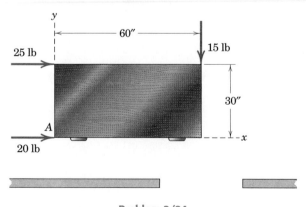

Problem 2/94

2/95 Under nonuniform and slippery road conditions, the four forces shown are exerted on the four drive wheels of the all-wheel-drive vehicle. Determine the resultant of this system and the x- and y-intercepts of its line of action. Note that the front and rear tracks are equal (i.e., $\overline{AB} = \overline{CD}$).

Ans. $\mathbf{R} = 903\mathbf{i} + 175\mathbf{j}$ N, $(0, -0.253)$ m, $(1.308, 0)$ m

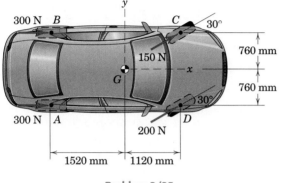

Problem 2/95

2/96 The rolling rear wheel of a front-wheel-drive automobile which is accelerating to the right is subjected to the five forces and one moment shown. The forces $A_x = 60$ lb and $A_y = 500$ lb are forces transmitted from the axle to the wheel, $F = 40$ lb is the friction force exerted by the road surface on the tire, $N = 600$ lb is the normal reaction force exerted by the road surface, and $W = 100$ lb is the weight of the wheel/tire unit. The couple $M = 2$ lb-ft is the bearing friction moment. Determine and locate the resultant of the system.

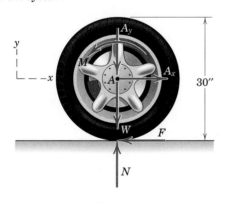

Problem 2/96

2/97 A rear-wheel-drive car is stuck in the snow between other parked cars as shown. In an attempt to free the car, three students exert forces on the car at points A, B, and C while the driver's actions result in a forward thrust of 40 lb acting parallel to the plane of rotation of each rear wheel. Treating the problem as two-dimensional, determine the equivalent force–couple system at the car center of mass G and locate the position x of the point on the car centerline through which the resultant passes. Neglect all forces not shown.

Ans. $\mathbf{R} = 185\mathbf{i} + 113.3\mathbf{j}$ lb
$M_G = 460$ lb-ft CCW
$x = 4.06$ ft

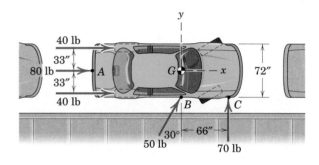

Problem 2/97

2/98 An exhaust system for a pickup truck is shown in the figure. The weights W_h, W_m, and W_t of the headpipe, muffler, and tailpipe are 10, 100, and 50 N, respectively, and act at the indicated points. If the exhaust-pipe hanger at point A is adjusted so that its tension F_A is 50 N, determine the required forces in the hangers at points B, C, and D so that the force–couple system at point O is zero. Why is a zero force–couple system at O desirable?

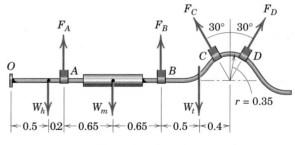

Dimensions in meters

Problem 2/98

SECTION B THREE-DIMENSIONAL FORCE SYSTEMS

2/7 RECTANGULAR COMPONENTS

Many problems in mechanics require analysis in three dimensions, and for such problems it is often necessary to resolve a force into its three mutually perpendicular components. The force $\mathbf{F}$ acting at point O in Fig. 2/16 has the *rectangular components* F_x, F_y, F_z, where

$$
\begin{aligned}
F_x &= F \cos \theta_x & F &= \sqrt{F_x^{\,2} + F_y^{\,2} + F_z^{\,2}} \\
F_y &= F \cos \theta_y & \mathbf{F} &= F_x \mathbf{i} + F_y \mathbf{j} + F_z \mathbf{k} \\
F_z &= F \cos \theta_z & \mathbf{F} &= F(\mathbf{i} \cos \theta_x + \mathbf{j} \cos \theta_y + \mathbf{k} \cos \theta_z)
\end{aligned}
\tag{2/11}
$$

The unit vectors $\mathbf{i}$, $\mathbf{j}$, and $\mathbf{k}$ are in the x-, y-, and z-directions, respectively. Using the direction cosines of $\mathbf{F}$, which are $l = \cos \theta_x$, $m = \cos \theta_y$, and $n = \cos \theta_z$, where $l^2 + m^2 + n^2 = 1$, we may write the force as

$$
\mathbf{F} = F(l\mathbf{i} + m\mathbf{j} + n\mathbf{k})
\tag{2/12}
$$

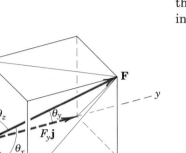

Figure 2/16

We may regard the right-side expression of Eq. 2/12 as the force magnitude F times a unit vector $\mathbf{n}_F$ which characterizes the direction of $\mathbf{F}$, or

$$
\mathbf{F} = F\mathbf{n}_F
\tag{2/12a}
$$

It is clear from Eqs. 2/12 and 2/12a that $\mathbf{n}_F = l\mathbf{i} + m\mathbf{j} + n\mathbf{k}$, which shows that the scalar components of the unit vector $\mathbf{n}_F$ are the direction cosines of the line of action of $\mathbf{F}$.

In solving three-dimensional problems, one must usually find the x, y, and z scalar components of a force. In most cases, the direction of a force is described (*a*) by two points on the line of action of the force or (*b*) by two angles which orient the line of action.

(a) Specification by two points on the line of action of the force.
If the coordinates of points A and B of Fig. 2/17 are known, the force $\mathbf{F}$ may be written as

$$
\mathbf{F} = F\mathbf{n}_F = F \frac{\overrightarrow{AB}}{AB} = F \frac{(x_2 - x_1)\mathbf{i} + (y_2 - y_1)\mathbf{j} + (z_2 - z_1)\mathbf{k}}{\sqrt{(x_2 - x_1)^2 + (y_2 - y_1)^2 + (z_2 - z_1)^2}}
$$

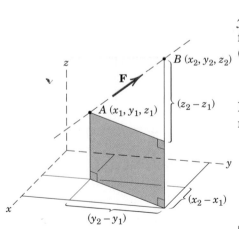

Figure 2/17

Thus the x, y, and z scalar components of $\mathbf{F}$ are the scalar coefficients of the unit vectors $\mathbf{i}$, $\mathbf{j}$, and $\mathbf{k}$, respectively.

(b) Specification by two angles which orient the line of action of the force. Consider the geometry of Fig. 2/18. We assume that the angles θ and ϕ are known. First resolve **F** into horizontal and vertical components.

$$F_{xy} = F \cos \phi$$
$$F_z = F \sin \phi$$

Then resolve the horizontal component F_{xy} into x- and y-components.

$$F_x = F_{xy} \cos \theta = F \cos \phi \cos \theta$$
$$F_y = F_{xy} \sin \theta = F \cos \phi \sin \theta$$

The quantities F_x, F_y, and F_z are the desired scalar components of **F**.

The choice of orientation of the coordinate system is arbitrary, with convenience being the primary consideration. However, we must use a right-handed set of axes in our three-dimensional work to be consistent with the right-hand-rule definition of the cross product. When we rotate from the x- to the y-axis through the 90° angle, the positive direction for the z-axis in a right-handed system is that of the advancement of a right-handed screw rotated in the same sense. This is equivalent to the right-hand rule.

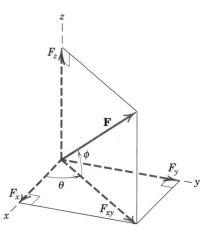

Figure 2/18

Dot Product

We can express the rectangular components of a force **F** (or any other vector) with the aid of the vector operation known as the *dot* or *scalar product* (see item 6 in Art. C/7 of Appendix C). The dot product of two vectors **P** and **Q**, Fig. 2/19a, is defined as the product of their magnitudes times the cosine of the angle α between them. It is written as

$$\mathbf{P} \cdot \mathbf{Q} = PQ \cos \alpha$$

We can view this product either as the orthogonal projection $P \cos \alpha$ of **P** in the direction of **Q** multiplied by Q, or as the orthogonal projection $Q \cos \alpha$ of **Q** in the direction of **P** multiplied by P. In either case the dot product of the two vectors is a scalar quantity. Thus, for instance, we can express the scalar component $F_x = F \cos \theta_x$ of the force **F** in Fig. 2/16 as $F_x = \mathbf{F} \cdot \mathbf{i}$, where **i** is the unit vector in the x-direction.

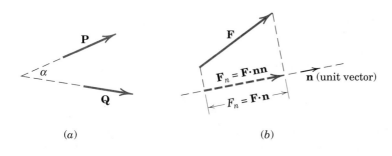

(a) (b)

Figure 2/19

In more general terms, if **n** is a unit vector in a specified direction, the projection of **F** in the **n**-direction, Fig. 2/19b, has the magnitude $F_n = \mathbf{F} \cdot \mathbf{n}$. If we want to express the projection in the **n**-direction as a vector quantity, then we multiply its scalar component, expressed by $\mathbf{F} \cdot \mathbf{n}$, by the unit vector **n** to give $\mathbf{F}_n = (\mathbf{F} \cdot \mathbf{n})\mathbf{n}$. We may write this as $\mathbf{F}_n = \mathbf{F} \cdot \mathbf{nn}$ without ambiguity because the term **nn** is not defined, and so the complete expression cannot be misinterpreted as $\mathbf{F} \cdot (\mathbf{nn})$.

If the direction cosines of **n** are α, β, and γ, then we may write **n** in vector component form like any other vector as

$$\mathbf{n} = \alpha\mathbf{i} + \beta\mathbf{j} + \gamma\mathbf{k}$$

where in this case its magnitude is unity. If the direction cosines of **F** with respect to reference axes x-y-z are l, m, and n, then the projection of **F** in the **n**-direction becomes

$$F_n = \mathbf{F} \cdot \mathbf{n} = F(l\mathbf{i} + m\mathbf{j} + n\mathbf{k}) \cdot (\alpha\mathbf{i} + \beta\mathbf{j} + \gamma\mathbf{k})$$
$$= F(l\alpha + m\beta + n\gamma)$$

because

$$\mathbf{i} \cdot \mathbf{i} = \mathbf{j} \cdot \mathbf{j} = \mathbf{k} \cdot \mathbf{k} = 1$$

and

$$\mathbf{i} \cdot \mathbf{j} = \mathbf{j} \cdot \mathbf{i} = \mathbf{i} \cdot \mathbf{k} = \mathbf{k} \cdot \mathbf{i} = \mathbf{j} \cdot \mathbf{k} = \mathbf{k} \cdot \mathbf{j} = 0$$

The latter two sets of equations are true because **i**, **j**, and **k** have unit length and are mutually perpendicular.

Angle between Two Vectors

If the angle between the force **F** and the direction specified by the unit vector **n** is θ, then from the dot-product definition we have $\mathbf{F} \cdot \mathbf{n} = Fn \cos \theta = F \cos \theta$, where $|\mathbf{n}| = n = 1$. Thus, the angle between **F** and **n** is given by

$$\theta = \cos^{-1} \frac{\mathbf{F} \cdot \mathbf{n}}{F} \qquad (2/13)$$

In general, the angle between any two vectors **P** and **Q** is

$$\theta = \cos^{-1} \frac{\mathbf{P} \cdot \mathbf{Q}}{PQ} \qquad (2/13a)$$

If a force **F** is perpendicular to a line whose direction is specified by the unit vector **n**, then $\cos \theta = 0$, and $\mathbf{F} \cdot \mathbf{n} = 0$. Note that this relationship does not mean that either **F** or **n** is zero, as would be the case with scalar multiplication where $(A)(B) = 0$ requires that either A or B (or both) be zero.

The dot-product relationship applies to nonintersecting vectors as well as to intersecting vectors. Thus, the dot product of the noninter-secting vectors **P** and **Q** in Fig. 2/20 is Q times the projection of **P**$'$ on **Q**, or $P'Q \cos \alpha = PQ \cos \alpha$ because **P**$'$ and **P** are the same when treated as free vectors.

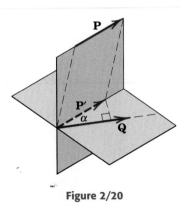

Figure 2/20

Sample Problem 2/10

A force **F** with a magnitude of 100 N is applied at the origin O of the axes x-y-z as shown. The line of action of **F** passes through a point A whose coordinates are 3 m, 4 m, and 5 m. Determine (a) the x, y, and z scalar components of **F**, (b) the projection F_{xy} of **F** on the x-y plane, and (c) the projection F_{OB} of **F** along the line OB.

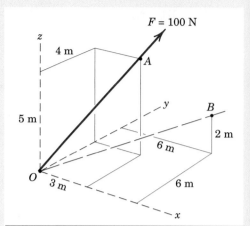

Solution.   **Part (a).**   We begin by writing the force vector **F** as its magnitude F times a unit vector $\mathbf{n}_{OA}$.

$$\mathbf{F} = F\mathbf{n}_{OA} = F\,\frac{\overrightarrow{OA}}{OA} = 100\left[\frac{3\mathbf{i} + 4\mathbf{j} + 5\mathbf{k}}{\sqrt{3^2 + 4^2 + 5^2}}\right]$$

$$= 100[0.424\mathbf{i} + 0.566\mathbf{j} + 0.707\mathbf{k}]$$

$$= 42.4\mathbf{i} + 56.6\mathbf{j} + 70.7\mathbf{k} \text{ N}$$

The desired scalar components are thus

① $\qquad\qquad F_x = 42.4 \text{ N} \qquad F_y = 56.6 \text{ N} \qquad F_z = 70.7 \text{ N} \qquad\qquad$ *Ans.*

Part (b).   The cosine of the angle θ_{xy} between **F** and the x-y plane is

$$\cos\theta_{xy} = \frac{\sqrt{3^2 + 4^2}}{\sqrt{3^2 + 4^2 + 5^2}} = 0.707$$

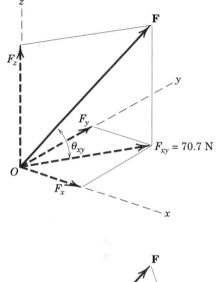

so that $F_{xy} = F\cos\theta_{xy} = 100(0.707) = 70.7 \text{ N}$ $\qquad\qquad$ *Ans.*

Part (c).   The unit vector $\mathbf{n}_{OB}$ along OB is

$$\mathbf{n}_{OB} = \frac{\overrightarrow{OB}}{OB} = \frac{6\mathbf{i} + 6\mathbf{j} + 2\mathbf{k}}{\sqrt{6^2 + 6^2 + 2^2}} = 0.688\mathbf{i} + 0.688\mathbf{j} + 0.229\mathbf{k}$$

The scalar projection of **F** on OB is

② $F_{OB} = \mathbf{F}\cdot\mathbf{n}_{OB} = (42.4\mathbf{i} + 56.6\mathbf{h} + 70.7\mathbf{k})\cdot(0.688\mathbf{i} + 0.688\mathbf{j} + 0.229\mathbf{k})$

$$= (42.4)(0.688) + (56.6)(0.688) + (70.7)(0.229)$$

$$= 84.4\text{N} \qquad\qquad\qquad\qquad\qquad\qquad\qquad \textit{Ans.}$$

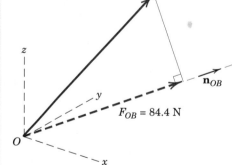

If we wish to express the projection as a vector, we write

$$\mathbf{F}_{OB} = \mathbf{F}\cdot\mathbf{n}_{OB}\mathbf{n}_{OB}$$

$$= 84.4(0.688\mathbf{i} + 0.688\mathbf{j} + 0.229\mathbf{k})$$

$$= 58.1\mathbf{i} + 58.1\mathbf{j} + 19.35\mathbf{k} \text{ N}$$

Helpful Hints

① In this example all scalar components are positive. Be prepared for the case where a direction cosine, and hence the scalar component, are negative.

② The dot product automatically finds the projection or scalar component of **F** along line OB as shown.

PROBLEMS

Introductory Problems

2/99 Express **F** as a vector in terms of the unit vectors **i**, **j**, and **k**. Determine the angle between **F** and the *y*-axis.

$$Ans.\ \mathbf{F} = 18.86\mathbf{i} - 23.6\mathbf{j} + 51.9\mathbf{k}\ N$$

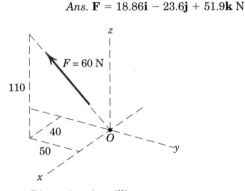

Dimensions in millimeters

Problem 2/99

2/100 The 70-m microwave transmission tower is steadied by three guy cables as shown. Cable *AB* carries a tension of 12 kN. Express the corresponding force on point *B* as a vector.

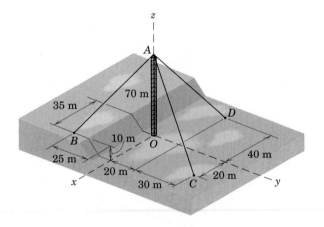

Problem 2/100

2/101 Express the force **F** as a vector in terms of the unit vectors **i**, **j**, and **k**. Determine the angles θ_x, θ_y, and θ_z which **F** makes with the positive *x*-, *y*-, and *z*-axes.

$$Ans.\ \mathbf{F} = -290\mathbf{i} + 507\mathbf{j} + 471\mathbf{k}\ lb$$
$$\theta_x = 112.7°,\ \theta_y = 47.5°,\ \theta_z = 51.1°$$

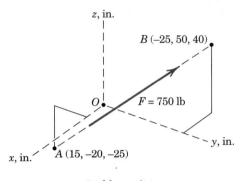

Problem 2/101

2/102 Express **F** as a vector in terms of the unit vectors **i**, **j**, and **k**. Determine the projection, both as a scalar and as a vector, of **F** onto line *OA*, which lies in the *x-y* plane.

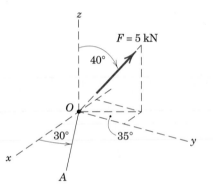

Problem 2/102

2/103 The force **F** has a magnitude of 300 lb and acts along the diagonal of the parallelepiped as shown. Express **F** in terms of its magnitude times the appropriate unit vector and determine its *x*-, *y*-, and *z*-components.

$$Ans.\ \mathbf{F} = 300 \left[\tfrac{1}{3}\mathbf{i} - \tfrac{2}{3}\mathbf{j} - \tfrac{2}{3}\mathbf{k}\right]\ lb$$
$$F_x = 100\ lb,\ F_y = -200\ lb,\ F_z = -200\ lb$$

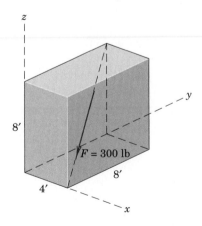

Problem 2/103

2/104 The turnbuckle is tightened until the tension in the cable AB equals 2.4 kN. Determine the vector expression for the tension **T** as a force acting on member AD. Also find the magnitude of the projection of **T** along the line AC.

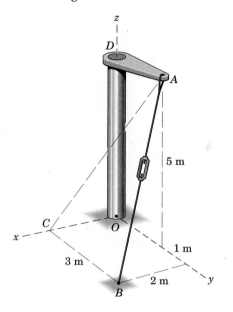

Problem 2/104

2/105 The rigid pole and cross-arm assembly is supported by the three cables shown. A turnbuckle at D is tightened until it induces a tension T in CD of 1.2 kN. Express **T** as a vector. Does it make any difference in the result which coordinate system is used?

Ans. **T** $= 0.321\mathbf{i} + 0.641\mathbf{j} - 0.962\mathbf{k}$ kN, No

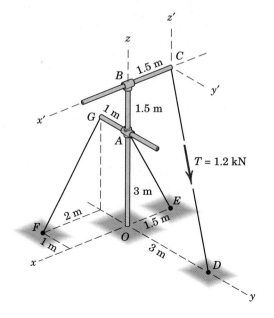

Problem 2/105

2/106 Use the result cited for Prob. 2/105 and determine the magnitude T_{GF} of the projection of **T** onto line GF.

Representative Problems

2/107 The tension in the supporting cable AB is 10 kN. Write the force which the cable exerts on the boom BC as a vector **T**. Determine the angles θ_x, θ_y, and θ_z which the line of action of **T** forms with the positive x-, y-, and z-axes.

Ans. **T** $= 10[0.406\mathbf{i} - 0.761\mathbf{j} + 0.507\mathbf{k}]$ kN
$\theta_x = 66.1°$, $\theta_y = 139.5°$, $\theta_z = 59.5°$

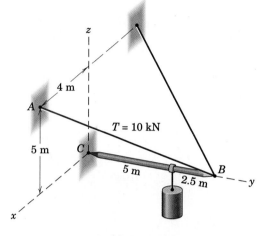

Problem 2/107

2/108 The cable BC carries a tension of 750 N. Write this tension as a force **T** acting on point B in terms of the unit vectors **i**, **j**, and **k**. The elbow at A forms a right angle.

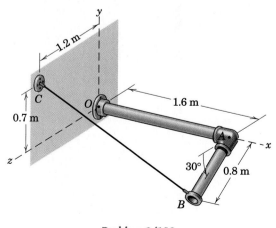

Problem 2/108

2/109 Derive the expression for the projection F_{DC} of the force **F** onto the line directed from D to C.

$$Ans. \ F_{DC} = \frac{(b^2 - a^2)F}{\sqrt{a^2 + b^2}\sqrt{a^2 + b^2 + c^2}}$$

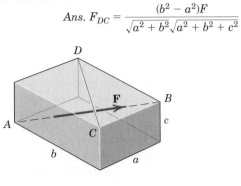

Problem 2/109

2/110 If M is the center of the face $ABCD$ of the rectangular solid, express **F** as a vector in terms of the unit vectors **i**, **j**, and **k**. Determine the scalar projection of **F** along line AE. The quantities F, a, b, and c are known.

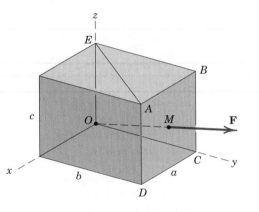

Problem 2/110

2/111 Determine the angle θ between the 200-lb force and line OC.

$$Ans. \ \theta = 54.9°$$

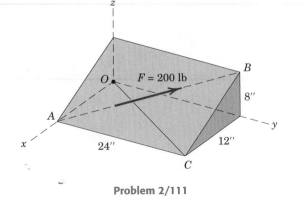

Problem 2/111

2/112 The rectangular plate is supported by hinges along its side BC and by the cable AE. If the cable tension is 300 N, determine the projection onto line BC of the force exerted on the plate by the cable. Note that E is the midpoint of the horizontal upper edge of the structural support.

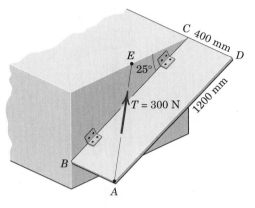

Problem 2/112

2/113 The access door is held in the 30° open position by the chain AB. If the tension in the chain is 100 N, determine the projection of the tension force onto the diagonal axis CD of the door.

$$Ans. \ T_{CD} = 46.0 \text{ N}$$

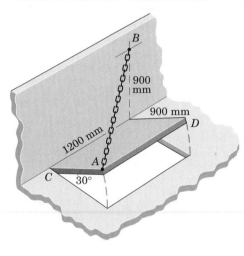

Problem 2/113

▶**2/114** The spring of constant $k = 15$ lb/in. is attached to the disk at point A and to the end fitting at point B as shown. The spring is unstretched when θ_A and θ_B are both zero. If the disk is rotated 15° clockwise and the end fitting is rotated 30° counterclockwise, determine a vector expression for the force **F** which the spring exerts at point A. The magnitude of the spring force is the constant k multiplied by the deflection (lengthening or shortening) of the spring.

Ans. $\mathbf{F} = -0.719\mathbf{i} + 9.48\mathbf{j} - 1.734\mathbf{k}$ lb

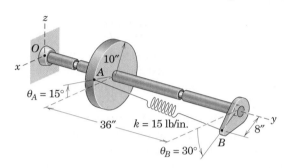

Problem 2/114

▶**2/115** Determine the x-, y-, and z-components of force **F** which acts on the tetrahedron as shown. The quantities a, b, c, and F are known, and M is the midpoint of edge AB.

Ans. $F_x = \dfrac{2acF}{\sqrt{a^2 + b^2}\sqrt{a^2 + b^2 + 4c^2}}$

$F_y = \dfrac{2bcF}{\sqrt{a^2 + b^2}\sqrt{a^2 + b^2 + 4c^2}}$

$F_z = F\sqrt{\dfrac{a^2 + b^2}{a^2 + b^2 + 4c^2}}$

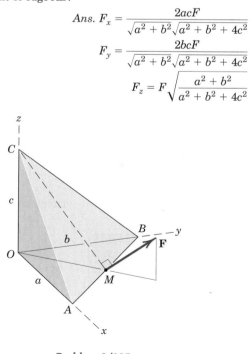

Problem 2/115

▶**2/116** A force **F** is applied to the surface of the sphere as shown. The angles θ and ϕ locate point P, and point M is the midpoint of ON. Express **F** in vector form, using the given x-, y-, and z-coordinates.

Ans. $\mathbf{F} = F\left[\dfrac{(2\sin\phi - 1)(\mathbf{i}\cos\theta + \mathbf{j}\sin\theta) + \mathbf{k}(2\cos\phi)}{\sqrt{5 - 4\sin\phi}}\right]$

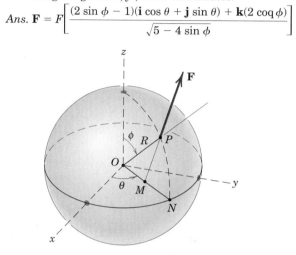

Problem 2/116

2/8 MOMENT AND COUPLE

In two-dimensional analyses it is often convenient to determine a moment magnitude by scalar multiplication using the moment-arm rule. In three dimensions, however, the determination of the perpendicular distance between a point or line and the line of action of the force can be a tedious computation. A vector approach with cross-product multiplication then becomes advantageous.

Moments in Three Dimensions

Consider a force **F** with a given line of action acting on a body, Fig. 2/21*a*, and any point *O* not on this line. Point *O* and the line of **F** establish a plane *A*. The moment **M**$_O$ of **F** about an axis through *O* normal to the plane has the magnitude $M_O = Fd$, where *d* is the perpendicular distance from *O* to the line of **F**. This moment is also referred to as the moment of **F** about the *point O*.

The vector **M**$_O$ is normal to the plane and is directed along the axis through *O*. We can describe both the magnitude and the direction of **M**$_O$ by the vector cross-product relation introduced in Art. 2/4. (Refer to item 7 in Art. C/7 of Appendix C.) The vector **r** runs from *O* to *any* point on the line of action of **F**. As described in Art. 2/4, the cross product of **r** and **F** is written **r** × **F** and has the magnitude $(r \sin \alpha)F$, which is the same as Fd, the magnitude of **M**$_O$.

The correct direction and sense of the moment are established by the right-hand rule, described previously in Arts. 2/4 and 2/5. Thus, with **r** and **F** treated as free vectors emanating from *O*, Fig. 2/21*b*, the thumb points in the direction of **M**$_O$ if the fingers of the right hand curl in the direction of rotation from **r** to **F** through the angle α. Therefore, we may write the moment of **F** about the axis through *O* as

$$\boxed{\mathbf{M}_O = \mathbf{r} \times \mathbf{F}} \qquad (2/14)$$

The order **r** × **F** of the vectors *must* be maintained because **F** × **r** would produce a vector with a sense opposite to that of **M**$_O$; that is, $\mathbf{F} \times \mathbf{r} = -\mathbf{M}_O$.

Evaluating the Cross Product

The cross-product expression for **M**$_O$ may be written in the determinant form

$$\boxed{\mathbf{M}_O = \begin{vmatrix} \mathbf{i} & \mathbf{j} & \mathbf{k} \\ r_x & r_y & r_z \\ F_x & F_y & F_z \end{vmatrix}} \qquad (2/15)$$

(Refer to item 7 in Art. C/7 of Appendix C if you are not already familiar with the determinant representation of the cross product.) Note the symmetry and order of the terms, and note that a *right-handed* coordinate system must be used. Expansion of the determinant gives

$$\mathbf{M}_O = (r_y F_z - r_z F_y)\mathbf{i} + (r_z F_x - r_x F_z)\mathbf{j} + (r_x F_y - r_y F_x)\mathbf{k}$$

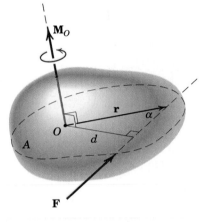

(*a*)

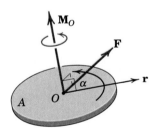

(*b*)

Figure 2/21

To gain more confidence in the cross-product relationship, examine the three components of the moment of a force about a point as obtained from Fig. 2/22. This figure shows the three components of a force **F** acting at a point A located relative to O by the vector **r**. The scalar magnitudes of the moments of these forces about the positive x-, y-, and z-axes through O can be obtained from the moment-arm rule, and are

$$M_x = r_y F_z - r_z F_y \qquad M_y = r_z F_x - r_x F_z \qquad M_z = r_x F_y - r_y F_x$$

which agree with the respective terms in the determinant expansion for the cross product **r** × **F**.

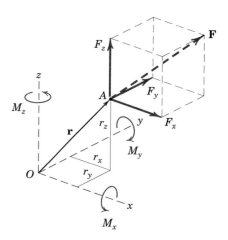

Figure 2/22

Moment about an Arbitrary Axis

We can now obtain an expression for the moment $\mathbf{M}_\lambda$ of **F** about *any* axis λ through O, as shown in Fig. 2/23. If **n** is a unit vector in the λ-direction, then we can use the dot-product expression for the component of a vector as described in Art. 2/7 to obtain $\mathbf{M}_O \cdot \mathbf{n}$, the component of $\mathbf{M}_O$ in the direction of λ. This scalar is the magnitude of the moment $\mathbf{M}_\lambda$ of **F** about λ.

To obtain the vector expression for the moment $\mathbf{M}_\lambda$ of **F** about λ, multiply the magnitude by the directional unit vector **n** to obtain

$$\mathbf{M}_\lambda = (\mathbf{r} \times \mathbf{F} \cdot \mathbf{n})\mathbf{n} \tag{2/16}$$

where **r** × **F** replaces $\mathbf{M}_O$. The expression **r** × **F** · **n** is known as a *triple scalar product* (see item 8 in Art. C/7, Appendix C). It need not be written (**r** × **F**) · **n** because a cross product cannot be formed by a vector and a scalar. Thus, the association **r** × (**F** · **n**) would have no meaning.

The triple scalar product may be represented by the determinant

$$|\mathbf{M}_\lambda| = M_\lambda = \begin{vmatrix} r_x & r_y & r_z \\ F_x & F_y & F_z \\ \alpha & \beta & \gamma \end{vmatrix} \tag{2/17}$$

where α, β, γ are the direction cosines of the unit vector **n**.

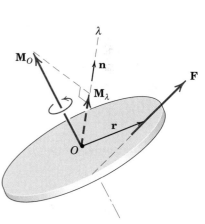

Figure 2/23

Varignon's Theorem in Three Dimensions

In Art. 2/4 we introduced Varignon's theorem in two dimensions. The theorem is easily extended to three dimensions. Figure 2/24 shows a system of concurrent forces $\mathbf{F}_1, \mathbf{F}_2, \mathbf{F}_3, \ldots$. The sum of the moments about O of these forces is

$$\mathbf{r} \times \mathbf{F}_1 + \mathbf{r} \times \mathbf{F}_2 + \mathbf{r} \times \mathbf{F}_3 + \cdots = \mathbf{r} \times (\mathbf{F}_1 + \mathbf{F}_2 + \mathbf{F}_3 + \cdots)$$

$$= \mathbf{r} \times \Sigma\mathbf{F}$$

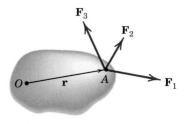

Figure 2/24

where we have used the distributive law for cross products. Using the symbol $\mathbf{M}_O$ to represent the sum of the moments on the left side of the above equation, we have

$$\mathbf{M}_O = \Sigma(\mathbf{r} \times \mathbf{F}) = \mathbf{r} \times \mathbf{R}$$

(2/18)

This equation states that the sum of the moments of a system of concurrent forces about a given point equals the moment of their sum about the same point. As mentioned in Art. 2/4, this principle has many applications in mechanics.

Couples in Three Dimensions

The concept of the couple was introduced in Art. 2/5 and is easily extended to three dimensions. Figure 2/25 shows two equal and opposite forces $\mathbf{F}$ and $-\mathbf{F}$ acting on a body. The vector $\mathbf{r}$ runs from *any* point B on the line of action of $-\mathbf{F}$ to *any* point A on the line of action of $\mathbf{F}$. Points A and B are located by position vectors $\mathbf{r}_A$ and $\mathbf{r}_B$ from *any* point O. The combined moment of the two forces about O is

$$\mathbf{M} = \mathbf{r}_A \times \mathbf{F} + \mathbf{r}_B \times (-\mathbf{F}) = (\mathbf{r}_A - \mathbf{r}_B) \times \mathbf{F}$$

However, $\mathbf{r}_A - \mathbf{r}_B = \mathbf{r}$, so that all reference to the moment center O disappears, and the moment of the couple becomes

$$\mathbf{M} = \mathbf{r} \times \mathbf{F}$$

(2/19)

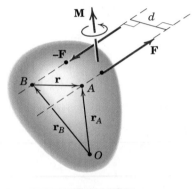

Figure 2/25

Thus, the moment of a couple is the *same about all points*. The magnitude of $\mathbf{M}$ is $M = Fd$, where d is the perpendicular distance between the lines of action of the two forces, as described in Art. 2/5.

The moment of a couple is a *free vector*, whereas the moment of a force about a point (which is also the moment about a defined axis through the point) is a *sliding vector* whose direction is along the axis through the point. As in the case of two dimensions, a couple tends to produce a pure rotation of the body about an axis normal to the plane of the forces which constitute the couple.

Couple vectors obey all of the rules which govern vector quantities. Thus, in Fig. 2/26 the couple vector $\mathbf{M}_1$ due to $\mathbf{F}_1$ and $-\mathbf{F}_1$ may be added

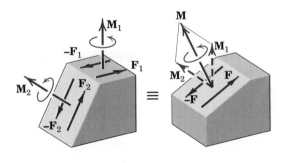

Figure 2/26

as shown to the couple vector $\mathbf{M}_2$ due to $\mathbf{F}_2$ and $-\mathbf{F}_2$ to produce the couple $\mathbf{M}$, which, in turn, can be produced by $\mathbf{F}$ and $-\mathbf{F}$.

In Art. 2/5 we learned how to replace a force by its equivalent force–couple system. You should also be able to carry out this replacement in three dimensions. The procedure is represented in Fig. 2/27, where the force $\mathbf{F}$ acting on a rigid body at point A is replaced by an equal force at point B and the couple $\mathbf{M} = \mathbf{r} \times \mathbf{F}$. By adding the equal and opposite forces $\mathbf{F}$ and $-\mathbf{F}$ at B, we obtain the couple composed of $-\mathbf{F}$ and the original $\mathbf{F}$. Thus, we see that the couple vector is simply the moment of the original force about the point to which the force is being moved. We emphasize that $\mathbf{r}$ is a vector which runs from B to *any* point on the line of action of the original force passing through A.

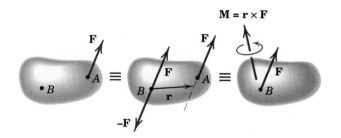

Figure 2/27

Sample Problem 2/11

Determine the moment of force **F** about point O (a) by inspection and (b) by the formal cross-product definition $\mathbf{M}_O = \mathbf{r} \times \mathbf{F}$.

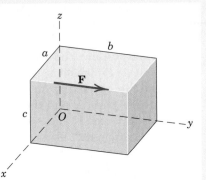

Solution. (a) Because **F** is parallel to the y-axis, **F** has no moment about that axis. It should be clear that the moment arm from the x-axis to the line of action of **F** is c and that the moment of **F** about the x-axis is negative. Similarly, the moment arm from the z-axis to the line of action of **F** is a and the moment of **F** about the z-axis is positive. So we have

$$\mathbf{M}_O = -cF\mathbf{i} + aF\mathbf{k} = F(-c\mathbf{i} + a\mathbf{k}) \qquad Ans.$$

(b) Formally,

① $$\mathbf{M}_O = \mathbf{r} \times \mathbf{F} = (a\mathbf{i} + c\mathbf{k}) \times F\mathbf{j} = aF\mathbf{k} - cF\mathbf{i}$$

$$= F(-c\mathbf{i} + a\mathbf{k}) \qquad Ans.$$

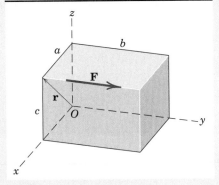

Helpful Hint

① Again we stress that **r** runs *from* the moment center *to* the line of action of **F**. Another permissible, but less convenient, position vector is $\mathbf{r} = a\mathbf{i} + b\mathbf{j} + c\mathbf{k}$.

Sample Problem 2/12

The turnbuckle is tightened until the tension in cable AB is 2.4 kN. Determine the moment about point O of the cable force acting on point A and the magnitude of this moment.

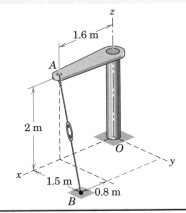

Solution. We begin by writing the described force as a vector.

$$\mathbf{T} = T\mathbf{n}_{AB} = 2.4\left[\frac{0.8\mathbf{i} + 1.5\mathbf{j} - 2\mathbf{k}}{\sqrt{0.8^2 + 1.5^2 + 2^2}}\right]$$

$$= 0.731\mathbf{i} + 1.371\mathbf{j} - 1.829\mathbf{k} \text{ kN}$$

The moment of this force about point O is

① $$\mathbf{M}_O = \mathbf{r}_{OA} \times \mathbf{T} = (1.6\mathbf{i} + 2\mathbf{k}) \times (0.731\mathbf{i} + 1.371\mathbf{j} - 1.829\mathbf{k})$$

$$= -2.74\mathbf{i} + 4.39\mathbf{j} + 2.19\mathbf{k} \text{ kN·m} \qquad Ans.$$

This vector has a magnitude

$$M_O = \sqrt{2.74^2 + 4.39^2 + 2.19^2} = 5.62 \text{ kN·m} \qquad Ans.$$

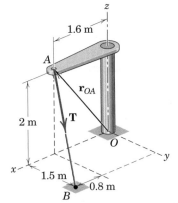

Helpful Hint

① The student should verify by inspection the signs of the moment components.

Sample Problem 2/13

A tension $\mathbf{T}$ of magnitude 10 kN is applied to the cable attached to the top A of the rigid mast and secured to the ground at B. Determine the moment M_z of $\mathbf{T}$ about the z-axis passing through the base O.

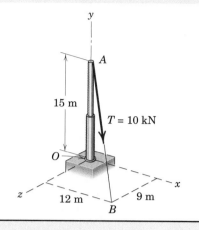

Solution (a).   The required moment may be obtained by finding the component along the z-axis of the moment $\mathbf{M}_O$ of $\mathbf{T}$ about point O. The vector $\mathbf{M}_O$ is normal to the plane defined by $\mathbf{T}$ and point O, as shown in the accompanying figure. In the use of Eq. 2/14 to find $\mathbf{M}_O$, the vector $\mathbf{r}$ is any vector from point O to the line of action of $\mathbf{T}$. The simplest choice is the vector from O to A, which is written as $\mathbf{r} = 15\mathbf{j}$ m. The vector expression for $\mathbf{T}$ is

$$\mathbf{T} = T\mathbf{n}_{AB} = 10 \left[\frac{12\mathbf{i} - 15\mathbf{j} + 9\mathbf{k}}{\sqrt{(12)^2 + (-15)^2 + (9)^2}} \right]$$

$$= 10(0.566\mathbf{i} - 0.707\mathbf{j} + 0.424\mathbf{k}) \text{ kN}$$

From Eq. 2/14,

$$[\mathbf{M}_O = \mathbf{r} \times \mathbf{F}] \qquad \mathbf{M}_O = 15\mathbf{j} \times 10(0.566\mathbf{i} - 0.707\mathbf{j} + 0.424\mathbf{k})$$

$$= 150(-0.566\mathbf{k} + 0.424\mathbf{i}) \text{ kN·m}$$

The value M_z of the desired moment is the scalar component of $\mathbf{M}_O$ in the z-direction or $M_z = \mathbf{M}_O \cdot \mathbf{k}$. Therefore,

$$M_z = 150(-0.566\mathbf{k} + 0.424\mathbf{i}) \cdot \mathbf{k} = -84.9 \text{ kN·m} \qquad \textit{Ans.}$$

The minus sign indicates that the vector $\mathbf{M}_z$ is in the negative z-direction. Expressed as a vector, the moment is $\mathbf{M}_z = -84.9\mathbf{k}$ kN·m.

Solution (b).   The force of magnitude T is resolved into components T_z and T_{xy} in the x-y plane. Since T_z is parallel to the z-axis, it can exert no moment about this axis. The moment M_z is, then, due only to T_{xy} and is $M_z = T_{xy}d$, where d is the perpendicular distance from T_{xy} to O. The cosine of the angle between T and T_{xy} is $\sqrt{15^2 + 12^2} / \sqrt{15^2 + 12^2 + 9^2} = 0.906$, and therefore,

$$T_{xy} = 10(0.906) = 9.06 \text{ kN}$$

The moment arm d equals $\overline{OA}$ multiplied by the sine of the angle between T_{xy} and OA, or

$$d = 15 \frac{12}{\sqrt{12^2 + 15^2}} = 9.37 \text{ m}$$

Hence, the moment of $\mathbf{T}$ about the z-axis has the magnitude

$$M_z = 9.06(9.37) = 84.9 \text{ kN·m} \qquad \textit{Ans.}$$

and is clockwise when viewed in the x-y plane.

Solution (c).   The component T_{xy} is further resolved into its components T_x and T_y. It is clear that T_y exerts no moment about the z-axis since it passes through it, so that the required moment is due to T_x alone. The direction cosine of $\mathbf{T}$ with respect to the x-axis is $12/\sqrt{9^2 + 12^2 + 15^2} = 0.566$ so that $T_x = 10(0.566) = 5.66$ kN. Thus,

$$M_z = 5.66(15) = 84.9 \text{ kN·m} \qquad \textit{Ans.}$$

Helpful Hints

① We could also use the vector from O to B for $\mathbf{r}$ and obtain the same result, but using vector OA is simpler.

② It is always helpful to accompany your vector operations with a sketch of the vectors so as to retain a clear picture of the geometry of the problem.

③ Sketch the x-y view of the problem and show d.

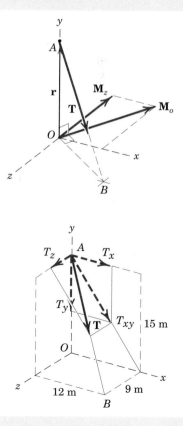

Sample Problem 2/14

Determine the magnitude and direction of the couple **M** which will replace the two given couples and still produce the same external effect on the block. Specify the two forces **F** and −**F**, applied in the two faces of the block parallel to the y-z plane, which may replace the four given forces. The 30-N forces act parallel to the y-z plane.

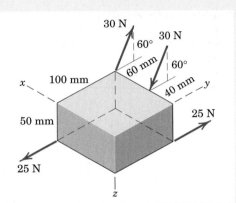

Solution. The couple due to the 30-N forces has the magnitude $M_1 = 30(0.06) = 1.80$ N·m. The direction of **M**$_1$ is normal to the plane defined by the two forces, and the sense, shown in the figure, is established by the right-hand convention. The couple due to the 25-N forces has the magnitude $M_2 = 25(0.10) = 2.50$ N·m with the direction and sense shown in the same figure. The two couple vectors combine to give the components

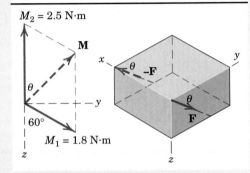

$$M_y = 1.80 \sin 60° = 1.559 \text{ N·m}$$

$$M_z = -2.50 + 1.80 \cos 60° = -1.600 \text{ N·m}$$

① Thus, $$M = \sqrt{(1.559)^2 + (-1.600)^2} = 2.23 \text{ N·m} \qquad Ans.$$

with $$\theta = \tan^{-1} \frac{1.559}{1.600} = \tan^{-1} 0.974 = 44.3° \qquad Ans.$$

The forces **F** and −**F** lie in a plane normal to the couple **M**, and their moment arm as seen from the right-hand figure is 100 mm. Thus, each force has the magnitude

$$[M = Fd] \qquad F = \frac{2.23}{0.10} = 22.3 \text{ N} \qquad Ans.$$

and the direction $\theta = 44.3°$.

Helpful Hint

① Bear in mind that the couple vectors are *free vectors* and therefore have no unique lines of action.

Sample Problem 2/15

A force of 40 lb is applied at A to the handle of the control lever which is attached to the fixed shaft OB. In determining the effect of the force on the shaft at a cross section such as that at O, we may replace the force by an equivalent force at O and a couple. Describe this couple as a vector **M**.

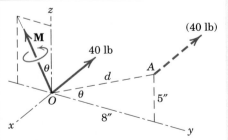

Solution. The couple may be expressed in vector notation as $\mathbf{M} = \mathbf{r} \times \mathbf{F}$, where $\mathbf{r} = \overrightarrow{OA} = 8\mathbf{j} + 5\mathbf{k}$ in. and $\mathbf{F} = -40\mathbf{i}$ lb. Thus,

$$\mathbf{M} = (8\mathbf{j} + 5\mathbf{k}) \times (-40\mathbf{i}) = -200\mathbf{j} + 320\mathbf{k} \text{ lb-in.}$$

Alternatively we see that moving the 40-lb force through a distance $d = \sqrt{5^2 + 8^2} = 9.43$ in. to a parallel position through O requires the addition of a couple **M** whose magnitude is

$$M = Fd = 40(9.43) = 377 \text{ lb-in.} \qquad Ans.$$

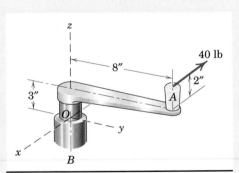

The couple vector is perpendicular to the plane in which the force is shifted, and its sense is that of the moment of the given force about O. The direction of **M** in the y-z plane is given by

$$\theta = \tan^{-1} \frac{5}{8} = 32.0° \qquad Ans.$$

PROBLEMS

Introductory Problems

2/117 Determine the moment of force **F** about point O.

Ans. $\mathbf{M}_O = F(c\mathbf{j} - b\mathbf{k})$

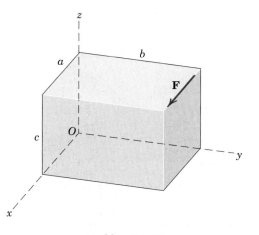

Problem 2/117

2/118 Determine the moment of force **F** about point A.

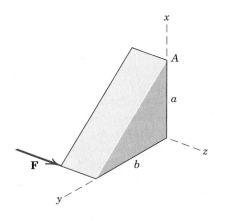

Problem 2/118

2/119 Determine the moment about O of the force of magnitude F for the case (a) when the force **F** is applied at A and for the case (b) when **F** is applied at B.

Ans. (a) $\mathbf{M}_O = FL\mathbf{i}$

(b) $\mathbf{M}_O = F(L\mathbf{i} + D\mathbf{k})$

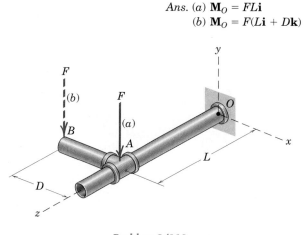

Problem 2/119

2/120 In opening a door which is equipped with a heavy-duty return mechanism, a person exerts a force **P** of magnitude 6 lb as shown. Force **P** and the normal n to the face of the door lie in a vertical plane. Compute the moment of **P** about the z-axis.

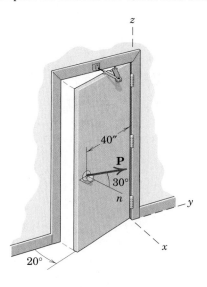

Problem 2/120

2/121 The two forces acting on the handles of the pipe wrenches constitute a couple **M**. Express the couple as a vector.

Ans. $\mathbf{M} = -75\mathbf{i} + 22.5\mathbf{j}$ N·m

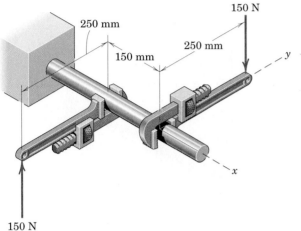

Problem 2/121

2/122 The bent bar has a mass ρ per unit of length. Determine the moment of the weight of the bar about point O.

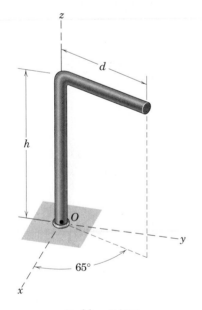

Problem 2/122

2/123 A helicopter is shown here with certain three-dimensional geometry given. During a ground test, a 400-N aerodynamic force is applied to the tail rotor at P as shown. Determine the moment of this force about point O of the airframe.

Ans. $\mathbf{M}_O = 480\mathbf{i} + 2400\mathbf{k}$ N·m

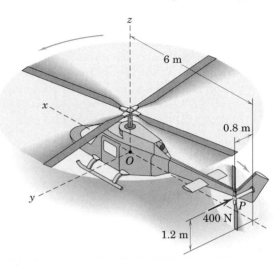

Problem 2/123

2/124 The 4-lb force is applied at point A of the crank assembly. Determine the moment of this force about point O.

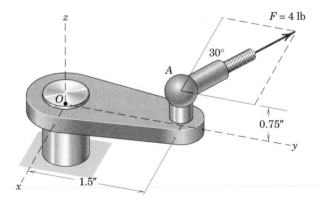

Problem 2/124

2/125 The right-angle pipe *OAB* of Prob. 2/108 is shown again here. Replace the 750-N tensile force which the cable exerts on point *B* by a force–couple system at point *O*.

Ans. $\mathbf{R} = -598\mathbf{i} + 411\mathbf{j} + 189.5\mathbf{k}$ N
$\mathbf{M}_O = -361\mathbf{i} - 718\mathbf{j} + 419\mathbf{k}$ N·m

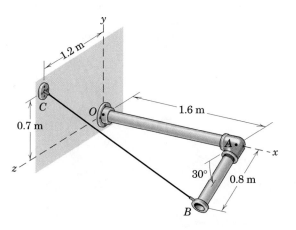

Problem 2/125

2/126 Determine the moment associated with the pair of 400-N forces applied to the T-shaped structure.

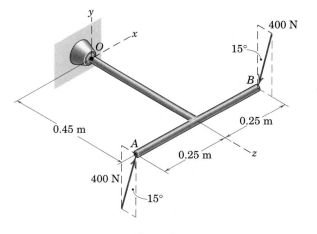

Problem 2/126

2/127 If the magnitude of the moment of **F** about line *CD* is 50 N·m, determine the magnitude of **F**.

Ans. $F = 228$ N

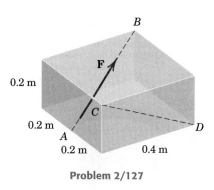

Problem 2/127

Representative Problems

2/128 A mechanic applies the horizontal 100-N force perpendicular to the wrench as indicated in the figure. Determine the moment of this force about the bolt center *O*. The wrench centerline lies in the *x-y* plane.

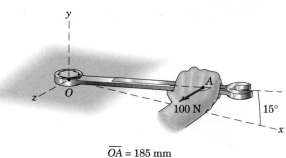

$\overline{OA} = 185$ mm

Problem 2/128

2/129 Two 1.2-lb thrusters on the nonrotating satellite are simultaneously fired as shown. Compute the moment associated with this couple and state about which satellite axes rotations will begin to occur.

Ans. $\mathbf{M} = -60\mathbf{i} + 48\mathbf{k}$ lb-in.

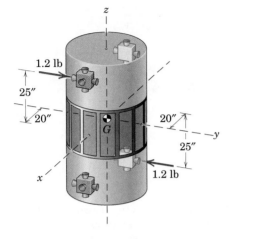

Problem 2/129

2/130 Compute the moment $\mathbf{M}_O$ of the 250-lb force about the axis O-O.

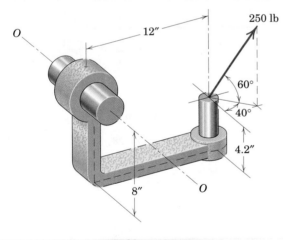

Problem 2/130

2/131 A space shuttle orbiter is subjected to thrusts from five of the engines of its reaction control system. Four of the thrusts are shown in the figure; the fifth is an 850-N upward thrust at the right rear, symmetric to the 850-N thrust shown on the left rear. Compute the moment of these forces about point G and show that the forces have the same moment about all points.

Ans. $\mathbf{M} = 3400\mathbf{i} - 51\,000\mathbf{j} - 51\,000\mathbf{k}$ N·m

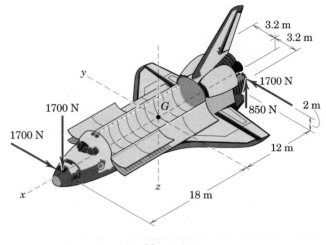

Problem 2/131

2/132 In picking up a load from position B, a cable tension $\mathbf{T}$ of magnitude 24 kN is developed. Calculate the moment which $\mathbf{T}$ produces about the base O of the construction crane.

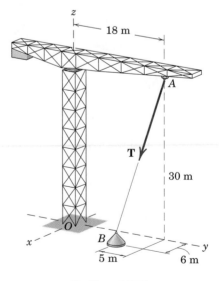

Problem 2/132

2/133 The specialty wrench shown is used for difficult-to-access bolts which have a low torque specification. Determine the moment about O of the 8-lb force applied at point A of the wrench handle. Discuss any shortcomings of this wrench.

Ans. $\mathbf{M}_O = 96\mathbf{i} - 72\mathbf{k}$ lb-in.

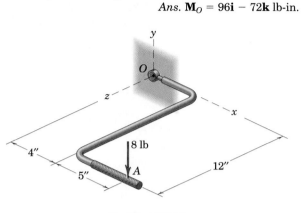

Problem 2/133

2/134 In order to decrease the undesirable moment about the x-axis, a mechanic uses his left hand to support the wrench handle at point B. What upward force F must he exert in order that there be no moment about the x-axis? What then is the net moment about O?

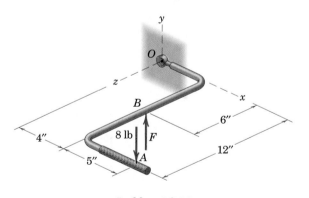

Problem 2/134

2/135 Determine the moment of the 100-lb force about point A by (a) using the vector cross-product relation and (b) resolving the force into its components and finding their respective moments.

Ans. $\mathbf{M}_A = -96.4\mathbf{i} - 173.2\mathbf{j} - 100\mathbf{k}$ lb-in.

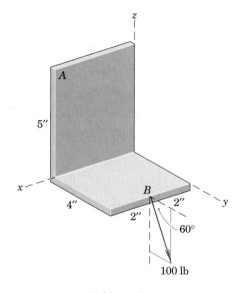

Problem 2/135

2/136 The specialty wrench shown in the figure is designed for access to the hold-down bolt on certain automobile distributors. For the configuration shown where the wrench lies in a vertical plane and a horizontal 200-N force is applied at A perpendicular to the handle, calculate the moment $\mathbf{M}_O$ applied to the bolt at O. For what value of the distance d would the z-component of $\mathbf{M}_O$ be zero?

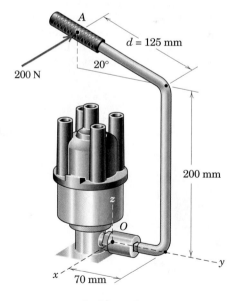

Problem 2/136

2/137 The moment **M** applied to the pulley wheel causes a tension $T = 80$ N in the cable which is secured to the wheel at D and to the ground at E. Determine the moment about O of this 80-N force as applied at C.

$$Ans. \mathbf{M}_O = -15.49\mathbf{i} + 4\mathbf{k} \text{ N·m}$$

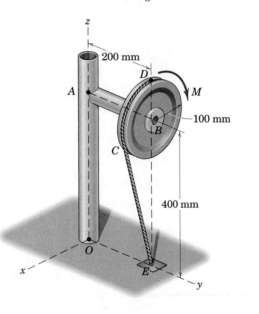

Problem 2/137

2/138 An 80-lb force is applied to the end of the wrench as the sprocket wheel is attached to an engine crankshaft. The force is perpendicular to the plane containing points O, A, and B. Determine the moment of this force about point O.

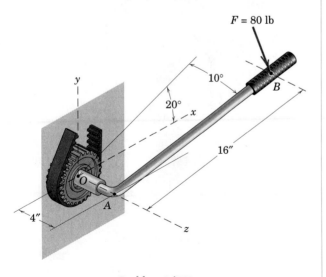

Problem 2/138

2/139 Using the principles to be developed in Chapter 3 on equilibrium, one can determine that the tension in cable AB is 143.4 N. Determine the moment about the x-axis of this tension force acting on point A. Compare your result with the moment of the weight W of the 15-kg uniform plate about the x-axis. What is the moment of the tension force acting at A about line OB?

$$Ans. M_x = 31.1 \text{ N·m}, (M_x)_W = -31.1 \text{ N·m}$$
$$M_{OB} = 0$$

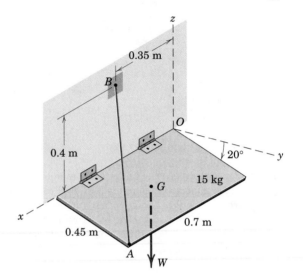

Problem 2/139

2/140 The rigid pole and cross-arm assembly of Prob. 2/105 is shown again here. Determine the vector expression for the moment of the 1.2-kN tension (*a*) about point *O* and (*b*) about the pole *z*-axis. Find each moment in two different ways.

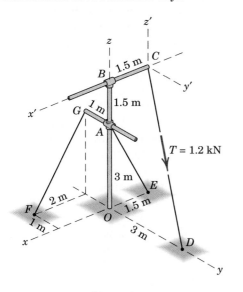

Problem 2/140

2/141 A 1.8-lb vertical force is applied to the knob of the window-opener mechanism when the crank *BC* is horizontal. Determine the moment of the force about point *A* and about line *AB*.

Ans. $\mathbf{M}_A = -5.40\mathbf{i} + 4.68\mathbf{j}$ lb-in.
$\mathbf{M}_{AB} = -4.05\mathbf{i} - 2.34\mathbf{k}$ lb-in.

2/142 Determine the vector expression for the moment $\mathbf{M}_O$ of the 600-N force about point *O*. The design specification for the bolt at *O* would require this result.

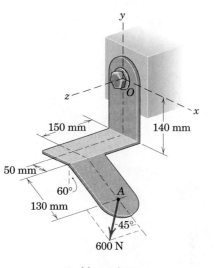

Problem 2/142

Problem 2/141

2/9 RESULTANTS

In Art. 2/6 we defined the resultant as the simplest force combination which can replace a given system of forces without altering the external effect on the rigid body on which the forces act. We found the magnitude and direction of the resultant force for the two-dimensional force system by a vector summation of forces, Eq. 2/9, and we located the line of action of the resultant force by applying the principle of moments, Eq. 2/10. These same principles can be extended to three dimensions.

In the previous article we showed that a force could be moved to a parallel position by adding a corresponding couple. Thus, for the system of forces $\mathbf{F}_1$, $\mathbf{F}_2$, $\mathbf{F}_3$ . . . acting on a rigid body in Fig. 2/28a, we may move each of them in turn to the arbitrary point O, provided we also introduce a couple for each force transferred. Thus, for example, we may move force $\mathbf{F}_1$ to O, provided we introduce the couple $\mathbf{M}_1 = \mathbf{r}_1 \times \mathbf{F}_1$, where $\mathbf{r}_1$ is a vector from O to any point on the line of action of $\mathbf{F}_1$. When all forces are shifted to O in this manner, we have a system of concurrent forces at O and a system of couple vectors, as represented in part b of the figure. The concurrent forces may then be added vectorially to produce a resultant force $\mathbf{R}$, and the couples may also be added to produce a resultant couple $\mathbf{M}$, Fig. 2/28c. The general force system, then, is reduced to

$$\mathbf{R} = \mathbf{F}_1 + \mathbf{F}_2 + \mathbf{F}_3 + \cdots = \Sigma\mathbf{F}$$
$$\mathbf{M} = \mathbf{M}_1 + \mathbf{M}_2 + \mathbf{M}_3 + \cdots = \Sigma(\mathbf{r} \times \mathbf{F})$$

(2/20)

The couple vectors are shown through point O, but because they are free vectors, they may be represented in any parallel positions. The magnitudes of the resultants and their components are

$$R_x = \Sigma F_x \qquad R_y = \Sigma F_y \qquad R_z = \Sigma F_z$$
$$R = \sqrt{(\Sigma F_x)^2 + (\Sigma F_y)^2 + (\Sigma F_z)^2}$$
$$\mathbf{M}_x = \Sigma(\mathbf{r} \times \mathbf{F})_x \qquad \mathbf{M}_y = \Sigma(\mathbf{r} \times \mathbf{F})_y \qquad \mathbf{M}_z = \Sigma(\mathbf{r} \times \mathbf{F})_z$$
$$M = \sqrt{M_x^2 + M_y^2 + M_z^2}$$

(2/21)

The cables of a suspension bridge exert a three-dimensional system of concentrated forces on this bridge tower.

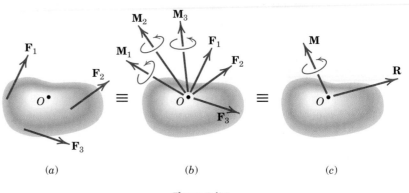

(a) (b) (c)

Figure 2/28

The point O selected as the point of concurrency for the forces is arbitrary, and the magnitude and direction of **M** depend on the particular point O selected. The magnitude and direction of **R**, however, are the same no matter which point is selected.

In general, any system of forces may be replaced by its resultant force **R** and the resultant couple **M**. In dynamics we usually select the mass center as the reference point. The change in the linear motion of the body is determined by the resultant force, and the change in the angular motion of the body is determined by the resultant couple. In statics, the body is in *complete equilibrium* when the resultant force **R** is zero and the resultant couple **M** is also zero. Thus, the determination of resultants is essential in both statics and dynamics.

We now examine the resultants for several special force systems.

Concurrent Forces. When forces are concurrent at a point, only the first of Eqs. 2/20 needs to be used because there are no moments about the point of concurrency.

Parallel Forces. For a system of parallel forces not all in the same plane, the magnitude of the parallel resultant force **R** is simply the magnitude of the algebraic sum of the given forces. The position of its line of action is obtained from the principle of moments by requiring that $\mathbf{r} \times \mathbf{R} = \mathbf{M}_O$. Here **r** is a position vector extending from the force–couple reference point O to the final line of action of **R**, and $\mathbf{M}_O$ is the sum of the moments of the individual forces about O. See Sample Problem 2/17 for an example of parallel-force systems.

Coplanar Forces. Article 2/6 was devoted to this force system.

Wrench Resultant. When the resultant couple vector **M** is parallel to the resultant force **R**, as shown in Fig. 2/29, the resultant is called a *wrench*. By definition a wrench is positive if the couple and force vectors point in the same direction and negative if they point in opposite directions. A common example of a positive wrench is found with the application of a screwdriver, to drive a right-handed screw. Any general force system may be represented by a wrench applied along a unique line of action. This reduction is illustrated in Fig. 2/30, where part a of the figure represents, for the general force system, the resultant force **R** acting at some point O and the corresponding resultant couple **M**. Although **M** is a free vector, for convenience we represent it as acting through O.

In part b of the figure, **M** is resolved into components $\mathbf{M}_1$ along the direction of **R** and $\mathbf{M}_2$ normal to **R**. In part c of the figure, the couple $\mathbf{M}_2$ is replaced by its equivalent of two forces **R** and $-\mathbf{R}$ separated by a distance

Positive wrench   Negative wrench

Figure 2/29

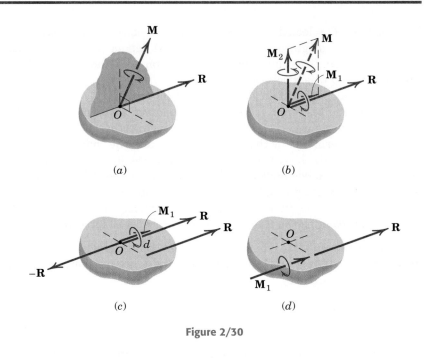

(a) (b)

(c) (d)

Figure 2/30

$d = M_2/R$ with $-\mathbf{R}$ applied at O to cancel the original $\mathbf{R}$. This step leaves the resultant $\mathbf{R}$, which acts along a new and unique line of action, and the parallel couple $\mathbf{M}_1$, which is a free vector, as shown in part d of the figure. Thus, the resultants of the original general force system have been transformed into a wrench (positive in this illustration) with its unique axis defined by the new position of $\mathbf{R}$.

We see from Fig. 2/30 that the axis of the wrench resultant lies in a plane through O normal to the plane defined by $\mathbf{R}$ and $\mathbf{M}$. The wrench is the simplest form in which the resultant of a general force system may be expressed. This form of the resultant, however, has limited application, because it is usually more convenient to use as the reference point some point O such as the mass center of the body or another convenient origin of coordinates not on the wrench axis.

Sample Problem 2/16

Determine the resultant of the force and couple system which acts on the rectangular solid.

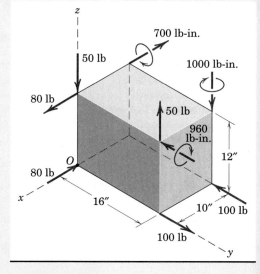

Solution.   We choose point O as a convenient reference point for the initial step of reducing the given forces to a force–couple system. The resultant force is

① $$\mathbf{R} = \Sigma\mathbf{F} = (80 - 80)\mathbf{i} + (100 - 100)\mathbf{j} + (50 - 50)\mathbf{k} = \mathbf{0} \text{ lb}$$

The sum of the moments about O is

② $$\mathbf{M}_O = [50(16) - 700]\mathbf{i} + [80(12) - 960]\mathbf{j} + [100(10) - 1000]\mathbf{k} \text{ lb-in.}$$
$$= 100\mathbf{i} \text{ lb-in.}$$

Hence, the resultant consists of a couple, which of course may be applied at any point on the body or the body extended.

Helpful Hints

① Since the force summation is zero, we conclude that the resultant, if it exists, must be a couple.

② The moments associated with the force pairs are easily obtained by using the $M = Fd$ rule and assigning the unit-vector direction by inspection. In many three-dimensional problems, this may be simpler than the $\mathbf{M} = \mathbf{r} \times \mathbf{F}$ approach.

Sample Problem 2/17

Determine the resultant of the system of parallel forces which act on the plate. Solve with a vector approach.

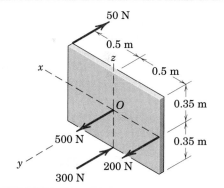

Solution.   Transfer of all forces to point O results in the force–couple system

$$\mathbf{R} = \Sigma\mathbf{F} = (200 + 500 - 300 - 50)\mathbf{j} = 350\mathbf{j} \text{ N}$$
$$\mathbf{M}_O = [50(0.35) - 300(0.35)]\mathbf{i} + [-50(0.50) - 200(0.50)]\mathbf{k}$$
$$= -87.5\mathbf{i} - 125\mathbf{k} \text{ N·m}$$

The placement of $\mathbf{R}$ so that it alone represents the above force–couple system is determined by the principle of moments in vector form

$$\mathbf{r} \times \mathbf{R} = \mathbf{M}_O$$
$$(x\mathbf{i} + y\mathbf{j} + z\mathbf{k}) \times 350\mathbf{j} = -87.5\mathbf{i} - 125\mathbf{k}$$
$$350x\mathbf{k} - 350z\mathbf{i} = -87.5\mathbf{i} - 125\mathbf{k}$$

From the one vector equation we may obtain the two scalar equations

$$350x = -125 \quad \text{and} \quad -350z = -87.5$$

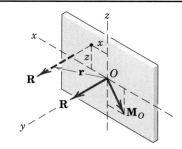

Hence, $x = -0.357$ m and $z = 0.250$ m are the coordinates through which the line of action of $\mathbf{R}$ must pass. The value of y may, of course, be any value, as permitted by the principle of transmissibility. Thus, as expected, the variable y
① drops out of the above vector analysis.

Helpful Hint

① You should also carry out a scalar solution to this problem.

Sample Problem 2/18

Replace the two forces and the negative wrench by a single force **R** applied at A and the corresponding couple **M**.

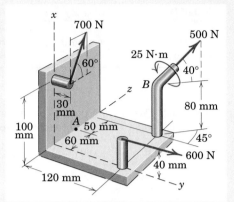

Solution. The resultant force has the components

$[R_x = \Sigma F_x]$ $R_x = 500 \sin 40° + 700 \sin 60° = 928$ N

$[R_y = \Sigma F_y]$ $R_y = 600 + 500 \cos 40° \cos 45° = 871$ N

$[R_z = \Sigma F_z]$ $R_z = 700 \cos 60° + 500 \cos 40° \sin 45° = 621$ N

Thus, $\mathbf{R} = 928\mathbf{i} + 871\mathbf{j} + 621\mathbf{k}$ N

and $R = \sqrt{(928)^2 + (871)^2 + (621)^2} = 1416$ N   *Ans.*

The couple to be added as a result of moving the 500-N force is

① $[\mathbf{M} = \mathbf{r} \times \mathbf{F}]$ $\mathbf{M}_{500} = (0.08\mathbf{i} + 0.12\mathbf{j} + 0.05\mathbf{k}) \times 500(\mathbf{i} \sin 40°$
$+ \mathbf{j} \cos 40° \cos 45° + \mathbf{k} \cos 40° \sin 45°)$

where **r** is the vector from A to B.

The term-by-term, or determinant, expansion gives

$$\mathbf{M}_{500} = 18.95\mathbf{i} - 5.59\mathbf{j} - 16.90\mathbf{k} \text{ N·m}$$

② The moment of the 600-N force about A is written by inspection of its x- and z-components, which gives

$$\mathbf{M}_{600} = (600)(0.060)\mathbf{i} + (600)(0.040)\mathbf{k}$$
$$= 36.0\mathbf{i} + 24.0\mathbf{k} \text{ N·m}$$

The moment of the 700-N force is easily obtained from the moments of the x- and z-components of the force. The result becomes

$$\mathbf{M}_{700} = (700 \cos 60°)(0.030)\mathbf{i} - [(700 \sin 60°)(0.060)$$
$$+ (700 \cos 60°)(0.100)]\mathbf{j} - (700 \sin 60°)(0.030)\mathbf{k}$$
$$= 10.5\mathbf{i} - 71.4\mathbf{j} - 18.19\mathbf{k} \text{ N·m}$$

③ Also, the couple of the given wrench may be written

$$\mathbf{M}' = 25.0(-\mathbf{i} \sin 40° - \mathbf{j} \cos 40° \cos 45° - \mathbf{k} \cos 40° \sin 45°)$$
$$= -16.07\mathbf{i} - 13.54\mathbf{j} - 13.54\mathbf{k} \text{ N·m}$$

Therefore, the resultant couple on adding together the **i**-, **j**-, and **k**-terms of the four **M**'s is

④ $\mathbf{M} = 49.4\mathbf{i} - 90.5\mathbf{j} - 24.6\mathbf{k} \text{ N·m}$

and $M = \sqrt{(49.4)^2 + (90.5)^2 + (24.6)^2} = 106.0 \text{ N·m}$ *Ans.*

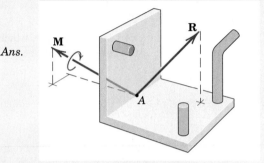

Helpful Hints

① *Suggestion:* Check the cross-product results by evaluating the moments about A of the components of the 500-N force directly from the sketch.

② For the 600-N and 700-N forces it is easier to obtain the components of their moments about the coordinate directions through A by inspection of the figure than it is to set up the cross-product relations.

③ The 25-N·m couple vector of the *wrench* points in the direction opposite to that of the 500-N force, and we must resolve it into its x-, y-, and z-components to be added to the other couple-vector components.

④ Although the resultant couple vector **M** in the sketch of the resultants is shown through A, we recognize that a couple vector is a free vector and therefore has no specified line of action.

Sample Problem 2/19

Determine the wrench resultant of the three forces acting on the bracket. Calculate the coordinates of the point P in the x-y plane through which the resultant force of the wrench acts. Also find the magnitude of the couple **M** of the wrench.

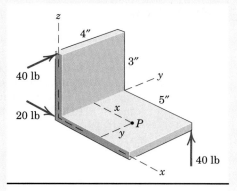

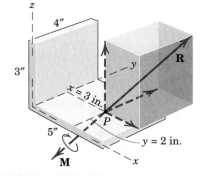

Solution.   The direction cosines of the couple **M** of the wrench must be the ① same as those of the resultant force **R**, assuming that the wrench is positive. The resultant force is

$$\mathbf{R} = 20\mathbf{i} + 40\mathbf{j} + 40\mathbf{k} \text{ lb} \qquad R = \sqrt{(20)^2 + (40)^2 + (40)^2} = 60 \text{ lb}$$

and its direction cosines are

$$\cos \theta_x = 20/60 = 1/3 \quad \cos \theta_y = 40/60 = 2/3 \quad \cos \theta_z = 40/60 = 2/3$$

The moment of the wrench couple must equal the sum of the moments of the given forces about point P through which **R** passes. The moments about P of the three forces are

$$(\mathbf{M})_{R_x} = 20y\mathbf{k} \text{ lb-in.}$$

$$(\mathbf{M})_{R_y} = -40(3)\mathbf{i} - 40x\mathbf{k} \text{ lb-in.}$$

$$(\mathbf{M})_{R_z} = 40(4 - y)\mathbf{i} - 40(5 - x)\mathbf{j} \text{ lb-in.}$$

and the total moment is

$$\mathbf{M} = (40 - 40y)\mathbf{i} + (-200 + 40x)\mathbf{j} + (-40x + 20y)\mathbf{k} \text{ lb-in.}$$

The direction cosines of **M** are

$$\cos \theta_x = (40 - 40y)/M$$

$$\cos \theta_y = (-200 + 40x)/M$$

$$\cos \theta_z = (-40x + 20y)/M$$

where M is the magnitude of **M**. Equating the direction cosines of **R** and **M** gives

$$40 - 40y = \frac{M}{3}$$

$$-200 + 40x = \frac{2M}{3}$$

$$-40x + 20y = \frac{2M}{3}$$

Solution of the three equations gives

$$M = -120 \text{ lb-in.} \qquad x = 3 \text{ in.} \qquad y = 2 \text{ in.} \qquad Ans.$$

We see that M turned out to be negative, which means that the couple vector is pointing in the direction opposite to **R**, which makes the wrench negative.

Helpful Hint

① We assume initially that the wrench is positive. If **M** turns out to be negative, then the direction of the couple vector is opposite to that of the resultant force.

PROBLEMS

Introductory Problems

2/143 Three forces act at point O. If it is known that the y-component of the resultant $\mathbf{R}$ is -5 kN and that the z-component is 6 kN, determine F_3, θ, and R.

Ans. $F_3 = 10.82$ kN, $\theta = 33.7°$, $R = 10.49$ kN

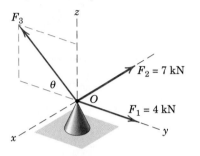

Problem 2/143

2/144 Three equal forces are exerted on the equilateral plate as shown. Reduce the force system to an equivalent force–couple system at point O. Show that $\mathbf{R}$ is perpendicular to $\mathbf{M}_O$.

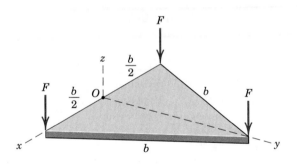

Problem 2/144

2/145 The thin rectangular plate is subjected to the four forces shown. Determine the equivalent force–couple system at O. What is the resultant of the system?

$$\text{Ans. } \mathbf{R} = \mathbf{0}, \mathbf{M}_O = Fb\left(1 + \frac{\sqrt{3}}{2}\right)\mathbf{i}$$

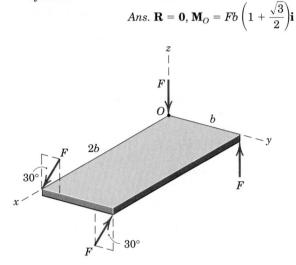

Problem 2/145

2/146 The spacecraft of Prob. 2/129 is repeated here. The plan is to fire four 1.2-lb thrusters as shown in order to spin up the spacecraft about its z-axis, but the thruster at A fails. Determine the equivalent force–couple system at G for the remaining three thrusters.

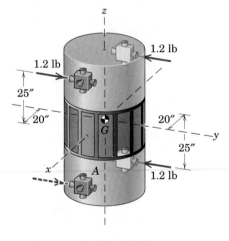

Problem 2/146

2/147 An oil tanker moves away from its docked position under the action of reverse thrust from screw A, forward thrust from screw B, and side thrust from the bow thruster C. Determine the equivalent force–couple system at the mass center G.

Ans. $\mathbf{R} = -8\mathbf{i}$ kN, $\mathbf{M}_G = 48\mathbf{j} + 820\mathbf{k}$ kN·m

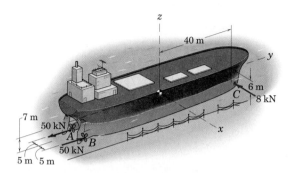

Problem 2/147

2/148 The pulley and gear are subjected to the loads shown. For these forces, determine the equivalent force–couple system at point O.

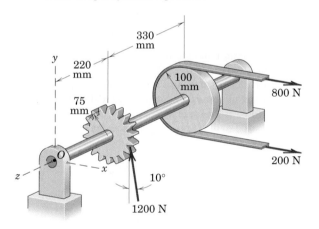

Problem 2/148

2/149 Determine the force–couple system at O which is equivalent to the two forces applied to the shaft AOB. Is $\mathbf{R}$ perpendicular to $\mathbf{M}_O$?

Ans. $\mathbf{R} = -266\mathbf{j} + 1085\mathbf{k}$ N
$\mathbf{M}_O = -48.9\mathbf{j} - 114.5\mathbf{k}$ N·m

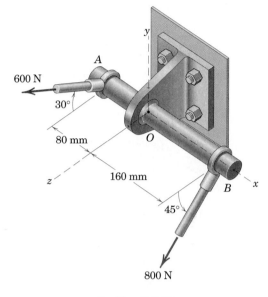

Problem 2/149

Representative Problems

2/150 The commercial airliner of Prob. 2/86 is redrawn here with three-dimensional information supplied. If engine 3 suddenly fails, determine the resultant of the three remaining engine thrust vectors, each of which has a magnitude of 90 kN. Specify the y- and z-coordinates of the point through which the line of action of the resultant passes. This information would be critical to the design criteria of performance with engine failure.

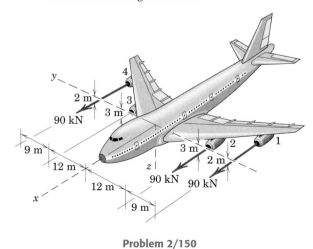

Problem 2/150

2/151 Represent the resultant of the force system acting on the pipe assembly by a single force **R** at A and a couple **M**.

Ans. $\mathbf{R} = 120\mathbf{i} - 180\mathbf{j} - 100\mathbf{k}$ N
$\mathbf{M}_A = 100\mathbf{j} + 50\mathbf{k}$ N·m

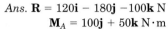
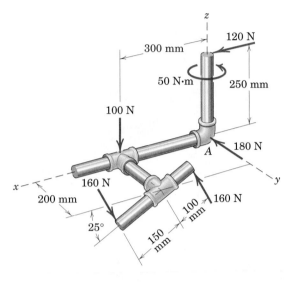

Problem 2/151

2/152 Two upward loads are exerted on the small three-dimensional truss. Reduce these two loads to a single force–couple system at point O. Show that **R** is perpendicular to $\mathbf{M}_O$. Then determine the point in the x-z plane through which the resultant passes.

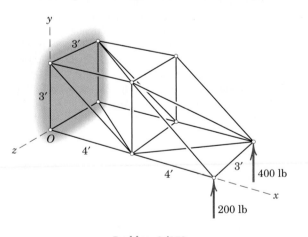

Problem 2/152

2/153 A 50-lb computer and a 20-lb laser printer rest on a horizontal tabletop. Determine the resultant of the system of two weights and specify the coordinates of the point P in the x-y plane through which the resultant passes. The two weights act at the points G_1 and G_2.

Ans. $\mathbf{R} = -70\mathbf{k}$ lb, $\mathbf{M}_O = -2800\mathbf{i} + 1440\mathbf{j}$ lb-ft
$(x, y) = (20.6, 40)$ in.

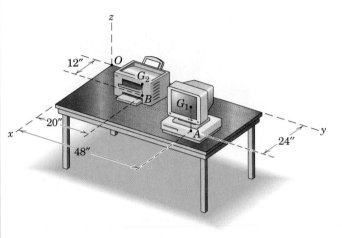

Problem 2/153

2/154 The motor mounted on the bracket is acted on by its 160-N weight, and its shaft resists the 120-N thrust and 25-N·m couple applied to it. Determine the resultant of the force system shown in terms of a force **R** at A and a couple **M**.

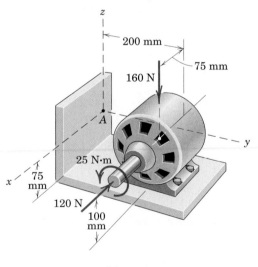

Problem 2/154

2/155 In tightening a bolt whose center is at point O, a person exerts a 40-lb force on the ratchet handle with his right hand. In addition, with his left hand he exerts a 20-lb force as shown in order to secure the socket onto the bolt head. Determine the equivalent force–couple system at O. Then find the point in the x-y plane through which the line of action of the single-force resultant passes.

Ans. $\mathbf{R} = -20\mathbf{j} - 40\mathbf{k}$ lb, $\mathbf{M}_O = -56\mathbf{i} - 320\mathbf{j}$ lb-in.
$$x = -6.4 \text{ in.}, y = 1.4 \text{ in.}$$

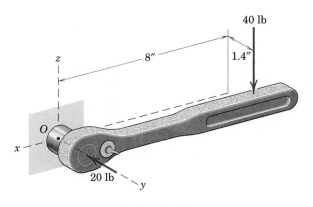

Problem 2/155

2/156 Replace the two forces acting on the pole by a wrench. Write the moment **M** associated with the wrench as a vector and specify the coordinates of the point P in the y-z plane through which the line of action of the wrench passes.

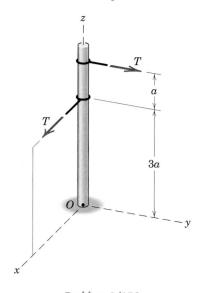

Problem 2/156

2/157 Replace the two forces acting on the frame by a wrench. Write the moment associated with the wrench as a vector and specify the coordinates of the point P in the y-z plane through which the line of action of the wrench passes. Note that the force of magnitude F is parallel to the x-axis.

Ans. $\mathbf{R} = F(\mathbf{i} - 3\mathbf{k})$, $\mathbf{M} = \dfrac{3aF}{10}(\mathbf{i} - 3\mathbf{k})$
$$y = \frac{a}{10}, z = 2a$$

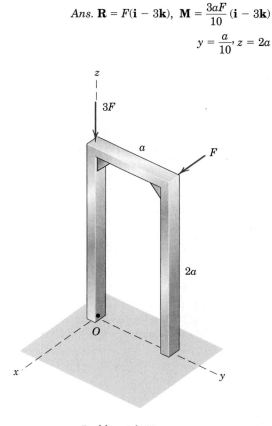

Problem 2/157

2/158 For the system of two forces in Prob. 2/149, determine the coordinates of the point in the x-z plane through which the line of action of the resultant of the system passes.

▶**2/159** The resultant of the two forces and couple may be represented by a wrench. Determine the vector expression for the moment **M** of the wrench and find the coordinates of the point P in the x-z plane through which the resultant force of the wrench passes.

Ans. $\mathbf{M} = 10\mathbf{i} + 10\mathbf{j}$ N·m
$x = z = 0.1$ m

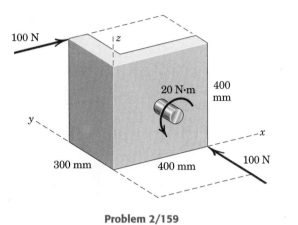
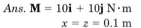

Problem 2/159

▶**2/160** For the position shown, the crankshaft of a small two-cylinder compressor is subjected to the 400-N and 800-N forces exerted by the connecting rods and the 200-N·m couple. Replace this loading system by a force–couple system at point A. Show that **R** is not perpendicular to $\mathbf{M}_A$. Then replace this force–couple system by a wrench. Determine the magnitude M of the moment of the wrench, the magnitude of the force **R** of the wrench, and the coordinates of the point in the x-z plane through which the line of action of the wrench passes.

Ans. $M = 85.8$ N·m, $R = 1108$ N
$x = 0.1158$ m, $z = -0.478$ m

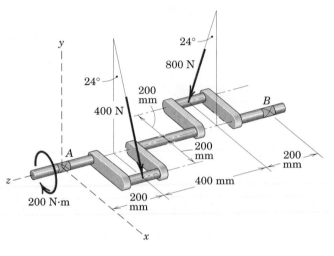

Problem 2/160

2/10   CHAPTER REVIEW

In Chapter 2 we have established the properties of forces, moments, and couples, and the correct procedures for representing their effects. Mastery of this material is essential for our study of equilibrium in the chapters which follow. Failure to correctly use the procedures of Chapter 2 is a common cause of errors in applying the principles of equilibrium. When difficulties arise, you should refer to this chapter to be sure that the forces, moments, and couples are correctly represented.

Forces

There is frequent need to represent forces as vectors, to resolve a single force into components along desired directions, and to combine two or more concurrent forces into an equivalent resultant force. Specifically, you should be able to:

1. Resolve a given force vector into its components along given directions, and express the vector in terms of the unit vectors along a given set of axes.

2. Express a force as a vector when given its magnitude and information about its line of action. This information may be in the form of two points along the line of action or angles which orient the line of action.

3. Use the dot product to compute the projection of a vector onto a specified line and the angle between two vectors.

4. Compute the resultant of two or more forces concurrent at a point.

Moments

The tendency of a force to rotate a body about an axis is described by a moment (or torque), which is a vector quantity. We have seen that finding the moment of a force is often facilitated by combining the moments of the components of the force. When working with moment vectors you should be able to:

1. Determine a moment by using the moment-arm rule.

2. Use the vector cross product to compute a moment vector in terms of a force vector and a position vector locating the line of action of the force.

3. Utilize Varignon's theorem to simplify the calculation of moments, in both scalar and vector forms.

4. Use the triple scalar product to compute the moment of a force vector about a given axis through a given point.

Couples

A couple is the combined moment of two equal, opposite, and non-collinear forces. The unique effect of a couple is to produce a pure twist or rotation regardless of where the forces are located. The couple is useful in replacing a force acting at a point by a force–couple system at a different point. To solve problems involving couples you should be able to:

1. Compute the moment of a couple, given the couple forces and either their separation distance or any position vectors locating their lines of action.

2. Replace a given force by an equivalent force–couple system, and vice versa.

Resultants

We can reduce an arbitrary system of forces and couples to a single resultant force applied at an arbitrary point, and a corresponding resultant couple. We can further combine this resultant force and couple into a wrench to give a single resultant force along a unique line of action, together with a parallel couple vector. To solve problems involving resultants you should be able to:

1. Compute the magnitude, direction, and line of action of the resultant of a system of coplanar forces if that resultant is a force; otherwise, compute the moment of the resultant couple.

2. Apply the principle of moments to simplify the calculation of the moment of a system of coplanar forces about a given point.

3. Replace a given general force system by a wrench along a specific line of action.

Equilibrium

You will use the preceding concepts and methods when you study equilibrium in the following chapters. Let us summarize the concept of equilibrium:

1. When the resultant force on a body is zero ($\Sigma\mathbf{F} = \mathbf{0}$), the body is in *translational* equilibrium. This means that its center of mass is either at rest or moving in a straight line with constant velocity.

2. In addition, if the resultant couple is zero ($\Sigma\mathbf{M} = \mathbf{0}$), the body is in *rotational* equilibrium, either having no rotational motion or rotating with a constant angular velocity.

3. When both resultants are zero, the body is in *complete* equilibrium.

REVIEW PROBLEMS

2/161 A die is being used to cut threads on a rod. If 15-lb forces are applied as shown, determine the magnitude F of the equal forces exerted on the $\frac{1}{4}$-in. rod by each of the four cutting surfaces so that their external effect on the rod is equivalent to that of the two 15-lb forces.

Ans. F = 300 lb

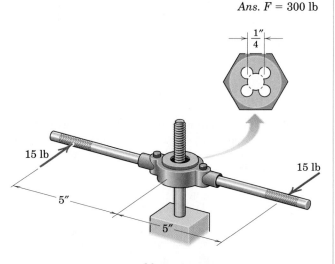

Problem 2/161

2/162 Using the principles of equilibrium to be developed in Chapter 3, you will soon be able to verify that the tension in cable AB is 85.8% of the weight of the cylinder of mass m, while the tension in cable AC is 55.5% of the suspended weight. Write each tension force acting on point A as a vector if the mass m is 60 kg.

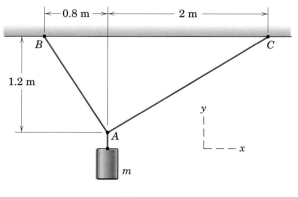

Problem 2/162

2/163 The control lever is subjected to a clockwise couple of 80 N·m exerted by its shaft at A and is designed to operate with a 200-N pull as shown. If the resultant of the couple and the force passes through A, determine the proper dimension x of the lever.

Ans. x = 266 mm

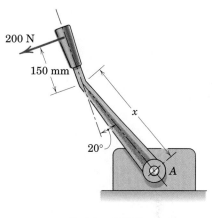

Problem 2/163

2/164 The blades of the portable fan generate a 1.2-lb thrust **T** as shown. Compute the moment M_O of this force about the rear support point O. For comparison, determine the moment about O due to the weight of the motor–fan unit AB, whose weight of 9 lb acts at G.

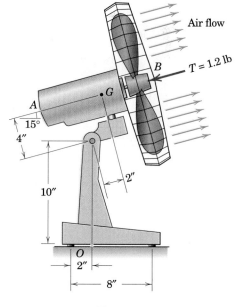

Problem 2/164

2/165 For the angular position $\theta = 60°$ of the crank OA, the gas pressure on the piston induces a compressive force P in the connecting rod along its centerline AB. If this force produces a moment of 720 N·m about the crank axis O, calculate P.

Ans. P = 9.18 kN

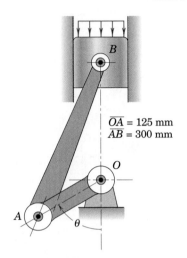

$\overline{OA} = 125$ mm
$\overline{AB} = 300$ mm

Problem 2/165

2/166 Calculate the moment M_O of the 250-N force about the base point O of the robot.

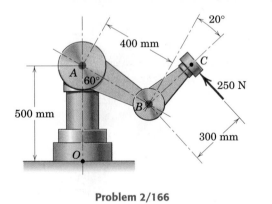

Problem 2/166

2/167 Represent the resultant of the three forces and couple by a force–couple system located at point A.

Ans. R = 4.75 kN, M_A = 21.1 kN·m CCW

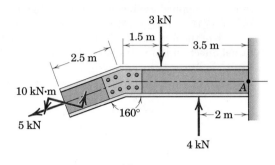

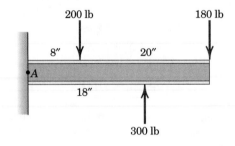

Problem 2/167

2/168 Reduce the given loading system to a force–couple system at point A. Then determine the distance x to the right of point A at which the resultant of the three forces acts.

Problem 2/168

2/169 The directions of rotation of the input shaft *A* and output shaft *B* of the worm-gear reducer are indicated by the curved dashed arrows. An input torque (couple) of 80 N·m is applied to shaft *A* in the direction of rotation. The output shaft *B* supplies a torque of 320 N·m to the machine which it drives (not shown). The shaft of the driven machine exerts an equal and opposite reacting torque on the output shaft of the reducer. Determine the resultant **M** of the two couples which act on the reducer unit and calculate the direction cosine of **M** with respect to the *x*-axis.

Ans. $\mathbf{M} = -320\mathbf{i} - 80\mathbf{j}$ N·m
$\cos \theta_x = -0.970$

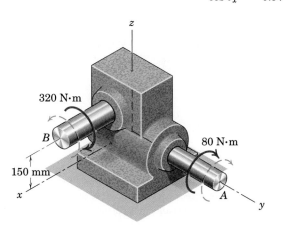

Problem 2/169

2/170 Determine the moment of the force **P** about point *A*.

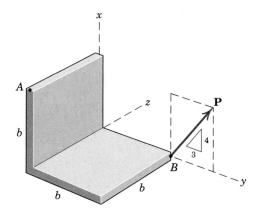

Problem 2/170

2/171 When the pole *OA* is in the position shown, the tension in cable *AB* is 3 kN. (*a*) Write the tension force exerted on the small collar at point *A* as a vector using the coordinates shown. (*b*) Determine the moment of this force about point *O* and state the moments about the *x*-, *y*-, and *z*-axes. (*c*) Determine the projection of this tension force onto line *AO*.

Ans. (*a*) $\mathbf{T}_{AB} = -2.05\mathbf{i} - 1.432\mathbf{j} - 1.663\mathbf{k}$ kN
(*b*) $\mathbf{M}_O = 7.63\mathbf{i} - 10.90\mathbf{j}$ kN·m
(*c*) $T_{AO} = 2.69$ kN

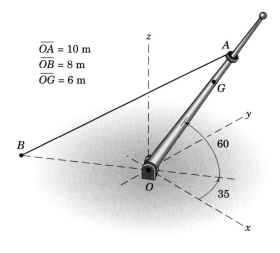

$\overline{OA} = 10$ m
$\overline{OB} = 8$ m
$\overline{OG} = 6$ m

Problem 2/171

2/172 A force **F** acts along the line *AB* inside the right circular cylindrical shell as shown. The quantities *r*, *h*, *θ*, and *F* are known. Using the *x*-, *y*-, and *z*-coordinates shown, express **F** as a vector.

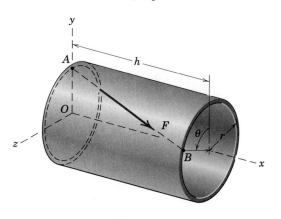

Problem 2/172

2/173 Three couples are formed by the three pairs of equal and opposite forces. Determine the resultant **M** of the three couples.

Ans. $\mathbf{M} = -20\mathbf{i} - 6.77\mathbf{j} - 37.2\mathbf{k}$ N·m

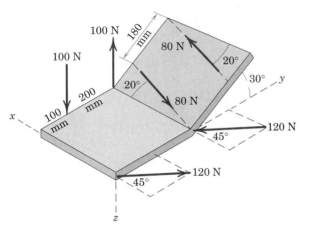

Problem 2/173

2/174 During a drilling operation, the small robotic device is subjected to an 800-N force at point C as shown. Replace this force by an equivalent force–couple system at point O.

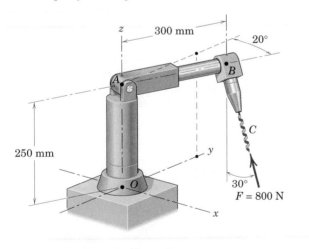

Problem 2/174

2/175 The combined action of the three forces on the base at O may be obtained by establishing their resultant through O. Determine the magnitudes of **R** and the accompanying couple **M**.

Ans. $R = 1093$ lb, $M = 9730$ lb-ft

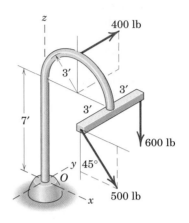

Problem 2/175

Computer-Oriented Problems

*2/176 Four forces are exerted on the eyebolt as shown. If the net effect on the bolt is a direct pull of 600 lb in the y-direction, determine the necessary values of T and θ.

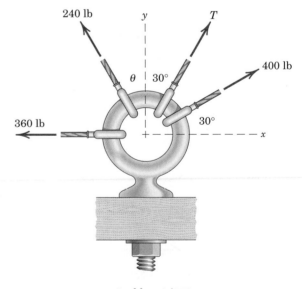

Problem 2/176

*2/177 The force **F** is directed from A toward D and D is allowed to move from B to C as measured by the variable s. Consider the projection of **F** onto line EF as a function of s. In particular, determine and plot the fraction n of the magnitude F which is projected as a function of s/d. Note that s/d varies from 0 to $2\sqrt{2}$.

$$Ans.\ n = \frac{\sqrt{2}\,\frac{s}{d} + 1}{\sqrt{5}\,\sqrt{\left(\frac{s}{d}\right)^2 + 5 - 2\sqrt{2}\,\frac{s}{d}}}$$

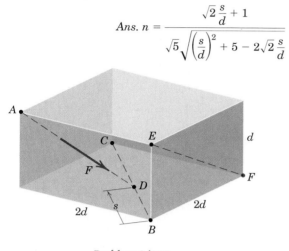

Problem 2/177

*2/178 The throttle-control lever OA rotates in the range $0 \le \theta \le 90°$. An internal torsional return spring exerts a restoring moment about O given by $M = K(\theta + \pi/4)$, where $K = 500$ N·mm/rad and θ is in radians. Determine and plot as a function of θ the tension T required to make the net moment about O zero. Use the two values $d = 60$ mm and $d = 160$ mm and comment on the relative design merits. The effects of the radius of the pulley at B are negligible.

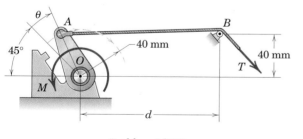

Problem 2/178

*2/179 With the 300-lb cylindrical part P in its grip, the robotic arm pivots about O through the range $-45° \le \theta \le 45°$ with the angle at A locked at 120°. Determine and plot (as a function of θ) the moment at O due to the combined effects of the 300-lb part P, the 120-lb weight of member OA (mass center at G_1), and the 50-lb weight of member AB (mass center at G_2). The end grip is included as a part of member AB. The lengths L_1 and L_2 are 3 ft and 2 ft, respectively. What is the maximum value of M_O and at what value of θ does this maximum occur?

$$Ans.\ M_O = 1230 \cos \theta + 650 \cos(60° - \theta)\ \text{lb-ft}$$
$$(M_O)_{max} = 1654\ \text{lb-ft at } \theta = 19.90°$$

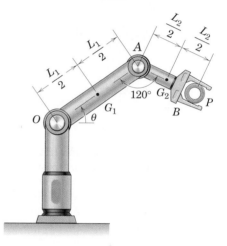

Problem 2/179

*2/180 A motor attached to the shaft at O causes the arm OA to rotate over the range $0 \leq \theta \leq 180°$. The unstretched length of the spring is 0.65 m, and it can support both tension and compression. If the net moment about O must be zero, determine and plot the required motor torque M as a function of θ.

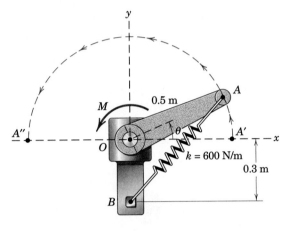

Problem 2/180

*2/181 A flagpole with attached light triangular frame is shown here for an arbitrary position during its raising. The 75-N tension in the ereating cable remains constant. Determine and plot the moment about the pivot O of the 75-N force for the range $0 \leq \theta \leq 90°$. Determine the maximum value of this moment and the elevation angle at which it occurs; comment on the physical significance of the latter. The effects of the diameter of the drum at D may be neglected.

$$Ans. \ \mathbf{M}_O = \frac{1350 \sin (\theta + 60°)}{\sqrt{45 + 36 \cos (\theta + 60°)}} \, \mathbf{k} \, \mathrm{N \cdot m}$$
$$(M_O)_{max} = 225 \ \mathrm{N \cdot m} \ \mathrm{at} \ \theta = 60°$$

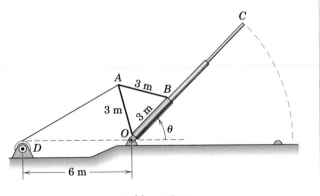

Problem 2/181

*2/182 The tension T in cable AB is maintained at a constant value of 120 N. Determine the moment $\mathbf{M}_O$ of this tension about point O over the range $0 \leq \theta \leq 90°$. Plot the x-, y-, and z-components of $\mathbf{M}_O$ as functions of θ.

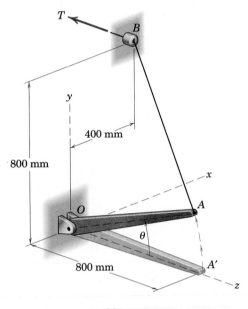

Problem 2/182

*2/183 The arm AB rotates in the range $0 \leq \theta \leq 180°$, and the spring is unstretched when $\theta = 90°$. Determine as a function of θ the moment about O of the spring force as applied at B. Plot the three scalar components of $\mathbf{M}_O$, and state the maximum absolute value of each component.

Ans. $\mathbf{M}_O = \dfrac{\sqrt{1.5 - 0.5 \sin \theta} - 1}{\sqrt{1.5 - 0.5 \sin \theta}} [-518\mathbf{i} + 259 \cos \theta \mathbf{k}]$ lb-ft

$$(M_{O_x})_{max} = 95.1 \text{ lb-ft at } \theta = 0, 180°$$

$$(M_{O_z})_{max} = 47.6 \text{ lb-ft at } \theta = 0, 180°$$

Problem 2/183

*2/184 The rectangular plate is tilted about its lower edge by a cable tensioned at a constant 600 N. Determine and plot the moment of this tension about the lower edge AB of the plate for the range $0 \leq \theta \leq 90°$.

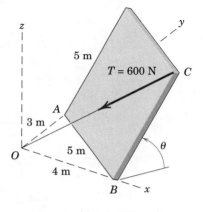

Problem 2/184

In many applications of mechanics, the sum of the forces acting on a body is zero, and a state of equilibrium exists. This apparatus is designed to hold a car body in equilibrium for a wide range of orientations during vehicle production.

3 EQUILIBRIUM

3/1   INTRODUCTION

Statics deals primarily with the description of the force conditions necessary and sufficient to maintain the equilibrium of engineering structures. This chapter on equilibrium, therefore, constitutes the most important part of statics, and the procedures developed here form the basis for solving problems in both statics and dynamics. We will make continual use of the concepts developed in Chapter 2 involving forces, moments, couples, and resultants as we apply the principles of equilibrium.

When a body is in equilibrium, the resultant of *all* forces acting on it is zero. Thus, the resultant force **R** and the resultant couple **M** are both zero, and we have the equilibrium equations

$$\boxed{\mathbf{R} = \Sigma \mathbf{F} = 0 \qquad \mathbf{M} = \Sigma \mathbf{M} = 0} \qquad (3/1)$$

These requirements are both necessary and sufficient conditions for equilibrium.

All physical bodies are three-dimensional, but we can treat many of them as two-dimensional when the forces to which they are subjected act in a single plane or can be projected onto a single plane. When this simplification is not possible, the problem must be treated as three-dimensional. We will follow the arrangement used in Chapter 2, and discuss in Section A the equilibrium of bodies subjected to two-dimensional force systems and in Section B the equilibrium of bodies subjected to three-dimensional force systems.

SECTION A EQUILIBRIUM IN TWO DIMENSIONS

3/2 SYSTEM ISOLATION AND THE FREE-BODY DIAGRAM

Before we apply Eqs. 3/1, we must define unambiguously the particular body or mechanical system to be analyzed and represent clearly and completely *all* forces acting *on* the body. Omission of a force which acts *on* the body in question, or inclusion of a force which does not act *on* the body, will give erroneous results.

A *mechanical system* is defined as a body or group of bodies which can be conceptually isolated from all other bodies. A system may be a single body or a combination of connected bodies. The bodies may be rigid or nonrigid. The system may also be an identifiable fluid mass, either liquid or gas, or a combination of fluids and solids. In statics we study primarily forces which act on rigid bodies at rest, although we also study forces acting on fluids in equilibrium.

Once we decide which body or combination of bodies to analyze, we then treat this body or combination as a single body *isolated* from all surrounding bodies. This isolation is accomplished by means of the ***free-body diagram***, which is a diagrammatic representation of the isolated system treated as a single body. The diagram shows all forces applied to the system by mechanical contact with other bodies, which are imagined to be removed. If appreciable body forces are present, such as gravitational or magnetic attraction, then these forces must also be shown on the free-body diagram of the isolated system. Only after such a diagram has been carefully drawn should the equilibrium equations be written. Because of its critical importance, we emphasize here that

> **the free-body diagram is the most important single step in the solution of problems in mechanics.**

Before attempting to draw a free-body diagram, we must recall the basic characteristics of force. These characteristics were described in Art. 2/2, with primary attention focused on the vector properties of force. Forces can be applied either by direct physical contact or by remote action. Forces can be either internal or external to the system under consideration. Application of force is accompanied by reactive force, and both applied and reactive forces may be either concentrated or distributed. The principle of transmissibility permits the treatment of force as a sliding vector as far as its external effects on a rigid body are concerned.

We will now use these force characteristics to develop conceptual models of isolated mechanical systems. These models enable us to write the appropriate equations of equilibrium, which can then be analyzed.

Modeling the Action of Forces

Figure 3/1 shows the common types of force application on mechanical systems for analysis in two dimensions. Each example shows the force exerted *on* the body to be *isolated,* by the body to be *removed.* Newton's third law, which notes the existence of an equal and opposite reaction to every action, must be carefully observed. The force exerted *on* the body in question *by* a contacting or supporting member is always in the sense to oppose the movement of the isolated body which would occur if the contacting or supporting body were removed.

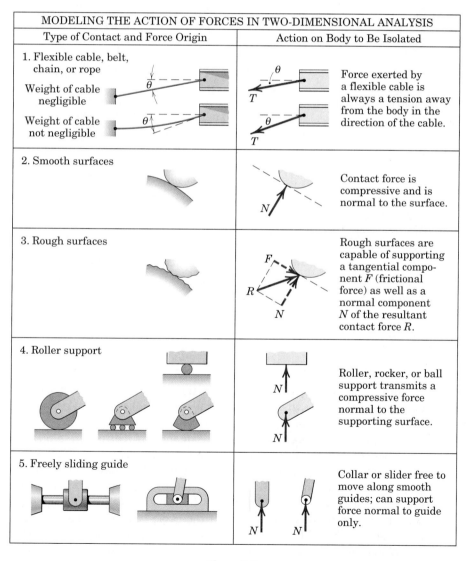

MODELING THE ACTION OF FORCES IN TWO-DIMENSIONAL ANALYSIS	
Type of Contact and Force Origin	Action on Body to Be Isolated
1. Flexible cable, belt, chain, or rope Weight of cable negligible Weight of cable not negligible	Force exerted by a flexible cable is always a tension away from the body in the direction of the cable.
2. Smooth surfaces	Contact force is compressive and is normal to the surface.
3. Rough surfaces	Rough surfaces are capable of supporting a tangential component F (frictional force) as well as a normal component N of the resultant contact force R.
4. Roller support	Roller, rocker, or ball support transmits a compressive force normal to the supporting surface.
5. Freely sliding guide	Collar or slider free to move along smooth guides; can support force normal to guide only.

Figure 3/1

MODELING THE ACTION OF FORCES IN TWO-DIMENSIONAL ANALYSIS (*cont.*)	
Type of Contact and Force Origin	Action on Body to Be Isolated
6. Pin connection	**Pin free to turn** A freely hinged pin connection is capable of supporting a force in any direction in the plane normal to the pin axis. We may either show two components R_x and R_y or a magnitude R and direction θ. A pin not free to turn also supports a couple M.
7. Built-in or fixed support	A built-in or fixed support is capable of supporting an axial force F, a transverse force V (shear force), and a couple M (bending moment) to prevent rotation.
8. Gravitational attraction	The resultant of gravitational attraction on all elements of a body of mass m is the weight $W = mg$ and acts toward the center of the earth through the center mass G.
9. Spring action	Spring force is tensile if spring is stretched and compressive if compressed. For a linearly elastic spring the stiffness k is the force required to deform the spring a unit distance.

Figure 3/1, continued

In Fig. 3/1, Example 1 depicts the action of a flexible cable, belt, rope, or chain on the body to which it is attached. Because of its flexibility, a rope or cable is unable to offer any resistance to bending, shear, or compression and therefore exerts only a tension force in a direction tangent to the cable at its point of attachment. The force exerted *by* the cable *on* the body to which it is attached is always *away* from the body. When the tension T is large compared with the weight of the cable, we may assume that the cable forms a straight line. When the cable weight is not negligible compared with its tension, the sag of the cable becomes important, and the tension in the cable changes direction and magnitude along its length.

When the smooth surfaces of two bodies are in contact, as in Example 2, the force exerted by one on the other is *normal* to the tangent to the surfaces and is compressive. Although no actual surfaces are perfectly smooth, we can assume this to be so for practical purposes in many instances.

When mating surfaces of contacting bodies are rough, as in Example 3, the force of contact is not necessarily normal to the tangent to the surfaces, but may be resolved into a *tangential* or *frictional component F* and a *normal component N*.

Example 4 illustrates a number of forms of mechanical support which effectively eliminate tangential friction forces. In these cases the net reaction is normal to the supporting surface.

Example 5 shows the action of a smooth guide on the body it supports. There cannot be any resistance parallel to the guide.

Example 6 illustrates the action of a pin connection. Such a connection can support force in any direction normal to the axis of the pin. We usually represent this action in terms of two rectangular components. The correct sense of these components in a specific problem depends on how the member is loaded. When not otherwise initially known, the sense is arbitrarily assigned and the equilibrium equations are then written. If the solution of these equations yields a positive algebraic sign for the force component, the assigned sense is correct. A negative sign indicates the sense is opposite to that initially assigned.

If the joint is free to turn about the pin, the connection can support only the force R. If the joint is not free to turn, the connection can also support a resisting couple M. The sense of M is arbitrarily shown here, but the true sense depends on how the member is loaded.

Example 7 shows the resultants of the rather complex distribution of force over the cross section of a slender bar or beam at a build-in or fixed support. The sense of the reactions F and V and the bending couple M in a given problem depends, of course, on how the member is loaded.

One of the most common forces is that due to gravitational attraction, Example 8. This force affects all elements of mass in a body and is, therefore, distributed throughout it. The resultant of the gravitational forces on all elements is the weight $W = mg$ of the body, which passes through the center of mass G and is directed toward the center of the earth for earthbound structures. The location of G is frequently obvious from the geometry of the body, particularly where there is symmetry. When the location is not readily apparent, it must be determined by experiment or calculations.

Similar remarks apply to the remote action of magnetic and electric forces. These forces of remote action have the same overall effect on a rigid body as forces of equal magnitude and direction applied by direct external contact.

Example 9 illustrates the action of a *linear* elastic spring and of a *nonlinear* spring with either hardening or softening characteristics. The force exerted by a linear spring, in tension or compression, is given by $F = kx$, where k is the *stiffness* of the spring and x is its deformation measured from the neutral or undeformed position.

The representations in Fig. 3/1 are *not* free-body diagrams, but are merely elements used to construct free-body diagrams. Study these nine conditions and identify them in the problem work so that you can draw the correct free-body diagrams.

Another view of the car-body lifting device shown in the chapter-opening photograph.

Construction of Free-Body Diagrams

The full procedure for drawing a free-body diagram which isolates a body or system consists of the following steps.

Step 1. Decide which system to isolate. The system chosen should usually involve one or more of the desired unknown quantities.

Step 2. Next isolate the chosen system by drawing a diagram which represents its *complete external boundary*. This boundary defines the isolation of the system from *all* other attracting or contacting bodies, which are considered removed. This step is often the most crucial of all. Make certain that you have *completely isolated* the system before proceeding with the next step.

Step 3. Identify all forces which act *on* the isolated system as applied *by* the removed contacting and attracting bodies, and represent them in their proper positions on the diagram of the isolated system. Make a systematic traverse of the entire boundary to identify all contact forces. Include body forces such as weights, where appreciable. Represent all known forces by vector arrows, each with its proper magnitude, direction, and sense indicated. Each unknown force should be represented by a vector arrow with the unknown magnitude or direction indicated by symbol. If the sense of the vector is also unknown, you must arbitrarily assign a sense. The subsequent calculations with the equilibrium equations will yield a positive quantity if the correct sense was assumed and a negative quantity if the incorrect sense was assumed. It is necessary to be *consistent* with the assigned characteristics of unknown forces throughout all of the calculations. If you are consistent, the solution of the equilibrium equations will reveal the correct senses.

Step 4. Show the choice of coordinate axes directly on the diagram. Pertinent dimensions may also be represented for convenience. Note, however, that the free-body diagram serves the purpose of focusing attention on the action of the external forces, and therefore the diagram should not be cluttered with excessive extraneous information. Clearly distinguish force arrows from arrows representing quantities other than forces. For this purpose a colored pencil may be used.

Completion of the foregoing four steps will produce a correct free-body diagram to use in applying the governing equations, both in statics and in dynamics. Be careful not to omit from the free-body diagram certain forces which may not appear at first glance to be needed in the calculations. It is only through *complete* isolation and a systematic representation of *all* external forces that a reliable accounting of the effects of all applied and reactive forces can be made. Very often a force which at first glance may not appear to influence a desired result does indeed have an influence. Thus, the only safe procedure is to include on the free-body diagram all forces whose magnitudes are not obviously negligible.

The free-body method is extremely important in mechanics because it ensures an accurate definition of a mechanical system and focuses attention on the exact meaning and application of the force laws of statics and dynamics. Review the foregoing four steps for constructing a free-body diagram while studying the sample free-body diagrams shown in Fig. 3/2 and the Sample Problems which appear at the end of the next article.

Examples of Free-Body Diagrams

Figure 3/2 gives four examples of mechanisms and structures together with their correct free-body diagrams. Dimensions and magnitudes are omitted for clarity. In each case we treat the entire system as

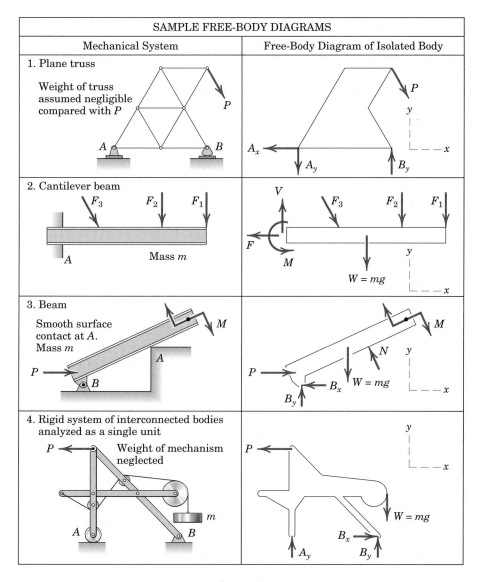

Figure 3/2

a single body, so that the internal forces are not shown. The characteristics of the various types of contact forces illustrated in Fig. 3/1 are used in the four examples as they apply.

In Example 1 the truss is composed of structural elements which, taken all together, constitute a rigid framework. Thus, we may remove the entire truss from its supporting foundation and treat it as a single rigid body. In addition to the applied external load P, the free-body diagram must include the reactions on the truss at A and B. The rocker at B can support a vertical force only, and this force is transmitted to the structure at B (Example 4 of Fig. 3/1). The pin connection at A (Example 6 of Fig. 3/1) is capable of supplying both a horizontal and a vertical force component to the truss. If the total weight of the truss members is appreciable compared with P and the forces at A and B, then the weights of the members must be included on the free-body diagram as external forces.

In this relatively simple example it is clear that the vertical component A_y must be directed down to prevent the truss from rotating clockwise about B. Also, the horizontal component A_x will be to the left to keep the truss from moving to the right under the influence of the horizontal component of P. Thus, in constructing the free-body diagram for this simple truss, we can easily perceive the correct sense of each of the components of force exerted *on* the truss *by* the foundation at A and can, therefore, represent its correct physical sense on the diagram. When the correct physical sense of a force or its component is not easily recognized by direct observation, it must be assigned arbitrarily, and the correctness of or error in the assignment is determined by the algebraic sign of its calculated value.

In Example 2 the cantilever beam is secured to the wall and subjected to three applied loads. When we isolate that part of the beam to the right of the section at A, we must include the reactive forces applied *to* the beam *by* the wall. The resultants of these reactive forces are shown acting on the section of the beam (Example 7 of Fig. 3/1). A vertical force V to counteract the excess of downward applied force is shown, and a tension F to balance the excess of applied force to the right must also be included. Then, to prevent the beam from rotating about A, a counterclockwise couple M is also required. The weight mg of the beam must be represented through the mass center (Example 8 of Fig. 3/1).

In the free-body diagram of Example 2, we have represented the somewhat complex system of forces which actually act on the cut section of the beam by the equivalent force–couple system in which the force is broken down into its vertical component V (shear force) and its horizontal component F (tensile force). The couple M is the bending moment in the beam. The free-body diagram is now complete and shows the beam in equilibrium under the action of six forces and one couple.

In Example 3 the weight $W = mg$ is shown acting through the center of mass of the beam, whose location is assumed known (Example 8 of Fig. 3/1). The force exerted by the corner A on the beam is normal to the smooth surface of the beam (Example 2 of Fig. 3/1). To perceive this action more clearly, visualize an enlargement of the contact point A, which would appear somewhat rounded, and consider the force exerted by this rounded corner on the straight surface of the beam, which is assumed to be smooth. If the contacting surfaces at the corner were not smooth, a tangential frictional component of force could exist. In addition to the applied force P and couple M, there is the pin connection at B, which exerts both an x- and a y-component of force on the beam. The positive senses of these components are assigned arbitrarily.

In Example 4 the free-body diagram of the entire isolated mechanism contains three unknown forces if the loads mg and P are known. Any one of many internal configurations for securing the cable leading from the mass m would be possible without affecting the external response of the mechanism as a whole, and this fact is brought out by the free-body diagram. This hypothetical example is used to show that the forces internal to a rigid assembly of members do not influence the values of the external reactions.

We use the free-body diagram in writing the equilibrium equations, which are discussed in the next article. When these equations are solved, some of the calculated force magnitudes may be zero. This would indicate that the assumed force does not exist. In Example 1 of Fig. 3/2, any of the reactions A_x, A_y, or B_y can be zero for specific values of the truss geometry and of the magnitude, direction, and sense of the applied load P. A zero reaction force is often difficult to identify by inspection, but can be determined by solving the equilibrium equations.

Similar comments apply to calculated force magnitudes which are negative. Such a result indicates that the actual sense is the opposite of the assumed sense. The assumed positive senses of B_x and B_y in Example 3 and B_y in Example 4 are shown on the free-body diagrams. The correctness of these assumptions is proved or disproved according to whether the algebraic signs of the computed forces are plus or minus when the calculations are carried out in an actual problem.

The isolation of the mechanical system under consideration is a crucial step in the formulation of the mathematical model. The most important aspect to the correct construction of the all-important free-body diagram is the clear-cut and unambiguous decision as to what is included and what is excluded. This decision becomes unambiguous only when the boundary of the free-body diagram represents a complete traverse of the body or system of bodies to be isolated, starting at some arbitrary point on the boundary and returning to that same point. The system within this closed boundary is the isolated free body, and all contact forces and all body forces transmitted to the system across the boundary must be accounted for.

The following exercises provide practice with drawing free-body diagrams. This practice is helpful before using such diagrams in the application of the principles of force equilibrium in the next article.

Even complex pulley systems such as the ones seen here are easily handled with a systematic equilibrium analysis.

FREE-BODY DIAGRAM EXERCISES

3/A In each of the five following examples, the body to be isolated is shown in the left-hand diagram, and an *incomplete* free-body diagram (FBD) of the isolated body is shown on the right. Add whatever forces are necessary in each case to form a complete free-body diagram. The weights of the bodies are negligible unless otherwise indicated. Dimensions and numerical values are omitted for simplicity.

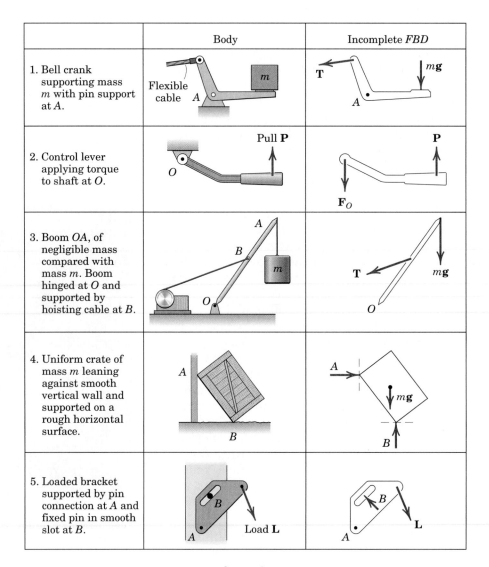

	Body	Incomplete *FBD*
1. Bell crank supporting mass m with pin support at A.		
2. Control lever applying torque to shaft at O.		
3. Boom OA, of negligible mass compared with mass m. Boom hinged at O and supported by hoisting cable at B.		
4. Uniform crate of mass m leaning against smooth vertical wall and supported on a rough horizontal surface.		
5. Loaded bracket supported by pin connection at A and fixed pin in smooth slot at B.		

Figure 3/A

3/B In each of the five following examples, the body to be isolated is shown in the left-hand diagram, and either a *wrong* or an *incomplete* free-body diagram (FBD) is shown on the right. Make whatever changes or additions are necessary in each case to form a correct and complete free-body diagram. The weights of the bodies are negligible unless otherwise indicated. Dimensions and numerical values are omitted for simplicity.

	Body	Wrong or Incomplete *FBD*
1. Lawn roller of mass m being pushed up incline θ.		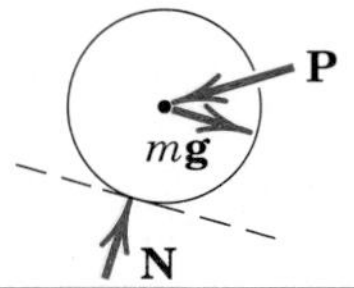
2. Prybar lifting body A having smooth horizontal surface. Bar rests on horizontal rough surface.	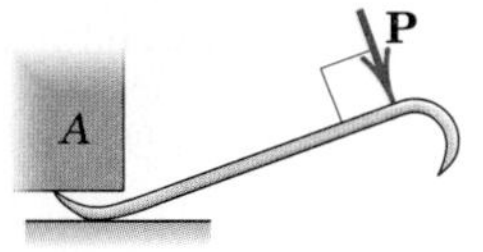	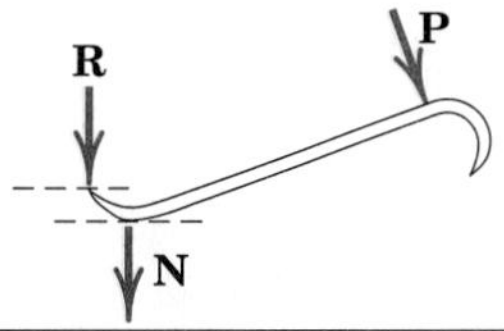
3. Uniform pole of mass m being hoisted into position by winch. Horizontal supporting surface notched to prevent slipping of pole.	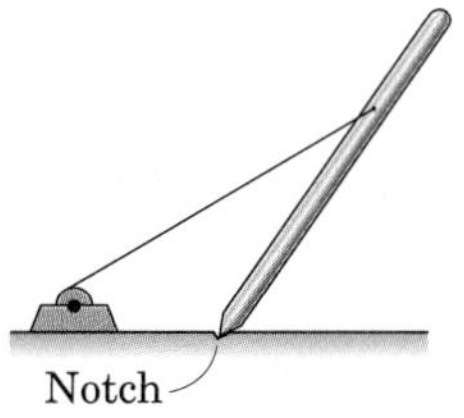	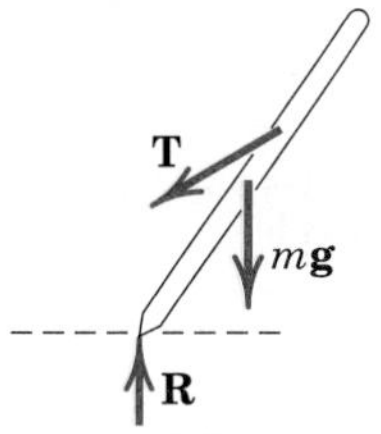
4. Supporting angle bracket for frame; pin joints.	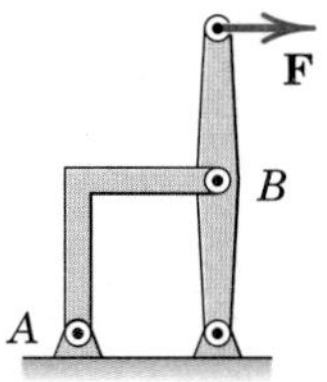	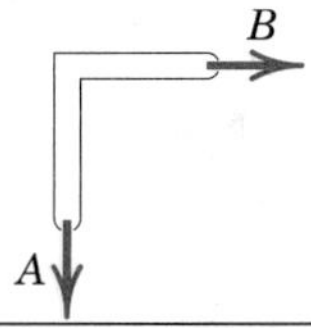
5. Bent rod welded to support at A and subjected to two forces and couple.	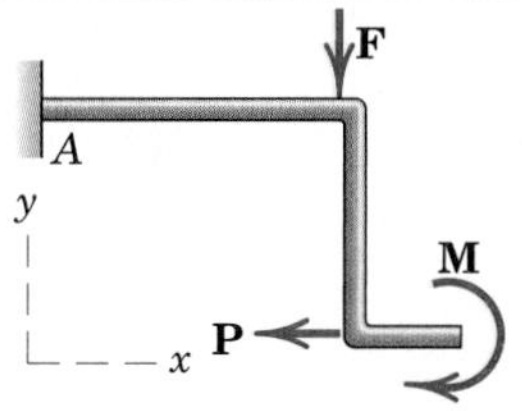	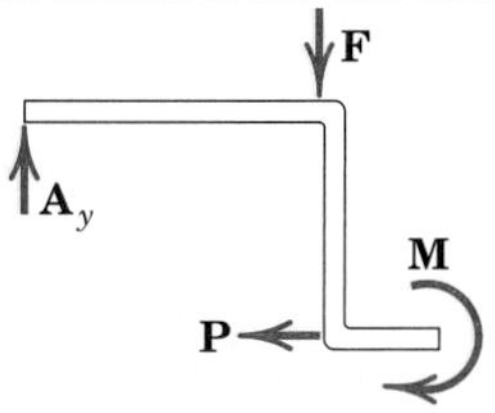

Figure 3/B

3/C Draw a complete and correct free-body diagram of each of the bodies designated in the statements. The weights of the bodies are significant only if the mass is stated.

All forces, known and unknown, should be labeled. (*Note*: The sense of some reaction components cannot always be determined without numerical calculation.)

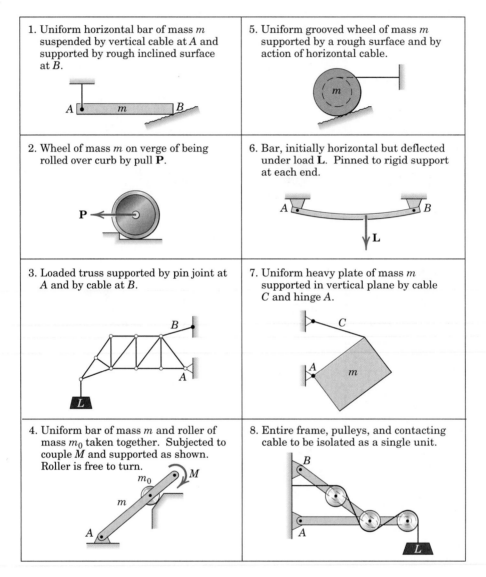

1. Uniform horizontal bar of mass m suspended by vertical cable at A and supported by rough inclined surface at B.

5. Uniform grooved wheel of mass m supported by a rough surface and by action of horizontal cable.

2. Wheel of mass m on verge of being rolled over curb by pull $\mathbf{P}$.

6. Bar, initially horizontal but deflected under load $\mathbf{L}$. Pinned to rigid support at each end.

3. Loaded truss supported by pin joint at A and by cable at B.

7. Uniform heavy plate of mass m supported in vertical plane by cable C and hinge A.

4. Uniform bar of mass m and roller of mass m_0 taken together. Subjected to couple M and supported as shown. Roller is free to turn.

8. Entire frame, pulleys, and contacting cable to be isolated as a single unit.

Figure 3/C

3/3   EQUILIBRIUM CONDITIONS

In Art. 3/1 we defined equilibrium as the condition in which the resultant of all forces and moments acting on a body is zero. Stated in another way, a body is in equilibrium if all forces and moments applied to it are in balance. These requirements are contained in the vector equations of equilibrium, Eqs. 3/1, which in two dimensions may be written in scalar form as

$$\Sigma F_x = 0 \qquad \Sigma F_y = 0 \qquad \Sigma M_O = 0 \qquad (3/2)$$

The third equation represents the zero sum of the moments of all forces about any point O on or off the body. Equations 3/2 are the necessary and sufficient conditions for complete equilibrium in two dimensions. They are necessary conditions because, if they are not satisfied, there can be no force or moment balance. They are sufficient because once they are satisfied, there can be no imbalance, and equilibrium is assured.

The equations relating force and acceleration for rigid-body motion are developed in *Vol. 2 Dynamics* from Newton's second law of motion. These equations show that the acceleration of the mass center of a body is proportional to the resultant force $\Sigma \mathbf{F}$ acting on the body. Consequently, if a body moves with constant velocity (zero acceleration), the resultant force on it must be zero, and the body may be treated as in a state of translational equilibrium.

For complete equilibrium in two dimensions, all three of Eqs. 3/2 must hold. However, these conditions are independent requirements, and one may hold without another. Take, for example, a body which slides along a horizontal surface with increasing velocity under the action of applied forces. The force–equilibrium equations will be satisfied in the vertical direction where the acceleration is zero, but not in the horizontal direction. Also, a body, such as a flywheel, which rotates about its fixed mass center with increasing angular speed is not in rotational equilibrium, but the two force–equilibrium equations will be satisfied.

Categories of Equilibrium

Applications of Eqs. 3/2 fall naturally into a number of categories which are easily identified. The categories of force systems acting on bodies in two-dimensional equilibrium are summarized in Fig. 3/3 and are explained further as follows.

Category 1, equilibrium of collinear forces, clearly requires only the one force equation in the direction of the forces (x-direction), since all other equations are automatically satisfied.

Category 2, equilibrium of forces which lie in a plane (x-y plane) and are concurrent at a point O, requires the two force equations only, since the moment sum about O, that is, about a z-axis through O, is necessarily zero. Included in this category is the case of the equilibrium of a particle.

Category 3, equilibrium of parallel forces in a plane, requires the one force equation in the direction of the forces (x-direction) and one moment equation about an axis (z-axis) normal to the plane of the forces.

CATEGORIES OF EQUILIBRIUM IN TWO DIMENSIONS		
Force System	Free-Body Diagram	Independent Equations
1. Collinear	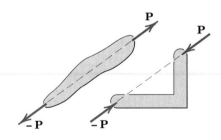	$\Sigma F_x = 0$
2. Concurrent at a point		$\Sigma F_x = 0$ $\Sigma F_y = 0$
3. Parallel		$\Sigma F_x = 0 \quad \Sigma M_z = 0$
4. General		$\Sigma F_x = 0 \quad \Sigma M_z = 0$ $\Sigma F_y = 0$

Figure 3/3

Category 4, equilibrium of a general system of forces in a plane (*x-y*), requires the two force equations in the plane and one moment equation about an axis (*z*-axis) normal to the plane.

Two- and Three-Force Members

You should be alert to two frequently occurring equilibrium situations. The first situation is the equilibrium of a body under the action of two forces only. Two examples are shown in Fig. 3/4, and we see that for such a *two-force member* to be in equilibrium, the forces must be *equal, opposite,* and *collinear.* The shape of the member does not affect this simple requirement. In the illustrations cited, we consider the weights of the members to be negligible compared with the applied forces.

The second situation is a *three-force member,* which is a body under the action of three forces, Fig. 3/5*a.* We see that equilibrium requires the lines of action of the three forces to be *concurrent.* If they were not concurrent, then one of the forces would exert a resultant moment about the point of intersection of the other two, which would violate the requirement of zero moment about every point. The only exception occurs when the three forces are parallel. In this case we may consider the point of concurrency to be at infinity.

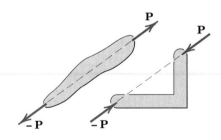

Two-force members

Figure 3/4

The principle of the concurrency of three forces in equilibrium is of considerable use in carrying out a graphical solution of the force equations. In this case the polygon of forces is drawn and made to close, as shown in Fig. 3/5b. Frequently, a body in equilibrium under the action of more than three forces may be reduced to a three-force member by a combination of two or more of the known forces.

Alternative Equilibrium Equations

In addition to Eqs. 3/2, there are two other ways to express the general conditions for the equilibrium of forces in two dimensions. The first way is illustrated in Fig. 3/6, parts (a) and (b). For the body shown in Fig. 3/6a, if $\Sigma M_A = 0$, then the resultant, if it still exists, cannot be a couple, but must be a force **R** passing through A. If now the equation $\Sigma F_x = 0$ holds, where the x-direction is arbitrary, it follows from Fig. 3/6b that the resultant force **R**, if it still exists, not only must pass through A, but also must be perpendicular to the x-direction as shown. Now, if $\Sigma M_B = 0$, where B is any point such that the line AB is not perpendicular to the x-direction, we see that **R** must be zero, and thus the body is in equilibrium. Therefore, an alternative set of equilibrium equations is

$$\Sigma F_x = 0 \qquad \Sigma M_A = 0 \qquad \Sigma M_B = 0$$

where the two points A and B must not lie on a line perpendicular to the x-direction.

A third formulation of the equilibrium conditions may be made for a coplanar force system. This is illustrated in Fig. 3/6, parts (c) and (d). Again, if $\Sigma M_A = 0$ for any body such as that shown in Fig. 3/6c, the resultant, if any, must be a force **R** through A. In addition, if $\Sigma M_B = 0$, the resultant, if one still exists, must pass through B as shown in Fig. 3/6d. Such a force cannot exist, however, if $\Sigma M_C = 0$, where C is not

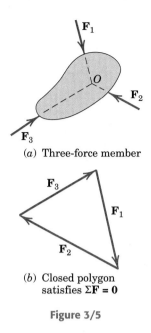

(a) Three-force member

(b) Closed polygon satisfies $\Sigma \mathbf{F} = \mathbf{0}$

Figure 3/5

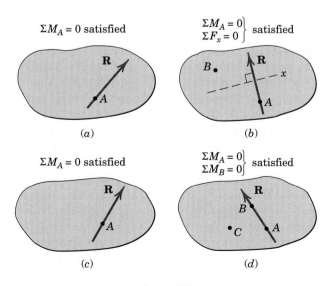

Figure 3/6

collinear with A and B. Thus, we may write the equations of equilibrium as

$$\Sigma M_A = 0 \qquad \Sigma M_B = 0 \qquad \Sigma M_C = 0$$

where A, B, and C are any three points not on the same straight line.

When equilibrium equations are written which are not independent, redundant information is obtained, and a correct solution of the equations will yield $0 = 0$. For example, for a general problem in two dimensions with three unknowns, three moment equations written about three points which lie on the same straight line are not independent. Such equations will contain duplicated information, and solution of two of them can at best determine two of the unknowns, with the third equation merely verifying the identity $0 = 0$.

Constraints and Statical Determinacy

The equilibrium equations developed in this article are both necessary and sufficient conditions to establish the equilibrium of a body. However, they do not necessarily provide all the information required to calculate all the unknown forces which may act on a body in equilibrium. Whether the equations are adequate to determine all the unknowns depends on the characteristics of the constraints against possible movement of the body provided by its supports. By *constraint* we mean the restriction of movement.

In Example 4 of Fig. 3/1 the roller, ball, and rocker provide constraint normal to the surface of contact, but none tangent to the surface. Thus, a tangential force cannot be supported. For the collar and slider of Example 5, constraint exists only normal to the guide. In Example 6 the fixed-pin connection provides constraint in both directions, but offers no resistance to rotation about the pin unless the pin is not free to turn. The fixed support of Example 7, however, offers constraint against rotation as well as lateral movement.

If the rocker which supports the truss of Example 1 in Fig. 3/2 were replaced by a pin joint, as at A, there would be one additional constraint beyond those required to support an equilibrium configuration with no freedom of movement. The three scalar conditions of equilibrium, Eqs. 3/2, would not provide sufficient information to determine all four unknowns, since A_x and B_x could not be solved for separately; only their sum could be determined. These two components of force would be dependent on the deformation of the members of the truss as influenced by their corresponding stiffness properties. The horizontal reactions A_x and B_x would also depend on any initial deformation required to fit the dimensions of the structure to those of the foundation between A and B. Thus, we cannot determine A_x and B_x by a rigid-body analysis.

Again referring to Fig. 3/2, we see that if the pin B in Example 3 were not free to turn, the support could transmit a couple to the beam through the pin. Therefore, there would be four unknown supporting reactions acting on the beam, namely, the force at A, the two components of force at B, and the couple at B. Consequently the three independent scalar equations of equilibrium would not provide enough information to compute all four unknowns.

A rigid body, or rigid combination of elements treated as a single body, which possesses more external supports or constraints than are necessary to maintain an equilibrium position is called *statically indeterminate*. Supports which can be removed without destroying the equilibrium condition of the body are said to be *redundant*. The number of redundant supporting elements present corresponds to the *degree of statical indeterminacy* and equals the total number of unknown external forces, minus the number of available independent equations of equilibrium. On the other hand, bodies which are supported by the minimum number of constraints necessary to ensure an equilibrium configuration are called *statically determinate*, and for such bodies the equilibrium equations are sufficient to determine the unknown external forces.

The problems on equilibrium in this article and throughout *Vol. 1 Statics* are generally restricted to statically determinate bodies where the constraints are just sufficient to ensure a stable equilibrium configuration and where the unknown supporting forces can be completely determined by the available independent equations of equilibrium.

We must be aware of the nature of the constraints before we attempt to solve an equilibrium problem. A body can be recognized as statically indeterminate when there are more unknown external reactions than there are available independent equilibrium equations for the force system involved. It is always well to count the number of unknown variables on a given body and to be certain that an equal number of independent equations can be written; otherwise, effort might be wasted in attempting an impossible solution with the aid of the equilibrium equations only. The unknown variables may be forces, couples, distances, or angles.

Adequacy of Constraints

In discussing the relationship between constraints and equilibrium, we should look further at the question of the adequacy of constraints. The existence of three constraints for a two-dimensional problem does not always guarantee a stable equilibrium configuration. Figure 3/7 shows four different types of constraints. In part *a* of the figure, point *A* of the rigid body is fixed by the two links and cannot move, and the third link prevents any rotation about *A*. Thus, this body is *completely fixed* with three *adequate (proper) constraints*.

In part *b* of the figure, the third link is positioned so that the force transmitted by it passes through point *A* where the other two constraint forces act. Thus, this configuration of constraints can offer no initial resistance to rotation about *A*, which would occur when external loads were applied to the body. We conclude, therefore, that this body is *incompletely fixed* under *partial constraints*.

The configuration in part *c* of the figure gives us a similar condition of incomplete fixity because the three parallel links could offer no initial resistance to a small vertical movement of the body as a result of external loads applied to it in this direction. The constraints in these two examples are often termed *improper*.

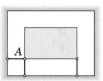

(*a*) Complete fixity
Adequate constraints

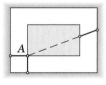

(*b*) Incomplete fixity
Partial constraints

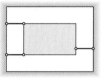

(*c*) Incomplete fixity
Partial constraints

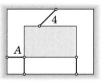

(*d*) **Excessive fixity**
Redundant constraint

Figure 3/7

In part *d* of Fig. 3/7 we have a condition of complete fixity, with link 4 acting as a fourth constraint which is unnecessary to maintain a fixed position. Link 4, then, is a *redundant constraint*, and the body is statically indeterminate.

As in the four examples of Fig. 3/7, it is generally possible by direct observation to conclude whether the constraints on a body in two-dimensional equilibrium are adequate (proper), partial (improper), or redundant. As indicated previously, the vast majority of problems in this book are statically determinate with adequate (proper) constraints.

KEY CONCEPTS

Approach to Solving Problems

The sample problems at the end of this article illustrate the application of free-body diagrams and the equations of equilibrium to typical statics problems. These solutions should be studied thoroughly. In the problem work of this chapter and throughout mechanics, it is important to develop a logical and systematic approach which includes the following steps:

1. Identify clearly the quantities which are known and unknown.

2. Make an unambiguous choice of the body (or system of connected bodies treated as a single body) to be isolated and draw its complete free-body diagram, labeling all external known and unknown but identifiable forces and couples which act on it.

3. Choose a convenient set of reference axes, always using right-handed axes when vector cross products are employed. Choose moment centers with a view to simplifying the calculations. Generally the best choice is one through which as many unknown forces pass as possible. Simultaneous solutions of equilibrium equations are frequently necessary, but can be minimized or avoided by a careful choice of reference axes and moment centers.

4. Identify and state the applicable force and moment principles or equations which govern the equilibrium conditions of the problem. In the following sample problems these relations are shown in brackets and precede each major calculation.

5. Match the number of independent equations with the number of unknowns in each problem.

6. Carry out the solution and check the results. In many problems engineering judgment can be developed by first making a reasonable guess or estimate of the result prior to the calculation and then comparing the estimate with the calculated value.

Sample Problem 3/1

Determine the magnitudes of the forces **C** and **T**, which, along with the other three forces shown, act on the bridge-truss joint.

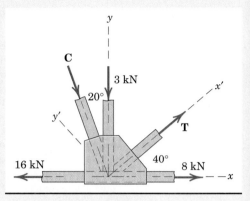

Solution.   The given sketch constitutes the free-body diagram of the isolated ① section of the joint in question and shows the five forces which are in equilibrium.

Solution 1 (scalar algebra).   For the x-y axes as shown we have

$[\Sigma F_x = 0]$ $8 + T \cos 40° + C \sin 20° - 16 = 0$

$0.766T + 0.342C = 8$ (a)

$[\Sigma F_y = 0]$ $T \sin 40° - C \cos 20° - 3 = 0$

$0.643T - 0.940C = 3$ (b)

Simultaneous solution of Eqs. (a) and (b) produces

$T = 9.09 \text{ kN}$ $C = 3.03 \text{ kN}$ *Ans.*

Solution II (scalar algebra).   To avoid a simultaneous solution, we may use axes ② x'-y' with the first summation in the y'-direction to eliminate reference to T. Thus,

$[\Sigma F_{y'} = 0]$ $-C \cos 20° - 3 \cos 40° - 8 \sin 40° + 16 \sin 40° = 0$

$C = 3.03 \text{ kN}$ *Ans.*

$[\Sigma F_{x'} = 0]$ $T + 8 \cos 40° - 16 \cos 40° - 3 \sin 40° - 3.03 \sin 20° = 0$

$T = 9.09 \text{ kN}$ *Ans.*

Solution III (vector algebra).   With unit vectors **i** and **j** in the x- and y-directions, the zero summation of forces for equilibrium yields the vector equation

$[\Sigma \mathbf{F} = \mathbf{0}]$ $8\mathbf{i} + (T \cos 40°)\mathbf{i} + (T \sin 40°)\mathbf{j} - 3\mathbf{j} + (C \sin 20°)\mathbf{i}$

$- (C \cos 20°)\mathbf{j} - 16\mathbf{i} = \mathbf{0}$

Equating the coefficients of the **i**- and **j**-terms to zero gives

$8 + T \cos 40° + C \sin 20° - 16 = 0$

$T \sin 40° - 3 - C \cos 20° = 0$

which are the same, of course, as Eqs. (a) and (b), which we solved above.

Solution IV (geometric).   The polygon representing the zero vector sum of the five forces is shown. Equations (a) and (b) are seen immediately to give the projections of the vectors onto the x- and y-directions. Similarly, projections onto the x'- and y'-directions give the alternative equations in Solution II.

A graphical solution is easily obtained. The known vectors are laid off head-③ to-tail to some convenient scale, and the directions of **T** and **C** are then drawn to close the polygon. The resulting intersection at point P completes the solution, thus enabling us to measure the magnitudes of **T** and **C** directly from the drawing to whatever degree of accuracy we incorporate in the construction.

Helpful Hints

① Since this is a problem of concurrent forces, no moment equation is necessary.

② The selection of reference axes to facilitate computation is always an important consideration. Alternatively in this example we could take a set of axes along and normal to the direction of **C** and employ a force summation normal to **C** to eliminate it.

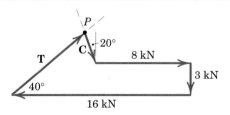

③ The known vectors may be added in any order desired, but they must be added before the unknown vectors.

Sample Problem 3/2

Calculate the tension T in the cable which supports the 1000-lb load with the pulley arrangement shown. Each pulley is free to rotate about its bearing, and the weights of all parts are small compared with the load. Find the magnitude of the total force on the bearing of pulley C.

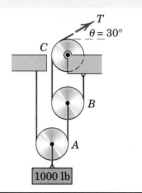

Solution. The free-body diagram of each pulley is drawn in its relative position to the others. We begin with pulley A, which includes the only known force. With the unspecified pulley radius designated by r, the equilibrium of moments about its center O and the equilibrium of forces in the vertical direction require

① $[\Sigma M_O = 0]$ $\qquad T_1 r - T_2 r = 0 \qquad T_1 = T_2$

$[\Sigma F_y = 0] \qquad T_1 + T_2 - 1000 = 0 \qquad 2T_1 = 1000 \qquad T_1 = T_2 = 500$ lb

From the example of pulley A we may write the equilibrium of forces on pulley B by inspection as

$$T_3 = T_4 = T_2/2 = 250 \text{ lb}$$

For pulley C the angle $\theta = 30°$ in no way affects the moment of T about the center of the pulley, so that moment equilibrium requires

$$T = T_3 \qquad \text{or} \qquad T = 250 \text{ lb} \qquad \qquad Ans.$$

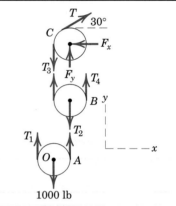

Equilibrium of the pulley in the x- and y-directions requires

$[\Sigma F_x = 0] \qquad\qquad 250 \cos 30° - F_x = 0 \qquad\qquad F_x = 217 \text{ lb}$

$[\Sigma F_y = 0] \qquad\qquad F_y + 250 \sin 30° - 250 = 0 \qquad F_y = 125 \text{ lb}$

$[F = \sqrt{F_x^2 + F_y^2}] \qquad F = \sqrt{(217)^2 + (125)^2} = 250 \text{ lb} \qquad\qquad Ans.$

Helpful Hint

① Clearly the radius r does not influence the results. Once we have analyzed a simple pulley, the results should be perfectly clear by inspection.

Sample Problem 3/3

The uniform 100-kg I-beam is supported initially by its end rollers on the horizontal surface at A and B. By means of the cable at C it is desired to elevate end B to a position 3 m above end A. Determine the required tension P, the reaction at A, and the angle θ made by the beam with the horizontal in the elevated position.

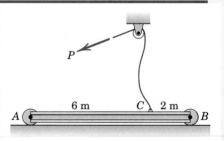

Solution. In constructing the free-body diagram, we note that the reaction on the roller at A and the weight are vertical forces. Consequently, in the absence of other horizontal forces, P must also be vertical. From Sample Problem 3/2 we see immediately that the tension P in the cable equals the tension P applied to the beam at C.

Moment equilibrium about A eliminates force R and gives

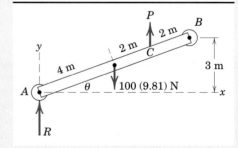

① $[\Sigma M_A = 0] \qquad P(6 \cos \theta) - 981(4 \cos \theta) = 0 \qquad P = 654 \text{ N} \qquad Ans.$

Equilibrium of vertical forces requires

$[\Sigma F_y = 0] \qquad\qquad 654 + R - 981 = 0 \qquad R = 327 \text{ N} \qquad\qquad Ans.$

The angle θ depends only on the specified geometry and is

$$\sin \theta = 3/8 \qquad \theta = 22.0° \qquad\qquad Ans.$$

Helpful Hint

① Clearly the equilibrium of this parallel force system is independent of θ.

Sample Problem 3/4

Determine the magnitude T of the tension in the supporting cable and the magnitude of the force on the pin at A for the jib crane shown. The beam AB is a standard 0.5-m I-beam with a mass of 95 kg per meter of length.

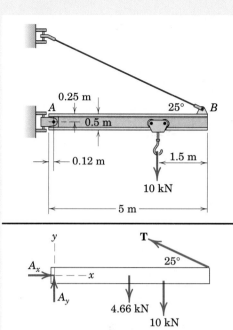

Algebraic solution. The system is symmetrical about the vertical x-y plane through the center of the beam, so the problem may be analyzed as the equilibrium of a coplanar force system. The free-body diagram of the beam is shown in the figure with the pin reaction at A represented in terms of its two rectangular components. The weight of the beam is $95(10^{-3})(5)9.81 = 4.66$ kN and acts through its center. Note that there are three unknowns A_x, A_y, and T, which may be found from the three equations of equilibrium. We begin with a moment equation about A, which eliminates two of the three unknowns from the equation. In applying the moment equation about A, it is simpler to consider the moments of the x- and y-components of $\mathbf{T}$ than it is to compute the perpendicular distance from $\mathbf{T}$ to A. Hence, with the counterclockwise sense as positive we write

①

②

$$[\Sigma M_A = 0] \qquad (T \cos 25°)0.25 + (T \sin 25°)(5 - 0.12)$$
$$- 10(5 - 1.5 - 0.12) - 4.66(2.5 - 0.12) = 0$$

from which $\qquad\qquad\qquad T = 19.61$ kN $\qquad\qquad\qquad\qquad$ *Ans.*

Equating the sums of forces in the x- and y-directions to zero gives

$$[\Sigma F_x = 0] \qquad A_x - 19.61 \cos 25° = 0 \qquad A_x = 17.77 \text{ kN}$$

$$[\Sigma F_y = 0] \qquad A_y + 19.61 \sin 25° - 4.66 - 10 = 0 \qquad A_y = 6.37 \text{ kN}$$

③ $\quad [A = \sqrt{A_x{}^2 + A_y{}^2}] \quad A = \sqrt{(17.77)^2 + (6.37)^2} = 18.88 \text{ kN} \qquad$ *Ans.*

Graphical solution. The principle that three forces in equilibrium must be concurrent is utilized for a graphical solution by combining the two known vertical forces of 4.66 and 10 kN into a single 14.66-kN force, located as shown on the modified free-body diagram of the beam in the lower figure. The position of this resultant load may easily be determined graphically or algebraically. The intersection of the 14.66-kN force with the line of action of the unknown tension $\mathbf{T}$ defines the point of concurrency O through which the pin reaction $\mathbf{A}$ must pass. The unknown magnitudes of $\mathbf{T}$ and $\mathbf{A}$ may now be found by adding the forces head-to-tail to form the closed equilibrium polygon of forces, thus satisfying their zero vector sum. After the known vertical load is laid off to a convenient scale, as shown in the lower part of the figure, a line representing the given direction of the tension $\mathbf{T}$ is drawn through the tip of the 14.66-kN vector. Likewise a line representing the direction of the pin reaction $\mathbf{A}$, determined from the concurrency established with the free-body diagram, is drawn through the tail of the 14.66-kN vector. The intersection of the lines representing vectors $\mathbf{T}$ and $\mathbf{A}$ establishes the magnitudes T and A necessary to make the vector sum of the forces equal to zero. These magnitudes are scaled from the diagram. The x- and y-components of $\mathbf{A}$ may be constructed on the force polygon if desired.

Free-body diagram

Helpful Hints

① The justification for this step is Varignon's theorem, explained in Art. 2/4. Be prepared to take full advantage of this principle frequently.

② The calculation of moments in two-dimensional problems is generally handled more simply by scalar algebra than by the vector cross product $\mathbf{r} \times \mathbf{F}$. In three dimensions, as we will see later, the reverse is often the case.

③ The direction of the force at A could be easily calculated if desired. However, in designing the pin A or in checking its strength, it is only the magnitude of the force that matters.

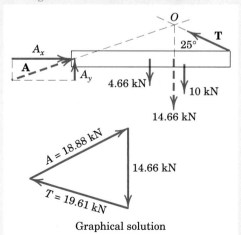

Graphical solution

PROBLEMS

Introductory Problems

3/1 Determine the force P required to maintain the 200-kg engine in the position for which $\theta = 30°$. The diameter of the pulley at B is negligible.

Ans. $P = 1759$ N

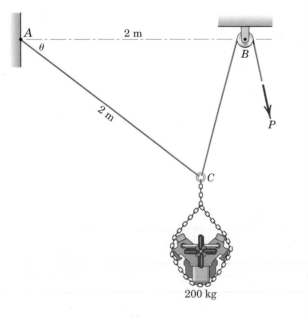

200 kg

Problem 3/1

3/2 The mass center G of the 1400-kg rear-engine car is located as shown in the figure. Determine the normal force under each tire when the car is in equilibrium. State any assumptions.

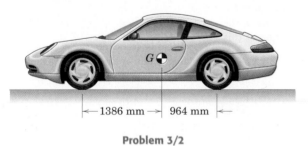

Problem 3/2

3/3 A carpenter carries a 12-lb 2-in. by 4-in. board as shown. What downward force does he feel on his shoulder at A?

Ans. $N_A = 18$ lb

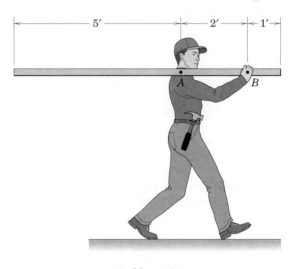

Problem 3/3

3/4 In the side view of a 70-kg television resting on a 24-kg cabinet, the mass centers are labeled G_1 and G_2. Determine the force reactions at A and B. (Note that the mass center of most televisions is located well forward because of the heavy nature of the front portion of picture tubes.)

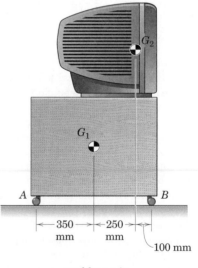

Problem 3/4

3/5 The roller stand is used to support portions of long boards as they are being cut on a table saw. If the board exerts a 25-N downward force on the roller C, determine the vertical reactions at A and D. Note that the connection at B is rigid, and that the feet A and D are fairly lengthy horizontal tubes with a nonslip coating.

Ans. $N_A = 8.45$ N, $N_D = 16.55$ N

Problem 3/5

3/6 The 450-kg uniform I-beam supports the load shown. Determine the reactions at the supports.

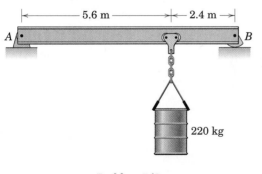

Problem 3/6

3/7 Calculate the force and moment reactions at the bolted base O of the overhead traffic-signal assembly. Each traffic signal weighs 80 lb, while the weights of members OC and AC are 110 lb and 120 lb, respectively.

Ans. $O_x = 0$, $O_y = 390$ lb, $M_O = 5040$ lb-ft CW

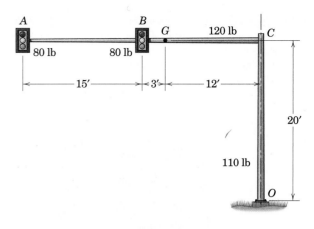

Problem 3/7

3/8 The 20-kg homogeneous smooth sphere rests on the two inclines as shown. Determine the contact forces at A and B.

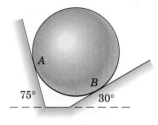

Problem 3/8

3/9 A 120-lb crate resets on the 60-lb pickup tailgate. Calculate the tension T in each of the two restraining cables, one of which is shown. The centers of gravity are at G_1 and G_2. The crate is located midway between the two cables.

Ans. $T = 131.2$ lb

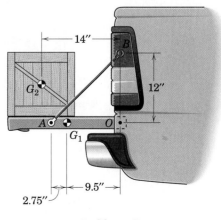

Problem 3/9

3/10 A portable electric generator has a mass of 160 kg with mass center at G. Determine the upward force F necessary to reduce the normal force at A to one-half its nominal ($F = 0$) value.

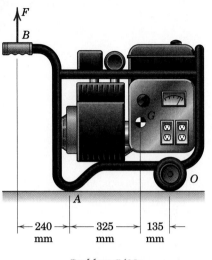

Problem 3/10

3/11 With what force magnitude T must the person pull on the cable in order to cause the scale A to read 500 lb? The weights of the pulleys and cables are negligible. State any assumptions.

Ans. T = 100 lb

Problem 3/11

3/12 The device shown is designed to aid in the removal of pull-tab tops from cans. If the user exerts a 40-N force at A, determine the tension T in the portion BC of the pull tab.

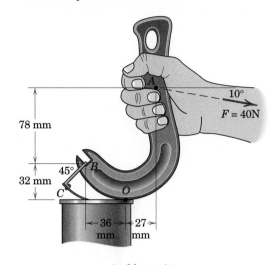

Problem 3/12

3/13 A woodcutter wishes to cause the tree trunk to fall uphill, even though the trunk is leaning downhill. With the aid of the winch W, what tension T in the cable will be required? The 1200-lb trunk has a center of gravity at G. The felling notch at O is sufficiently large so that the resisting moment there is negligible.

Ans. T = 81.2 lb

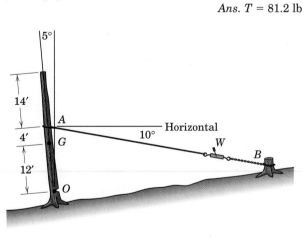

Problem 3/13

3/14 To facilitate shifting the position of a lifting hook when it is not under load, the sliding hanger shown is used. The projections at A and B engage the flanges of a box beam when a load is supported, and the hook projects through a horizontal slot in the beam. Compute the forces at A and B when the hook supports a 300-kg mass.

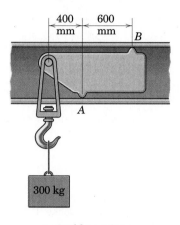

Problem 3/14

3/15 Three cables are joined at the junction ring C. Determine the tensions in cables AC and BC caused by the weight of the 30-kg cylinder.

Ans. $T_{AC} = 215$ N, $T_{BC} = 264$ N

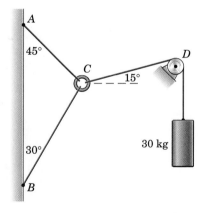

Problem 3/15

3/16 A 700-N axial force is required to remove the pulley from its shaft. What force F must be exerted on the handle of each of the two prybars? Friction at the contact points B and E is sufficient to prevent slipping; friction at the pulley contact points C and F is negligible.

Problem 3/16

3/17 The uniform beam has a mass of 50 kg per meter of length. Compute the reactions at the support O. The force loads shown lie in a vertical plane.

Ans. $O_x = -0.7$ kN, $O_y = 5.98$ kN, $M_O = 9.12$ kN·m

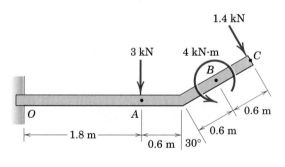

Problem 3/17

Representative Problems

3/18 A pipe P is being bent by the pipe bender as shown. If the hydraulic cylinder applies a force of magnitude $F = 24$ kN to the pipe at C, determine the magnitude of the roller reactions at A and B.

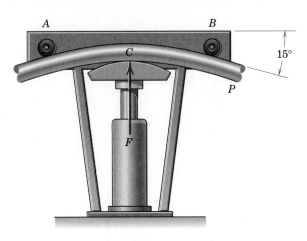

Problem 3/18

3/19 The uniform 15-m pole has a mass of 150 kg and is supported by its smooth ends against the vertical walls and by the tension T in the vertical cable. Compute the reactions at A and B.

Ans. $A = B = 327$ N

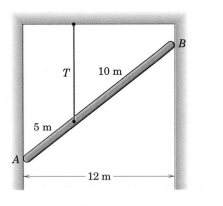

Problem 3/19

3/20 Determine the reactions at A and E if $P = 500$ N. What is the maximum value which P may have for static equilibrium? Neglect the weight of the structure compared with the applied loads.

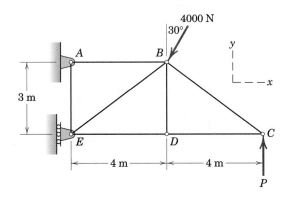

Problem 3/20

3/21 While digging a small hole prior to planting a tree, a homeowner encounters rocks. If he exerts a horizontal 50-lb force on the prybar as shown, what is the horizontal force exerted on rock C? Note that a small ledge on rock C supports a vertical force reaction there. Neglect friction at B. Complete solutions (a) including and (b) excluding the weight of the prybar.

Ans. (a) $F_C = 380$ lb, (b) $F_C = 325$ lb

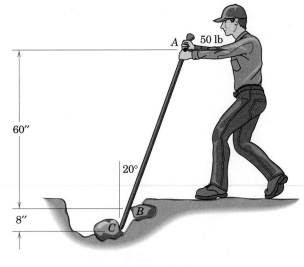

Problem 3/21

3/22 Determine the force P required to begin rolling the uniform cylinder of mass m over the obstruction of height h.

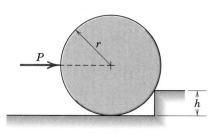

Problem 3/22

3/23 A 35-N axial force at B is required to open the spring-loaded plunger of the water nozzle. Determine the required force F applied to the handle at A and the magnitude of the pin reaction at O. Note that the plunger passes through a vertically-elongated hole in the handle at B, so that negligible vertical force is transmitted there.

Ans. $F = 13.98$ N, $O = 48.8$ N

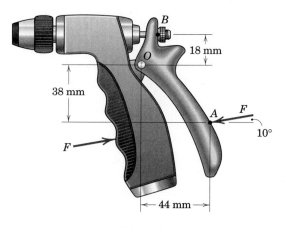

Problem 3/23

3/24 A person holds a 60-lb suitcase by its handle as indicated in the figure. Determine the tension in each of the four identical links AB.

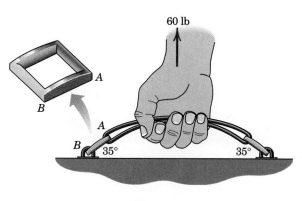

Problem 3/24

3/25 A block placed under the head of the claw hammer as shown greatly facilitates the extraction of the nail. If a 50-lb pull on the handle is required to pull the nail, calculate the tension T in the nail and the magnitude A of the force exerted by the hammer head on the block. The contacting surfaces at A are sufficiently rough to prevent slipping.

Ans. $T = 200$ lb, $A = 188.8$ lb

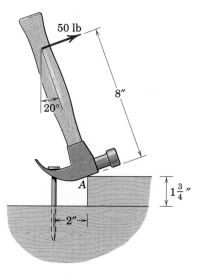

Problem 3/25

3/26 The indicated location of the center of gravity of the 3600-lb pickup truck is for the unladen condition. If a load whose center of gravity is $x = 16$ in. behind the rear axle is added to the truck, determine the load weight W_L for which the normal forces under the front and rear wheels are equal.

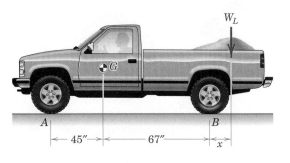

Problem 3/26

3/27 The wall-mounted 2.5-kg light fixture has its mass center at G. Determine the reactions at A and B and also calculate the moment supported by the adjustment thumbscrew at C. (Note that the lightweight frame ABC has about 250 mm of horizontal tubing, directed into and out of the paper, at both A and B.)

Ans. $B_x = 32.0$ N
$A_x = 32.0$ N, $A_y = 24.5$ N
$M_C = 2.45$ N·m CW

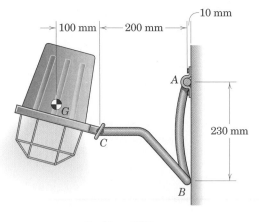

Problem 3/27

3/28 To test the validity of aerodynamic assumptions made in the design of the aircraft, its model is being tested in a wind tunnel. The support bracket is connected to a force and moment balance, which is zeroed when there is no airflow. Under test conditions, the lift L, drag D, and pitching moment M_G act as shown. The force balance records the lift, drag, and a moment M_P. Determine M_G in terms of L, D, and M_P.

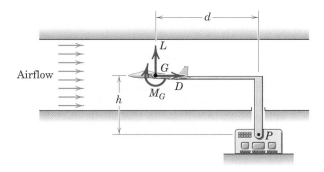

Problem 3/28

3/29 The chain binder is used to secure loads of logs, lumber, pipe, and the like. If the tension T_1 is 2 kN when $\theta = 30°$, determine the force P required on the lever and the corresponding tension T_2 for this position. Assume that the surface under A is perfectly smooth.

Ans. $P = 166.7$ N, $T_2 = 1.917$ kN

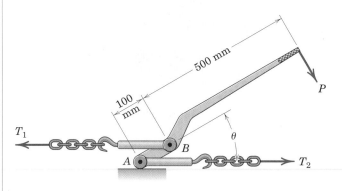

Problem 3/29

3/30 The device shown is designed to apply pressure when bonding laminate to each side of a countertop near an edge. If a 30-lb force is applied to the handle, determine the force which each roller exerts on its corresponding surface.

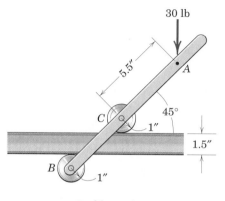

Problem 3/30

3/31 The two light pulleys are fastened together and form an integral unit. They are prevented from turning about their bearing at O by a cable wound securely around the smaller pulley and fastened to point A. Calculate the magnitude R of the force supported by the bearing O for the applied 2-kN load.

Ans. R = 4.38 kN

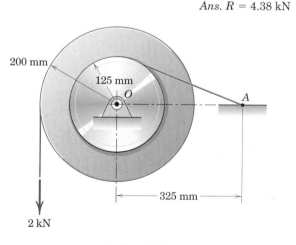

Problem 3/31

3/32 In a procedure to evaluate the strength of the triceps muscle, a person pushes down on a load cell with the palm of his hand as indicated in the figure. If the load-cell reading is 35 lb, determine the vertical tensile force F generated by the triceps muscle. The lower arm weighs 3.2 lb with mass center at G. State any assumptions.

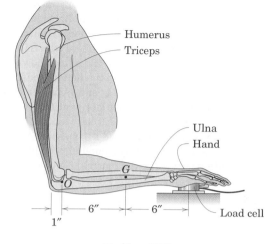

Problem 3/32

3/33 A person is performing slow arm curls with a 20-lb weight as indicated in the figure. The brachialis muscle group (consisting of the biceps and brachialis muscles) is the major factor in this exercise. Determine the magnitude F of the brachialis-muscle-group force and the magnitude E of the elbow joint reaction at point E for the forearm position shown in the figure. Take the dimensions shown to locate the effective points of application of the two muscle groups; these points are 8 in. directly above E and 2 in. directly to the right of E. Include the 3.2-lb forearm weight which acts at point G. State any assumptions.

Ans. F = 154.2 lb, *E* = 131.8 lb

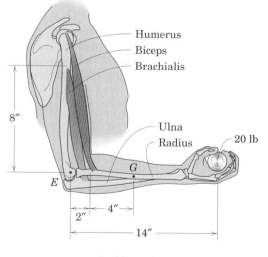

Problem 3/23

3/34 A woman is holding a 3.6-kg sphere in her hand with the entire arm held horizontally as shown in the figure. A tensile force in the deltoid muscle prevents the arm from rotating about the shoulder joint O; this force acts at the 21° angle shown. Determine the force exerted by the deltoid muscle on the upper arm at A and the x- and y-components of the force reaction at the shoulder joint O. The mass of the upper arm is $m_U = 1.9$ kg, the mass of the lower arm is $m_L = 1.1$ kg, and the mass of the hand is $m_H = 0.4$ kg; all the corresponding weights act at the locations shown in the figure.

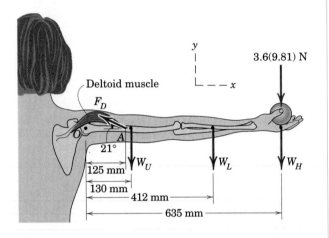

Problem 3/34

3/35 With his weight W equally distributed on both feet, a man begins to slowly rise from a squatting position as indicated in the figure. Determine the tensile force F in the patellar tendon and the magnitude of the force reaction at point O, which is the contact area between the tibia and the femur. Note that the line of action of the patellar tendon force is along its midline. Neglect the weight of the lower leg.

Ans. $F = 2.25W$, $O = 2.67W$

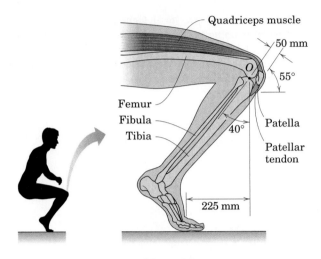

Problem 3/35

3/36 The elements of an on-off mechanism for a table lamp are shown in the figure. The electrical switch S requires a 0.9-lb force in order to depress it. What corresponding force F must be exerted on the handle at A?

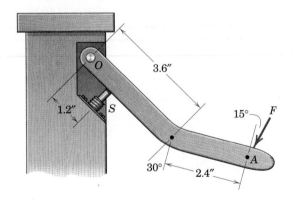

Problem 3/36

3/37 The uniform 18-kg bar OA is held in the position shown by the smooth pin at O and the cable AB. Determine the tension T in the cable and the magnitude and direction of the external pin reaction at O.

Ans. $T = 99.5$ N, $O = 246$ N, 70.3° CCW from x-axis

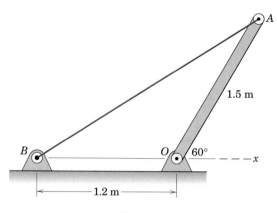

Problem 3/37

3/38 A person attempts to move a 40-lb shop vacuum by pulling on the hose as indicated. What force F will cause the unit to tip clockwise if wheel A is against an obstruction?

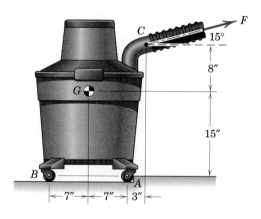

Problem 3/38

3/39 The exercise machine is designed with a lightweight cart which is mounted on small rollers so that it is free to move along the inclined ramp. Two cables are attached to the cart—one for each hand. If the hands are together so that the cables are parallel and if each cable lies essentially in a vertical plane, determine the force P which each hand must exert on its cable in order to maintain an equilibrium position. The mass of the person is 70 kg, the ramp angle θ is 15°, and the angle β is 18°. In addition, calculate the force R which the ramp exerts on the cart.

Ans. $P = 45.5$ N, $R = 691$ N

Problem 3/39

3/40 The device shown is used to test automobile-engine valve springs. The torque wrench is directly connected to arm OB. The specification for the automotive intake-valve spring is that 83 lb of force should reduce its length from 2 in. (unstressed length) to $1\frac{11}{16}$ in. What is the corresponding reading M on the torque wrench, and what force F exerted on the torque-wrench handle is required to produce this reading? Neglect the small effects of changes in the angular position of arm OB.

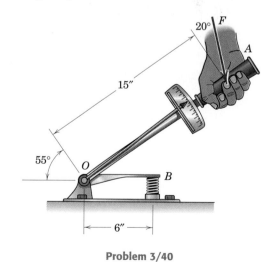

Problem 3/40

3/41 During an engine test on the ground, a propeller thrust $T = 3000$ N is generated on the 1800-kg airplane with mass center at G. The main wheels at B are locked and do not skid; the small tail wheel at A has no brake. Compute the percent change n in the normal forces at A and B as compared with their "engine-off" values.

Ans. $n_A = -32.6\%, n_B = 2.28\%$

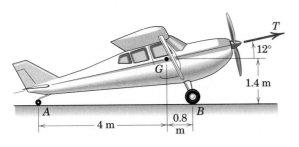

Problem 3/41

3/42 A rocker arm with rollers at A and B is shown in the position when the valve is open and the valve spring is fully compressed. In this position, the spring force is 900 N. Determine the force which the rocker arm exerts on the camshaft C. Also calculate the magnitude of the force supported by the rocker-arm shaft O.

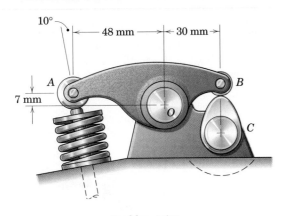

Problem 3/42

3/43 The hook wrench or pin spanner is used to turn shafts and collars. If a moment of 80 N·m is required to turn the 200-mm-diameter collar about its center O under the action of the applied force P, determine the contact force R on the smooth surface at A. Engagement of the pin at B may be considered to occur at the periphery of the collar.

Ans. $R = 1047$ N

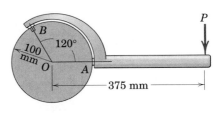

Problem 3/43

3/44 The dolly shown is useful in the handling of 55-gallon drums. Determine the force F necessary to hold a drum in the position shown. You may neglect the weight of the dolly in comparison with the 500-lb weight of the drum, whose center of gravity is at G. There is sufficient friction to prevent slipping at the contact point P.

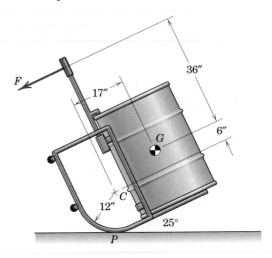

Problem 3/44

3/45 In sailing at a constant speed with the wind, the sailboat is driven by a 1000-lb force against its mainsail and a 400-lb force against its staysail as shown. The total resistance due to fluid friction through the water is the force R. Determine the resultant of the lateral forces perpendicular to motion applied to the hull by the water.

Ans. M = 8000 lb-ft

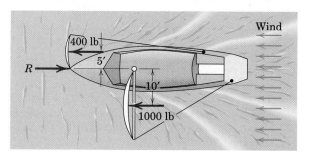

Problem 3/45

3/46 Estimate the force F required to lift the rear tires of the race car off the ground. You may assume that part of CD of the dolly jack is horizontal. The mass of the car and the driver combined is 700 kg with mass center at G. The driver applies the brakes during the jacking. State any additional assumptions.

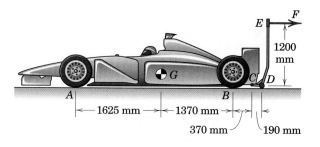

Problem 3/46

3/47 A portion of the shifter mechanism for a manual car transmission is shown in the figure. For the 4-lb force exerted on the shift knob, determine the corresponding force P exerted by the shift link BC on the transmission (not shown). Neglect friction in the ball-and-socket joint at O, in the joint at B, and in the slip tube near support D. Note that a soft rubber bushing at D allows the slip tube to self-align with link BC.

Ans. P = 13.14 lb

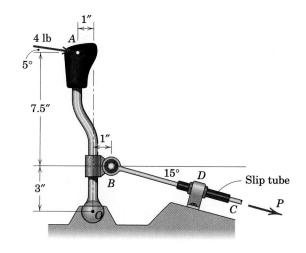

Problem 3/47

3/48 The small sailboat may be tipped at its moorings as shown to effect repairs below the waterline. One attached rope is passed under the keel and secured to the dock. The other rope is attached to the mast and is used to tip the boat. The boat shown has a displacement (which equals the total mass) of 5000 kg with mass center at G. The metacenter M is the point on the centerline of the boat through which the vertical resultant of the buoyant forces passes, and $\overline{GM}$ = 0.8 m. Calculate the tension T required to hold the boat in the position shown.

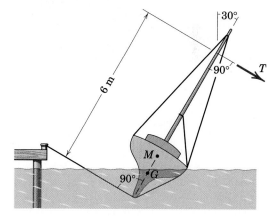

Problem 3/48

3/49 A torque (moment) of 24 N·m is required to turn the bolt about its axis. Determine P and the forces between the smooth hardened jaws of the wrench and the corners A and B of the hexagonal head. Assume that the wrench fits easily on the bolt so that contact is made at corners A and B only.

Ans. $P = 200$ N, $A = 2870$ N, $B = 3070$ N

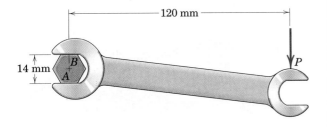

Problem 3/49

3/50 Determine the moment M which must be applied to the shaft in order to hold the homogeneous hemisphere in an arbitrary angular position as measured by the angle θ. The radii of gear A, gear B, and the hemisphere are r_A, r_B, and r, respectively. Assume the friction in all bearings to be negligible.

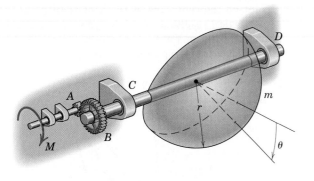

Problem 3/50

3/51 The car complete with driver weighs 1700 lb and without the two airfoils has a 50%–50% front–rear weight distribution at a certain speed at which there is no lift on the car. It is estimated that at this speed each of the airfoils A_1 and A_2 will generate 400 lb of downward force L and 50 lb of drag force D on the car. Specify the vertical reactions N_A and N_B under the two pairs of wheels at that speed when the airfoils are added. Assume that the addition of the airfoils does not affect the drag and zero-lift conditions of the car body itself and that the engine has sufficient power for equilibrium at that speed. The weight of the airfoils may be neglected.

Ans. $N_A = 1201$ lb (48.0%), $N_B = 1299$ lb (52.0%)

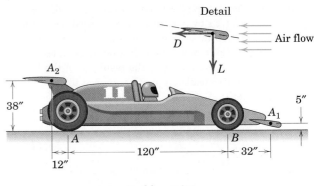

Problem 3/51

3/52 To test the deflection of the uniform 200-lb beam the 120-lb boy exerts a pull of 40 lb on the rope rigged as shown. Compute the force supported by the pin at the hinge O.

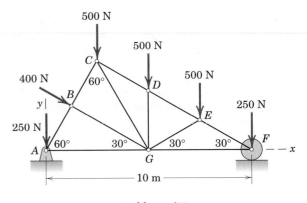

Problem 3/52

3/53 Determine the external reactions at A and F for the roof truss loaded as shown. The vertical loads represent the effect of the supported roofing materials, while the 400-N force represents a wind load.

Ans. $A_x = 346$ N, $A_y = 1100$ N, $F_y = 1100$ N

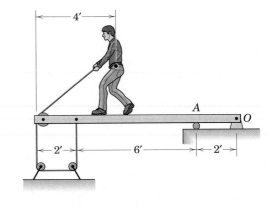

Problem 3/53

3/54 The member *OBC* and sheave at *C* together have a mass of 500 kg, with a combined center of mass at *G*. Calculate the magnitude of the force supported by the pin connection at *O* when the 3-kN load is applied. The collar at *A* can provide support in the horizontal direction only.

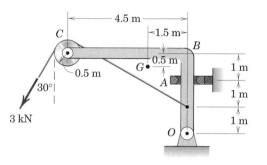

Problem 3/54

3/55 It is desired that a person be able to begin closing the van hatch from the open position shown with a 10-lb vertical force *P*. As a design exercise, determine the necessary force in each of the two hydraulic struts *AB*. The mass center of the 90-lb door is 1.5 in. directly below point *A*. Treat the problem as two-dimensional.

Ans. F = 187.1 lb

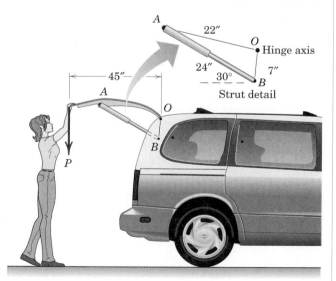

Problem 3/55

3/56 The man pushes the lawn mower at a steady speed with a force *P* which is parallel to the incline. The mass of the mower with attached grass bag is 50 kg with mass center at *G*. If $\theta = 15°$, determine the normal forces N_B and N_C under each pair of wheels *B* and *C*. Neglect friction. Compare with the normal forces for the conditions of $\theta = 0$ and $P = 0$.

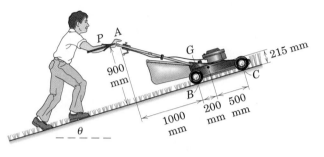

Problem 3/56

3/57 Determine the tension *T* in the turnbuckle for the pulley–cable system in terms of the mass *m* of the body which it supports. Neglect the mass of the pulleys and cable.

Ans. $T = \frac{2}{7}mg$

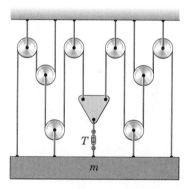

Problem 3/57

3/58 The cargo door for an airplane of circular fuselage section consists of the uniform quarter-circular segment AB of mass m. A detent in the hinge at A holds the door open in the position shown. Determine the moment exerted by the hinge on the door.

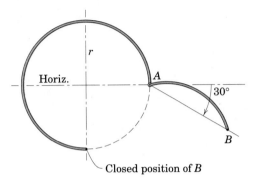

Problem 3/58

3/59 Pulley A delivers a steady torque (moment) of 100 N·m to a pump through its shaft at C. The tension in the lower side of the belt is 600 N. The driving motor B has a mass of 100 kg and rotates clockwise. As a design consideration, determine the magnitude R of the force on the supporting pin at O.

Ans. $R = 1.167$ kN

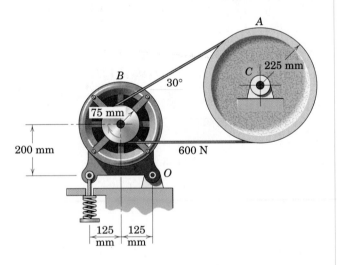

Problem 3/59

3/60 Certain elements of an in-refrigerator ice-cube maker are shown in the figure. (A "cube" has the form of a cylindrical segment!) Once the cube freezes and a small heater (not shown) forms a thin film of water between the cube and supporting surface, a motor rotates the ejector arm OA to remove the cube. If there are eight cubes and eight arms, determine the required torque M as a function of θ. The weight of eight cubes is 0.56 lb, and the center-of-gravity distance $\bar{r} = 0.55r$. Neglect friction, and assume that the resultant of the distributed normal force acting on the cube passes through point O.

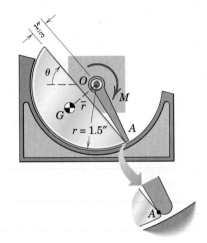

Problem 3/60

SECTION B   EQUILIBRIUM IN THREE DIMENSIONS

3/4   EQUILIBRIUM CONDITIONS

We now extend our principles and methods developed for two-dimensional equilibrium to the case of three-dimensional equilibrium. In Art. 3/1 the general conditions for the equilibrium of a body were stated in Eqs. 3/1, which require that the resultant force and resultant couple on a body in equilibrium be zero. These two vector equations of equilibrium and their scalar components may be written as

$$\Sigma \mathbf{F} = \mathbf{0} \quad \text{or} \quad \begin{cases} \Sigma F_x = 0 \\ \Sigma F_y = 0 \\ \Sigma F_z = 0 \end{cases}$$

$$\Sigma \mathbf{M} = \mathbf{0} \quad \text{or} \quad \begin{cases} \Sigma M_x = 0 \\ \Sigma M_y = 0 \\ \Sigma M_z = 0 \end{cases} \tag{3/3}$$

The first three scalar equations state that there is no resultant force acting on a body in equilibrium in any of the three coordinate directions. The second three scalar equations express the further equilibrium requirement that there be no resultant moment acting on the body about any of the coordinate axes or about axes parallel to the coordinate axes. These six equations are both necessary and sufficient conditions for complete equilibrium. The reference axes may be chosen arbitrarily as a matter of convenience, the only restriction being that a right-handed coordinate system should be chosen when vector notation is used.

The six scalar relationships of Eqs. 3/3 are independent conditions because any of them can be valid without the others. For example, for a car which accelerates on a straight and level road in the x-direction, Newton's second law tells us that the resultant force on the car equals its mass times its acceleration. Thus $\Sigma F_x \neq 0$, but the remaining two force–equilibrium equations are satisfied because all other acceleration components are zero. Similarly, if the flywheel of the engine of the accelerating car is rotating with increasing angular speed about the x-axis, it is not in rotational equilibrium about this axis. Thus, for the flywheel alone, $\Sigma M_x \neq 0$ along with $\Sigma F_x \neq 0$, but the remaining four equilibrium equations for the flywheel would be satisfied for its mass-center axes.

In applying the vector form of Eqs. 3/3, we first express each of the forces in terms of the coordinate unit vectors $\mathbf{i}$, $\mathbf{j}$, and $\mathbf{k}$. For the first equation, $\Sigma \mathbf{F} = \mathbf{0}$, the vector sum will be zero only if the coefficients of $\mathbf{i}$, $\mathbf{j}$, and $\mathbf{k}$ in the expression are, respectively, zero. These three sums, when each is set equal to zero, yield precisely the three scalar equations of equilibrium, $\Sigma F_x = 0$, $\Sigma F_y = 0$, and $\Sigma F_z = 0$.

For the second equation, $\Sigma \mathbf{M} = \mathbf{0}$, where the moment sum may be taken about any convenient point O, we express the moment of each force as the cross product $\mathbf{r} \times \mathbf{F}$, where $\mathbf{r}$ is the position vector from O to any point on the line of action of the force $\mathbf{F}$. Thus $\Sigma \mathbf{M} = \Sigma(\mathbf{r} \times \mathbf{F}) = \mathbf{0}$. When the coefficients of $\mathbf{i}$, $\mathbf{j}$, and $\mathbf{k}$ in the resulting moment equation are set equal to zero, respectively, we obtain the three scalar moment equations $\Sigma M_x = 0$, $\Sigma M_y = 0$, and $\Sigma M_z = 0$.

Free-Body Diagrams

The summations in Eqs. 3/3 include the effects of *all* forces on the body under consideration. We learned in the previous article that the free-body diagram is the only reliable method for disclosing all forces and moments which should be included in our equilibrium equations. In three dimensions the free-body diagram serves the same essential purpose as it does in two dimensions and should *always* be drawn. We have our choice either of drawing a pictorial view of the isolated body with all external forces represented or of drawing the orthogonal projections of the free-body diagram. Both representations are illustrated in the sample problems at the end of this article.

The correct representation of forces on the free-body diagram requires a knowledge of the characteristics of contacting surfaces. These characteristics were described in Fig. 3/1 for two-dimensional problems, and their extension to three-dimensional problems is represented in Fig. 3/8 for the most common situations of force transmission. The representations in both Figs. 3/1 and 3/8 will be used in three-dimensional analysis.

The essential purpose of the free-body diagram is to develop a reliable picture of the physical action of all forces (and couples if any) acting on a body. So it is helpful to represent the forces in their correct physical sense whenever possible. In this way, the free-body diagram becomes a closer model to the actual physical problem than it would be if the forces were arbitrarily assigned or always assigned in the same mathematical sense as that of the assigned coordinate axis.

For example, in part 4 of Fig. 3/8, the correct sense of the unknowns R_x and R_y may be known or perceived to be in the sense opposite to those of the assigned coordinate axes. Similar conditions apply to the sense of couple vectors, parts 5 and 6, where their sense by the right-hand rule may be assigned opposite to that of the respective coordinate direction. By this time, you should recognize that a negative answer for an unknown force or couple vector merely indicates that its physical action is in the sense opposite to that assigned on the free-body diagram. Frequently, of course, the correct physical sense is not known initially, so that an arbitrary assignment on the free-body diagram become necessary.

Categories of Equilibrium

Application of Eqs. 3/3 falls into four categories which we identify with the aid of Fig. 3/9. These categories differ in the number and type (force or moment) of independent equilibrium equations required to solve the problem.

Category 1, equilibrium of forces all concurrent at point O, requires all three force equations, but no moment equations because the moment of the forces about any axis through O is zero.

Category 2, equilibrium of forces which are concurrent with a line, requires all equations except the moment equation about that line, which is automatically satisfied.

MODELING THE ACTION OF FORCES IN THREE-DIMENSIONAL ANALYSIS

Type of Contact and Force Origin	Action on Body to Be Isolated
1. Member in contact with smooth surface, or ball-supported member 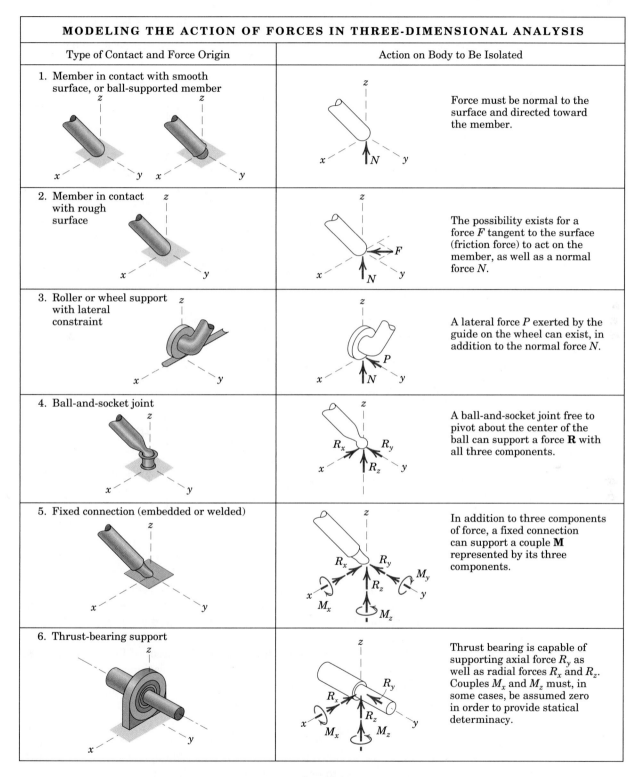	Force must be normal to the surface and directed toward the member.
2. Member in contact with rough surface	The possibility exists for a force F tangent to the surface (friction force) to act on the member, as well as a normal force N.
3. Roller or wheel support with lateral constraint	A lateral force P exerted by the guide on the wheel can exist, in addition to the normal force N.
4. Ball-and-socket joint	A ball-and-socket joint free to pivot about the center of the ball can support a force **R** with all three components.
5. Fixed connection (embedded or welded)	In addition to three components of force, a fixed connection can support a couple **M** represented by its three components.
6. Thrust-bearing support	Thrust bearing is capable of supporting axial force R_y as well as radial forces R_x and R_z. Couples M_x and M_z must, in some cases, be assumed zero in order to provide statical determinacy.

Figure 3/8

CATEGORIES OF EQUILIBRIUM IN THREE DIMENSIONS

Force System	Free-Body Diagram	Independent Equations
1. Concurrent at a point	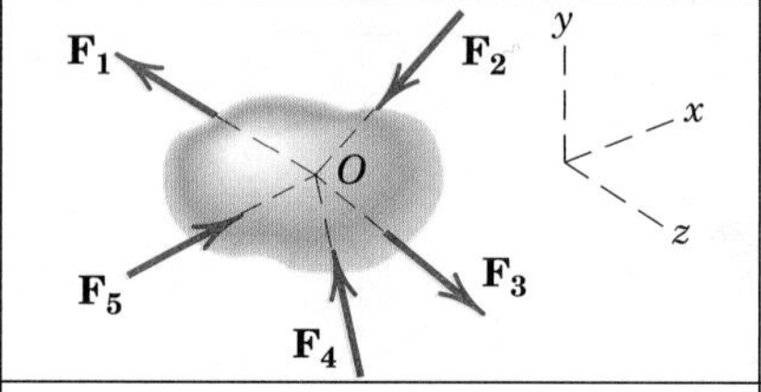	$\Sigma F_x = 0$ $\Sigma F_y = 0$ $\Sigma F_z = 0$
2. Concurrent with a line	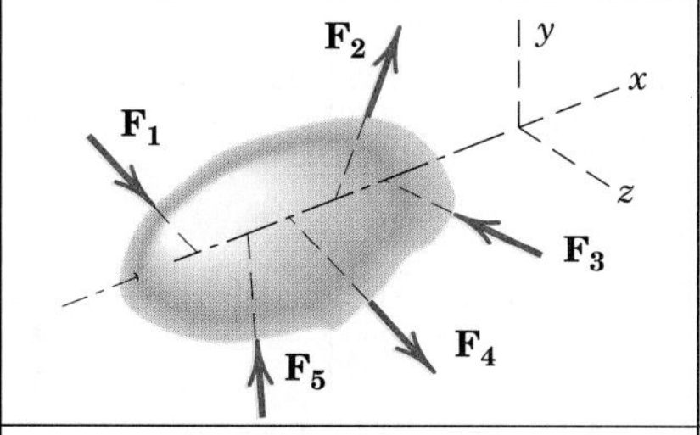	$\Sigma F_x = 0 \quad \Sigma M_y = 0$ $\Sigma F_y = 0 \quad \Sigma M_z = 0$ $\Sigma F_z = 0$
3. Parallel	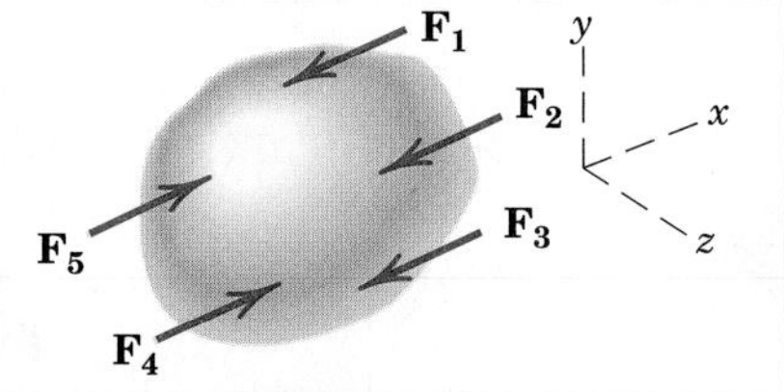	$\Sigma F_x = 0 \quad \Sigma M_y = 0$ $\quad\quad\quad\; \Sigma M_z = 0$
4. General	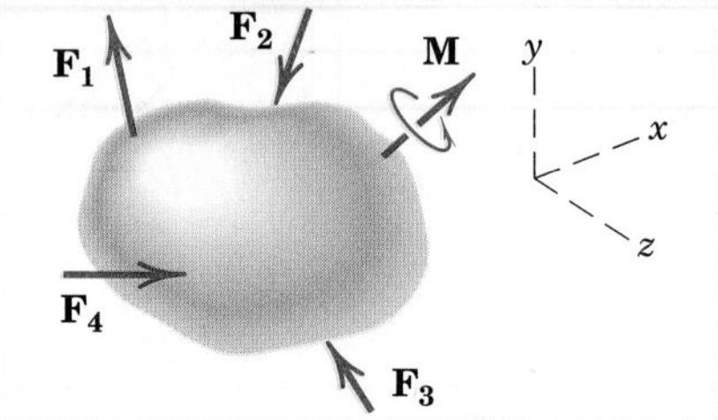	$\Sigma F_x = 0 \quad \Sigma M_x = 0$ $\Sigma F_y = 0 \quad \Sigma M_y = 0$ $\Sigma F_z = 0 \quad \Sigma M_z = 0$

Figure 3/9

Category 3, equilibrium of parallel forces, requires only one force equation, the one in the direction of the forces (x-direction as shown), and two moment equations about the axes (y and z) which are normal to the direction of the forces.

Category 4, equilibrium of a general system of forces, requires all three force equations and all three moment equations.

The observations contained in these statements are generally quite evident when a given problem is being solved.

Constraints and Statical Determinacy

The six scalar relations of Eqs. 3/3, although necessary and sufficient conditions to establish equilibrium, do not necessarily provide all of the information required to calculate the unknown forces acting in a three-dimensional equilibrium situation. Again, as we found with two dimensions, the question of adequacy of information is decided by the characteristics of the constraints provided by the supports. An analytical criterion for determining the adequacy of constraints is available, but it is beyond the scope of this treatment.* In Fig. 3/10, however, we cite four examples of constraint conditions to alert the reader to the problem.

Part *a* of Fig. 3/10 shows a rigid body whose corner point *A* is completely fixed by the links 1, 2, and 3. Links 4, 5, and 6 prevent rotations about the axes of links 1, 2, and 3, respectively, so that the body is *completely fixed* and the constraints are said to be *adequate*. Part *b* of the figure shows the same number of constraints, but we see that they provide no resistance to a moment which might be applied about axis *AE*. Here the body is *incompletely fixed* and only *partially constrained*.

Similarly, in Fig. 3/10*c* the constraints provide no resistance to an unbalanced force in the *y*-direction, so here also is a case of incomplete fixity with partial constraints. In Fig. 3/10*d*, if a seventh constraining link were imposed on a system of six constraints placed properly for complete fixity, more supports would be provided than would be necessary to establish the equilibrium position, and link 7 would be *redundant*. The body would then be *statically indeterminate* with such a seventh link in place. With only a few exceptions, the supporting constraints for rigid bodies in equilibrium in this book are adequate, and the bodies are statically determinate.

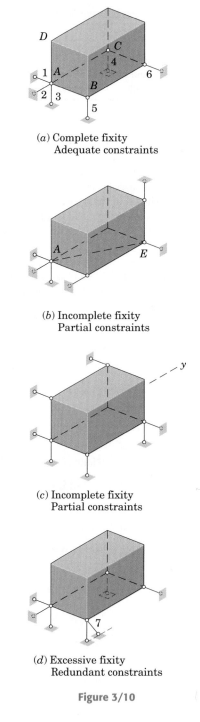

(*a*) Complete fixity
Adequate constraints

(*b*) Incomplete fixity
Partial constraints

(*c*) Incomplete fixity
Partial constraints

(*d*) Excessive fixity
Redundant constraints

Figure 3/10

© Thomas Ebert/Laif/Redux

The equilibrium of structural components such as these shell-like panels is an issue both during and after construction. This structure will be used to house airships.

*See the first author's *Statics, 2nd Edition SI Version*, 1975, Art. 16.

Sample Problem 3/5

The uniform 7-m steel shaft has a mass of 200 kg and is supported by a ball-and-socket joint at A in the horizontal floor. The ball end B rests against the smooth vertical walls as shown. Compute the forces exerted by the walls and the floor on the ends of the shaft.

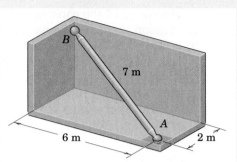

Solution. The free-body diagram of the shaft is first drawn where the contact forces acting on the shaft at B are shown normal to the wall surfaces. In addition to the weight $W = mg = 200(9.81) = 1962$ N, the force exerted by the floor on the ball joint at A is represented by its x-, y-, and z-components. These components are shown in their correct physical sense, as should be evident from the requirement that A be held in place. The vertical position of B is found from $7 = \sqrt{2^2 + 6^2 + h^2}$, $h = 3$ m. Right-handed coordinate axes are assigned as shown.

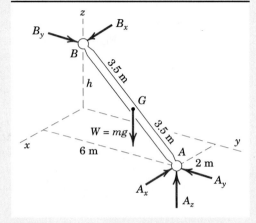

Vector solution. We will use A as a moment center to eliminate reference to the forces at A. The position vectors needed to compute the moments about A are

$$\mathbf{r}_{AG} = -1\mathbf{i} - 3\mathbf{j} + 1.5\mathbf{k} \text{ m} \qquad \text{and} \qquad \mathbf{r}_{AB} = -2\mathbf{i} - 6\mathbf{j} + 3\mathbf{k} \text{ m}$$

where the mass center G is located halfway between A and B.

The vector moment equation gives

$$[\Sigma \mathbf{M}_A = 0] \qquad \mathbf{r}_{AB} \times (\mathbf{B}_x + \mathbf{B}_y) + \mathbf{r}_{AG} \times \mathbf{W} = 0$$

$$(-2\mathbf{i} - 6\mathbf{j} + 3\mathbf{k}) \times (B_x\mathbf{i} + B_y\mathbf{j}) + (-\mathbf{i} - 3\mathbf{j} + 1.5\mathbf{k}) \times (-1962\mathbf{k}) = 0$$

$$\begin{vmatrix} \mathbf{i} & \mathbf{j} & \mathbf{k} \\ -2 & -6 & 3 \\ B_x & B_y & 0 \end{vmatrix} + \begin{vmatrix} \mathbf{i} & \mathbf{j} & \mathbf{k} \\ -1 & -3 & 1.5 \\ 0 & 0 & -1962 \end{vmatrix} = 0$$

$$(-3B_y + 5890)\mathbf{i} + (3B_x - 1962)\mathbf{j} + (-2B_y + 6B_x)\mathbf{k} = 0$$

Equating the coefficients of $\mathbf{i}$, $\mathbf{j}$, and $\mathbf{k}$ to zero and solving give

$$B_x = 654 \text{ N} \qquad \text{and} \qquad B_y = 1962 \text{ N} \qquad \qquad \textit{Ans.}$$

The forces at A are easily determined by

$$[\Sigma \mathbf{F} = 0] \qquad (654 - A_x)\mathbf{i} + (1962 - A_y)\mathbf{j} + (-1962 + A_z)\mathbf{k} = 0$$

and

$$A_x = 654 \text{ N} \qquad A_y = 1962 \text{ N} \qquad A_z = 1962 \text{ N}$$

Finally

$$A = \sqrt{A_x{}^2 + A_y{}^2 + A_z{}^2}$$

$$= \sqrt{(654)^2 + (1962)^2 + (1962)^2} = 2850 \text{ N} \qquad \qquad \textit{Ans.}$$

Scalar solution. Evaluating the scalar moment equations about axes through A parallel, respectively, to the x- and y-axes, gives

$$[\Sigma M_{A_x} = 0] \qquad 1962(3) - 3B_y = 0 \qquad B_y = 1962 \text{ N}$$

$$[\Sigma M_{A_y} = 0] \qquad -1962(1) + 3B_x = 0 \qquad B_x = 654 \text{ N}$$

The force equations give, simply,

$$[\Sigma F_x = 0] \qquad -A_x + 654 = 0 \qquad A_x = 654 \text{ N}$$

$$[\Sigma F_y = 0] \qquad -A_y + 1962 = 0 \qquad A_y = 1962 \text{ N}$$

$$[\Sigma F_z = 0] \qquad A_z - 1962 = 0 \qquad A_z = 1962 \text{ N}$$

Helpful Hints

① We could, of course, assign all of the unknown components of force in the positive mathematical sense, in which case A_x and A_y would turn out to be negative upon computation. The free-body diagram describes the physical situation, so it is generally preferable to show the forces in their correct physical senses wherever possible.

② Note that the third equation $-2B_y + 6B_x = 0$ merely checks the results of the first two equations. This result could be anticipated from the fact that an equilibrium system of forces concurrent with a line requires only two moment equations (Category 2 under *Categories of Equilibrium*).

③ We observe that a moment sum about an axis through A parallel to the z-axis merely gives us $6B_x - 2B_y = 0$, which serves only as a check as noted previously. Alternatively we could have first obtained A_z from $\Sigma F_z = 0$ and then taken our moment equations about axes through B to obtain A_x and A_y.

Sample Problem 3/6

A 200-N force is applied to the handle of the hoist in the direction shown. The bearing A supports the thrust (force in the direction of the shaft axis), while bearing B supports only radial load (load normal to the shaft axis). Determine the mass m which can be supported and the total radial force exerted on the shaft by each bearing. Assume neither bearing to be capable of supporting a moment about a line normal to the shaft axis.

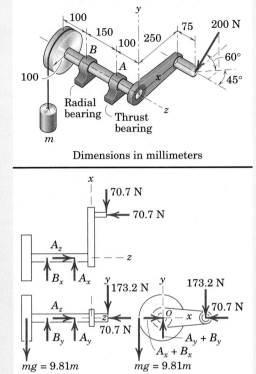

Dimensions in millimeters

Solution. The system is clearly three-dimensional with no lines or planes of symmetry, and therefore the problem must be analyzed as a general space system of forces. A scalar solution is used here to illustrate this approach, although a solution using vector notation would also be satisfactory. The free-body diagram of the shaft, lever, and drum considered a single body could be shown by a

① space view if desired, but is represented here by its three orthogonal projections.

The 200-N force is resolved into its three components, and each of the three views shows two of these components. The correct directions of A_x and B_x may be seen by inspection by observing that the line of action of the resultant of the two 70.7-N forces passes between A and B. The correct sense of the forces A_y and B_y cannot be determined until the magnitudes of the moments are obtained, so they are arbitrarily assigned. The x-y projection of the bearing forces is shown in terms of the sums of the unknown x- and y-components. The addition of A_z and the weight $W = mg$ completes the free-body diagrams. It should be noted that the three views represent three two-dimensional problems related by the corresponding components of the forces.

② From the x-y projection

$$[\Sigma M_O = 0] \qquad 100(9.81m) - 250(173.2) = 0 \qquad m = 44.1 \text{ kg} \qquad Ans.$$

From the x-z projection

$$[\Sigma M_A = 0] \qquad 150B_x + 175(70.7) - 250(70.7) = 0 \qquad B_x = 35.4 \text{ N}$$

$$[\Sigma F_x = 0] \qquad A_x + 35.4 - 70.7 = 0 \qquad A_x = 35.4 \text{ N}$$

③ The y-z view gives

$$[\Sigma M_A = 0] \qquad 150B_y + 175(173.2) - 250(44.1)(9.81) = 0 \qquad B_y = 520 \text{ N}$$

$$[\Sigma F_y = 0] \qquad A_y + 520 - 173.2 - (44.1)(9.81) = 0 \qquad A_y = 86.8 \text{ N}$$

$$[\Sigma F_z = 0] \qquad A_z = 70.7 \text{ N}$$

The total radial forces on the bearings become

$$[A_r = \sqrt{A_x^2 + A_y^2}] \qquad A_r = \sqrt{(35.4)^2 + (86.8)^2} = 93.5 \text{ N} \qquad Ans.$$

④ $$[B = \sqrt{B_x^2 + B_y^2}] \qquad B = \sqrt{(35.4)^2 + (520)^2} = 521 \text{ N} \qquad Ans.$$

Helpful Hints

① If the standard three views of orthographic projection are not entirely familiar, then review and practice them. Visualize the three views as the images of the body projected onto the front, top, and end surfaces of a clear plastic box placed over and aligned with the body.

② We could have started with the x-z projection rather than with the x-y projection.

③ The y-z view could have followed immediately after the x-y view since the determination of A_y and B_y may be made after m is found.

④ Without the assumption of zero moment supported by each bearing about a line normal to the shaft axis, the problem would be statically indeterminate.

Sample Problem 3/7

The welded tubular frame is secured to the horizontal x-y plane by a ball-and-socket joint at A and receives support from the loose-fitting ring at B. Under the action of the 2-kN load, rotation about a line from A to B is prevented by the cable CD, and the frame is stable in the position shown. Neglect the weight of the frame compared with the applied load and determine the tension T in the cable, the reaction at the ring, and the reaction components at A.

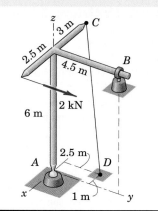

Solution. The system is clearly three-dimensional with no lines or planes of symmetry, and therefore the problem must be analyzed as a general space system of forces. The free-body diagram is drawn, where the ring reaction is shown in terms of its two components. All unknowns except $\mathbf{T}$ may be eliminated by a moment sum about the line AB. The direction of AB is specified by the unit

① vector $\mathbf{n} = \dfrac{1}{\sqrt{6^2 + 4.5^2}}(4.5\mathbf{j} + 6\mathbf{k}) = \frac{1}{5}(3\mathbf{j} + 4\mathbf{k})$. The moment of $\mathbf{T}$ about AB

is the component in the direction of AB of the vector moment about the point A and equals $\mathbf{r}_1 \times \mathbf{T} \cdot \mathbf{n}$. Similarly the moment of the applied load F about AB is $\mathbf{r}_2 \times \mathbf{F} \cdot \mathbf{n}$. With $\overline{CD} = \sqrt{46.2}$ m, the vector expressions for $\mathbf{T}$, $\mathbf{F}$, $\mathbf{r}_1$, and $\mathbf{r}_2$ are

$$\mathbf{T} = \frac{T}{\sqrt{46.2}}(2\mathbf{i} + 2.5\mathbf{j} - 6\mathbf{k}) \qquad \mathbf{F} = 2\mathbf{j} \text{ kN}$$

② $$\mathbf{r}_1 = -\mathbf{i} + 2.5\mathbf{j} \text{ m} \qquad \mathbf{r}_2 = 2.5\mathbf{i} + 6\mathbf{k} \text{ m}$$

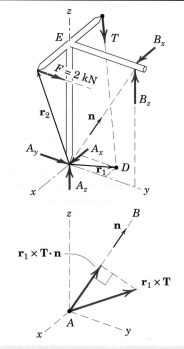

The moment equation now becomes

$$[\Sigma M_{AB} = 0] \quad (-\mathbf{i} + 2.5\mathbf{j}) \times \frac{T}{\sqrt{46.2}}(2\mathbf{i} + 2.5\mathbf{j} - 6\mathbf{k}) \cdot \tfrac{1}{5}(3\mathbf{j} + 4\mathbf{k})$$

$$+ (2.5\mathbf{i} + 6\mathbf{k}) \times (2\mathbf{j}) \cdot \tfrac{1}{5}(3\mathbf{j} + 4\mathbf{k}) = \mathbf{0}$$

Completion of the vector operations gives

$$-\frac{48T}{\sqrt{46.2}} + 20 = 0 \qquad T = 2.83 \text{ kN} \qquad\qquad \textit{Ans.}$$

and the components of T become

$$T_x = 0.833 \text{ kN} \qquad T_y = 1.042 \text{ kN} \qquad T_z = -2.50 \text{ kN}$$

We may find the remaining unknowns by moment and force summations as follows:

$$[\Sigma M_z = 0] \qquad 2(2.5) - 4.5B_x - 1.042(3) = 0 \qquad B_x = 0.417 \text{ kN} \qquad \textit{Ans.}$$

$$[\Sigma M_x = 0] \qquad 4.5B_z - 2(6) - 1.042(6) = 0 \qquad B_z = 4.06 \text{ kN} \qquad \textit{Ans.}$$

$$[\Sigma F_x = 0] \qquad A_x + 0.417 + 0.833 = 0 \qquad A_x = -1.250 \text{ kN} \qquad \textit{Ans.}$$

③ $$[\Sigma F_y = 0] \qquad A_y + 2 + 1.042 = 0 \qquad A_y = -3.04 \text{ kN} \qquad \textit{Ans.}$$

$$[\Sigma F_z = 0] \qquad A_z + 4.06 - 2.50 = 0 \qquad A_z = -1.556 \text{ kN} \qquad \textit{Ans.}$$

Helpful Hints

① The advantage of using vector notation in this problem is the freedom to take moments directly about any axis. In this problem this freedom permits the choice of an axis that eliminates five of the unknowns.

② Recall that the vector $\mathbf{r}$ in the expression $\mathbf{r} \times \mathbf{F}$ for the moment of a force is a vector from the moment center to *any* point on the line of action of the force. Instead of $\mathbf{r}_1$, an equally simple choice would be the vector $\overrightarrow{AC}$.

③ The negative signs associated with the A-components indicate that they are in the opposite direction to those shown on the free-body diagram.

PROBLEMS

Introductory Problems

3/61 Determine the tensions in cables AB, AC, and AD.

$Ans.$ $T_{AB} = 569$ N, $T_{AC} = 376$ N, $T_{AD} = 467$ N

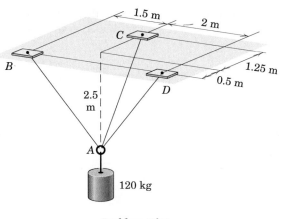

Problem 3/61

3/62 A uniform steel plate 18 in. square weighing 68 lb is suspended in the horizontal plane by the three vertical wires as shown. Calculate the tension in each wire.

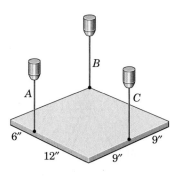

Problem 3/62

3/63 The horizontal steel shaft weights 620 lb and is suspended by a vertical cable from A and by a second cable BC which lies in a vertical transverse plane and loops underneath the shaft. Calculate the tensions T_1 and T_2 in the cables.

$Ans.$ $T_1 = 155$ lb, $T_2 = 260$ lb

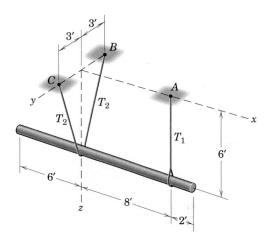

Problem 3/63

3/64 Two steel I-beams, each with a mass of 100 kg, are welded together at right angles and lifted by vertical cables so that the beams remain in a horizontal plane. Compute the tension in each of the cables A, B, and C.

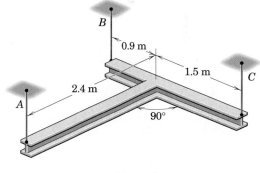

Problem 3/64

3/65 The chain-supported portion of a light fixture is in the shape of part of a spherical shell. If the mass of the glass unit is m, determine the tension T in each of the three chains.

Ans. $T = 0.373mg$

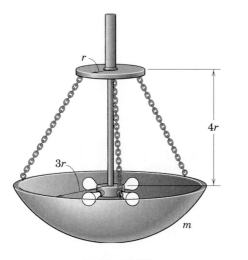

Problem 3/65

3/66 An overhead view of a car is shown in the figure. Two different locations C and D are considered for a single jack. In each case, the entire right side of the car is lifted just off the ground. Determine the normal reaction forces at A and B and the vertical jacking force required for the case of each jacking location. Consider the 1600-kg car to be rigid. The mass center G is on the midline of the car.

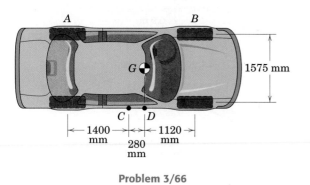

Problem 3/66

3/67 The industrial door is a uniform rectangular panel weighing 1200 lb and rolls along the fixed rail D on its hanger-mounted wheels A and B. The door is maintained in a vertical plane by the floor-mounted guide roller C, which bears against the bottom edge. For the position shown compute the horizontal side thrust on each of the wheels A and B, which must be accounted for in the design of the brackets.

Ans. $A_x = 48$ lb, $B_x = 12$ lb

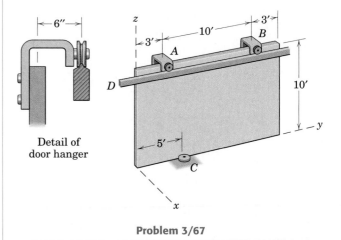

Problem 3/67

3/68 A uniform steel ring 60 in. in diameter and weighing 600 lb is lifted by the three cables, each 50 in. long, attached at points A, B, and C as shown. Compute the tension in each cable.

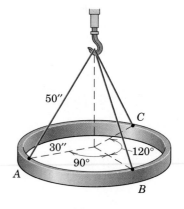

Problem 3/68

3/69 A three-legged stool is subjected to the load L as shown. determine the vertical force reaction under each leg. Neglect the weight of the stool.

Ans. $N_A = 0.533L$, $N_B = N_C = 0.233L$

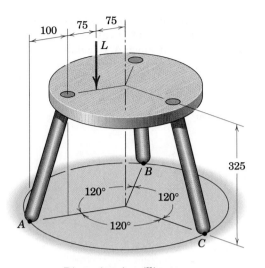

Dimensions in millimeters

Problem 3/69

3/70 Determine the compression in each of the three legs of the tripod subjected to the vertical 2-kN force. The weight of the legs is negligible compared with the applied load. Solve by using the force–equilibrium equation $\Sigma \mathbf{F} = \mathbf{0}$.

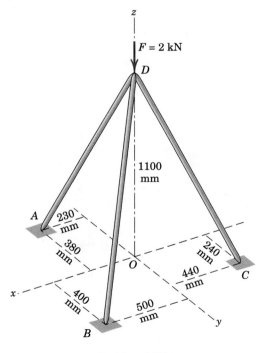

Problem 3/70

Representative Problems

3/71 One of the vertical walls supporting end B of the 200-kg uniform shaft of Sample Problem 3/5 is turned through a 30° angle as shown here. End A is still supported by the ball-and-socket connection in the horizontal x-y plane. Calculate the magnitudes of the forces $\mathbf{P}$ and $\mathbf{R}$ exerted on the ball end B of the shaft by the vertical walls C and D, respectively.

Ans. $P = 1584$ N, $R = 755$ N

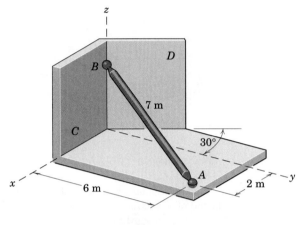

Problem 3/71

3/72 The 9-m steel boom has a mass of 600 kg with center of mass at midlength. It is supported by a ball-and-socket joint at A and the two cables under tensions T_1 and T_2. The cable which supports the 2000-kg load leads through a sheave (pulley) at B and is secured to the vertical x-y plane at F. Calculate the magnitude of the tension T_1. (*Hint:* Write a moment equation which eliminates all unknowns except T_1.)

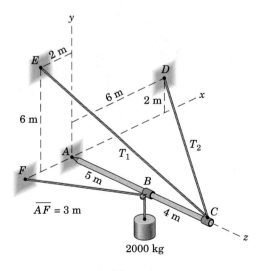

Problem 3/72

3/73 The smooth homogeneous sphere rests in the 120° groove and bears against the end plate, which is normal to the direction of the groove. Determine the angle θ, measured from the horizontal, for which the reaction on each side of the groove equals the force supported by the end plate.

Ans. $\theta = 30°$

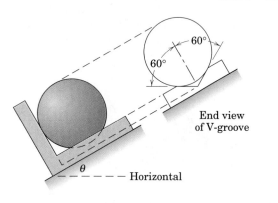

Problem 3/73

3/74 The small tripod-like stepladder is useful for supporting one end of a walking board. If F denotes the magnitude of the downward load from such a board (not shown), determine the reaction at each of the three feet A, B, and C. Neglect friction.

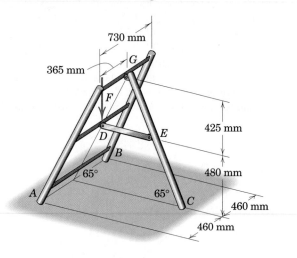

Problem 3/74

3/75 The mass center of the 30-kg door is in the center of the panel. If the weight of the door is supported entirely by the lower hinge A, calculate the magnitude of the total force supported by the hinge at B.

Ans. $B = 190.2$ N

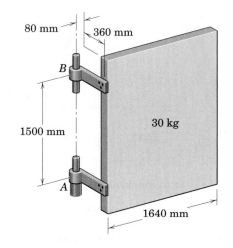

Problem 3/75

3/76 As part of a check on its design, a lower A-arm (part of an automobile suspension) is supported by bearings at A and B and subjected to the pair of 900-N forces at C and D. The suspension spring, not shown for clarity, exerts a force F_S at E as shown, where E is in plane $ABCD$. Determine the magnitude F_S of the spring force and the magnitudes F_A and F_B of the bearing forces at A and B which are perpendicular to the hinge axis AB.

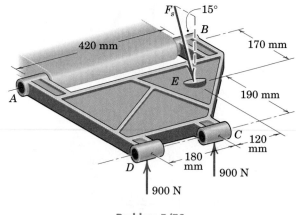

Problem 3/76

3/77 The rigid unit of post, bracket, and motor weighs 60 lb with its mass center at G located 12 in. from the vertical centerline of the post. The post is welded to the fixed base at A. The motor, which drives a machine through a flexible shaft, turns in the direction indicated and delivers a torque of 1500 lb-in. In addition, a 50-lb force is applied to the bracket as shown. Determine the vector expressions for the total force $\mathbf{R}$ and moment $\mathbf{M}$ applied to the post at A by the supporting base. (*Caution:* Be careful to assign the torque (couple) which acts on the motor shaft in its correct sense consistent with Newton's third law.)

Ans. $\mathbf{R} = 50\mathbf{i} + 60\mathbf{k}$ lb, $\mathbf{M} = -280\mathbf{j} + 150\mathbf{k}$ lb-in.

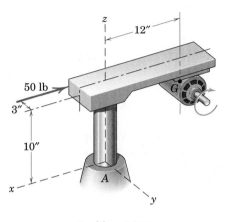

Problem 3/77

3/78 Determine the magnitudes of the force $\mathbf{R}$ and couple $\mathbf{M}$ exerted by the nut and bolt on the loaded bracket at O to maintain equilibrium.

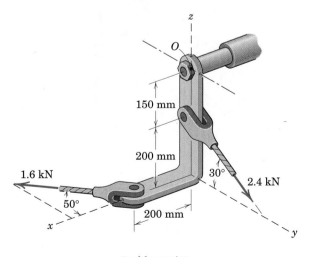

Problem 3/78

3/79 The 480-lb V-8 engine is supported on an engine stand and rotated 90° from its upright position so that its center of gravity G is in the position shown. Determine the vertical reaction at each roller of the stand. Neglect the weight of the stand itself.

Ans. $N_A = 282$ lb, $N_B = 34.8$ lb, $N_C = 162.8$ lb

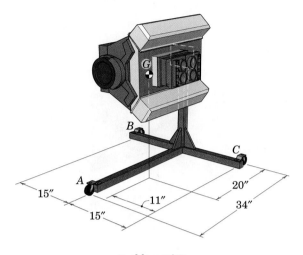

Problem 3/79

3/80 During a test, the left engine of the twin-engine airplane is revved up and a 500-lb thrust is generated. The main wheels at B and C are braked in order to prevent motion. Determine the change (compared with the nominal values with both engines off) in the normal reaction forces at A, B, and C.

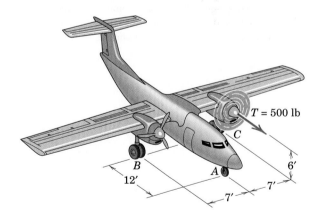

Problem 3/80

3/81 The 25-kg rectangular access door is held in the 90° open position by the single prop CD. Determine the force F in the prop and the magnitude of the force normal to the hinge axis AB in each of the small hinges A and B.

Ans. $F = 140.5$ N, $A_n = 80.6$ N, $B_n = 95.4$ N

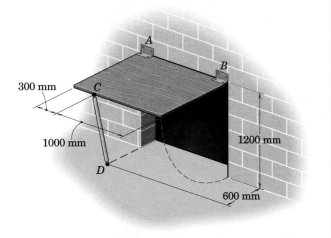

Problem 3/81

3/82 One of the three landing pads for a proposed Mars lander is shown in the figure. As part of a design check on the distribution of force in the landing struts, compute the force in each of the struts AC, BC, and CD when the lander is resting on a horizontal surface on Mars. The arrangement is symmetrical with respect to the x-z plane. The mass of the lander is 600 kg. (Assume equal support by the pads and consult Table D/2 in Appendix D as needed.)

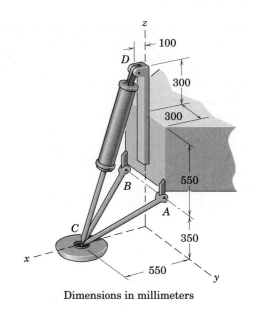

Dimensions in millimeters

Problem 3/82

3/83 Gear C drives the V-belt pulley D at a constant speed. For the belt tensions shown calculate the gear-tooth force P and the magnitudes of the total forces supported by the bearings at A and B.

Ans. $P = 70.9$ N, $A = 83.3$ N, $B = 208$ N

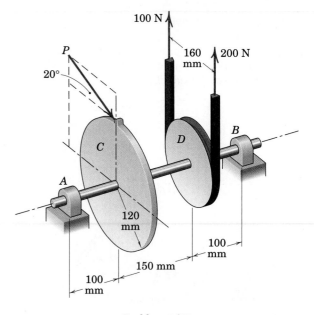

Problem 3/83

3/84 The spring of modulus $k = 900$ N/m is stretched a distance $\delta = 60$ mm when the mechanism is in the position shown. Calculate the force P_{min} required to initiate rotation about the hinge axis BC, and determine the corresponding magnitudes of the bearing forces which are perpendicular to BC. What is the normal reaction force at D if $P = P_{min}/2$?

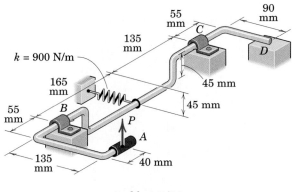

Problem 3/84

3/85 A force P of 200 N on the handle of the cable reel is required to wind up the underground cable as it comes from the manhole. The drum diameter is 1000 mm. For the horizontal position of the crank handle shown, calculate the magnitudes of the bearing forces at A and B. Neglect the weight of the drum.

Ans. $A = 116.7$ N, $B = 313$ N

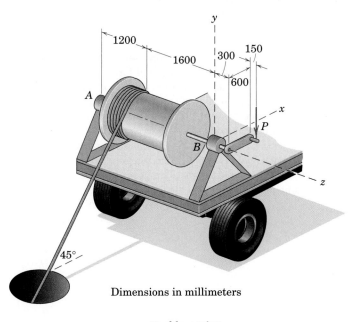

Dimensions in millimeters

Problem 3/85

3/86 The shaft, lever, and handle are welded together and constitute a single rigid body. Their combined mass is 28 kg with mass center at G. The assembly is mounted in bearings A and B, and rotation is prevented by link CD. Determine the forces exerted on the shaft by bearings A and B while the 30-N·m couple is applied to the handle as shown. Would these forces change if the couple were applied to the shaft AB rather than to the handle?

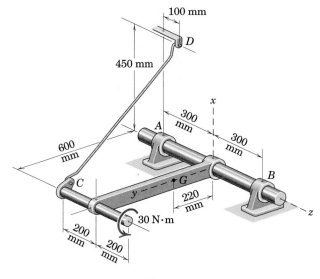

Problem 3/86

3/87 Each of the two legs of the welded frame weighs 100 lb. A wire from C to D prevents the frame from rotating out of the horizontal plane about an axis through its bearing at B and its ball-and-socket joint at A. Calculate the tension T in the wire and the magnitude of the total force supported by the connection at A.

Ans. $T = 245$ lb, $A = 122.5$ lb

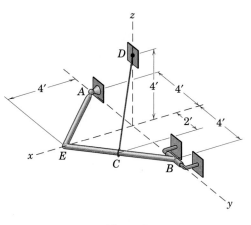

Problem 3/87

3/88 Consider the rudder assembly of a radio-controlled model airplane. For the 15° position shown in the figure, the net pressure acting on the left side of the rectangular rudder area is $p = 4(10^{-5})$ N/mm². Determine the required force P in the control rod DE and the horizontal components of the reactions at hinges A and B which are parallel to the rudder surface. Assume the aerodynamic pressure to be uniform.

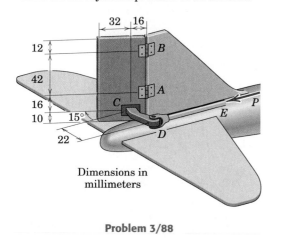

Dimensions in millimeters

Problem 3/88

3/89 The rigid pole and cross-arms of Prob. 2/105 are shown again here. Determine the tensions T_{AE} and T_{GF} in the two supporting cables resulting from the 1.2-kN tension in cable CD. Assume the absence of any resisting moments on the base of the pole at O about the x- and y-axes, but not about the z-axis.

Ans. $T_{AE} = 4.30$ kN, $T_{GF} = 3.47$ kN

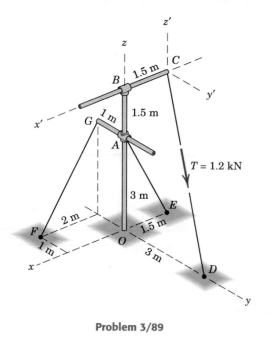

Problem 3/89

3/90 A homogeneous door of mass m, height h, and width w is leaned against a wall for painting. Small wooden strips are placed beneath corners A, B, and C. There is negligible friction at C, but friction at A and B is sufficient to prevent slipping. Determine the y- and z-components of the force reactions at A and B and the force normal to the wall at C.

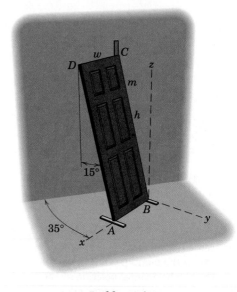

Problem 3/90

3/91 The upper ends of the vertical coil springs in the stock racecar can be moved up and down by means of a screw mechanism not shown. This adjustment permits a change in the downward force at each wheel as an optimum handling setup is sought. Initially, scales indicate the normal forces to be 800 lb, 800 lb, 1000 lb, and 1000 lb at A, B, C, and D, respectively, If the top of the right rear spring at A is lowered so that the scale at A reads an additional 100 lb, determine the corresponding changes in the normal forces at B, C, and D. Neglect the effects of the small attitude changes (pitch and roll angles) caused by the spring adjustment. The front wheels are the same distance apart as the rear wheels.

Ans. $\Delta N_B = -100$ lb, $\Delta N_C = 100$ lb
$\Delta N_D = -100$ lb

Simplified spring detail

Problem 3/91

3/92 A uniform bar of length b and mass m is suspended at its ends by two wires, each of length b, from points A and B in the horizontal plane a distance b apart. A couple M is applied to the bar causing it to rotate about a vertical axis to the equilibrium position shown. Derive an expression for the height h which it rises from its original equilibrium position where it hangs freely with no applied moment. What value of M is required to raise the bar the maximum amount b?

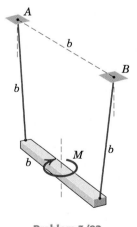

Problem 3/92

3/93 The boom AB lies in the vertical y-z plane and is supported by a ball-and-socket joint at B and by the two cables at A. Calculate the tension in each cable resulting from the 20-kN force acting in the horizontal

plane and applied at the midpoint M of the boom. Neglect the weight of the boom.

Ans. $T_1 = 33.0$ kN, $T_2 = 22.8$ kN

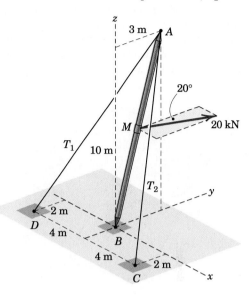

Problem 3/93

3/94 A rectangular sign over a store has a mass of 100 kg, with the center of mass in the center of the rectangle. The support against the wall at point C may be treated as a ball-and-socket joint. At corner D support is provided in the y-direction only. Calculate the tensions T_1 and T_2 in the supporting wires, the total force supported at C, and the lateral force R supported at D.

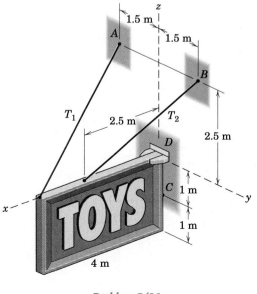

Problem 3/94

▶**3/95** The uniform rectangular panel *ABCD* has a mass of 40 kg and is hinged at its corners *A* and *B* to the fixed vertical surface. A wire from *E* to *D* keeps edges *BC* and *AD* horizontal. Hinge *A* can support thrust along the hinge axis *AB*, whereas hinge *B* supports force normal to the hinge axis only. Find the tension *T* in the wire and the magnitude *B* of the force supported by hinge *B*.

Ans. $T = 277$ N, $B = 169.9$ N

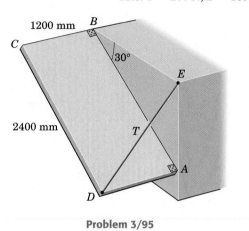

Problem 3/95

▶**3/96** The vertical plane containing the utility cable turns 30° at the vertical pole *OC*. The tensions T_1 and T_2 are both 950 N. In order to prevent long-term leaning of the pole, guy wires *AD* and *BE* are utilized. If the two guy wires are adjusted so as to carry equal tensions *T* which together reduce the moment at *O* to zero, determine the net horizontal reaction at *O*. Determine the required value of *T*. Neglect the weight of the pole.

Ans. $O = 144.9$ N, $T = 471$ N

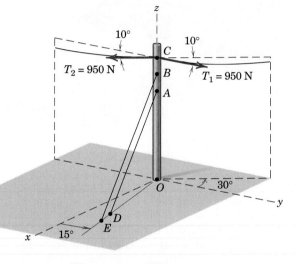

$\overline{OA}$ = 9 m	$\overline{OD}$ = 8 m
$\overline{OB}$ = 11 m	$\overline{OE}$ = 10 m
$\overline{OC}$ = 13 m	

Problem 3/96

3/5 CHAPTER REVIEW

In Chapter 3 we have applied our knowledge of the properties of forces, moments, and couples studied in Chapter 2 to solve problems involving rigid bodies in equilibrium. Complete equilibrium of a body requires that the vector resultant of all forces acting on it be zero ($\Sigma \mathbf{F} = \mathbf{0}$) and the vector resultant of all moments on the body about a point (or axis) also be zero ($\Sigma \mathbf{M} = \mathbf{0}$). We are guided in all of our solutions by these two requirements, which are easily comprehended physically.

Frequently, it is not the theory but its application which presents difficulty. The crucial steps in applying our principles of equilibrium should be quite familiar by now. They are:

1. Make an unequivocal decision as to which system (a body or collection of bodies) in equilibrium is to be analyzed.

2. Isolate the system in question from all contacting bodies by drawing its *free-body diagram* showing *all* forces and couples acting *on* the isolated system from external sources.

3. Observe the principle of action and reaction (Newton's third law) when assigning the sense of each force.

4. Label all forces and couples, known and unknown.

5. Choose and label reference axes, always choosing a right-handed set when vector notation is used (which is usually the case for three-dimensional analysis).

6. Check the adequacy of the constraints (supports) and match the number of unknowns with the number of available independent equations of equilibrium.

When solving an equilibrium problem, we should first check to see that the body is statically determinate. If there are more supports than are necessary to hold the body in place, the body is statically indeterminate, and the equations of equilibrium by themselves will not enable us to solve for all of the external reactions. In applying the equations of equilibrium, we choose scalar algebra, vector algebra, or graphical analysis according to both preference and experience; vector algebra is particularly useful for many three-dimensional problems.

The algebra of a solution can be simplified by the choice of a moment axis which eliminates as many unknowns as possible or by the choice of a direction for a force summation which avoids reference to certain unknowns. A few moments of thought to take advantage of these simplifications can save appreciable time and effort.

The principles and methods covered in Chapter 2 and 3 constitute the most basic part of statics. They lay the foundation for what follows not only in statics but in dynamics as well.

REVIEW PROBLEMS

3/97 Calculate the magnitude of the force supported by the pin at *B* for the bell crank loaded and supported as shown.

Ans. B = 60.0 lb

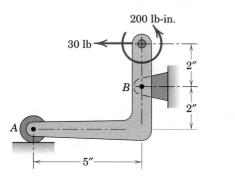

Problem 3/97

3/98 A 50-kg acrobat pedals her unicycle across the taut but slightly elastic cable. If the deflection at the center of the 18-m span is 75 mm, determine the tension in the cable. Neglect the effects of the weights of the cable and unicycle.

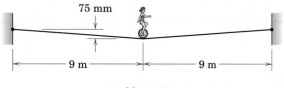

Problem 3/98

3/99 The uniform 5-m bar with a mass of 100 kg is hinged at *O* and prevented from rotating in the vertical plane beyond the 30° position by the fixed roller at *A*. Calculate the magnitude of the total force supported by the pin at *O*.

Ans. O = 1769 N

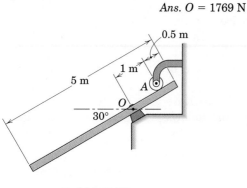

Problem 3/99

3/100 The tool shown is used for straightening twisted members as wooden framing is completed. If the force *P* = 30 lb is applied to the handle as shown, determine the normal forces applied to the installed stud at points *A* and *B*. Ignore friction.

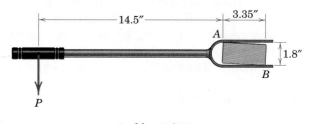

Problem 3/100

3/101 The device shown in the figure is useful for lifting drywall panels into position prior to fastening to the stud wall. Estimate the magnitude *P* of the force required to lift the 25-kg panel. State any assumptions.

Ans. P = 351 N

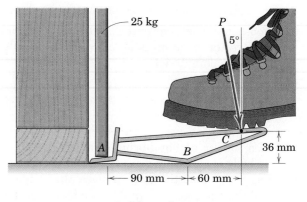

Problem 3/101

3/102 The designers of an aircraft landing-gear system wish to cause the forces in both struts *AB* and *CD* to act along their respective lengths. What angle θ should they specify for strut *AB*? The weights of all members are small compared with the forces which act on the system shown. Treat as two-dimensional.

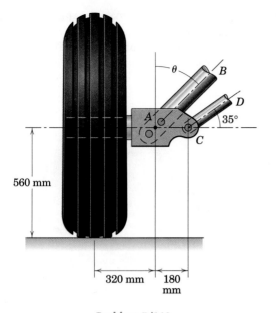

Problem 3/102

3/103 A freeway sign measuring 12 ft by 6 ft is supported by the single mast as shown. The sign, supporting framework, and mast together weigh 600 lb, with center of gravity 10 ft away from the vertical centerline of the mast. When the sign is subjected to the direct blast of a 75-mi/hr wind, an average pressure difference of 17.5 lb/ft² is developed between the front and back sides of the sign, with the resultant of the wind-pressure forces acting at the center of the sign. Determine the magnitudes of the force and moment reactions at the base of the mast. Such results would be instrumental in the design of the base.

Ans. R = 1396 lb, *M* = 25,600 lb-ft

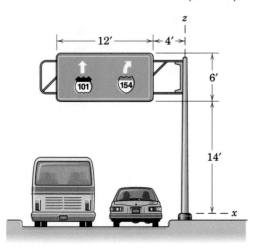

Problem 3/103

3/104 If the weight of the boom is negligible compared with the applied 30-kN load, determine the cable tensions T_1 and T_2 and the force acting at the ball joint at A.

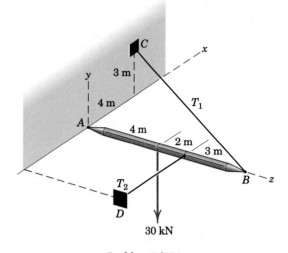

Problem 3/104

3/105 Magnetic tape under a tension of 10 N at D passes around the guide pulleys and through the erasing head at C at constant speed. As a result of a small amount of friction in the bearings of the pulleys, the tape at E is under a tension of 11 N. Determine the tension T in the supporting spring at B. The plate lies in a horizontal plane and is mounted on a precision needle bearing at A.

Ans. T = 10.62 N

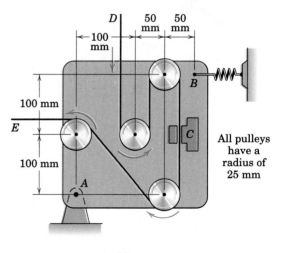

Problem 3/105

3/106 The curved arm BC and attached cables AB and AC support a power line which lies in the vertical y-z plane. The tangents to the power line at the insulator below A make $15°$ angles with the horizontal y-axis. If the tension in the power line at the insulator is 1.3 kN, calculate the total force supported by the bolt at D on the pole bracket. The weight of the arm BC can be neglected compared with the other forces, and it can be assumed that the bolt at E supports horizontal force only.

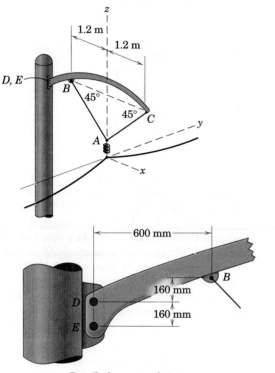

Detail of arm attachment

Problem 3/106

3/107 Determine the tension T required to hold the rectangular solid in the position shown. The 125-kg rectangular solid is homogeneous. Friction at D is negligible.

Ans. $T = 1053$ N

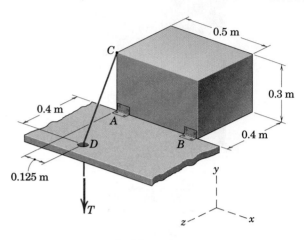

Problem 3/107

3/108 A vertical force P on the foot pedal of the bell crank is required to produce a tension T of 400 N in the vertical control rod. Determine the corresponding bearing reactions at A and B.

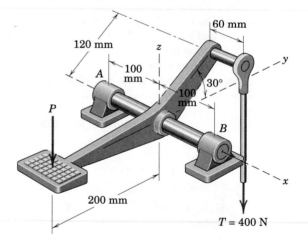

Problem 3/108

3/109 The wind blowing normal to the plane of the rectangular sign exerts a uniform pressure of 175 N/m^2 as indicated in the figure. Determine the changes in the forces exerted at A and B by each support. There are two symmetrically placed I-beam uprights, and the width of the sign is 3 m.

Ans. $\Delta A = 1451$ N down, $\Delta B = 1451$ N up

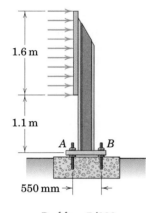

Problem 3/109

3/110 Each of the three uniform 1200-mm bars has a mass of 20 kg. The bars are welded together into the configuration shown and suspended by three vertical wires. Bars AB and BC lie in the horizontal x-y plane, and the third bar lies in a plane parallel to the x-z plane. Compute the tension in each wire.

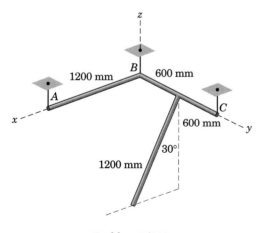

Problem 3/110

3/111 A wheel of mass m and radius r with its mass center G at the geometric center rests in a small depression of width b. Determine the minimum couple M applied to the wheel at the location shown in order to roll the wheel out of the depression. Assume no slipping occurs. What is the influence of r_0, r, and θ?

Ans. $M = \frac{1}{2}mgb$

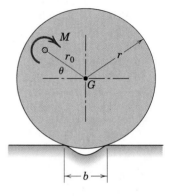

Problem 3/111

3/112 The 900-lb boom with center of gravity at G is held in the position shown by a ball-and-socket joint at O and the two cables AB and AC. Determine the two cable tensions and the x-, y-, and z-components of the force reaction at O.

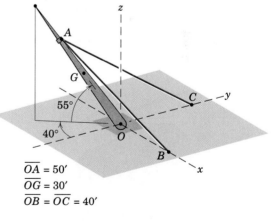

$\overline{OA} = 50'$
$\overline{OG} = 30'$
$\overline{OB} = \overline{OC} = 40'$

Problem 3/112

3/113 The stand is used for storing and dispensing cable. The combined weight of the stand, reel, and cable is 325 lb with center of gravity at G. Determine the normal forces at all four feet (A, B, C, and D) if (a) the cable comes off at the top of the reel and (b) the cable comes off at the bottom of the reel. Note that the cable departs the reel at its horizontal midpoint in both cases. The effective radius of the reel is 6 in.

Ans. (a) $N_A = N_D = 47.1$ lb, $N_B = N_C = 109.0$ lb
(b) $N_A = N_D = 59.6$ lb, $N_B = N_C = 96.5$ lb

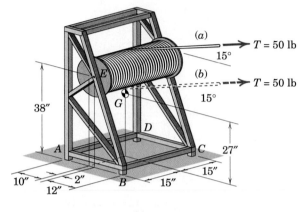

Problem 3/113

3/114 The drum and shaft are welded together and have a mass of 50 kg with mass center at G. The shaft is subjected to a torque (couple) of 120 N·m, and the drum is prevented from rotating by the cord wrapped securely around it and attached to point C. Calculate the magnitudes of the forces supported by bearings A and B.

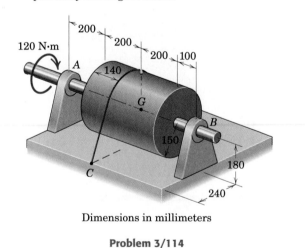

Dimensions in millimeters

Problem 3/114

▶**3/115** The L-shaped bar is supported by a ball-and-socket joint at O [case (a)] and the two cables as shown. Explain why this configuration is improperly constrained. (b) The ball-and-socket joint is now replaced by the universal joint which can support, in addition to three force reactions, a moment about the y-axis but no moments about the x- and z-axes. Plot the two cable tensions, the magnitude of the force reaction at O, and the moment reaction at O as functions of the position x of the 100-lb cylinder over the range $0.5 \leq x \leq 4.5$ ft. Explain any unusual characteristics of these plots. Neglect the weight of the bar throughout.

Ans. $T_{AD} = 208$ lb, $T_{BC} = 128.0$ lb
$O = 303$ lb, all constant
$M_{O_y} = -100x + 250$ lb-ft

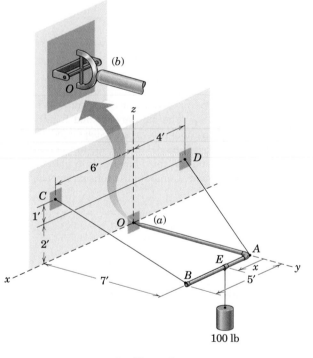

Problem 3/115

⬛ *Computer-Oriented Problems*

*3/116 Determine and plot the tension ratio T/mg required to hold the uniform slender bar in equilibrium for any angle θ from just above zero to just under 45°. The bar AB of mass m is uniform.

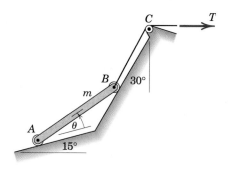

Problem 3/116

*3/117 The jib crane is designed for a maximum capacity of 10 kN, and its uniform I-beam has a mass of 200 kg. (*a*) Plot the magnitude R of the force on the pin at A as a function of x through its operating range of $x = 0.2$ m to $x = 3.8$ m. On the same set of axes, plot the x- and y-components of the pin reaction at A. (*b*) Determine the minimum value of R and the corresponding value of x. (*c*) For what value of R should the pin at A be designed? (Use $g = 10$ m/s^2.)

$Ans.$ (*a*) $R = \frac{1}{3} \{400x^2 - 400x + 1072\}^{1/2}$
(*b*) $R_{min} = 10.39$ kN at $x = 0.5$ m
(*c*) $R_{max} = 24.3$ kN at $x = 3.8$ m

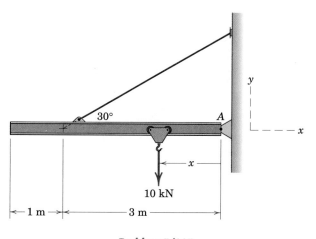

Problem 3/117

*3/118 Two traffic signals are attached to the 36-ft support cable at equal intervals as shown. Determine the equilibrium configuration angles α, β, and γ, as well as the tension in each cable segment.

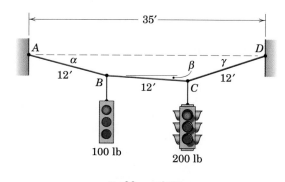

Problem 3/118

*3/119 The two traffic signals of Prob. 3/118 are now repositioned so that segment BC of the 36-ft support cable is 10 ft in length and is horizontal. Specify the necessary lengths AB and CD and the tensions in all three cable segments.

$Ans.$ $\overline{AB} = 17.01$ ft, $\overline{CD} = 8.99$ ft
$T_{AB} = 503$ lb, $T_{BC} = 493$ lb, $T_{CD} = 532$ lb

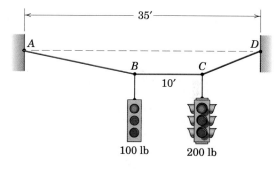

Problem 3/119

*3/120 The vertical position of the two 20-kg cylinders is controlled by the torque M applied to the central shaft. The cords attached to the ends of the arms pass through smooth holes in a fixed surface at A and B. Plot M as a function of θ from $\theta = 0°$ to $\theta = 180°$. Determine the maximum value of M and the corresponding value of θ. Take $g = 10$ m/s^2 for your calculations.

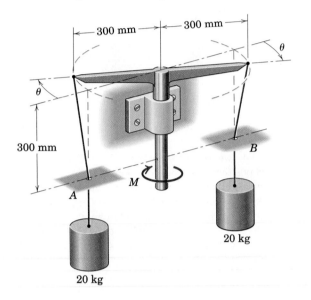

Problem 3/120

*3/121 The horizontal boom is supported by the cables AB and CD and by a ball-and-socket joint at O. To determine the influence on the reaction at O of the position of the vertical load L along the boom, we may neglect the weight of the boom. If R represents the magnitude of the total force at O, determine by calculus the minimum ratio R/L and the corresponding value of x. Then write a computer program for R/L and plot the results for $0 < x < 6$ m as a check on your calculations.

$$Ans. \; R/L = \sqrt{47x^2/162 - x/3 + 1}$$
$$(R/L)_{\min} = 0.951 \text{ at } x = 0.574 \text{ m}$$

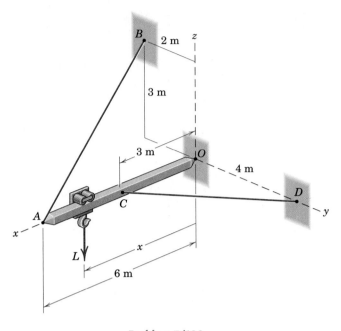

Problem 3/121

*3/122 The mass center of the 10-kg arm OC is located at G, and the spring of constant $k = 1.2$ kN/m is unstretched when $\theta = 0$. Plot the applied moment M required for static equilibrium over the range $0 \leq \theta \leq 180°$. Determine the value of θ for which $M = 0$ (if any) and the minimum and maximum values of M along with the corresponding values of θ at which these extremes occur. The motion of this mechanism occurs in a vertical plane. Take M to be positive when counterclockwise. The value of m_2 is 3 kg.

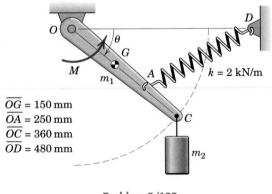

$$\overline{OG} = 150 \text{ mm}$$
$$\overline{OA} = 250 \text{ mm}$$
$$\overline{OC} = 360 \text{ mm}$$
$$\overline{OD} = 480 \text{ mm}$$

Problem 3/122

*3/123 The basic features of a small backhoe are shown in the illustration. Member *BE* (complete with hydraulic cylinder *CD* and bucket-control links *DF* and *DE*) weighs 500 lb with mass center at G_1. The bucket and its load of clay weigh 350 lb with mass center at G_2. To disclose the operational design characteristics of the backhoe, determine and plot the force *T* in the hydraulic cylinder *AB* as a function of the angular position θ of member *BE* over the range $0 \leq \theta \leq 90°$. For what value of θ is the force *T* equal to zero? Member *OH* is fixed for this exercise; note that its controlling hydraulic cylinder (hidden) extends from near point *O* to pin *I*. Similarly, the bucket-control hydraulic cylinder *CD* is held at a fixed length.

Ans. $T = 0$ at $\theta = 1.729°$

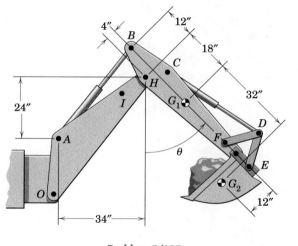

Problem 3/123

*3/124 The system of Prob. 3/107 is shown again here, only now the 125-kg homogeneous rectangular solid is depicted as having rotated an angle θ about the hinge axis *AB*. Determine and plot the following quantities as functions of θ over the range $0 \leq \theta \leq 60°$: *T*, A_y, A_z, B_x, B_y, and B_z. The hinge at *A* cannot exert an axial thrust. Assume all hinge force components to be in the positive coordinate directions.

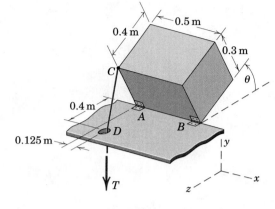

Problem 3/124

*3/125 The vertical pole, utility cable, and two guy wires from Prob. 3/96 are shown again here. As part of a design study, the following conditions are considered. The tension T_2 is a constant 1000 N, and its 10° angle is fixed. The 10° angle for T_1 is also fixed, but the magnitude of T_1 is allowed to vary from 0 to 2000 N. For each value of T_1, determine and plot the magnitude of the equal tensions *T* in cables *AD* and *BE* and the angle θ for which the moment at *O* will be zero. State the values of *T* and θ for $T_1 = 1000$ N.

Ans. $T = 495$ N, $\theta = 15°$

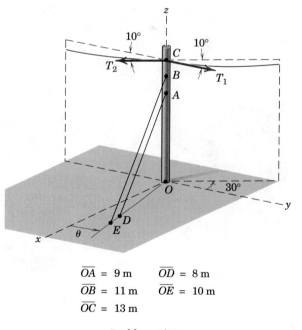

$\overline{OA} = 9$ m	$\overline{OD} = 8$ m
$\overline{OB} = 11$ m	$\overline{OE} = 10$ m
$\overline{OC} = 13$ m	

Problem 3/125

This view from Lisbon, Portugal, shows a variety of structures. In all cases, however, the engineers had to calculate the force supported by each major component of the overall structure.

STRUCTURES

4/1   INTRODUCTION

In Chapter 3 we studied the equilibrium of a single rigid body or a system of connected members treated as a single rigid body. We first drew a free-body diagram of the body showing all forces external to the isolated body and then we applied the force and moment equations of equilibrium. In Chapter 4 we focus on the determination of the forces internal to a structure, that is, forces of action and reaction between the connected members. An engineering structure is any connected system of members built to support or transfer forces and to safely withstand the loads applied to it. To determine the forces internal to an engineering structure, we must dismember the structure and analyze separate free-body diagrams of individual members or combinations of members. This analysis requires careful application of Newton's third law, which states that each action is accompanied by an equal and opposite reaction.

In Chapter 4 we analyze the internal forces acting in several types of structures, namely, trusses, frames, and machines. In this treatment we consider only *statically determinate* structures, which do not have more supporting constraints than are necessary to maintain an equilibrium configuration. Thus, as we have already seen, the equations of equilibrium are adequate to determine all unknown reactions.

The analysis of trusses, frames and machines, and beams under concentrated loads constitutes a straightforward application of the material developed in the previous two chapters. The basic procedure developed in Chapter 3 for isolating a body by constructing a correct free-body diagram is essential for the analysis of statically determinate structures.

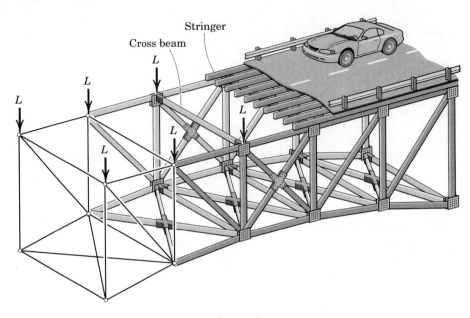

Figure 4/1

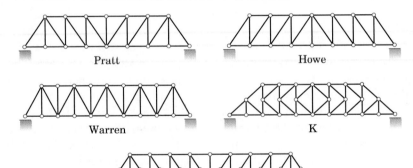

Commonly Used Bridge Trusses

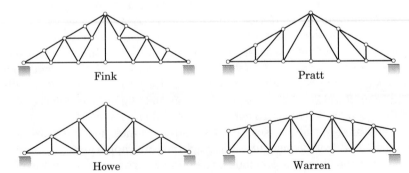

Commonly Used Roof Trusses

Figure 4/2

4/2   PLANE TRUSSES

A framework composed of members joined at their ends to form a rigid structure is called a *truss*. Bridges, roof supports, derricks, and other such structures are common examples of trusses. Structural members commonly used are I-beams, channels, angles, bars, and special shapes which are fastened together at their ends by welding, riveted connections, or large bolts or pins. When the members of the truss lie essentially in a single plane, the truss is called a *plane truss*.

For bridges and similar structures, plane trusses are commonly utilized in pairs with one truss assembly placed on each side of the structure. A section of a typical bridge structure is shown in Fig. 4/1. The combined weight of the roadway and vehicles is transferred to the longitudinal stringers, then to the cross beams, and finally, with the weights of the stringers and cross beams accounted for, to the upper joints of the two plane trusses which form the vertical sides of the structure. A simplified model of the truss structure is indicated at the left side of the illustration; the forces L represent the joint loadings.

Several examples of commonly used trusses which can be analyzed as plane trusses are shown in Fig. 4/2.

Simple Trusses

The basic element of a plane truss is the triangle. Three bars joined by pins at their ends, Fig. 4/3a, constitute a rigid frame. The term *rigid* is used to mean noncollapsible and also to mean that deformation of the members due to induced internal strains is negligible. On the other hand, four or more bars pin-jointed to form a polygon of as many sides constitute a nonrigid frame. We can make the nonrigid frame in Fig. 4/3b rigid, or stable, by adding a diagonal bar joining A and D or B and C and thereby forming two triangles. We can extend the structure by adding additional units of two end-connected bars, such as DE and CE or AF and DF, Fig. 4/3c, which are pinned to two fixed joints. In this way the entire structure will remain rigid.

Structures built from a basic triangle in the manner described are known as *simple trusses*. When more members are present than are needed to prevent collapse, the truss is statically indeterminate. A statically indeterminate truss cannot be analyzed by the equations of equilibrium alone. Additional members or supports which are not necessary for maintaining the equilibrium configuration are called *redundant*.

To design a truss we must first determine the forces in the various members and then select appropriate sizes and structural shapes to withstand the forces. Several assumptions are made in the force analysis of simple trusses. First, we assume all members to be *two-force members*. A two-force member is one in equilibrium under the action of two forces only, as defined in general terms with Fig. 3/4 in Art. 3/3. Each member of a truss is normally a straight link joining the two points of application of force. The two forces are applied at the ends of the member and are necessarily equal, opposite, and *collinear* for equilibrium.

The member may be in tension or compression, as shown in Fig. 4/4. When we represent the equilibrium of a portion of a two-force member, the tension T or compression C acting on the cut section is the same

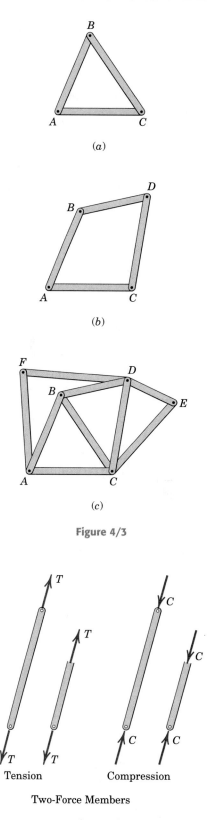

(a)

(b)

(c)

Figure 4/3

Tension   Compression

Two-Force Members

Figure 4/4

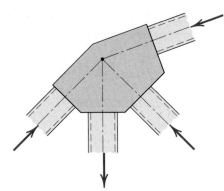

Figure 4/5

for all sections. We assume here that the weight of the member is small compared with the force it supports. If it is not, or if we must account for the small effect of the weight, we can replace the weight W of the member by two forces, each $W/2$ if the member is uniform, with one force acting at each end of the member. These forces, in effect, are treated as loads externally applied to the pin connections. Accounting for the weight of a member in this way gives the correct result for the average tension or compression along the member but will not account for the effect of bending of the member.

Truss Connections and Supports

When welded or riveted connections are used to join structural members, we may usually assume that the connection is a pin joint if the centerlines of the members are concurrent at the joint as in Fig. 4/5.

We also assume in the analysis of simple trusses that all external forces are applied at the pin connections. This condition is satisfied in most trusses. In bridge trusses the deck is usually laid on cross beams which are supported at the joints, as shown in Fig. 4/1.

For large trusses, a roller, rocker, or some kind of slip joint is used at one of the supports to provide for expansion and contraction due to temperature changes and for deformation from applied loads. Trusses and frames in which no such provision is made are statically indeterminate, as explained in Art. 3/3. Figure 3/1 shows examples of such joints.

Two methods for the force analysis of simple trusses will be given. Each method will be explained for the simple truss shown in Fig. 4/6*a*. The free-body diagram of the truss as a whole is shown in Fig. 4/6*b*. The external reactions are usually determined first, by applying the equilibrium equations to the truss as a whole. Then the force analysis of the remainder of the truss is performed.

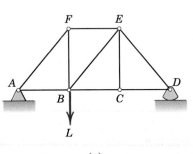

(*a*)

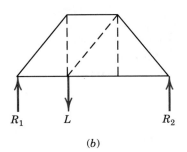

(*b*)

Figure 4/6

4/3 Method of Joints

This method for finding the forces in the members of a truss consists of satisfying the conditions of equilibrium for the forces acting on the connecting pin of each joint. The method therefore deals with the equilibrium of concurrent forces, and only two independent equilibrium equations are involved.

We begin the analysis with any joint where at least one known load exists and where not more than two unknown forces are present. The solution may be started with the pin at the left end. Its free-body diagram is shown in Fig. 4/7. With the joints indicated by letters, we usually designate the force in each member by the two letters defining the ends of the member. The proper directions of the forces should be evident by inspection for this simple case. The free-body diagrams of portions of members AF and AB are also shown to clearly indicate the mechanism of the action and reaction. The member AB actually makes contact on the left side of the pin, although the force AB is drawn from the right side and is shown acting away from the pin. Thus, if we consistently draw the force arrows on the *same* side of the pin as the member, then tension (such as AB) will always be indicated by an arrow *away*

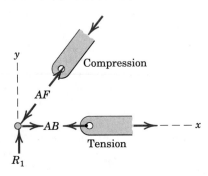

Figure 4/7

from the pin, and compression (such as *AF*) will always be indicated by an arrow *toward* the pin. The magnitude of *AF* is obtained from the equation $\Sigma F_y = 0$ and *AB* is then found from $\Sigma F_x = 0$.

Joint *F* may be analyzed next, since it now contains only two unknowns, *EF* and *BF*. Proceeding to the next joint having no more than two unknowns, we subsequently analyze joints *B*, *C*, *E*, and *D* in that order. Figure 4/8 shows the free-body diagram of each joint and its corresponding force polygon, which represents graphically the two equilibrium conditions $\Sigma F_x = 0$ and $\Sigma F_y = 0$. The numbers indicate the order in which the joints are analyzed. We note that, when joint *D* is finally reached, the computed reaction R_2 must be in equilibrium with the forces in members *CD* and *ED*, which were determined previously from the two neighboring joints. This requirement provides a check on the correctness of our work. Note that isolation of joint *C* shows that the force in *CE* is zero when the equation $\Sigma F_y = 0$ is applied. The force in

This New York City bridge structure suggests that members of a simple truss need not be straight.

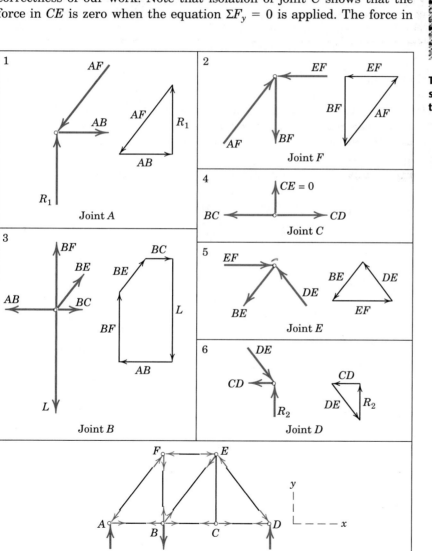

Figure 4/8

this member would not be zero, of course, if an external vertical load were applied at C.

It is often convenient to indicate the tension T and compression C of the various members directly on the original truss diagram by drawing arrows away from the pins for tension and toward the pins for compression. This designation is illustrated at the bottom of Fig. 4/8.

Sometimes we cannot initially assign the correct direction of one or both of the unknown forces acting on a given pin. If so, we may make an arbitrary assignment. A negative computed force value indicates that the initially assumed direction is incorrect.

Internal and External Redundancy

If a plane truss has more external supports than are necessary to ensure a stable equilibrium configuration, the truss as a whole is statically indeterminate, and the extra supports constitute *external* redundancy. If a truss has more internal members than are necessary to prevent collapse when the truss is removed from its supports, then the extra members constitute *internal* redundancy and the truss is again statically indeterminate.

For a truss which is statically determinate externally, there is a definite relation between the number of its members and the number of its joints necessary for internal stability without redundancy. Because we can specify the equilibrium of each joint by two scalar force equations, there are in all $2j$ such equations for a truss with j joints. For the entire truss composed of m two-force members and having the maximum of three unknown support reactions, there are in all $m + 3$ unknowns (m tension or compression forces and three reactions). Thus, for any plane truss, the equation $m + 3 = 2j$ will be satisfied if the truss is statically determinate internally.

A *simple* plane truss, formed by starting with a triangle and adding two new members to locate each new joint with respect to the existing structure, satisfies the relation automatically. The condition holds for the initial triangle, where $m = j = 3$, and m increases by 2 for each added joint while j increases by 1. Some other (nonsimple) statically determinate trusses, such as the K-truss in Fig. 4/2, are arranged differently, but can be seen to satisfy the same relation.

This equation is a necessary condition for stability but it is not a sufficient condition, since one or more of the m members can be arranged in such a way as not to contribute to a stable configuration of the entire truss. If $m + 3 > 2j$, there are more members than independent equations, and the truss is statically indeterminate internally with redundant members present. If $m + 3 < 2j$, there is a deficiency of internal members, and the truss is unstable and will collapse under load.

Special Conditions

We often encounter several special conditions in the analysis of trusses. When two collinear members are under compression, as indicated in Fig. 4/9a, it is necessary to add a third member to maintain

Harbour Bridge in Sydney, Australia

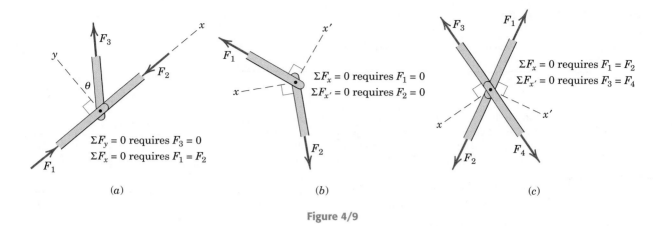

Figure 4/9

alignment of the two members and prevent buckling. We see from a force summation in the y-direction that the force F_3 in the third member must be zero and from the x-direction that $F_1 = F_2$. This conclusion holds regardless of the angle θ and holds also if the collinear members are in tension. If an external force with a component in the y-direction were applied to the joint, then F_3 would no longer be zero.

When two noncollinear members are joined as shown in Fig. 4/9b, then in the absence of an externally applied load at this joint, the forces in both members must be zero, as we can see from the two force summations.

When two pairs of collinear members are joined as shown in Fig. 4/9c, the forces in each pair must be equal and opposite. This conclusion follows from the force summations indicated in the figure.

Truss panels are frequently cross-braced as shown in Fig. 4/10a. Such a panel is statically indeterminate if each brace can support either tension or compression. However, when the braces are flexible members incapable of supporting compression, as are cables, then only the tension member acts and we can disregard the other member. It is usually evident from the asymmetry of the loading how the panel will deflect. If the deflection is as indicated in Fig. 4/10b, then member AB should be retained and CD disregarded. When this choice cannot be made by inspection, we may arbitrarily select the member to be retained. If the assumed tension turns out to be positive upon calculation, then the choice was correct. If the assumed tension force turns out to be negative, then the opposite member must be retained and the calculation redone.

We can avoid simultaneous solution of the equilibrium equations for two unknown forces at a joint by a careful choice of reference axes. Thus, for the joint indicated schematically in Fig. 4/11 where L is known and F_1 and F_2 are unknown, a force summation in the x-direction eliminates reference to F_1 and a force summation in the x'-direction eliminates reference to F_2. When the angles involved are not easily found, then a simultaneous solution of the equations using one set of reference directions for both unknowns may be preferable.

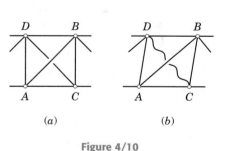

(a)   (b)

Figure 4/10

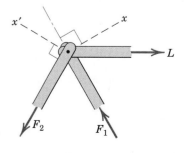

Figure 4/11

Sample Problem 4/1

Compute the force in each member of the loaded cantilever truss by the method of joints.

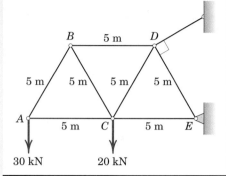

Solution.   If it were not desired to calculate the external reactions at D and E, the analysis for a cantilever truss could begin with the joint at the loaded end. However, this truss will be analyzed completely, so the first step will be to compute the external forces at D and E from the free-body diagram of the truss as a whole. The equations of equilibrium give

$$[\Sigma M_E = 0] \qquad\qquad 5T - 20(5) - 30(10) = 0 \qquad\qquad T = 80 \text{ kN}$$

$$[\Sigma F_x = 0] \qquad\qquad 80 \cos 30° - E_x = 0 \qquad\qquad E_x = 69.3 \text{ kN}$$

$$[\Sigma F_y = 0] \qquad 80 \sin 30° + E_y - 20 - 30 = 0 \qquad E_y = 10 \text{ kN}$$

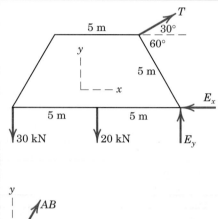

Next we draw free-body diagrams showing the forces acting on each of the connecting pins. The correctness of the assigned directions of the forces is verified when each joint is considered in sequence. There should be no question about the correct direction of the forces on joint A. Equilibrium requires

$$[\Sigma F_y = 0] \qquad 0.866AB - 30 = 0 \qquad AB = 34.6 \text{ kN } T \qquad\qquad Ans.$$

$$[\Sigma F_x = 0] \qquad AC - 0.5(34.6) = 0 \qquad AC = 17.32 \text{ kN } C \qquad\qquad Ans.$$

① where T stands for tension and C stands for compression.

Joint B must be analyzed next, since there are more than two unknown forces on joint C. The force BC must provide an upward component, in which case BD must balance the force to the left. Again the forces are obtained from

$$[\Sigma F_y = 0] \qquad 0.866BC - 0.866(34.6) = 0 \qquad BC = 34.6 \text{ kN } C \qquad Ans.$$

$$[\Sigma F_x = 0] \qquad BD - 2(0.5)(34.6) = 0 \qquad BD = 34.6 \text{ kN } T \qquad Ans.$$

Joint C now contains only two unknowns, and these are found in the same way as before:

$$[\Sigma F_y = 0] \qquad 0.866CD - 0.866(34.6) - 20 = 0$$

$$CD = 57.7 \text{ kN } T \qquad\qquad Ans.$$

$$[\Sigma F_x = 0] \qquad CE - 17.32 - 0.5(34.6) - 0.5(57.7) = 0$$

$$CE = 63.5 \text{ kN } C \qquad\qquad Ans.$$

Finally, from joint E there results

$$[\Sigma F_y = 0] \qquad 0.866DE = 10 \qquad DE = 11.55 \text{ kN } C \qquad Ans.$$

and the equation $\Sigma F_x = 0$ checks.

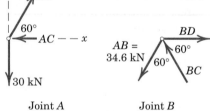

Joint A Joint B

Helpful Hint

① It should be stressed that the tension/compression designation refers to the *member*, not the joint. Note that we draw the force arrow on the same side of the joint as the member which exerts the force. In this way tension (arrow away from the joint) is distinguished from compression (arrow toward the joint).

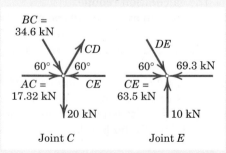

Joint C Joint E

Sample Problem 4/2

The simple truss shown supports the two loads, each of magnitude L. Determine the forces in members DE, DF, DG, and CD.

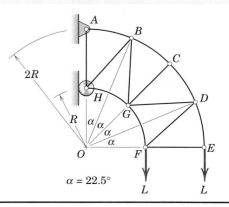

$$\alpha = 22.5°$$

Solution. First of all, we note that the curved members of this simple truss are all two-force members, so that the effect of each curved member within the truss is the same as that of a straight member.

We can begin with joint E because there are only two unknown member forces acting there. With reference to the free-body diagram and accompanying geometry for joint E, we note that $\beta = 180° - 11.25° - 90° = 78.8°$.

① $[\Sigma F_y = 0]$ $DE \sin 78.8° - L = 0$ $DE = 1.020L \ T$ *Ans.*

     $[\Sigma F_x = 0]$ $EF - DE \cos 78.8° = 0$ $EF = 0.1989L \ C$

We must now move to joint F, as there are still three unknown members at joint D. From the geometric diagram,

$$\gamma = \tan^{-1}\left[\frac{2R \sin 22.5°}{2R \cos 22.5° - R}\right] = 42.1°$$

From the free-body diagram of joint F,

$[\Sigma F_x = 0]$ $-GF \cos 67.5° + DF \cos 42.1° - 0.1989L = 0$

$[\Sigma F_y = 0]$ $GF \sin 67.5° + DF \sin 42.1° - L = 0$

Simultaneous solution of these two equations yields

$$GF = 0.646L \ T \qquad DF = 0.601L \ T \qquad \textit{Ans.}$$

For member DG, we move to the free-body diagram of joint D and the accompanying geometry.

$$\delta = \tan^{-1}\left[\frac{2R \cos 22.5° - 2R \cos 45°}{2R \sin 45° - 2R \sin 22.5°}\right] = 33.8°$$

$$\epsilon = \tan^{-1}\left[\frac{2R \sin 22.5° - R \sin 45°}{2R \cos 22.5° - R \cos 45°}\right] = 2.92°$$

Then from joint D:

$[\Sigma F_x = 0]$ $-DG \cos 2.92° - CD \sin 33.8° - 0.601L \sin 47.9° + 1.020L \cos 78.8° = 0$

$[\Sigma F_y = 0]$ $-DG \sin 2.92° + CD \cos 33.8° - 0.601L \cos 47.9° - 1.020L \sin 78.8° = 0$

The simultaneous solution is

$$CD = 1.617L \ T \qquad DG = -1.147L \text{ or } DG = 1.147L \ C \qquad \textit{Ans.}$$

Note that ϵ is shown exaggerated in the accompanying figures.

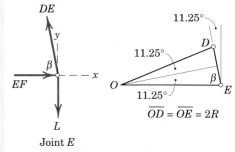

Joint E

Helpful Hint

① Rather than calculate and use the angle $\beta = 78.8°$ in the force equations, we could have used the $11.25°$ angle directly.

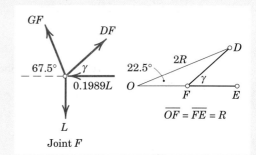

Joint F

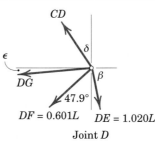

Joint D

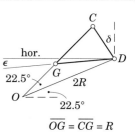

PROBLEMS

Introductory Problems

4/1 Determine the force in each member of the loaded truss. Explain why knowledge of the lengths of the members is unnecessary.

Ans. AB = 2000 lb *C, AC* = 1732 lb *T*
BC = 3460 lb *C*

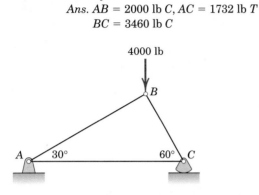

Problem 4/1

4/2 Determine the force in each member of the loaded truss. Identify any zero-force members by inspection.

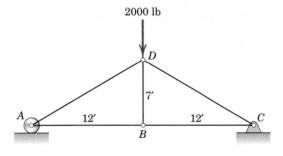

Problem 4/2

4/3 Determine the force in each member of the loaded truss.

Ans. AB = 3000 N *T, AC* = 4240 N *C, CD* = 4240 N *T*
AD = 3000 N *C, BC* = 6000 N *T*

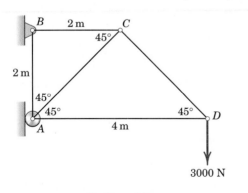

Problem 4/3

4/4 Determine the force in each member of the loaded truss.

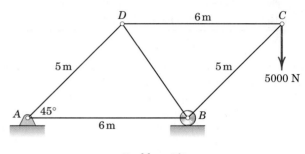

Problem 4/4

4/5 Determine the force in each member of the loaded truss.

Ans. AB = 5.63 kN *C, AF* = 3.38 kN *T*
BC = 4.13 kN *C, BE* = 0.901 kN *T*
BF = 4 kN *T,*   *CD* = 6.88 kN *C*
CE = 5.50 kN *T, DE* = 4.13 kN *T*
EF = 3.38 kN *T*

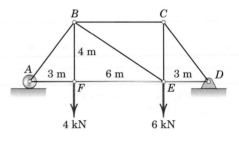

Problem 4/5

4/6 Calculate the force in each member of the loaded truss. All triangles are isosceles.

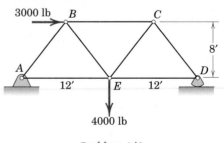

Problem 4/6

4/7 Determine the force in member AC of the loaded truss. The two quarter-circular members act as two-force members.

$$Ans.\ AC = \frac{L}{2}\ T$$

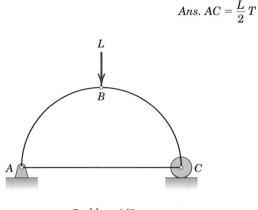

Problem 4/7

4/8 Determine the force in each member of the loaded truss. Make use of the symmetry of the truss and of the loading.

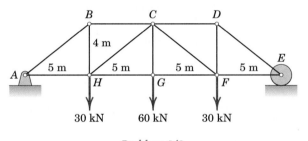

Problem 4/8

4/9 Determine the force in each member of the loaded truss.

$$Ans.\ AB = 14.42\ kN\ T,\ AC = 2.07\ kN\ C,\ AD = 0$$
$$BC = 6.45\ kN\ T,\ BD = 12.89\ kN\ C$$

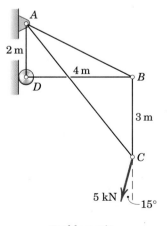

Problem 4/9

4/10 Determine the forces in members BE and CE of the loaded truss.

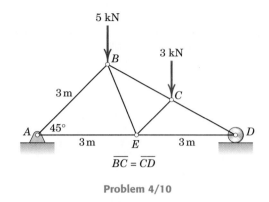

Problem 4/10

Representing Problems

4/11 Calculate the forces in members CG and CF for the truss shown.

$$Ans.\ CG = 2.24\ kN\ T,\ CF = 1\ kN\ C$$

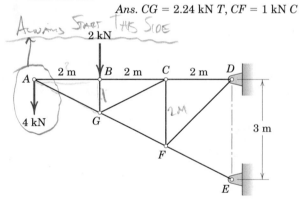

Problem 4/11

4/12 Each member of the truss is a uniform 20-ft bar weighing 400 lb. Calculate the average tension or compression in each member due to the weights of the members.

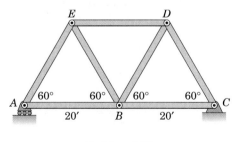

Problem 4/12

4/13 A drawbridge is being raised by a cable *EI*. The four joint loadings shown result from the weight of the roadway. Determine the forces in members *EF*, *DE*, *DF*, *CD*, and *FG*.

> Ans. *EF* = 3.87 kips *C*, *DE* = 4.61 kips *T*
> *DF* = 4.37 kips *C*, *CD* = 2.73 kips *T*
> *FG* = 7.28 kips *C*

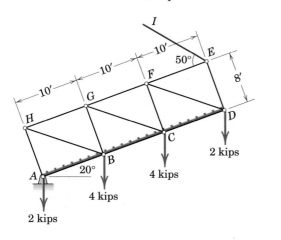

Problem 4/13

4/14 The truss is composed of equilateral triangles of sides *a* and is loaded and supported as shown. Determine the forces in members *EF*, *DE*, and *DF*.

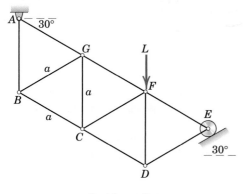

Problem 4/14

4/15 Determine the forces in members *BC* and *BG* of the loaded truss.

> Ans. *BC* = 3.46 kN *C*, *BG* = 1.528 kN *T*

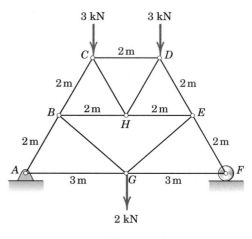

Problem 4/15

4/16 Determine the forces in members *BI*, *CI*, and *HI* for the loaded truss. All angles are 30°, 60°, or 90°.

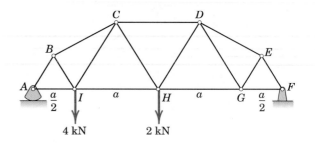

Problem 4/16

4/17 Determine the forces in members *AF*, *BE*, *BF*, and *CE* of the loaded truss.

Ans. AF = 6.13 kN *T*, *BE* = 5.59 kN *T*
BF = 6.50 kN *C*, *CE* = 5 kN *C*

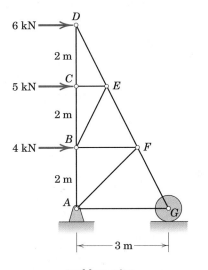

Problem 4/17

4/18 The signboard truss is designed to support a horizontal wind load of 800 lb. A separate analysis shows that $\frac{5}{8}$ of this force is transmitted to the center connection at *C* and the rest is equally divided between *D* and *B*. Calculate the forces in members *BE* and *BC*.

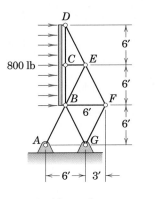

Problem 4/18

4/19 A snow load transfers the forces shown to the upper joints of a Pratt roof truss. Neglect any horizontal reactions at the supports and solve for the forces in all members.

Ans. AB = DE = BC = CD = 3.35 kN *C*
AH = EF = 3 kN *T*, *BH = DF* = 1 kN *C*
CF = CH = 1.414 kN *T*, *FG = GH* = 2 kN *T*

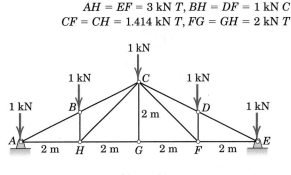

Problem 4/19

4/20 The loading of Prob. 4/19 is shown applied to a Howe roof truss. Neglect any horizontal reactions at the supports and solve for the forces in all members. Compare with the results of Prob. 4/19.

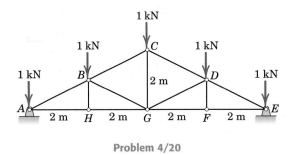

Problem 4/20

4/21 Determine the force in each member of the pair of trusses which support the 5000-lb load at their common joint *C*.

Ans. AB = BC = BG = 0, *AG = CG* = 2890 lb *C*

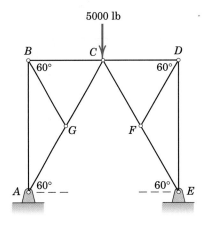

Problem 4/21

4/22 Determine the force in member *BF* of the loaded truss.

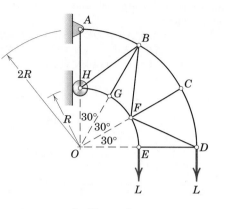

Problem 4/22

4/23 The rectangular frame is composed of four perimeter two-force members and two cables *AC* and *BD* which are incapable of supporting compression. Determine the forces in all members due to the load *L* in position (*a*) and then in position (*b*).

Ans. (*a*) $AB = AD = BD = 0, BC = L\ C$

$$AC = \frac{5L}{3}\ T, CD = \frac{4L}{3}\ C$$

(*b*) $AB = AD = BC = BD = 0$

$$AC = \frac{5L}{3}\ T, CD = \frac{4L}{3}\ C$$

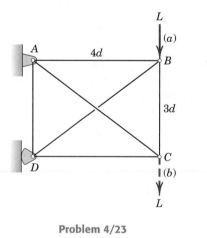

Problem 4/23

4/24 Verify the fact that each of the trusses contains one or more elements of redundancy and propose two separate changes, either one of which would remove the redundancy and produce complete statical determinacy. All members can support compression as well as tension.

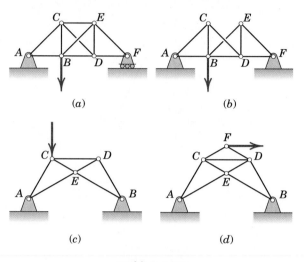

Problem 4/24

4/25 Determine the forces in members *AC* and *AD* of the loaded truss.

Ans. $AC = AD = 779$ lb *C*

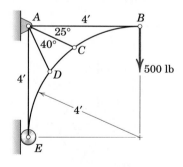

Problem 4/25

4/26 Analysis of the wind acting on a small Hawaiian church, which withstood the 165-mi/hr winds of Hurricane Iniki in 1992, showed the forces transmitted to each roof truss panel to be as shown. Treat the structure as a symmetrical simple truss and neglect any horizontal component of the support reaction at A. Identify the truss member which supports the largest force, tension or compression, and calculate this force.

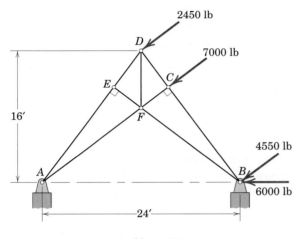

Problem 4/26

▶4/27 The tower for a transmission line is modeled by the truss shown. The crossed members in the center sections of the truss may be assumed to be capable of supporting tension only. For the loads of 1.8 kN applied in the vertical plane, compute the forces induced in members AB, DB, and CD.

Ans. $AB = 3.89$ kN C, $DB = 0$, $CD = 0.932$ kN C

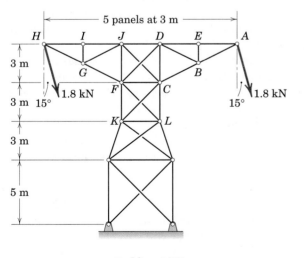

Problem 4/27

▶4/28 Determine the force in member CM of the loaded truss.

Ans. $CM = 3.41L$ T

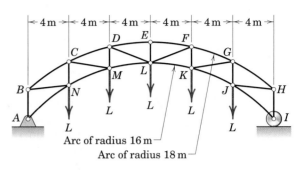

Problem 4/28

4/4 METHOD OF SECTIONS

When analyzing plane trusses by the method of joints, we need only two of the three equilibrium equations because the procedures involve concurrent forces at each joint. We can take advantage of the third or moment equation of equilibrium by selecting an entire section of the truss for the free body in equilibrium under the action of a nonconcurrent system of forces. This *method of sections* has the basic advantage that the force in almost any desired member may be found directly from an analysis of a section which has cut that member. Thus, it is not necessary to proceed with the calculation from joint to joint until the member in question has been reached. In choosing a section of the truss, we note that, in general, not more than three members whose forces are unknown should be cut, since there are only three available independent equilibrium relations.

Illustration of the Method

The method of sections will now be illustrated for the truss in Fig. 4/6, which was used in the explanation of the method of joints. The truss is shown again in Fig. 4/12a for ready reference. The external reactions are first computed as with the method of joints, by considering the truss as a whole.

Let us determine the force in the member *BE*, for example. An imaginary section, indicated by the dashed line, is passed through the truss, cutting it into two parts, Fig. 4/12b. This section has cut three members whose forces are initially unknown. In order for the portion of the truss on each side of the section to remain in equilibrium, it is necessary to apply to each cut member the force which was exerted on it by the member cut away. For simple trusses composed of straight two-force members, these forces, either tensile or compressive, will always be in the directions of the respective members. The left-hand section is in equilibrium under the action of the applied load L, the end reaction R_1, and the three forces exerted on the cut members by the right-hand section which has been removed.

We can usually draw the forces with their proper senses by a visual approximation of the equilibrium requirements. Thus, in balancing the moments about point B for the left-hand section, the force EF is clearly to the left, which makes it compressive, because it acts toward the cut section of member EF. The load L is greater than the reaction R_1, so that the force BE must be up and to the right to supply the needed upward component for vertical equilibrium. Force BE is therefore tensile, since it acts away from the cut section.

With the approximate magnitudes of R_1 and L in mind we see that the balance of moments about point E requires that BC be to the right. A casual glance at the truss should lead to the same conclusion when it is realized that the lower horizontal member will stretch under the tension caused by bending. The equation of moments about joint B eliminates three forces from the relation, and EF can be determined directly. The force BE is calculated from the equilibrium equation for the y-direction. Finally, we determine BC by balancing moments about point E. In this

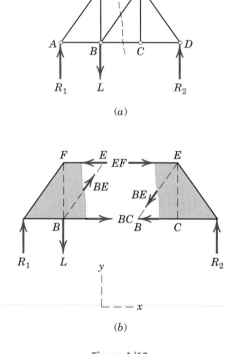

(a)

(b)

Figure 4/12

way each of the three unknowns has been determined independently of the other two.

The right-hand section of the truss, Fig. 4/12b, is in equilibrium under the action of R_2 and the same three forces in the cut members applied in the directions opposite to those for the left section. The proper sense for the horizontal forces can easily be seen from the balance of moments about points B and E.

Additional Considerations

It is essential to understand that in the method of sections an entire portion of the truss is considered a single body in equilibrium. Thus, the forces in members internal to the section are not involved in the analysis of the section as a whole. To clarify the free body and the forces acting externally on it, the cutting section is preferably passed through the members and not the joints. We may use either portion of a truss for the calculations, but the one involving the smaller number of forces will usually yield the simpler solution.

In some cases the methods of sections and joints can be combined for an efficient solution. For example, suppose we wish to find the force in a central member of a large truss. Furthermore, suppose that it is not possible to pass a section through this member without passing through at least four unknown members. It may be possible to determine the forces in nearby members by the method of sections and then progress to the unknown member by the method of joints. Such a combination of the two methods may be more expedient than exclusive use of either method.

The moment equations are used to great advantage in the method of sections. One should choose a moment center, either on or off the section, through which as many unknown forces as possible pass.

It is not always possible to assign the proper sense of an unknown force when the free-body diagram of a section is initially drawn. Once an arbitrary assignment is made, a positive answer will verify the assumed sense and a negative result will indicate that the force is in the sense opposite to that assumed. An alternative notation preferred by some is to assign all unknown forces arbitrarily as positive in the tension direction (away from the section) and let the algebraic sign of the answer distinguish between tension and compression. Thus, a plus sign would signify tension and a minus sign compression. On the other hand, the advantage of assigning forces in their correct sense on the free-body diagram of a section wherever possible is that doing so emphasizes the physical action of the forces more directly, and this practice is the one which is preferred here.

© Photodisc/Media Bakery

Many simple trusses are periodic in that there are repeated and identical structural sections.

Sample Problem 4/3

Calculate the forces induced in members KL, CL, and CB by the 20-ton load on the cantilever truss.

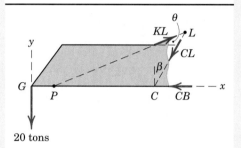

Solution. Although the vertical components of the reactions at A and M are statically indeterminate with the two fixed supports, all members other than AM are statically determinate. We may pass a section directly through members KL, CL, and CB and analyze the portion of the truss to the left of this section as a ① statically determinate rigid body.

The free-body diagram of the portion of the truss to the left of the section is shown. A moment sum about L quickly verifies the assignment of CB as compression, and a moment sum about C quickly discloses that KL is in tension. The direction of CL is not quite so obvious until we observe that KL and CB intersect at a point P to the right of G. A moment sum about P eliminates reference to KL and CB and shows that CL must be compressive to balance the moment of the 20-ton force about P. With these considerations in mind the solution becomes straightforward, as we now see how to solve for each of the three unknowns independently of the other two.

② Summing moments about L requires finding the moment arm $\overline{BL} = 16 + (26 - 16)/2 = 21$ ft. Thus,

$$[\Sigma M_L = 0] \qquad 20(5)(12) - CB(21) = 0 \qquad CB = 57.1 \text{ tons } C \qquad \textit{Ans.}$$

Next we take moments about C, which requires a calculation of $\cos \theta$. From the given dimensions we see $\theta = \tan^{-1}(5/12)$ so that $\cos \theta = 12/13$. Therefore,

$$[\Sigma M_C = 0] \qquad 20(4)(12) - \tfrac{12}{13}KL(16) = 0 \qquad KL = 65 \text{ tons } T \qquad \textit{Ans.}$$

Finally, we may find CL by a moment sum about P, whose distance from C is given by $\overline{PC}/16 = 24/(26 - 16)$ or $\overline{PC} = 38.4$ ft. We also need β, which is given by $\beta = \tan^{-1}(\overline{CB}/\overline{BL}) = \tan^{-1}(12/21) = 29.7°$ and $\cos \beta = 0.868$. We now have

③ $[\Sigma M_p = 0] \qquad 20(48 - 38.4) - CL(0.868)(38.4) = 0$

$$CL = 5.76 \text{ tons } C \qquad \textit{Ans.}$$

Helpful Hints

① We note that analysis by the method of joints would necessitate working with eight joints in order to calculate the three forces in question. Thus, the method of sections offers a considerable advantage in this case.

② We could have started with moments about C or P just as well.

③ We could also have determined CL by a force summation in either the x- or y-direction.

Sample Problem 4/4

Calculate the force in member DJ of the Howe roof truss illustrated. Neglect any horizontal components of force at the supports.

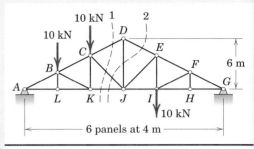

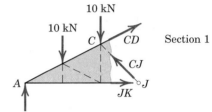

Solution. It is not possible to pass a section through DJ without cutting four members whose forces are unknown. Although three of these cut by section 2 are concurrent at J and therefore the moment equation about J could be used to obtain DE, the force in DJ cannot be obtained from the remaining two equilibrium principles. It is necessary to consider first the adjacent section 1 before analyzing section 2.

The free-body diagram for section 1 is drawn and includes the reaction of 18.33 kN at A, which is previously calculated from the equilibrium of the truss as a whole. In assigning the proper directions for the forces acting on the three cut members, we see that a balance of moments about A eliminates the effects of CD and JK and clearly requires that CJ be up and to the left. A balance of moments about C eliminates the effect of the three forces concurrent at C and indicates that JK must be to the right to supply sufficient counterclockwise moment. Again it should be fairly obvious that the lower chord is under tension because of the bending tendency of the truss. Although it should also be apparent that the top chord is under compression, for purposes of illustration the force in CD will
① be arbitrarily assigned as tension.

By the analysis of section 1, CJ is obtained from

$$[\Sigma M_A = 0] \qquad 0.707CJ(12) - 10(4) - 10(8) = 0 \qquad CJ = 14.14 \text{ kN } C$$

In this equation the moment of CJ is calculated by considering its horizontal and vertical components acting at point J. Equilibrium of moments about J requires

$$[\Sigma M_J = 0] \qquad 0.894CD(6) + 18.33(12) - 10(4) - 10(8) = 0$$

$$CD = -18.63 \text{ kN}$$

The moment of CD about J is calculated here by considering its two components
② as acting through D. The minus sign indicates that CD was assigned in the wrong direction.

Hence, $\qquad\qquad CD = 18.63 \text{ kN } C$

③ From the free-body diagram of section 2, which now includes the known value of CJ, a balance of moments about G is seen to eliminate DE and JK. Thus,

$$[\Sigma M_G = 0] \qquad 12DJ + 10(16) + 10(20) - 18.33(24) - 14.14(0.707)(12) = 0$$

$$DJ = 16.67 \text{ kN } T \qquad\qquad\qquad\qquad Ans.$$

Again the moment of CJ is determined from its components considered to be acting at J. The answer for DJ is positive, so that the assumed tensile direction is correct.

An alternative approach to the entire problem is to utilize section 1 to determine CD and then use the method of joints applied at D to determine DJ.

Helpful Hints

① There is no harm in assigning one or more of the forces in the wrong direction, as long as the calculations are consistent with the assumption. A negative answer will show the need for reversing the direction of the force.

② If desired, the direction of CD may be changed on the free-body diagram and the algebraic sign of CD reversed in the calculations, or else the work may be left as it stands with a note stating the proper direction.

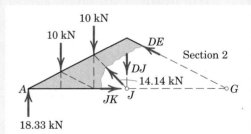

③ Observe that a section through members CD, DJ, and DE could be taken which would cut only three unknown members. However, since the forces in these three members are all concurrent at D, a moment equation about D would yield no information about them. The remaining two force equations would not be sufficient to solve for the three unknowns.

PROBLEMS

Introductory Problems

4/29 Determine the forces in members *CG* and *GH*.

Ans. CG = 0, GH = 27 kN T

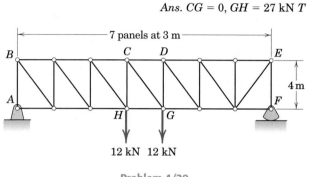

Problem 4/29

4/30 Determine the force in member *AE* of the loaded truss.

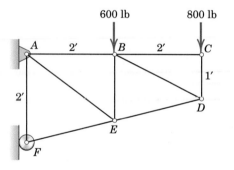

Problem 4/30

4/31 Determine the force in member *BC* of the loaded truss.

Ans. BC = 4.81 kips T

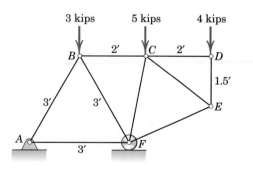

Problem 4/31

4/32 Determine the forces in members *GH* and *CG* for the truss loaded and supported as shown. Does the statical indeterminacy of the supports affect your calculation?

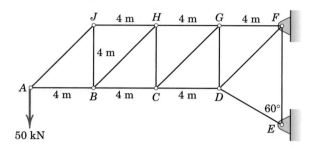

Problem 4/32

4/33 Determine the force in member *DG* of the loaded truss.

Ans. DG = L T

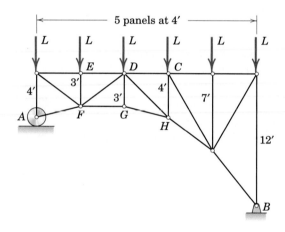

Problem 4/33

4/34 Determine the force in member *BE* of the loaded truss.

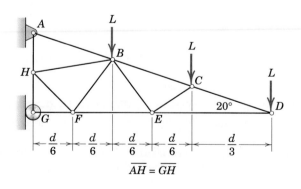

$$\overline{AH} = \overline{GH}$$

Problem 4/34

Representative Problems

4/35 Determine the forces in members *DE* and *DL*.

Ans. DE = 24 kN *T, DL* = 33.9 kN *C*

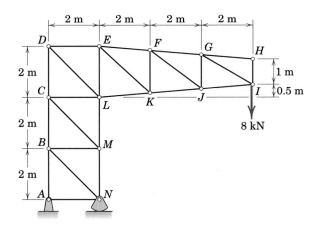

Problem 4/35

4/36 Calculate the forces in members *BC*, *BE*, and *EF*. Solve for each force from an equilibrium equation which contains that force as the only unknown.

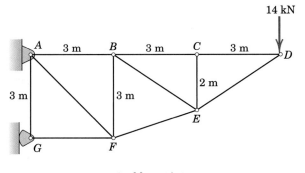

Problem 4/36

4/37 Calculate the forces in members *BC*, *CD*, and *CG* of the loaded truss composed of equilateral triangles, each of side length 8 m.

Ans. BC = 1.155 kN *T, CD* = 5.20 kN *T*
CG = 4.04 kN *C*

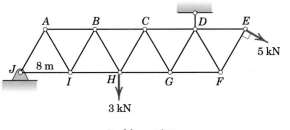

Problem 4/37

4/38 Determine the forces in members *BC* and *FG* of the loaded symmetrical truss. Show that this calculation can be accomplished by using one section and two equations, each of which contains only one of the two unknowns. Are the results affected by the statical indeterminacy of the supports at the base?

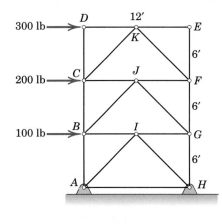

Problem 4/38

4/39 The truss shown is composed of 45° right triangles. The crossed members in the center two panels are slender tie rods incapable of supporting compression. Retain the two rods which are under tension and compute the magnitudes of their tensions. Also find the force in member *MN*.

Ans. FN = *GM* = 84.8 kN *T, MN* = 20 kN *T*

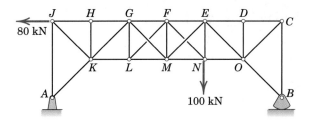

Problem 4/39

4/40 Determine the force in member *BF*.

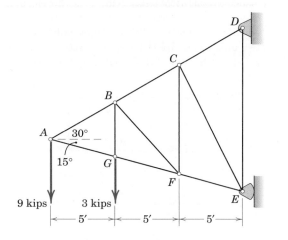

9 kips   3 kips

Problem 4/40

4/41 Determine the forces in members *CD*, *CJ*, and *DJ*.
Ans. $CD = 0.562L$ C, $CJ = 1.562L$ T
$DJ = 1.250L$ C

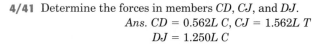
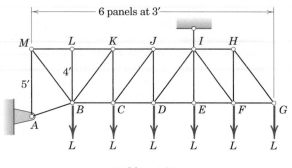

Problem 4/41

4/42 Compute the force in member *HN* of the loaded truss.

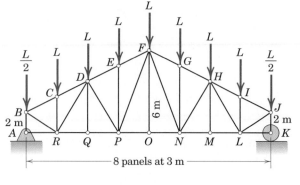

Problem 4/42

4/43 Determine the forces in members *DE*, *DL*, *LM*, and *EL* of the loaded symmetrical truss.
Ans. $DE = 4.80L$ C, $DL = 0.0446L$ T, $LM = 4.54L$ T
$EL = 3.80L$ T

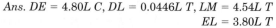
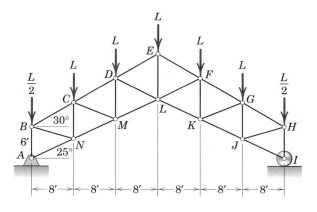

Problem 4/43

4/44 Determine the forces in members *DQ* and *CQ* of the loaded symmetrical truss.

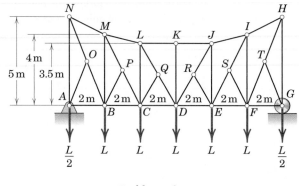

Problem 4/44

4/45 Calculate the forces in members *CB*, *CG*, and *FG* for the loaded truss without first calculating the force in any other member.
Ans. $CB = 56.2$ kN C, $CG = 13.87$ kN T
$FG = 19.62$ kN T

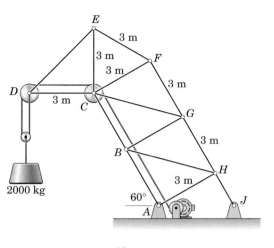

Problem 4/45

4/46 The hinged frames *ACE* and *DFB* are connected by two hinged bars, *AB* and *CD*, which cross without being connected. Compute the force in *AB*.

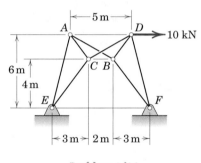

Problem 4/46

4/47 Determine the force in member *JM* of the loaded truss.

Ans. JM = 0.0901*L T*

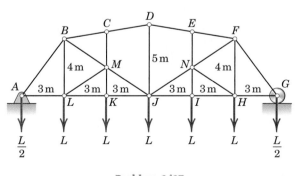

Problem 4/47

4/48 Determine the forces in members *DE*, *EI*, *FI*, and *HI* of the arched roof truss.

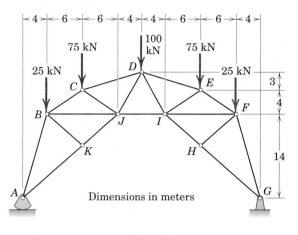

Dimensions in meters

Problem 4/48

4/49 Determine the force in member *GK* of the loaded symmetrical truss.

Ans. GK = 2.13*L T*

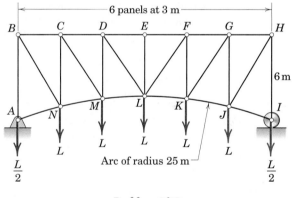

Problem 4/49

4/50 Determine the force in member *CL* of the loaded truss. The radius of curvature of the upper chord *BCDEFG* is 30 m.

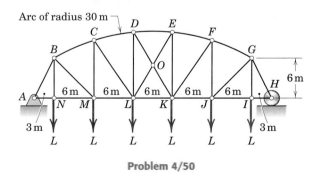

Problem 4/50

▶**4/51** Determine the force in member *DK* of the loaded overhead sign truss.

Ans. DK = 1 kip T

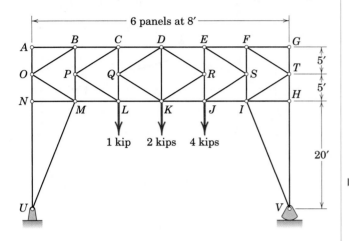

1 kip   2 kips   4 kips

Problem 4/51

▶**4/52** Determine the force in member *DG* of the compound truss. The joints all lie on radial lines subtending angles of 15° as indicated, and the curved members act as two-force members. Distance $\overline{OC} = \overline{OA} = \overline{OB} = R$.

Ans. DG = 0.569L C

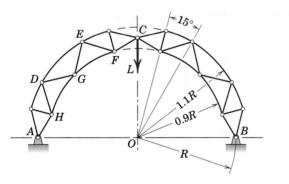

Problem 4/52

▶**4/53** Determine the force in member *CK* of the loaded truss.

Ans. CK = 9290 N C

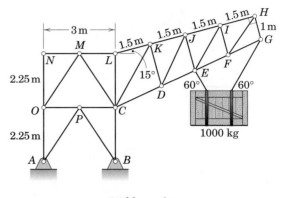

Problem 4/53

▶**4/54** A design model for a transmission-line tower is shown in the figure. Members *GH*, *FG*, *OP*, and *NO* are insulated cables; all other members are steel bars. For the loading shown, compute the forces in members *FI*, *FJ*, *EJ*, *EK*, and *ER*. Use a combination of methods if desired.

Ans. FI = ER = 0, FJ = 7.81 kN T
EJ = 3.61 kN C, EK = 22.4 kN C

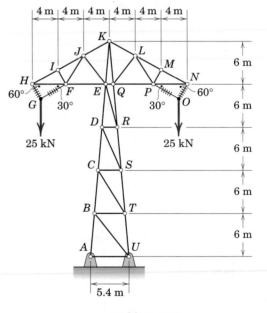

Problem 4/54

4/5   SPACE TRUSSES

A *space truss* is the three-dimensional counterpart of the plane truss described in the three previous articles. The idealized space truss consists of rigid links connected at their ends by ball-and-socket joints (such a joint is illustrated in Fig. 3/8 in Art. 3/4). Whereas a triangle of pin-connected bars forms the basic noncollapsible unit for the plane truss, a space truss, on the other hand, requires six bars joined at their ends to form the edges of a tetrahedron as the basic noncollapsible unit. In Fig. 4/13a the two bars AD and BD joined at D require a third support CD to keep the triangle ADB from rotating about AB. In Fig. 4/13b the supporting base is replaced by three more bars AB, BC, and AC to form a tetrahedron not dependent on the foundation for its own rigidity.

We may form a new rigid unit to extend the structure with three additional concurrent bars whose ends are attached to three fixed joints on the existing structure. Thus, in Fig. 4/13c the bars AF, BF, and CF are attached to the foundation and therefore fix point F in space. Likewise point H is fixed in space by the bars AH, DH, and CH. The three additional bars CG, FG, and HG are attached to the three fixed points C, F, and H and therefore fix G in space. The fixed point E is similarly created. We see now that the structure is entirely rigid. The two applied loads shown will result in forces in all of the members. A space truss formed in this way is called a *simple* space truss.

Ideally there must be point support, such as that given by a ball-and-socket joint, at the connections of a space truss to prevent bending in the members. As in riveted and welded connections for plane trusses, if the centerlines of joined members intersect at a point, we can justify the assumption of two-force members under simple tension and compression.

Statically Determinate Space Trusses

When a space truss is supported externally so that it is statically determinate as an entire unit, a relationship exists between the number of its joints and the number of its members necessary for internal stability without redundancy. Because the equilibrium of each joint is specified by three scalar force equations, there are in all $3j$ such equations for a space truss with j joints. For the entire truss composed of m members there are m unknowns (the tensile or compressive forces in the members) plus six unknown support reactions in the general case of a statically determinate space structure. Thus, for any space truss, the equation $m + 6 = 3j$ will be satisfied if the truss is statically determinate internally. A *simple* space truss satisfies this relation automatically. Starting with the initial tetrahedron, for which the equation holds, the structure is extended by adding three members and one joint at a time, thus preserving the equality.

As in the case of the plane truss, this relation is a necessary condition for stability, but it is not a sufficient condition, since one or more of the m members can be arranged in such a way as not to contribute to a stable configuration of the entire truss. If $m + 6 > 3j$, there are more members than there are independent equations, and the truss is statically indeterminate internally with redundant members present.

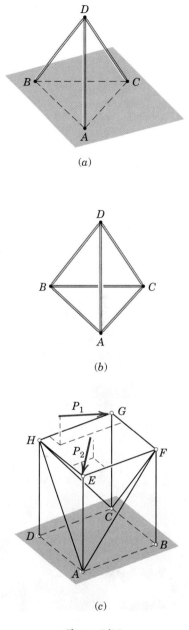

(a)

(b)

(c)

Figure 4/13

If $m + 6 < 3j$, there is a deficiency of internal members, and the truss is unstable and subject to collapse under load. This relationship between the number of joints and the number of members is very helpful in the preliminary design of a stable space truss, since the configuration is not as obvious as with a plane truss, where the geometry for statical determinacy is generally quite apparent.

Method of Joints for Space Trusses

The method of joints developed in Art. 4/3 for plane trusses may be extended directly to space trusses by satisfying the complete vector equation

$$\Sigma \mathbf{F} = \mathbf{0} \tag{4/1}$$

for each joint. We normally begin the analysis at a joint where at least one known force acts and not more than three unknown forces are present. Adjacent joints on which not more than three unknown forces act may then be analyzed in turn.

This step-by-step joint technique tends to minimize the number of simultaneous equations to be solved when we must determine the forces in all members of the space truss. For this reason, although it is not readily reduced to a routine, such an approach is recommended. As an alternative procedure, however, we may simply write $3j$ joint equations by applying Eq. 4/1 to all joints of the space frame. The number of unknowns will be $m + 6$ if the structure is noncollapsible when removed from its supports and those supports provide six external reactions. If, in addition, there are no redundant members, then the number of equations ($3j$) equals the number of unknowns ($m + 6$), and the entire system of equations can be solved simultaneously for the unknowns. Because of the large number of coupled equations, a computer solution is usually required. With this latter approach, it is not necessary to begin at a joint where at least one known and no more than three unknown forces act.

Method of Sections for Space Trusses

The method of sections developed in the previous article may also be applied to space trusses. The two vector equations

$$\Sigma \mathbf{F} = \mathbf{0} \qquad \text{and} \qquad \Sigma \mathbf{M} = \mathbf{0}$$

must be satisfied for any section of the truss, where the zero moment sum will hold for all moment axes. Because the two vector equations are equivalent to six scalar equations, we conclude that, in general, a section should not be passed through more than six members whose forces are unknown. The method of sections for space trusses is not widely used, however, because a moment axis can seldom be found which eliminates all but one unknown, as in the case of plane trusses.

Vector notation for expressing the terms in the force and moment equations for space trusses is of considerable advantage and is used in the sample problem which follows.

This space truss is used to support observation equipment in Bottrop, Germany.

Sample Problem 4/5

The space truss consists of the rigid tetrahedron $ABCD$ anchored by a ball-and-socket connection at A and prevented from any rotation about the x-, y-, or z-axes by the respective links 1, 2, and 3. The load L is applied to joint E, which is rigidly fixed to the tetrahedron by the three additional links. Solve for the forces in the members at joint E and indicate the procedure for the determination of the forces in the remaining members of the truss.

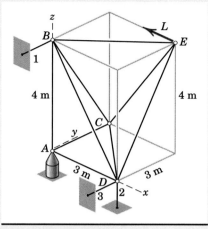

Solution. We note first that the truss is supported with six properly placed constraints, which are the three at A and the links 1, 2, and 3. Also, with $m = 9$ members and $j = 5$ joints, the condition $m + 6 = 3j$ for a sufficiency of members to provide a noncollapsible structure is satisfied.

① The external reactions at A, B, and D can be calculated easily as a first step, although their values will be determined from the solution of all forces on each of the joints in succession.

② We start with a joint on which at least one known force and not more than three unknown forces act, which in this case is joint E. The free-body diagram of joint E is shown with all force vectors arbitrarily assumed in their positive tension directions (away from the joint). The vector expressions for the three unknown forces are

$$\mathbf{F}_{EB} = \frac{F_{EB}}{\sqrt{2}}(-\mathbf{i} - \mathbf{j}), \; \mathbf{F}_{EC} = \frac{F_{EC}}{5}(-3\mathbf{i} - 4\mathbf{k}), \; \mathbf{F}_{ED} = \frac{F_{ED}}{5}(-3\mathbf{j} - 4\mathbf{k})$$

Equilibrium of joint E requires

$$[\Sigma \mathbf{F} = 0] \qquad \mathbf{L} + \mathbf{F}_{EB} + \mathbf{F}_{EC} + \mathbf{F}_{ED} = 0 \qquad \text{or}$$

$$-L\mathbf{i} + \frac{F_{EB}}{\sqrt{2}}(-\mathbf{i} - \mathbf{j}) + \frac{F_{EC}}{5}(-3\mathbf{i} + 4\mathbf{k}) + \frac{F_{ED}}{5}(-3\mathbf{j} - 4\mathbf{k}) = 0$$

Rearranging terms gives

$$\left(-L - \frac{F_{EB}}{\sqrt{2}} - \frac{3F_{EC}}{5}\right)\mathbf{i} + \left(-\frac{F_{EB}}{\sqrt{2}} - \frac{3F_{ED}}{5}\right)\mathbf{j} + \left(-\frac{4F_{EC}}{5} - \frac{4F_{ED}}{5}\right)\mathbf{k} = 0$$

Equating the coefficients of the $\mathbf{i}$-, $\mathbf{j}$-, and $\mathbf{k}$-unit vectors to zero gives the three equations

$$\frac{F_{EB}}{\sqrt{2}} + \frac{3F_{EC}}{5} = -L \qquad \frac{F_{EB}}{\sqrt{2}} + \frac{3F_{ED}}{5} = 0 \qquad F_{EC} + F_{ED} = 0$$

Solving the equations gives us

$$F_{EB} = -L/\sqrt{2} \qquad F_{EC} = -5L/6 \qquad F_{ED} = 5L/6 \qquad \textit{Ans.}$$

Thus, we conclude that F_{EB} and F_{EC} are compressive forces and F_{ED} is tension.

Unless we have computed the external reactions first, we must next analyze joint C with the known value of F_{EC} and the three unknowns F_{CB}, F_{CA}, and F_{CD}. The procedure is identical with that used for joint E. Joints B, D, and A are then analyzed in the same way and in that order, which limits the scalar unknowns to three for each joint. The external reactions computed from these analyses must, of course, agree with the values which can be determined initially from an analysis of the truss as a whole.

Helpful Hints

① *Suggestion:* Draw a free-body diagram of the truss as a whole and verify that the external forces acting on the truss are $\mathbf{A}_x = L\mathbf{i}$, $\mathbf{A}_y = L\mathbf{j}$, $\mathbf{A}_z = (4L/3)\mathbf{k}$, $\mathbf{B}_y = 0$, $\mathbf{D}_y = -L\mathbf{j}$, $\mathbf{D}_z = -(4L/3)\mathbf{k}$.

② With this assumption, a negative numerical value for a force indicates compression.

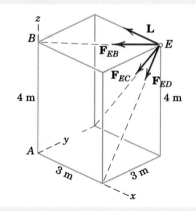

PROBLEMS

(In the following problems, use plus for tension and minus for compression.)

4/55 Determine the forces in members AB, AC, and AD. Point M is the centroid of triangle BCD.

Ans. $T_{AB} = 1607$ lb, $T_{AC} = 0$, $T_{AD} = -1607$ lb

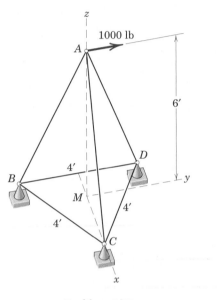

Problem 4/55

4/56 The base of an automobile jackstand forms an equilateral triangle of side length 10 in. and is centered under the collar A. Model the structure as one with a ball and socket at each joint and determine the forces in members BC, BD, and CD. Neglect any horizontal reaction components under the feet B, C, and D.

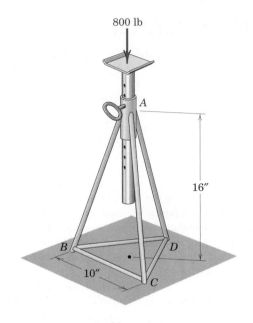

Problem 4/56

4/57 The space truss in the form of a tetrahedron is supported by ball-and-socket connections at its base points A and B and is prevented from rotating about AB by the vertical tie bar CD. After noting the vertical components of the reactions under the symmetrical truss at A and B, draw a free-body diagram of the triangular configuration of links BDE and determine the x-component of the force exerted by the foundation on the truss at B.

Ans. $B_x = P$

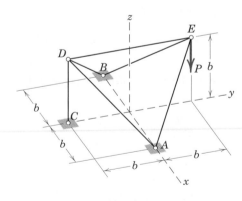

Problem 4/57

4/58 The rectangular space truss 16 m in height is erected on a horizontal square base 12 m on a side. Guy wires are attached to the structure at E and G as shown and are tightened until the tension T in each wire is 9 kN. Calculate the force F in each of the diagonal members.

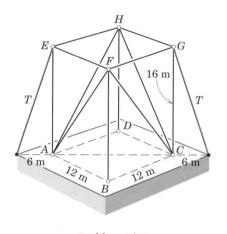

Problem 4/58

4/59 The tetrahedral space truss has a horizontal base ABC in the form of an isosceles triangle and legs AD, BD, and CD which support the mass m from point D. Each vertex of the base is suspended by a vertical wire from an overhead support. Calculate the forces induced in members AC and AB.

$$\text{Ans. } F_{AC} = -\frac{5}{54}mg, \quad F_{AB} = -\frac{4}{27}mg$$

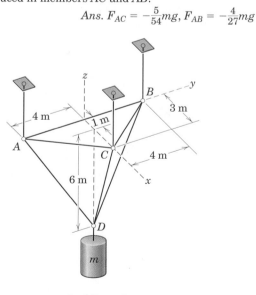

Problem 4/59

4/60 For the space truss shown, check the sufficiency of the supports and also the number of and arrangement of the members to ensure statical determinacy, both external and internal. Determine the forces in members AE, BE, BF, and CE.

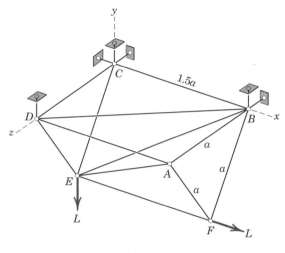

Problem 4/60

4/61 Determine the force in member BD of the regular pyramid with square base.

$$\text{Ans. } DB = -2.00L$$

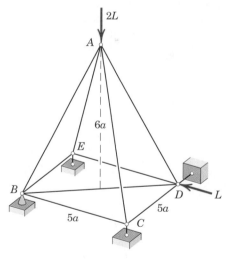

Problem 4/61

4/62 A space truss is being designed with the elements shown. How many more members are needed to make the truss stable internally? Name the members (by specifying their end points) which would produce one possible configuration of internal stability.

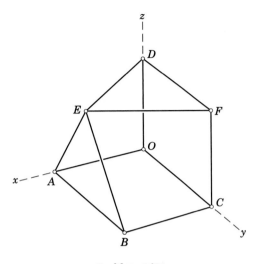

Problem 4/62

4/63 The space truss shown is secured to the fixed supports at A, B, and E and is loaded by the force L which has equal x- and y-components but no vertical z-component. Show that there is a sufficient number of members to provide internal stability and that their placement is adequate for this purpose. Next determine the forces in members CD, BC, and CE.

$$Ans. \ F_{BC} = \frac{L\sqrt{2}}{4}, F_{CD} = 0, F_{CE} = -\frac{L\sqrt{3}}{2}$$

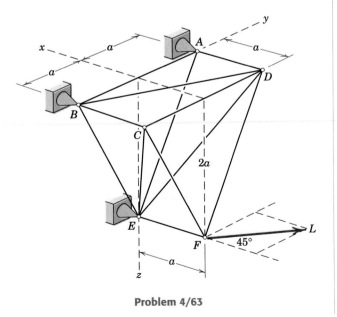

Problem 4/63

4/64 The pyramidal truss section $BCDEF$ is symmetric about the vertical x-z plane as shown. Cables AE, AF, and AB support a 5-kN load. Determine the force in member BE.

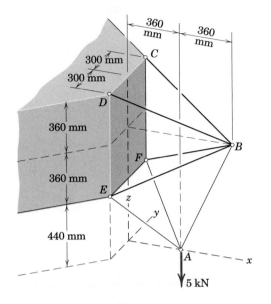

Problem 4/64

▶**4/65** The lengthy boom of an overhead construction crane, a portion of which is shown, is an example of a periodic structure—one which is composed of repeated and identical structural units. Use the method of sections to find the forces in members FJ and GJ.

$$Ans. \ FJ = 0, GJ = -70.8 \text{ kN}$$

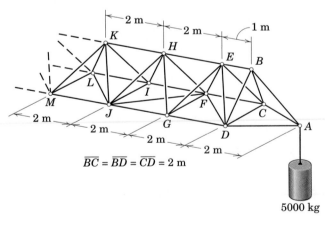

$$\overline{BC} = \overline{BD} = \overline{CD} = 2 \text{ m}$$

Problem 4/65

▶**4/66** A space truss consists of two pyramids on identical square bases in the horizontal *x-y* plane with common side *DG*. The truss is loaded at the vertex *A* by the downward force *L* and is supported by the vertical reactions shown at its corners. All members except the two base diagonals are of the same length *b*. Take advantage of the two vertical planes of symmetry and determine the forces in *AB* and *DA*. (Note that link *AB* prevents the two pyramids from hinging about *DG*.)

$$Ans. \; AB = -\frac{\sqrt{2}L}{4}$$

$$DA = -\frac{\sqrt{2}L}{8}$$

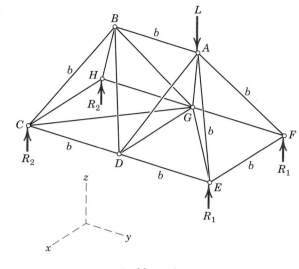

Problem 4/66

4/6   FRAMES AND MACHINES

A structure is called a *frame* or *machine* if at least one of its individual members is a *multiforce member*. A multiforce member is defined as one with three or more forces acting on it, or one with two or more forces and one or more couples acting on it. Frames are structures which are designed to support applied loads and are usually fixed in position. Machines are structures which contain moving parts and are designed to transmit input forces or couples to output forces or couples.

Because frames and machines contain multiforce members, the forces in these members in general will *not* be in the directions of the members. Therefore, we cannot analyze these structures by the methods developed in Arts. 4/3, 4/4, and 4/5 because these methods apply to simple trusses composed of two-force members where the forces are in the directions of the members.

Interconnected Rigid Bodies with Multiforce Members

In Chapter 3 we discussed the equilibrium of multiforce bodies, but we concentrated on the equilibrium of a *single* rigid body. In this present article we focus on the equilibrium of *interconnected* rigid bodies which include multiforce members. Although most such bodies may be analyzed as two-dimensional systems, there are numerous examples of frames and machines which are three-dimensional.

The forces acting on each member of a connected system are found by isolating the member with a free-body diagram and applying the equations of equilibrium. The *principle of action and reaction* must be carefully observed when we represent the forces of interaction on the separate free-body diagrams. If the structure contains more members or supports than are necessary to prevent collapse, then, as in the case of trusses, the problem is statically indeterminate, and the principles of equilibrium, although necessary, are not sufficient for solution. Although many frames and machines are statically indeterminate, we will consider in this article only those which are statically determinate.

If the frame or machine constitutes a rigid unit by itself when removed from its supports, like the A-frame in Fig. 4/14*a*, the analysis is best begun by establishing all the forces external to the structure treated as a single rigid body. We then dismember the structure and consider the equilibrium of each part separately. The equilibrium equations for the several parts will be related through the terms involving

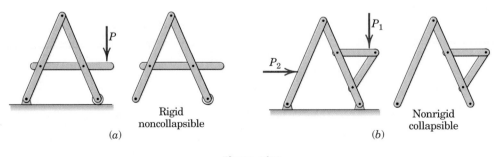

Rigid
noncollapsible

(*a*)

Nonrigid
collapsible

(*b*)

Figure 4/14

the forces of interaction. If the structure is not a rigid unit by itself but depends on its external supports for rigidity, as illustrated in Fig. 4/14*b*, then the calculation of the external support reactions cannot be completed until the structure is dismembered and the individual parts are analyzed.

Force Representation and Free-Body Diagrams

In most cases the analysis of frames and machines is facilitated by representing the forces in terms of their rectangular components. This is particularly so when the dimensions of the parts are given in mutually perpendicular directions. The advantage of this representation is that the calculation of moment arms is simplified. In some three-dimensional problems, particularly when moments are evaluated about axes which are not parallel to the coordinate axes, use of vector notation is advantageous.

It is not always possible to assign the proper sense to every force or its components when drawing the free-body diagrams, and it becomes necessary to make an arbitrary assignment. In any event, it is *absolutely necessary* that a force be *consistently* represented on the diagrams for interacting bodies which involve the force in question. Thus, for two bodies connected by the pin A, Fig. 4/15*a*, the force components must be consistently represented in *opposite* directions on the separate free-body diagrams.

For a ball-and-socket connection between members of a space frame, we must apply the action-and-reaction principle to all three components as shown in Fig. 4/15*b*. The assigned directions may prove to be wrong when the algebraic signs of the components are determined upon calculation. If A_x, for instance, should turn out to be negative, it is actually acting in the direction opposite to that originally represented. Accordingly, we would need to reverse the direction of the force on *both* members and to reverse the sign of its force terms in the equations. Or we may leave the representation as originally made, and the proper sense of the force will be understood from the negative sign. If we choose to use vector notation in labeling the forces, then we must be careful to use a plus sign for an action and a minus sign for the corresponding reaction, as shown in Fig. 4/16.

We may occasionally need to solve two or more equations simultaneously in order to separate the unknowns. In most instances, however, we can avoid simultaneous solutions by careful choice of the member or group of members for the free-body diagram and by a careful choice of moment axes which will eliminate undesired terms from the equations. The method of solution described in the foregoing paragraphs is illustrated in the following sample problems.

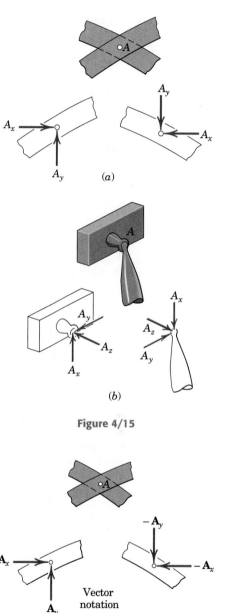

Figure 4/15

Figure 4/16

Sample Problem 4/6

The frame supports the 400-kg load in the manner shown. Neglect the weights of the members compared with the forces induced by the load and compute the horizontal and vertical components of all forces acting on each of the members.

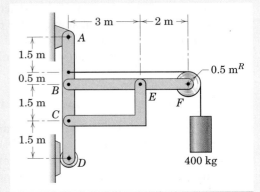

Solution. We observe first that the three supporting members which constitute the frame form a rigid assembly that can be analyzed as a single unit. We also observe that the arrangement of the external supports makes the frame statically determinate.

From the free-body diagram of the entire frame we determine the external reactions. Thus,

$$[\Sigma M_A = 0] \qquad 5.5(0.4)(9.81) - 5D = 0 \qquad D = 4.32 \text{ kN}$$

$$[\Sigma F_x = 0] \qquad A_x - 4.32 = 0 \qquad A_x = 4.32 \text{ kN}$$

$$[\Sigma F_y = 0] \qquad A_y - 3.92 = 0 \qquad A_y = 3.92 \text{ kN}$$

Next we dismember the frame and draw a separate free-body diagram of each member. The diagrams are arranged in their approximate relative positions to aid in keeping track of the common forces of interaction. The external reactions just obtained are entered onto the diagram for AD. Other known forces are the 3.92-kN forces exerted by the shaft of the pulley on the member BF, as obtained from the free-body diagram of the pulley. The cable tension of 3.92 kN is also shown acting on AD at its attachment point.

Next, the components of all unknown forces are shown on the diagrams. Here we observe that CE is a two-force member. The force components on CE have equal and opposite reactions, which are shown on BF at E and on AD at C. We may not recognize the actual sense of the components at B at first glance, so they may be arbitrarily but consistently assigned.

The solution may proceed by use of a moment equation about B or E for member BF, followed by the two force equations. Thus,

$$[\Sigma M_B = 0] \qquad 3.92(5) - \tfrac{1}{2}E_x(3) = 0 \qquad E_x = 13.08 \text{ kN} \qquad \textit{Ans.}$$

$$[\Sigma F_y = 0] \qquad B_y + 3.92 - 13.08/2 = 0 \qquad B_y = 2.62 \text{ kN} \qquad \textit{Ans.}$$

$$[\Sigma F_x = 0] \qquad B_x + 3.92 - 13.08 = 0 \qquad B_x = 9.15 \text{ kN} \qquad \textit{Ans.}$$

Positive numerical values of the unknowns mean that we assumed their directions correctly on the free-body diagrams. The value of $C_x = E_x = 13.08$ kN obtained by inspection of the free-body diagram of CE is now entered onto the diagram for AD, along with the values of B_x and B_y just determined. The equations of equilibrium may now be applied to member AD as a check, since all the forces acting on it have already been computed. The equations give

$$[\Sigma M_C = 0] \qquad 4.32(3.5) + 4.32(1.5) - 3.92(2) - 9.15(1.5) = 0$$

$$[\Sigma F_x = 0] \qquad 4.32 - 13.08 + 9.15 + 3.92 + 4.32 = 0$$

$$[\Sigma F_y = 0] \qquad -13.08/2 + 2.62 + 3.92 = 0$$

Helpful Hints

① We see the frame corresponds to the category illustrated in Fig. 4/14a.

② Without this observation, the problem solution would be much longer, because the three equilibrium equations for member BF would contain four unknowns: B_x, B_y, E_x, and E_y. Note that the direction of the line joining the two points of force application, and not the shape of the member, determines the direction of the forces acting on a two-force member.

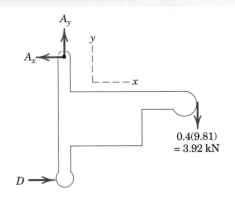

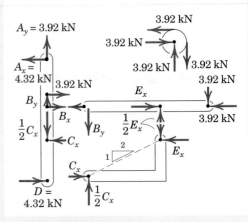

Sample Problem 4/7

Neglect the weight of the frame and compute the forces acting on all of its members.

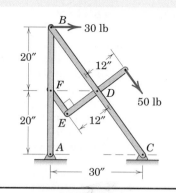

Solution. ① We note first that the frame is not a rigid unit when removed from its supports since *BDEF* is a movable quadrilateral and not a rigid triangle. Consequently the external reactions cannot be completely determined until the individual members are analyzed. However, we can determine the vertical components of the reactions at *A* and *C* from the free-body diagram of the frame ② as a whole. Thus,

$$[\Sigma M_C = 0] \qquad 50(12) + 30(40) - 30A_y = 0 \qquad A_y = 60 \text{ lb} \qquad \textit{Ans.}$$

$$[\Sigma F_y = 0] \qquad C_y - 50(4/5) - 60 = 0 \qquad C_y = 100 \text{ lb} \qquad \textit{Ans.}$$

Next we dismember the frame and draw the free-body diagram of each part. Since *EF* is a two-force member, the direction of the force at *E* on *ED* and at *F* on *AB* is known. We assume that the 30-lb force is applied to the pin as a part of ③ member *BC*. There should be no difficulty in assigning the correct directions for forces *E*, *F*, *D*, and B_x. The direction of B_y, however, may not be assigned by inspection and therefore is arbitrarily shown as downward on *AB* and upward on *BC*.

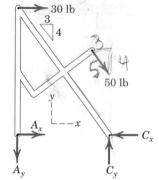

Member ED. The two unknowns are easily obtained by

$$[\Sigma M_D = 0] \qquad 50(12) - 12E = 0 \qquad E = 50 \text{ lb} \qquad \textit{Ans.}$$

$$[\Sigma F = 0] \qquad D - 50 - 50 = 0 \qquad D = 100 \text{ lb} \qquad \textit{Ans.}$$

Member EF. Clearly *F* is equal and opposite to *E* with the magnitude of 50 lb.

Member AB. Since *F* is now known, we solve for B_x, A_x, and B_y from

$$[\Sigma M_A = 0] \qquad 50(3/5)(20) - B_x(40) = 0 \qquad B_x = 15 \text{ lb} \qquad \textit{Ans.}$$

$$[\Sigma F_x = 0] \qquad A_x + 15 - 50(3/5) = 0 \qquad A_x = 15 \text{ lb} \qquad \textit{Ans.}$$

$$[\Sigma F_y = 0] \qquad 50(4/5) - 60 - B_y = 0 \qquad B_y = -20 \text{ lb} \qquad \textit{Ans.}$$

The minus sign shows that we assigned B_y in the wrong direction.

Member BC. ④ The results for B_x, B_y, and *D* are now transferred to *BC*, and the remaining unknown C_x is found from

$$[\Sigma F_x = 0] \qquad 30 + 100(3/5) - 15 - C_x = 0 \qquad C_x = 75 \text{ lb} \qquad \textit{Ans.}$$

We may apply the remaining two equilibrium equations as a check. Thus,

$$[\Sigma F_y = 0] \qquad 100 + (-20) - 100(4/5) = 0$$

$$[\Sigma M_C = 0] \qquad (30 - 15)(40) + (-20)(30) = 0$$

Helpful Hints

① We see that this frame corresponds to the category illustrated in Fig. 4/14*b*.

② The directions of A_x and C_x are not obvious initially and can be assigned arbitrarily to be corrected later if necessary.

③ Alternatively the 30-lb force could be applied to the pin considered a part of *BA*, with a resulting change in the reaction B_x.

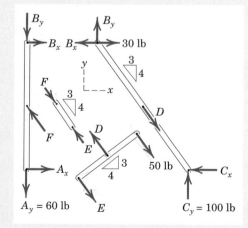

④ Alternatively we could have returned to the free-body diagram of the frame as a whole and found C_x.

Sample Problem 4/8

The machine shown is designed as an overload protection device which re-leases the load when it exceeds a predetermined value T. A soft metal shear pin S is inserted in a hole in the lower half and is acted on by the upper half. When the total force on the pin exceeds its strength, it will break. The two halves then rotate about A under the action of the tensions in BD and CD, as shown in the second sketch, and rollers E and F release the eye bolt. Determine the maximum allowable tension T if the pin S will shear when the total force on it is 800 N. Also compute the corresponding force on the hinge pin A.

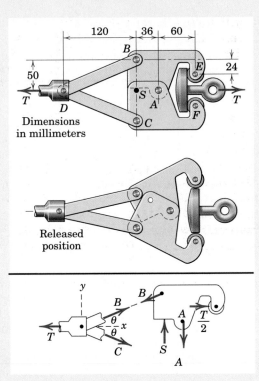

Dimensions in millimeters

Released position

Solution. Because of symmetry we analyze only one of the two hinged mem-bers. The upper part is chosen, and its free-body diagram along with that for the connection at D is drawn. Because of symmetry the forces at S and A have no ① x-components. The two-force members BD and CD exert forces of equal magni-tude $B = C$ on the connection at D. Equilibrium of the connection gives

$[\Sigma F_x = 0]$ $\qquad$ $B \cos \theta + C \cos \theta - T = 0$ $\qquad$ $2B \cos \theta = T$

$$B = T/(2 \cos \theta)$$

From the free-body diagram of the upper part we express the equilibrium of moments about point A. Substituting $S = 800$ N and the expression for B gives

② $[\Sigma M_A = 0]$ $\qquad$ $\dfrac{T}{2 \cos \theta} (\cos \theta)(50) + \dfrac{T}{2 \cos \theta} (\sin \theta)(36) - 36(800) - \dfrac{T}{2} (26) = 0$

Substituting $\sin \theta / \cos \theta = \tan \theta = 5/12$ and solving for T give

$$T\left(25 + \frac{5(36)}{2(12)} - 13\right) = 28\ 800$$

$$T = 1477 \text{ N} \qquad \text{or} \qquad T = 1.477 \text{ kN} \qquad\qquad \textit{Ans.}$$

Finally, equilibrium in the y-direction gives us

$[\Sigma F_y = 0]$ $\qquad$ $S - B \sin \theta - A = 0$

$$800 - \frac{1477}{2(12/13)} \frac{5}{13} - A = 0 \qquad A = 492 \text{ N} \qquad\qquad \textit{Ans.}$$

Helpful Hints

① It is always useful to recognize sym-metry. Here it tells us that the forces acting on the two parts behave as mirror images of each other with re-spect to the x-axis. Thus, we cannot have an action on one member in the plus x-direction and its reaction on the other member in the negative x-direction. Consequently the forces at S and A have no x-components.

② Be careful not to forget the moment of the y component of B. Note that our units here are newton-millimeters.

Sample Problem 4/9

In the particular position shown, the excavator applies a 20-kN force parallel to the ground. There are two hydraulic cylinders AC to control the arm OAB and a single cylinder DE to control arm $EBIF$. (*a*) Determine the force in the hydraulic cylinders AC and the pressure p_{AC} against their pistons, which have an effective diameter of 95 mm. (*b*) Also determine the force in hydraulic cylinder DE and the pressure p_{DE} against its 105-mm-diameter piston. Neglect the weights of the members compared with the effects of the 20-kN force.

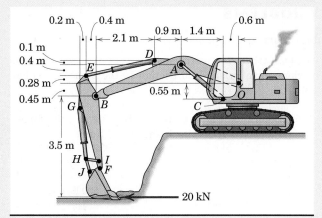

Solution.   (*a*) We begin by constructing a free-body diagram of the entire arm assembly. Note that we include only the dimensions necessary for this portion of the problem—details of the cylinders DE and GH are unnecessary at this time.

$[\Sigma M_O = 0]$ $-20\ 000(3.95) - 2F_{AC} \cos 41.3°(0.68) + 2F_{AC} \sin 41.3°(2) = 0$

$$F_{AC} = 48\ 800 \text{ N or } 48.8 \text{ kN} \qquad \qquad Ans.$$

① From $F_{AC} = p_{AC}A_{AC}$, $p_{AC} = \dfrac{F_{AC}}{A_{AC}} = \dfrac{48\ 800}{\left(\pi \dfrac{0.095^2}{4}\right)} = 6.89(10^6)$ Pa or 6.89 MPa   *Ans.*

(*b*) For cylinder DF, we "cut" the assembly at a location which makes the desired cylinder force external to our free-body diagram. This means isolating the vertical arm $EBIF$ along with the bucket and its applied force.

$[\Sigma M_B = 0]$ $-20\ 000(3.5) + F_{DE} \cos 11.31°(0.73) + F_{DE} \sin 11.31°(0.4) = 0$

$$F_{DE} = 88\ 100 \text{ N or } 88.1 \text{ kN} \qquad \qquad Ans.$$

$$p_{DE} = \frac{F_{DE}}{A_{DE}} = \frac{88\ 100}{\left(\pi \dfrac{0.105^2}{4}\right)} = 10.18(10^6) \text{ Pa or } 10.18 \text{ MPa} \qquad Ans.$$

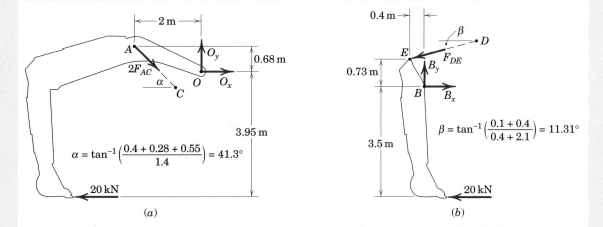

(*a*)

(*b*)

PROBLEMS

Introductory Problems

(Unless otherwise instructed, neglect the mass of the various members and all friction in the problems which follow.)

4/67 Determine the force in member *BD* and the forces supported by all pins in the loaded frame.

Ans. BD = 1932 lb, *A* = 1617 lb
B = *D* = 1932 lb

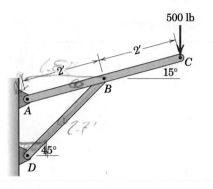

Problem 4/67

4/68 Determine the force supported by each pin of the loaded truss.

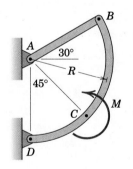

Problem 4/68

4/69 For an 80-N squeeze on the handles of the pliers, determine the force *F* applied to the round rod by each jaw. In addition, calculate the force supported by the pin at *A*.

Ans. F = 217 N, *A* = 297 N

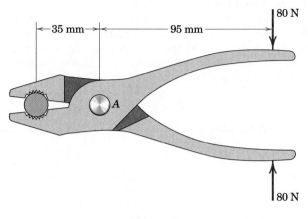

Problem 4/69

4/70 Compute the force supported by the pin at *A* for the slip-joint pliers under a grip of 30 lb.

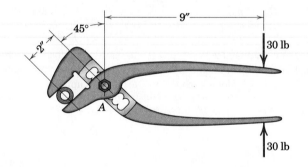

Problem 4/70

4/71 Determine the force supported by the roller at *E*.

Ans. E = 150 N

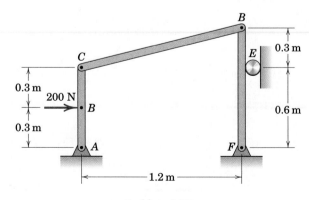

Problem 4/71

4/72 Calculate the magnitude of the force acting on the pin at D. Pin C is fixed in DE and bears against the smooth slot in the triangular plate.

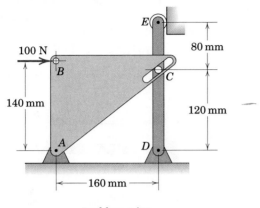

100 N

E

80 mm

B

140 mm

C

120 mm

A

D

← 160 mm →

Problem 4/72

4/73 Determine the reaction at the roller F for the frame loaded as shown.

Ans. F = 800 N

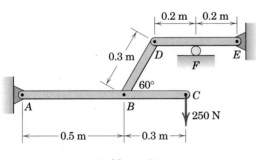

0.2 m | 0.2 m

0.3 m | D

E

F

60°

A

B

C

↓ 250 N

← 0.5 m → ← 0.3 m →

Problem 4/73

4/74 Given the values of the load L and dimension R, for what value of the couple M will the force in link CH be zero?

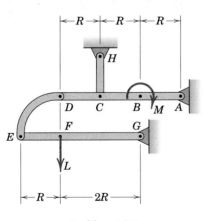

←R→←R→←R→

H

D   C   B   M   A

F   G

E

↓L

←R→← 2R →

Problem 4/74

4/75 The handheld press is useful for such tasks as squeezing rivets or punching holes. What force P is applied to the sheet metal at E for the 60-N forces applied to the handles?

Ans. P = 2050 N

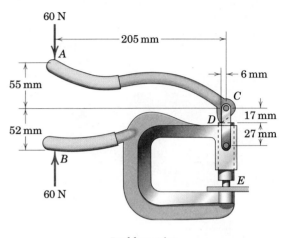

60 N

A

← 205 mm →

55 mm

6 mm

C

52 mm

D

17 mm

27 mm

B

60 N

E

Problem 4/75

4/76 The device shown is used to straighten bowed decking boards just prior to final nailing to the joists. There is a lower bracket (not shown) at O which fixes the part OA to a joist, so that the pivot A may be considered fixed. For a given force P exerted perpendicular to the handle ABC as shown, determine the corresponding normal force N applied to the bent board near point B. Neglect friction.

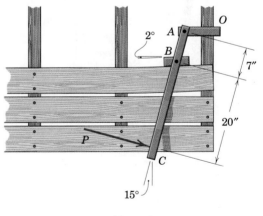

O

2°   A

B

7″

20″

P

C

15°

Problem 4/76

4/77 The "jaws-of-life" device is utilized by rescuers to pry apart wreckage, thus helping to free accident victims. If a pressure of 500 lb/in.^{2} is developed behind the piston P of area 20 in.^{2}, determine the vertical force R which is exerted by the jaw tips on the wreckage for the position shown. Note that link AB and its counterpart are both horizontal in the figure for this position.

Ans. R = 1111 lb

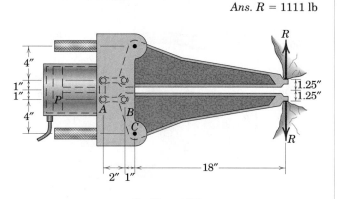

Problem 4/77

4/78 Determine the moment M which must be applied at A to keep the frame in static equilibrium in the position shown. Also calculate the magnitude of the pin reaction at A.

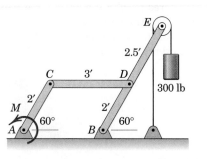

Problem 4/78

Representative Problems

4/79 The wingnut B of the collapsible bucksaw is tightened until the tension in rod AB is 200 N. Determine the force in the saw blade EF and the magnitude F of the force supported by pin C.

Ans. EF = 100 N T, F = 300 N

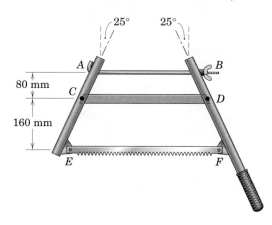

Problem 4/79

4/80 The elements of a floor jack are shown in the figure. The figure $CDFE$ is a parallelogram. Calculate the force in the hydraulic cylinder AB corresponding to the 2000-lb load supported as shown. What is the force in link EF?

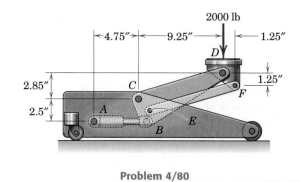

Problem 4/80

4/81 The device shown is used for lifting 55-gallon drums. Determine the magnitude of the force exerted at *B*.

Ans. B = 1.855 kN

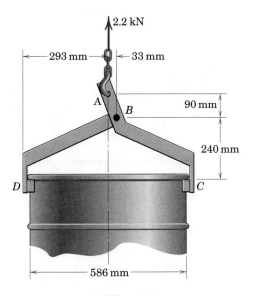

Problem 4/81

4/82 Calculate the *x*- and *y*-components of the force **C** which member *BC* exerts on member *ACD*. The cables are wrapped securely around the two pulleys, which are fastened together.

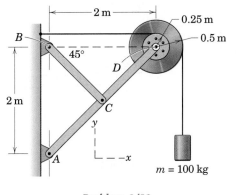

Problem 4/82

4/83 Determine the magnitude of the pin reaction at *A* and the magnitude and direction of the force reaction at the rollers. The pulleys at *C* and *D* are small.

Ans. A = 999 N, *F* = 314 N up

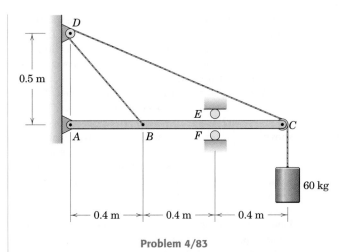

Problem 4/83

4/84 If a force *F* = 15 N is required to release the spring-loaded clamps, what are the normal reactions at *A* and *B* if *F* = 0?

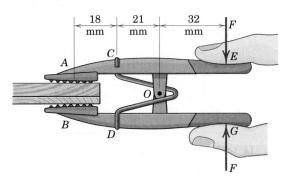

Problem 4/84

4/85 The figure illustrates a common problem associated with simple structures. Under the loadings L, the rafters can rotate, the ridge beam at A can lower, and the walls BC and DE can rotate outward, as shown in part b of the figure. This phenomenon is sometimes clearly observed in old wooden farm structures as a central sagging of the ridge beam when viewed from the side. A simple remedy is shown in part a of the figure. A chain or cable is stretched between fasteners at B and E, and the turnbuckle F is tightened until a proper tension is achieved, thereby preventing the outward tilting of the walls. For given values of the dimension d and the point loads L (which result from the distributed loads of the rafter and roofing weights and any additional loads such as snow), calculate the tension T required so that there are no outward forces on the walls at B and E. Assume that the support of the rafters at the ridge beam is purely horizontal and that all joints are free to rotate.

Ans. $T = \frac{24}{35}L$

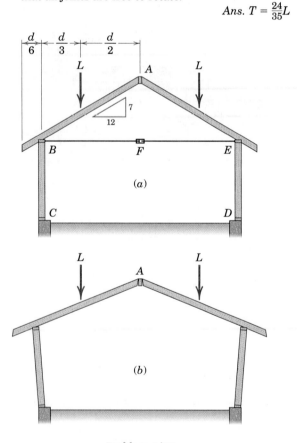

Problem 4/85

4/86 The ramp is used as passengers board a small commuter airplane. The total mass of the ramp and six passengers is 750 kg with mass center at G. Determine the force in the hydraulic cylinder AB and the magnitude of the pin reaction at C.

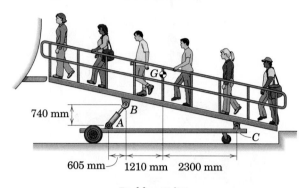

Problem 4/86

4/87 A small bolt cutter operated by hand for cutting small bolts and rods is shown in the sketch. For a hand grip $P = 150$ N, determine the force Q developed by each jaw on the rod to be cut.

Ans. $Q = 2.7$ kN

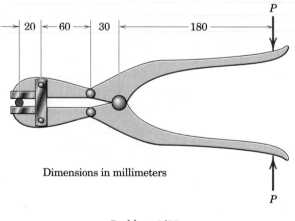

Dimensions in millimeters

Problem 4/87

4/88 The clamp shown in the figure is frequently used in welding operations. Determine the clamping force on the two metal pieces at E and the magnitudes of the forces supported by pins A, B, and D.

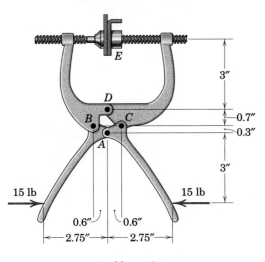

Problem 4/88

4/89 The clamp is adjusted so that it exerts a pair of 200-N compressive forces on the boards between its swivel grips. Determine the force in the threaded shaft BC and the magnitude of the pin reaction at D.
Ans. BC = 375 N C, D = 425 N

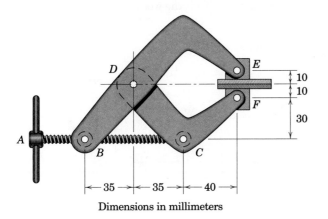

Dimensions in millimeters

Problem 4/89

4/90 When the crank AB is vertical, the beam CD is horizontal and the cable makes a 20° angle with the horizontal. Compute the moment M required for equilibrium of the frame.

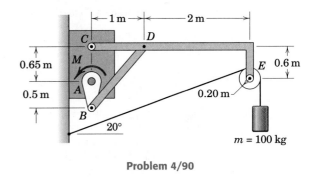

Problem 4/90

4/91 An 18-lb force is applied to the handle OAB of the cork puller. Determine the extraction force F exerted on the cork.
Ans. F = 54.2 lb

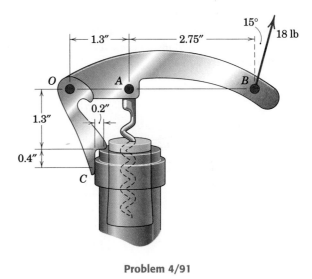

Problem 4/91

4/92 The illustration represents the framework for a storage-shelf unit, with the intermediate shelves not shown. The variable x represents different bracing configurations. Determine and plot, over the range $0 \le \dfrac{x}{h} \le 1$, the force in the brace EF corresponding to the applied force P. What happens if $x = 0$?

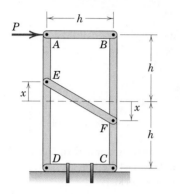

Problem 4/92

4/93 The simple crane supports the 6000-lb load. Determine the tension T in the cable and the magnitude of the pin reaction at O.

Ans. $T = 23{,}300$ lb, $O = 23{,}100$ lb

Problem 4/93

4/94 The dual-grip clamp shown in the figure is used to provide added clamping force with a positive action. If the vertical screw is tightened to produce a clamping force of 3 kN and then the horizontal screw is tightened until the force in the screw at A is doubled, find the total reaction R on the pin at B.

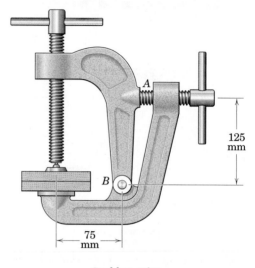

Problem 4/94

4/95 Determine the vertical clamping force at E in terms of the force P applied to the handle of the toggle clamp, which holds the workpiece F in place.

Ans. $E = 7.2P$

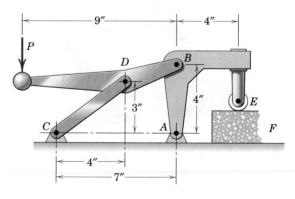

Problem 4/95

4/96 Compute the force in link *AB* of the lifting tongs which cross without touching.

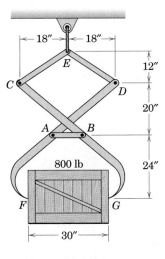

Problem 4/96

4/97 The elements of a spring-loaded mechanism for a car-trunk lid are shown in the figure. For the equilibrium position shown, determine the moment M_A which each of two torsional springs at *A* must exert on link *AB*. The mass of the trunk lid is 18 kg with mass center at *G*. The effects of the weights of the other members may be neglected.

Ans. $M_A = 9.94$ N·m CCW

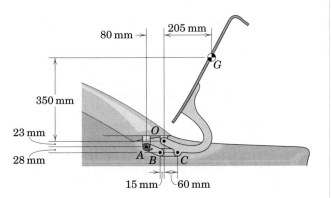

Problem 4/97

4/98 The machine shown is used for moving heavy items such as pallets of bricks around construction sites. For the horizontal boom position shown, determine the force in each of the two hydraulic cylinders *AB*. The mass of the boom is 1500 kg with mass center at G_1, and the mass of the cube of bricks is 2000 kg with mass center at G_2.

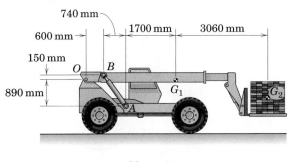

Problem 4/98

4/99 The forklift area of the machine of Prob. 4/98 is shown with additional dimensional detail. Determine the force in the single hydraulic cylinder *CD*. The mass of the cube of bricks is 2000 kg with mass center at G_2. You may neglect the effects of the mass of the forklift components.

Ans. $F_{CD} = 127.8$ kN

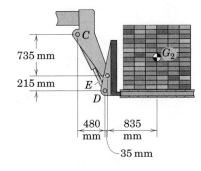

Problem 4/99

4/100 Determine the vertical clamping force at *E* in terms of the force *P* applied to the handle of the toggle clamp.

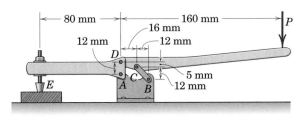

Problem 4/100

4/101 The device shown is used to drag loaded wooden pallets across warehouse floors. The wood board shown is one of several members that comprise the base of the pallet. For the 900-lb force applied by a forklift, determine the magnitude of the force supported by pin C and the normal gripping forces at A and B.

Ans. $C = 1229$ lb, $A_n = B_n = 692$ lb

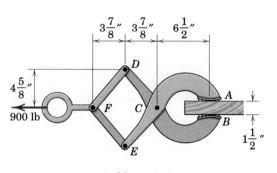

Problem 4/101

4/102 A modification of the pallet puller of Prob. 4/101 is shown here. For the same net 900-lb force as in Prob. 4/101, determine the magnitude of the force supported by pin C and the normal gripping forces at A and B.

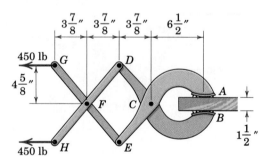

Problem 4/102

4/103 The figure shows a wheel puller which is designed to remove a V-belt pulley P from its tight-fitting shaft S by tightening of the central screw. If the pulley starts to slide off the shaft when the compression in the screw has reached 1.2 kN, calculate the magnitude of the force supported by each jaw at A. The adjusting screws D support horizontal force and keep the side arms parallel with the central screw.

Ans. $A = 0.626$ kN

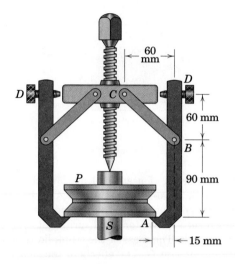

Problem 4/103

4/104 The elements of a front-hinged automobile-hood assembly are shown in the figure. The light linkages BC and CD and the gas-pressurized strut EF hold the hood in the open position shown. In this position, the hood is free to rotate clockwise about pin O; pin A is locked until the hood has been lowered to a nearly closed horizontal position. For a hood weight of 80 lb with center of gravity at G, determine the minimum compression force C in the strut which will maintain the open-hood position. Note that there are two links OA spaced across the front of the car, but only one set of the remaining links, located on the inside of the right-front fender.

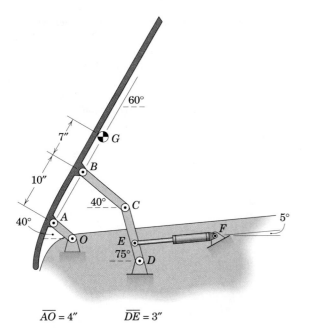

$\overline{AO} = 4''$ $\qquad \overline{DE} = 3''$

$\overline{BC} = \overline{CD} = 9''$

Problem 4/104

4/105 The elements of a rear suspension for a front-wheel-drive car are shown in the figure. Determine the magnitude of the force at each joint if the normal force F exerted on the tire has a magnitude of 3600 N.

Ans. $A = 4550$ N, $B = 4410$ N
$C = D = 1898$ N, $E = F = 5920$ N

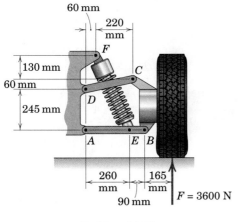

Problem 4/105

4/106 A 250-N force is applied to the foot-operated air pump. The return spring S exerts a 3-N·m moment on member OBA for this position. Determine the corresponding compression force C in the cylinder BD. If the diameter of the piston in the cylinder is 45 mm, estimate the air pressure generated for these conditions. State any assumptions.

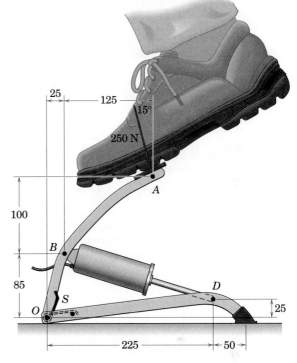

Dimensions in millimeters

Problem 4/106

4/107 The truck shown is used to deliver food to aircraft. The elevated unit weighs 2000 lb with center of gravity at G. Determine the required force in the hydraulic cylinder AB.

Ans. $F_{AB} = 6710$ lb

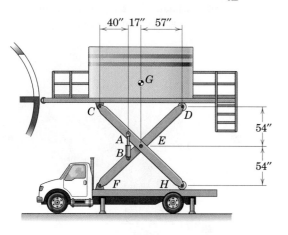

Problem 4/107

4/108 A lifting device for transporting 135-kg steel drums is shown. Calculate the magnitude of the force exerted on the drum at E and F.

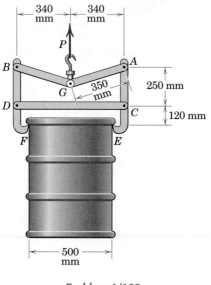

Problem 4/108

4/109 Determine the compression force C exerted on the can for an applied force $P = 50$ N when the can crusher is in the position shown. Note that there are two links AB and two links AOD, with one pair of linkages on each side of the stationary portion of the crusher. Also, pin B is on the vertical centerline of the can. Finally, note that small square projections E of the moving jaw move in recessed slots of the fixed frame.

Ans. $C = 249$ N

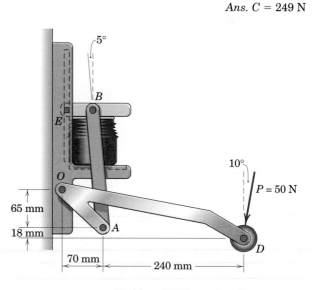

Problem 4/109

4/110 Determine the compression force C exerted on the can for an applied force $P = 50$ N when the can crusher is in the position shown. Point B is centered on the bottom of the can.

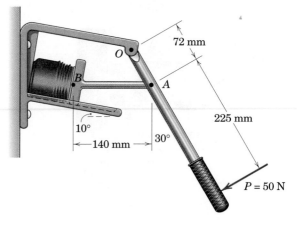

Problem 4/110

4/111 Determine the force in cylinder *AB* due to the combined weight of the bucket and operator. The combined mass is 180 kg with mass center at *G*.

Ans. $F_{AB} = 9200$ N

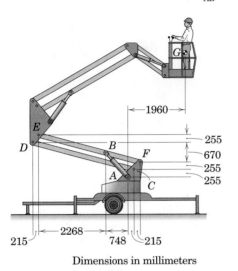

Dimensions in millimeters

Problem 4/111

4/112 Additional detail in the bucket area of the cherry-picker of Prob. 4/111 is given in the figure. Solve for the force in cylinder *HJ* due to the effects of the 180-kg combined mass of the bucket and operator, with mass center at *G*.

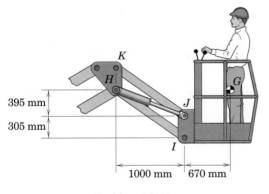

Problem 4/112

4/113 The car hoist allows the car to be driven onto the platform, after which the rear wheels are raised. If the loading from both rear wheels is 1500 lb, determine the force in the hydraulic cylinder *AB*. Neglect the weight of the platform itself. Member *BCD* is a right-angle bell crank pinned to the ramp at *C*.

Ans. AB = 3970 lb C

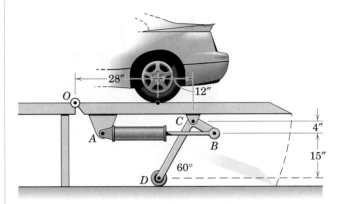

Problem 4/113

4/114 The aircraft landing gear consists of a spring- and hydraulically-loaded piston and cylinder *D* and the two pivoted links *OB* and *CB*. If the gear is moving along the runway at a constant speed with the wheel supporting a stabilized constant load of 24 kN, calculate the total force which the pin at *A* supports.

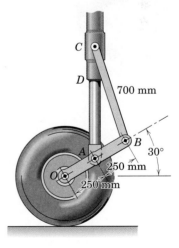

Problem 4/114

4/115 Determine the force in the hydraulic cylinder AB and the magnitude of the pin reaction at O for the position shown. The bucket and its load have a combined weight of 4000 lb with center of gravity at G. You may neglect the effect of the weights of the other members.

$Ans.\ F_{AB} = 17{,}140$ lb, $O = 16{,}590$ lb

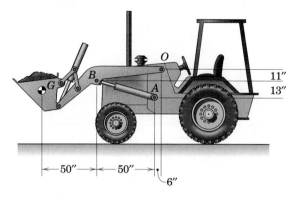

Problem 4/115

4/116 Consider the additional dimensional detail for the front-end loader of Prob. 4/115. Determine the force in the hydraulic cylinder CE. The weight of the bucket and its load is 4000 lb with center of gravity at G. You may ignore the effects of the weights of other members.

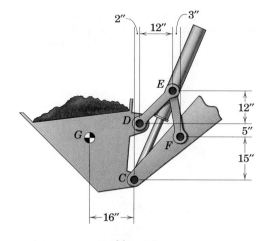

Problem 4/116

4/117 An adjustable tow bar connecting the tractor unit H with the landing gear J of a large aircraft is shown in the figure. Adjusting the height of the hook F at the end of the tow bar is accomplished by the hydraulic cylinder CD activated by a small hand pump (not shown). For the nominal position shown of the triangular linkage ABC, calculate the force P supplied by the cylinder to the pin C to position the tow bar. The rig has a total weight of 100 lb and is supported by the tractor hitch at E.

$Ans.\ P = 60.8$ lb

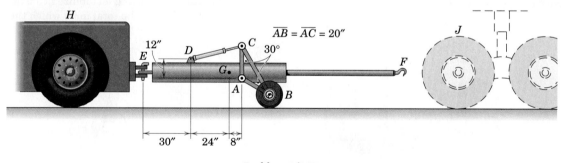

Problem 4/117

4/118 An automatic window-positioning system for a greenhouse is shown in the figure. Window *OA* has a mass of 50 kg with mass center at *G*. A motor (not shown) drives the gear *C*, which in turn moves the positioning rack *AB*. For the window position shown, determine the motor torque *M* which must be supplied to the pinion *C*. Neglect the weight of bar *AB*.

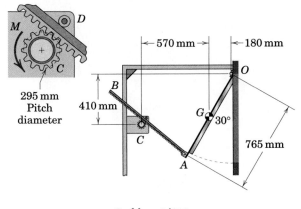

Problem 4/118

4/119 The pruning mechanism of a pole saw is shown as it cuts a branch *S*. For the particular position drawn, the actuating cord is parallel to the pole and carries a tension of 30 lb. Determine the shearing force *P* applied to the branch by the cutter and the total force supported by the pin at *E*. The force exerted by the light return spring at *C* is small and may be neglected.

Ans. P = 338 lb, *E* = 75.1 lb

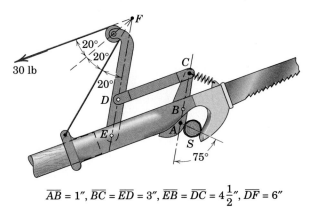

$$\overline{AB} = 1'', \overline{BC} = \overline{ED} = 3'', \overline{EB} = \overline{DC} = 4\tfrac{1}{2}'', \overline{DF} = 6''$$

Problem 4/119

4/120 The figure shows a special rig designed to erect vertical sections of a construction tower. The assembly *A* has a mass of 1.5 Mg and is elevated by the platform *B*, which itself has a mass of 2 Mg. The platform is guided up the fixed vertical column by rollers and is activated by the hydraulic cylinder *CD* and links *EDF* and *FH*. For the particular position shown, calculate the force *R* exerted by the hydraulic cylinder at *D* and the magnitude of the force supported by the pin at *E*.

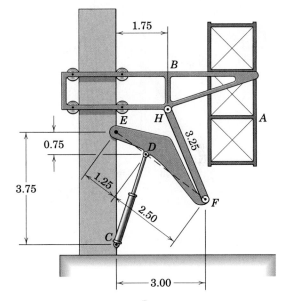

Dimensions in meters

Problem 4/120

4/121 Determine the force acting on member *ABC* at connection *A* for the loaded space frame shown. Each connection may be treated as a ball-and-socket joint.

Ans. A = 4.25 kN

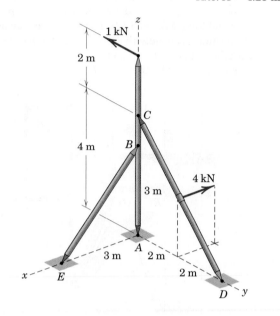

Problem 4/121

4/122 In the schematic representation of an actual structure, *T* represents a turnbuckle, *C* and *D* are non-thrust-bearing hinges whose axes are along the line *CD*, and *B*, *E*, and *F* are ball-and-socket joints. Determine the tension *T* in the turnbuckle and the force in member *EF*.

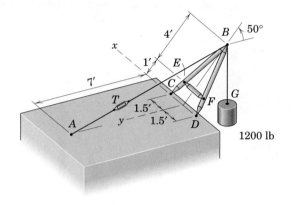

Problem 4/122

4/7   CHAPTER REVIEW

In Chapter 4 we have applied the principles of equilibrium to two classes of problems: (*a*) simple trusses and (*b*) frames and machines. No new theory was needed, since we merely drew the necessary free-body diagrams and applied our familiar equations of equilibrium. The structures dealt with in Chapter 4, however, have given us the opportunity to further develop our appreciation for a systematic approach to mechanics problems.

The most essential features of the analysis of these two classes of structures are reviewed in the following statements.

(a) Simple Trusses

1. Simple trusses are composed of two-force members joined at their ends and capable of supporting tension or compression. Each internal force, therefore, is always in the direction of the line joining the endpoints of its member.

2. Simple trusses are built from the basic rigid (noncollapsible) unit of the triangle for plane trusses and the tetrahedron for space trusses. Additional units of a truss are formed by adding new members, two for plane trusses and three for space trusses, attached to existing joints and joined at their ends to form a new joint.

3. The joints of simple trusses are assumed to be pin connections for plane trusses and ball-and-socket connections for space trusses. Thus, the joints can transmit force but not moment.

4. External loads are assumed to be applied only at the joints.

5. Trusses are statically determinate externally when the external constraints are not in excess of those required to maintain an equilibrium position.

6. Trusses are statically determinate internally when constructed in the manner described in item (2), where internal members are not in excess of those required to prevent collapse.

7. The *method of joints* utilizes the force equations of equilibrium for each joint. Analysis normally begins at a joint where at least one force is known and not more than two forces are unknown for plane trusses or not more than three forces are unknown for space trusses.

8. The *method of sections* utilizes a free body of an entire section of a truss containing two or more joints. In general, the method involves the equilibrium of a nonconcurrent system of forces. The moment equation of equilibrium is especially useful when the method of sections is used. In general, the forces acting on a section which cuts more than three unknown members of a plane truss cannot be solved for completely because there are only three independent equations of equilibrium.

9. The vector representing a force acting on a joint or a section is drawn on the same side of the joint or section as the member which transmits the force. With this convention, tension is indicated when the force arrow is away from the joint or section, and compression is indicated when the arrow points toward the joint or section.

10. When the two diagonal members which brace a quadrilateral panel are flexible members incapable of supporting compression, only the one in tension is retained in the analysis, and the panel remains statically determinate.

11. When two joined members under load are collinear and a third member with a different direction is joined with their connection, the force in the third member must be zero unless an external force is applied at the joint with a component normal to the collinear members.

(b) Frames and Machines

1. Frames and machines are structures which contain one or more multiforce members. A multiforce member is one which has acting on it three or more forces, or two or more forces and one or more couples.

2. Frames are structures designed to support loads, generally under static conditions. Machines are structures which transform input forces and moments to output forces and moments and generally involve moving parts. Some structures may be classified as either a frame or a machine.

3. Only frames and machines which are statically determinate externally and internally are considered here.

4. If a frame or machine as a whole is a rigid (noncollapsible) unit when its external supports are removed, then we begin the analysis by computing the external reactions on the entire unit. If a frame or machine as a whole is a nonrigid (collapsible) unit when its external supports are removed, then the analysis of the external reactions cannot be completed until the structure is dismembered.

5. Forces acting in the internal connections of frames and machines are calculated by dismembering the structure and constructing a separate free-body diagram of each part. The principle of action and reaction must be *strictly* observed; otherwise, error will result.

6. The force and moment equations of equilibrium are applied to the members as needed to compute the desired unknowns.

REVIEW PROBLEMS

4/123 The support apparatus for a power line for a light-rapid-transit train is shown. If a tension of 500 lb exists in cable *EDF*, determine the force *F* supported by the bolt at *B*.

Ans. F = 1478 lb

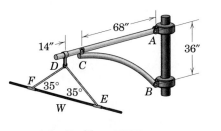

Problem 4/123

4/124 Determine the force in each member of the loaded truss.

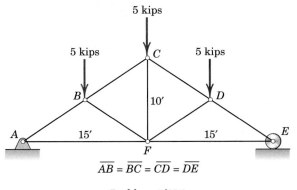

$$\overline{AB} = \overline{BC} = \overline{CD} = \overline{DE}$$

Problem 4/124

4/125 The specialty tool is used for installing and removing snap rings. Determine the spreading force applied at *G* and *H* if *P* = 50 N.

Ans. G = 181.8 N

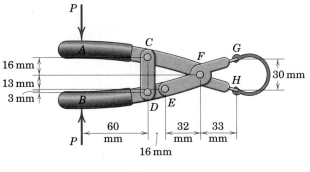

Problem 4/125

4/126 Determine the forces in members *CH* and *CF*.

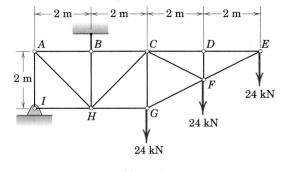

Problem 4/126

4/127 Calculate the force in member *BG* using a free-body diagram of the rigid member *ABC*.

Ans. BG = 1800 lb *C*

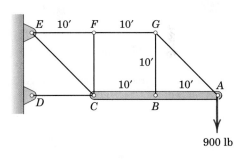

Problem 4/127

4/128 Show that the force in the horizontal member *BD* is independent of its position *x* within the triangular truss. Each side of the overall truss supports a centered vertical load *L* as shown by the two dashed vectors, and their loads are then distributed to the joints as shown by the solid vectors.

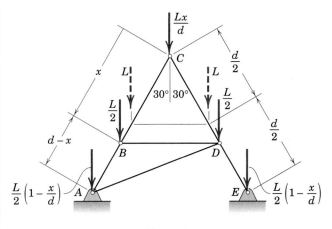

Problem 4/128

4/129 The basic structural shape and loading of Prob. 4/128 is now treated as the loaded frame shown in the figure. Determine the force in the horizontal member BD as a function of its position x within the frame.

$$\text{Ans. } BD = \frac{0.289Ld}{x}$$

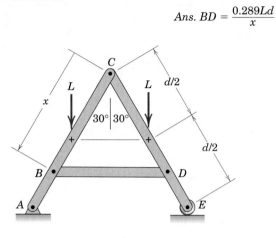

Problem 4/129

4/130 Determine the forces in members CH, AH, and CD of the loaded truss.

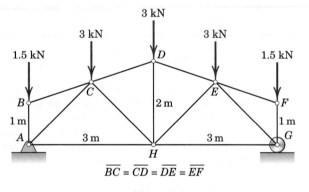

$$\overline{BC} = \overline{CD} = \overline{DE} = \overline{EF}$$

Problem 4/130

4/131 The nose-wheel assembly is raised by the application of a torque M to link BC through the shaft at B. If the arm and wheel AO have a combined weight of 100 lb with center of gravity at G, find the value of M necessary to lift the wheel when D is directly under B, at which position angle θ is 30°.

$$\text{Ans. } M = 1250 \text{ lb-in.}$$

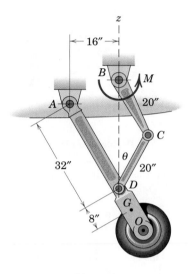

Problem 4/131

4/132 Determine the force in member BF of the loaded truss.

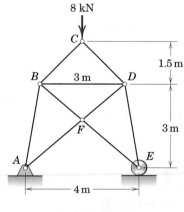

Problem 4/132

4/133 Determine the forces in members AB, BI, and CI of the simple truss. Note that all curved members are two-force members.

Ans. AB = 2.26L T, BI = L T, CI = 0.458L T

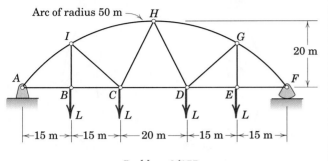

Arc of radius 50 m

Problem 4/133

4/134 The structure of Prob. 4/133 is modified in that the four curved members are replaced by the two members AIH and HGF. Instrumentation indicates the tension in members CH and DH to be $0.5L$ each. Determine the forces in members AB, BI, and CI. Is the problem solvable without the information about CH?

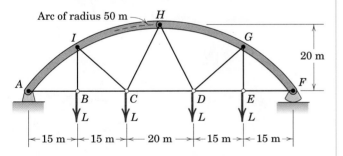

Arc of radius 50 m

Problem 4/134

4/135 An antitorque wrench is designed for use by a crewman of a spacecraft where he has no stable platform against which to push as he tightens a bolt. The pin A fits into an adjacent hole in the structure which contains the bolt to be turned. Successive oscillations of the gear and handle unit turn the socket in one direction through the action of a ratchet mechanism. The reaction against the pin A provides the "antitorque" characteristic of the tool. For a gripping force $P = 150$ N, determine the torque M transmitted to the bolt and the external reaction R against the pin A normal to the line AB. (One side of the tool is used for tightening and the opposite side for loosening a bolt.)

Ans. M = 7.88 N·m, R = 137.0 N

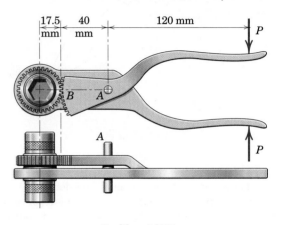

Problem 4/135

4/136 Determine the forces in members DM and DN of the loaded symmetrical truss.

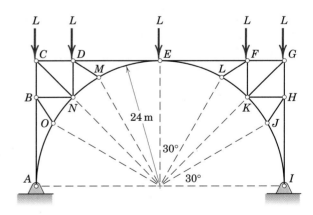

Problem 4/136

4/137 The depicted structure is under consideration as the upper portion of a transmission-line tower and is supported at points F, G, H, and I. Point C is directly above the center of rectangle $FGHI$. Determine the force in member CD.

Ans. CD = 2.4L T

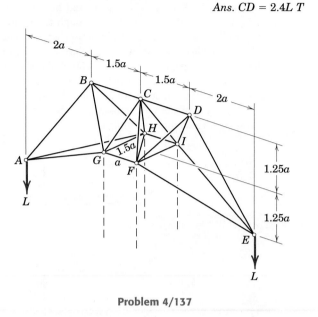

Problem 4/137

4/138 The structure shown is being considered as part of a large cylindrical vessel which must support external loads. Strain-gage instrumentation indicates that the compressive force in member BE is $0.8L$. Determine the forces in members AB and DE. Make use of symmetry.

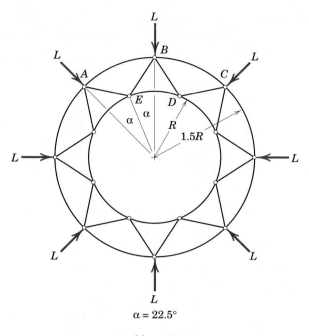

$\alpha = 22.5°$

Problem 4/138

4/139 Determine the forces in members AD and DG.

Ans. AD = 0.625L C, DG = 2.5L C

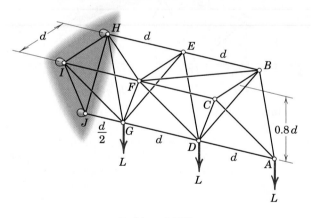

Problem 4/139

▶**4/140** Each of the landing struts for a planet exploration spacecraft is designed as a space truss symmetrical about the vertical x-z plane as shown. For a landing force $F = 2.2$ kN, calculate the corresponding force in member BE. The assumption of static equilibrium for the truss is permissible if the mass of the truss is very small. Assume equal loads in the symmetrically placed members.

Ans. $F_{BE} = 1.620$ kN

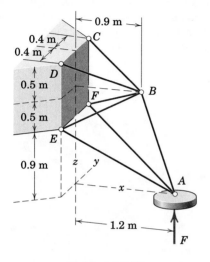

Problem 4/140

Computer-Oriented Problems*

***4/141** For a given force P on the handle of the toggle clamp the clamping force C increases to very large values as the angle θ decreases. For $P = 30$ lb determine the relationship between C and θ and plot it as a function of θ from $\theta = 2°$ to $\theta = 30°$. Assume that the shaft slides freely in its guide.

Ans. $C = \dfrac{60}{\sin \theta}$ lb

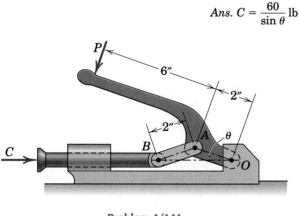

Problem 4/141

***4/142** The truck with bed-mounted crane is used in the delivery of cubes of bricks. For the given position of the carrier C along the boom, determine and plot the force F in the hydraulic cylinder AB as a function of the elevation angle θ for $0 \le \theta \le 75°$. Find the maximum value of F and the corresponding angle θ. Neglect the mass of the boom compared with that of the 1500-kg pallet of bricks.

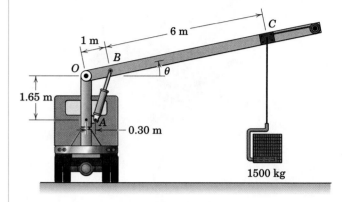

Problem 4/142

***4/143** The type of marine crane shown is utilized for both dockside and offshore operations. Determine and plot the force in member BC as a function of the boom angle θ for $0 \le \theta \le 80°$ and state the value of this force for $\theta = 40°$. Neglect the radius of all pulleys and the weight of the boom.

Ans. $BC = 190.5$ kN at $\theta = 40°$

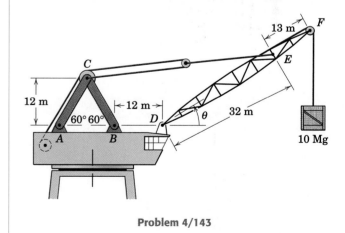

Problem 4/143

*4/144 Boom *OA* of the large waterfront crane can rotate about the vertical member *OB*. Determine and plot the forces in members *BC* and *BD* as functions of θ over the range −90° ≤ θ ≤ 90°. The weight of the crate is 3 tons. The 45° elevation angle of the boom remains constant; the cables for control of boom elevation angle have been omitted for clarity.

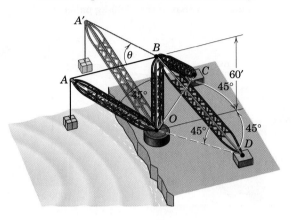

Problem 4/144

*4/145 A door-opening mechanism is shown in the figure. The spring-loaded hinges at *O* provide a moment $K_T\theta$ which tends to close the door, where θ is the door-opening angle and the torsional spring constant $K_T = 500$ lb-in./rad. The motor unit at *A* provides a variable moment *M* so that the slowly opening door is always in quasi-static equilibrium. Determine the moment *M* and the pin force at *B* as functions of θ for the range 0 ≤ θ ≤ 90°. State the value of *M* for θ = 45°.

Ans. M = 285 lb-in. at θ = 45°

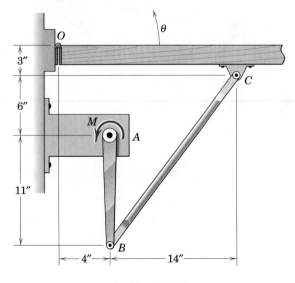

Problem 4/145

*4/146 The lift shown is used to elevate motorcycles during manufacture and repair. The hydraulic cylinder *OM* remains at the 60° angle as the elevation angle θ varies. The figures *ACDE* and *FHIJ* (*J* hidden) are parallelograms. There is a ball-and-socket joint at *M* to allow rotation of the frame *BKMNG*. If the total load *L* = 800 lb, determine and plot the required axial cylinder force *R* as a function of θ over the range 0 ≤ θ ≤ 90°. State the maximum value of *R* and the angle at which the maximum occurs. Over the same range of θ, plot the side force *S* applied to the cylinder rod at *M* and state its maximum absolute value.

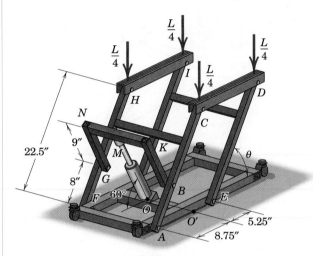

Problem 4/146

*4/147 The uniform 30-kg ventilation door *OAP* is opened by the mechanism shown. Plot the required force in the cylinder *DE* as a function of the door opening angle θ over the range $0 \le \theta \le \theta_{max}$, where θ_{max} is the maximum opening. Determine the minimum and maximum values of this force and the angles at which these extremes occur. Note that the cylinder is not horizontal when $\theta = 0$.

Ans. $(F_{DE})_{max} = 3580$ N at $\theta = 0$
$(F_{DE})_{min} = 0$ at $\theta_{max} = 65.9°$

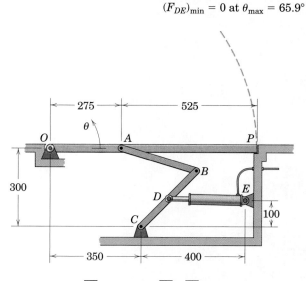

$\overline{AB} = 300$ $\qquad$ $\overline{CD} = \overline{DB} = 150$

Dimensions in millimeters

Problem 4/147

*4/148 The machine shown is used to help load luggage into airliners. The combined mass of the conveyor and luggage is 100 kg with mass center at *G*. Determine and plot the force in the hydraulic cylinder as a function of θ over the range $5° \le \theta \le 30°$ and state the maximum value over this range.

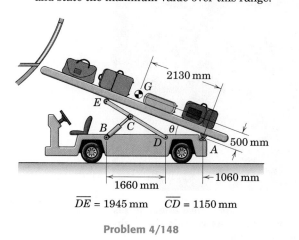

$\overline{DE} = 1945$ mm $\qquad$ $\overline{CD} = 1150$ mm

Problem 4/148

*4/149 The "jaws of life" device is used by rescuers to pry apart wreckage. A pressure of 35 MPa ($35(10^6)$ N/m²) is developed behind the piston of 50-mm radius. Begin by determining the prying force *R*, the force in link *AB*, and the horizontal force reaction at *C* for the condition shown on the left. Then develop expressions for and plot those quantities as functions of the jaw angle θ (shown on the right) over the range $0 \le \theta \le 45°$. State the minimum value of *R* and the value of θ for which this extreme occurs.

Ans. At $\theta = 0$: $R = 75\ 000$ N, $AB = 211\ 000$ N
$C_x = -85\ 400$ N
$R_{min} = 49\ 400$ N at $\theta = 23.2°$

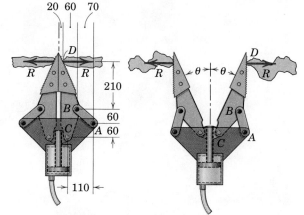

Dimensions in millimeters

Problem 4/149

When forces are continuously distributed over a region of a structure, the cumulative effect of this distribution must be determined. The designers of high-performance sailboats consider both air-pressure distributions on the sails and water-pressure distributions on the hull.

DISTRIBUTED FORCES

5/1   INTRODUCTION

In the previous chapters we treated all forces as concentrated along their lines of action and at their points of application. This treatment provided a reasonable model for those forces. Actually, "concentrated" forces do not exist in the exact sense, since every external force applied mechanically to a body is distributed over a finite contact area, however small.

The force exerted by the pavement on an automobile tire, for instance, is applied to the tire over its entire area of contact, Fig. 5/1a, which may be appreciable if the tire is soft. When analyzing the forces acting on the car as a whole, if the dimension b of the contact area is negligible compared with the other pertinent dimensions, such as the distance between wheels, then we may replace the actual distributed contact forces by their resultant R treated as a concentrated force. Even the force of contact between a hardened steel ball and its race in a loaded ball bearing, Fig. 5/1b, is applied over a finite though extremely small contact area. The forces applied to a two-force member of a truss, Fig. 5/1c, are applied over an actual area of contact of the pin against the hole and internally across the cut section as shown. In these and other similar examples we may treat the forces as concentrated when analyzing their external effects on bodies as a whole.

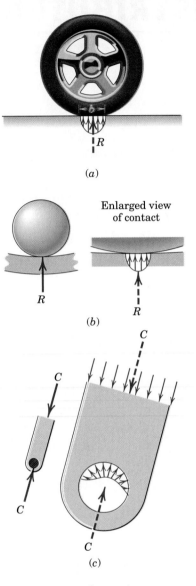

(a)

Enlarged view
of contact

R

(b)

C

C

C

C

Figure 5/1

If, on the other hand, we want to find the distribution of *internal* forces in the material of the body near the contact location, where the internal stresses and strains may be appreciable, then we must not treat the load as concentrated but must consider the actual distribution. This problem will not be discussed here because it requires a knowledge of the properties of the material and belongs in more advanced treatments of the mechanics of materials and the theories of elasticity and plasticity.

When forces are applied over a region whose dimensions are not negligible compared with other pertinent dimensions, then we must account for the actual manner in which the force is distributed. We do this by summing the effects of the distributed force over the entire region using mathematical integration. This requires that we know the intensity of the force at any location. There are three categories of such problems.

(1) Line Distribution. When a force is distributed along a line, as in the continuous vertical load supported by a suspended cable, Fig. 5/2a, the intensity w of the loading is expressed as force per unit length of line, newtons per meter (N/m) or pounds per foot (lb/ft).

(2) Area Distribution. When a force is distributed over an area, as with the hydraulic pressure of water against the inner face of a section of dam, Fig. 5/2b, the intensity is expressed as force per unit area. This intensity is called *pressure* for the action of fluid forces and *stress* for the internal distribution of forces in solids. The basic unit for pressure or stress in SI is the newton per square meter (N/m^2), which is also called the *pascal* (Pa). This unit, however, is too small for most applications (6895 Pa = 1 lb/in.^{2}). The kilopascal (kPa), which equals 10^3 Pa, is more commonly used for fluid pressure, and the megapascal, which equals 10^6 Pa, is used for stress. In the U.S. customary system of units, both fluid pressure and mechanical stress are commonly expressed in pounds per square inch (lb/in.^{2}).

(3) Volume Distribution. A force which is distributed over the volume of a body is called a *body force*. The most common body force is the force of gravitational attraction, which acts on all elements of mass in a body. The determination of the forces on the supports of the heavy cantilevered structure in Fig. 5/2c, for example, would require accounting for the distribution of gravitational force throughout the structure. The intensity of gravitational force is the *specific weight* ρg, where ρ is the density (mass per unit volume) and g is the acceleration due to gravity. The units for ρg are $(kg/m^3)(m/s^2) = N/m^3$ in SI units and lb/ft^3 or lb/in.^{3} in the U.S. customary system.

The body force due to the gravitational attraction of the earth (weight) is by far the most commonly encountered distributed force. Section A of this chapter treats the determination of the point in a body through which the resultant gravitational force acts, and discusses the associated geometric properties of lines, areas, and volumes. Section B treats distributed forces which act on and in beams and flexible cables and distributed forces which fluids exert on exposed surfaces.

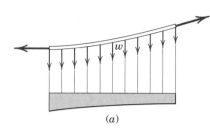

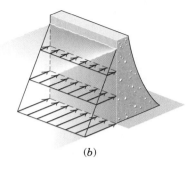

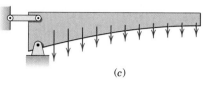

(a)

(b)

(c)

Figure 5/2

SECTION A CENTERS OF MASS AND CENTROIDS

5/2   CENTER OF MASS

Consider a three-dimensional body of any size and shape, having a mass m. If we suspend the body, as shown in Fig. 5/3, from any point such as A, the body will be in equilibrium under the action of the tension in the cord and the resultant W of the gravitational forces acting on all particles of the body. This resultant is clearly collinear with the cord. Assume that we mark its position by drilling a hypothetical hole of negligible size along its line of action. We repeat the experiment by suspending the body from other points such as B and C, and in each instance we mark the line of action of the resultant force. For all practical purposes these lines of action will be concurrent at a single point G, which is called the *center of gravity* of the body.

An exact analysis, however, would account for the slightly differing directions of the gravity forces for the various particles of the body, because those forces converge toward the center of attraction of the earth. Also, because the particles are at different distances from the earth, the intensity of the force field of the earth is not exactly constant over the body. As a result, the lines of action of the gravity-force resultants in the experiments just described will not be quite concurrent, and therefore no unique center of gravity exists in the exact sense. This is of no practical importance as long as we deal with bodies whose dimensions are small compared with those of the earth. We therefore assume a uniform and parallel force field due to the gravitational attraction of the earth, and this assumption results in the concept of a unique center of gravity.

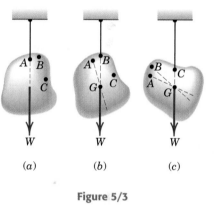

(a)   (b)   (c)

Figure 5/3

Determining the Center of Gravity

To determine mathematically the location of the center of gravity of any body, Fig. 5/4a, we apply the *principle of moments* (see Art. 2/6) to the parallel system of gravitational forces. The moment of the resultant gravitational force W about any axis equals the sum of the moments

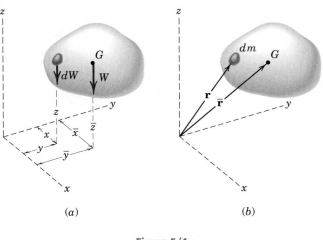

Figure 5/4

about the same axis of the gravitational forces dW acting on all particles treated as infinitesimal elements of the body. The resultant of the gravitational forces acting on all elements is the weight of the body and is given by the sum $W = \int dW$. If we apply the moment principle about the y-axis, for example, the moment about this axis of the elemental weight is $x\ dW$, and the sum of these moments for all elements of the body is $\int x\ dW$. This sum of moments must equal $W\bar{x}$, the moment of the sum. Thus, $\bar{x}W = \int x\ dW$.

With similar expressions for the other two components, we may express the coordinates of the center of gravity G as

$$\bar{x} = \frac{\int x\ dW}{W} \qquad \bar{y} = \frac{\int y\ dW}{W} \qquad \bar{z} = \frac{\int z\ dW}{W} \qquad \textbf{(5/1a)}$$

To visualize the physical moments of the gravity forces appearing in the third equation, we may reorient the body and attached axes so that the z-axis is horizontal. It is essential to recognize that the numerator of each of these expressions represents the *sum of the moments*, whereas the product of W and the corresponding coordinate of G represents the *moment of the sum*. This moment principle finds repeated use throughout mechanics.

With the substitution of $W = mg$ and $dW = g\ dm$, the expressions for the coordinates of the center of gravity become

$$\bar{x} = \frac{\int x\ dm}{m} \qquad \bar{y} = \frac{\int y\ dm}{m} \qquad \bar{z} = \frac{\int z\ dm}{m} \qquad \textbf{(5/1b)}$$

Equations 5/1b may be expressed in vector form with the aid of Fig. 5/4b, in which the elemental mass and the mass center G are located by their respective position vectors $\mathbf{r} = x\mathbf{i} + y\mathbf{j} + z\mathbf{k}$ and $\bar{\mathbf{r}} = \bar{x}\mathbf{i} + \bar{y}\mathbf{j} + \bar{z}\mathbf{k}$. Thus, Eqs. 5/1$b$ are the components of the single vector equation

$$\bar{\mathbf{r}} = \frac{\int \mathbf{r}\, dm}{m}$$

(5/2)

The density ρ of a body is its mass per unit volume. Thus, the mass of a differential element of volume dV becomes $dm = \rho\, dV$. If ρ is not constant throughout the body but can be expressed as a function of the coordinates of the body, we must account for this variation when calculating the numerators and denominators of Eqs. 5/1b. We may then write these expressions as

$$\bar{x} = \frac{\int x\rho\, dV}{\int \rho\, dV} \qquad \bar{y} = \frac{\int y\rho\, dV}{\int \rho\, dV} \qquad \bar{z} = \frac{\int z\rho\, dV}{\int \rho\, dV}$$

(5/3)

Center of Mass versus Center of Gravity

Equations 5/1b, 5/2, and 5/3 are independent of gravitational effects since g no longer appears. They therefore define a unique point in the body which is a function solely of the distribution of mass. This point is called the *center of mass*, and clearly it coincides with the center of gravity as long as the gravity field is treated as uniform and parallel.

It is meaningless to speak of the center of gravity of a body which is removed from the gravitational field of the earth, since no gravitational forces would act on it. The body would, however, still have its unique center of mass. We will usually refer henceforth to the center of mass rather than to the center of gravity. Also, the center of mass has a special significance in calculating the dynamic response of a body to unbalanced forces. This class of problems is discussed at length in *Vol. 2 Dynamics*.

In most problems the calculation of the position of the center of mass may be simplified by an intelligent choice of reference axes. In general the axes should be placed so as to simplify the equations of the boundaries as much as possible. Thus, polar coordinates will be useful for bodies with circular boundaries.

Another important clue may be taken from considerations of symmetry. Whenever there exists a line or plane of symmetry in a homogeneous body, a coordinate axis or plane should be chosen to coincide with this line or plane. The center of mass will always lie on such a line or plane, since the moments due to symmetrically located elements will always cancel, and the body may be considered composed of pairs of these elements. Thus, the center of mass G of the homogeneous right-circular cone of Fig. 5/5a will lie somewhere on its central axis, which is a line of symmetry. The center of mass of the half right-circular cone lies on its plane of symmetry, Fig. 5/5b. The center of mass of the half ring in Fig. 5/5c lies in both of its planes of symmetry and therefore is situated on

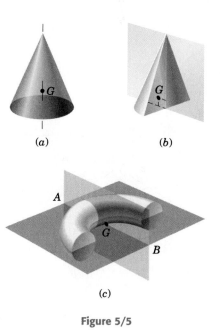

(a)

(b)

(c)

Figure 5/5

line AB. It is easiest to find the location of G by using symmetry when it exists.

5/3 CENTROIDS OF LINES, AREAS, AND VOLUMES

When the density ρ of a body is uniform throughout, it will be a constant factor in both the numerators and denominators of Eqs. 5/3 and will therefore cancel. The remaining expressions define a purely geometrical property of the body, since any reference to its mass properties has disappeared. The term *centroid* is used when the calculation concerns a geometrical shape only. When speaking of an actual physical body, we use the term *center of mass*. If the density is uniform throughout the body, the positions of the centroid and center of mass are identical, whereas if the density varies, these two points will, in general, not coincide.

The calculation of centroids falls within three distinct categories, depending on whether we can model the shape of the body involved as a line, an area, or a volume.

(1) Lines. For a slender rod or wire of length L, cross-sectional area A, and density ρ, Fig. 5/6, the body approximates a line segment, and $dm = \rho A\, dL$. If ρ and A are constant over the length of the rod, the coordinates of the center of mass also become the coordinates of the centroid C of the line segment, which, from Eqs. 5/1b, may be written

$$\bar{x} = \frac{\int x\, dL}{L} \qquad \bar{y} = \frac{\int y\, dL}{L} \qquad \bar{z} = \frac{\int z\, dL}{L} \qquad \textbf{(5/4)}$$

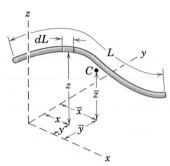

Figure 5/6

Note that, in general, the centroid C will not lie on the line. If the rod lies on a single plane, such as the x-y plane, only two coordinates need to be calculated.

(2) Areas. When a body of density ρ has a small but constant thickness t, we can model it as a surface area A, Fig. 5/7. The mass of an element becomes $dm = \rho t\, dA$. Again, if ρ and t are constant over the entire area, the coordinates of the center of mass of the body also become the coordinates of the centroid C of the surface area, and from Eqs. 5/1b the coordinates may be written

$$\bar{x} = \frac{\int x\, dA}{A} \qquad \bar{y} = \frac{\int y\, dA}{A} \qquad \bar{z} = \frac{\int z\, dA}{A} \qquad \textbf{(5/5)}$$

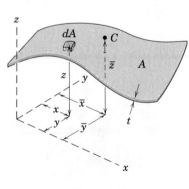

Figure 5/7

The numerators in Eqs. 5/5 are called the *first moments of area.** If the surface is curved, as illustrated in Fig. 5/7 with the shell segment, all three coordinates will be involved. The centroid C for the curved surface will in general not lie on the surface. If the area is a flat surface in,

---

*Second moments of areas (moments of first moments) appear later in our discussion of area moments of inertia in Appendix A.

say, the x-y plane, only the coordinates of C in that plane need to be calculated.

(3) Volumes. For a general body of volume V and density ρ, the element has a mass $dm = \rho\, dV$. The density ρ cancels if it is constant over the entire volume, and the coordinates of the center of mass also become the coordinates of the centroid C of the body. From Eqs. 5/3 or 5/1b they become

$$\bar{x} = \frac{\int x\, dV}{V} \qquad \bar{y} = \frac{\int y\, dV}{V} \qquad \bar{z} = \frac{\int z\, dV}{V} \qquad \textbf{(5/6)}$$

Choice of Element for Integration

The principal difficulty with a theory often lies not in its concepts but in the procedures for applying it. With mass centers and centroids the concept of the moment principle is simple enough; the difficult steps are the choice of the differential element and setting up the integrals. The following five guidelines will be useful.

(1) Order of Element. Whenever possible, a first-order differential element should be selected in preference to a higher-order element so that only one integration will be required to cover the entire figure. Thus, in Fig. 5/8a a first-order horizontal strip of area $dA = l\, dy$ will require only one integration with respect to y to cover the entire figure. The second-order element $dx\, dy$ will require two integrations, first with respect to x and second with respect to y, to cover the figure. As a further example, for the solid cone in Fig. 5/8b we choose a first-order element in the form of a circular slice of volume $dV = \pi r^2\, dy$. This choice requires only one integration, and thus is preferable to choosing a third-order element $dV = dx\, dy\, dz$, which would require three awkward integrations.

(2) Continuity. Whenever possible, we choose an element which can be integrated in one continuous operation to cover the figure. Thus, the horizontal strip in Fig. 5/8a would be preferable to the vertical strip in Fig. 5/9, which, if used, would require two separate integrals because of the discontinuity in the expression for the height of the strip at $x = x_1$.

(3) Discarding Higher-Order Terms. Higher-order terms may always be dropped compared with lower-order terms (see Art. 1/7). Thus, the vertical strip of area under the curve in Fig. 5/10 is given by the

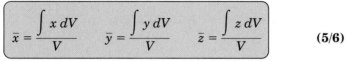

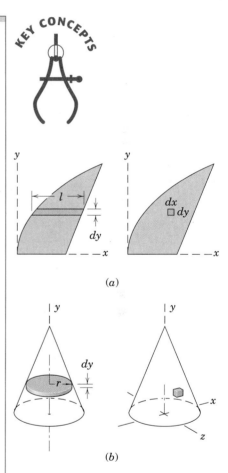

(a)

(b)

Figure 5/8

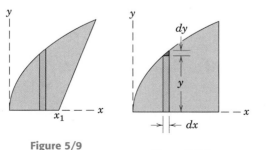

Figure 5/9

Figure 5/10

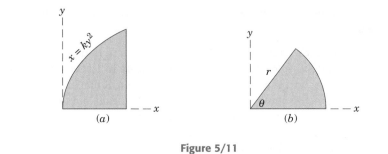

Figure 5/11

first-order term $dA = y\,dx$, and the second-order triangular area $\frac{1}{2}dx\,dy$ is discarded. In the limit, of course, there is no error.

(4) Choice of Coordinates. As a general rule, we choose the coordinate system which best matches the boundaries of the figure. Thus, the boundaries of the area in Fig. 5/11a are most easily described in rectangular coordinates, whereas the boundaries of the circular sector of Fig. 5/11b are best suited to polar coordinates.

(5) Centroidal Coordinate of Element. When a first- or second-order differential element is chosen, it is essential to use the *coordinate of the centroid of the element* for the moment arm in expressing the moment of the differential element. Thus, for the horizontal strip of area in Fig. 5/12a, the moment of dA about the y-axis is $x_c\,dA$, where x_c is the x-coordinate of the centroid C of the element. Note that x_c is *not* the x which describes either boundary of the area. In the y-direction for this element the moment arm y_c of the centroid of the element is the same, in the limit, as the y-coordinates of the two boundaries.

As a second example, consider the solid half-cone of Fig. 5/12b with the semicircular slice of differential thickness as the element of volume. The moment arm for the element in the x-direction is the distance x_c to the centroid of the face of the element and not the x-distance to the boundary of the element. On the other hand, in the z-direction the moment arm z_c of the centroid of the element is the same as the z-coordinate of the element.

With these examples in mind, we rewrite Eqs. 5/5 and 5/6 in the form

$$\bar{x} = \frac{\displaystyle\int x_c\,dA}{A} \qquad \bar{y} = \frac{\displaystyle\int y_c\,dA}{A} \qquad \bar{z} = \frac{\displaystyle\int z_c\,dA}{A} \tag{5/5a}$$

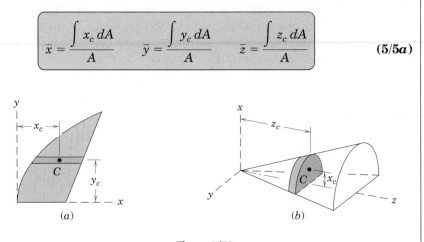

Figure 5/12

and

$$\bar{x} = \frac{\int x_c \, dV}{V} \qquad \bar{y} = \frac{\int y_c \, dV}{V} \qquad \bar{z} = \frac{\int z_c \, dV}{V} \qquad \textbf{(5/6a)}$$

It is *essential* to recognize that the subscript c serves as a reminder that the moment arms appearing in the numerators of the integral expressions for moments are *always* the coordinates of the *centroids* of the particular elements chosen.

At this point you should be certain to understand clearly the principle of moments, which was introduced in Art. 2/4. You should recognize the physical meaning of this principle as it is applied to the system of parallel weight forces depicted in Fig. 5/4a. Keep in mind the equivalence between the moment of the resultant weight W and the sum (integral) of the moments of the elemental weights dW, to avoid mistakes in setting up the necessary mathematics. Recognition of the principle of moments will help in obtaining the correct expression for the moment arm x_c, y_c, or z_c of the centroid of the chosen differential element.

Keeping in mind the physical picture of the principle of moments, we will recognize that Eqs. 5/4, 5/5, and 5/6, which are geometric relationships, are descriptive also of homogeneous physical bodies, because the density ρ cancels. If the density of the body in question is not constant but varies throughout the body as some function of the coordinates, then it will not cancel from the numerator and denominator of the mass-center expressions. In this event, we must use Eqs. 5/3 as explained earlier.

Sample Problems 5/1 through 5/5 which follow have been carefully chosen to illustrate the application of Eqs. 5/4, 5/5, and 5/6 for calculating the location of the centroid for line segments (slender rods), areas (thin flat plates), and volumes (homogeneous solids). The five integration considerations listed above are illustrated in detail in these sample problems.

Section C/10 of Appendix C contains a table of integrals which includes those needed for the problems in this and subsequent chapters. A summary of the centroidal coordinates for some of the commonly used shapes is given in Tables D/3 and D/4, Appendix D.

Sample Problem 5/1

Centroid of a circular arc. Locate the centroid of a circular arc as shown in the figure.

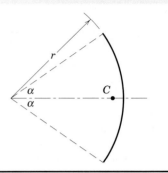

Solution. Choosing the axis of symmetry as the x-axis makes $\bar{y} = 0$. A differential element of arc has the length $dL = r\,d\theta$ expressed in polar coordinates, ① and the x-coordinate of the element is $r\cos\theta$.

Applying the first of Eqs. 5/4 and substituting $L = 2\alpha r$ give

$$[L\bar{x} = \int x\,dL] \qquad (2\alpha r)\bar{x} = \int_{-\alpha}^{\alpha} (r\cos\theta)\,r\,d\theta$$

$$2\alpha r\bar{x} = 2r^2 \sin\alpha$$

$$\bar{x} = \frac{r\sin\alpha}{\alpha} \qquad\qquad\qquad Ans.$$

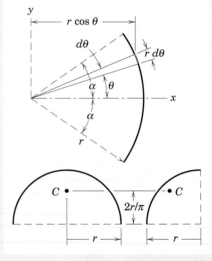

For a semicircular arc $2\alpha = \pi$, which gives $\bar{x} = 2r/\pi$. By symmetry we see immediately that this result also applies to the quarter-circular arc when the measurement is made as shown.

Helpful Hint

① It should be perfectly evident that polar coordinates are preferable to rectangular coordinates to express the length of a circular arc.

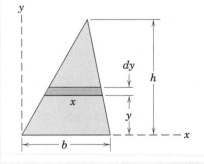

Sample Problem 5/2

Centroid of a triangular area. Determine the distance $\bar{h}$ from the base of a triangle of altitude h to the centroid of its area.

Solution. The x-axis is taken to coincide with the base. A differential strip of ① area $dA = x\,dy$ is chosen. By similar triangles $x/(h - y) = b/h$. Applying the second of Eqs. 5/5a gives

$$[A\bar{y} = \int y_c\,dA] \qquad \frac{bh}{2}\bar{y} = \int_0^h y\,\frac{b(h - y)}{h}\,dy = \frac{bh^2}{6}$$

and

$$\bar{y} = \frac{h}{3} \qquad\qquad\qquad Ans.$$

This same result holds with respect to either of the other two sides of the triangle considered a new base with corresponding new altitude. Thus, the centroid lies at the intersection of the medians, since the distance of this point from any side is one-third the altitude of the triangle with that side considered the base.

Helpful Hint

① We save one integration here by using the first-order element of area. Recognize that dA must be expressed in terms of the integration variable y; hence, $x = f(y)$ is required.

Sample Problem 5/3

Centroid of the area of a circular sector.   Locate the centroid of the area of a circular sector with respect to its vertex.

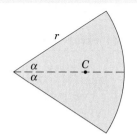

Solution I.   The x-axis is chosen as the axis of symmetry, and $\bar{y}$ is therefore automatically zero. We may cover the area by moving an element in the form of a partial circular ring, as shown in the figure, from the center to the outer periphery. The radius of the ring is r_0 and its thickness is dr_0, so that its area is

① $dA = 2r_0\alpha \, dr_0$.

② The x-coordinate to the centroid of the element from Sample Problem 5/1 is $x_c = r_0 \sin \alpha/\alpha$, where r_0 replaces r in the formula. Thus, the first of Eqs. 5/5a gives

$$[A\bar{x} = \int x_c \, dA] \qquad \frac{2\alpha}{2\pi}(\pi r^2)\bar{x} = \int_0^r \left(\frac{r_0 \sin \alpha}{\alpha}\right)(2r_0\alpha \, dr_0)$$

$$r^2\alpha\bar{x} = \frac{2}{3}r^3 \sin \alpha$$

$$\bar{x} = \frac{2}{3}\frac{r \sin \alpha}{\alpha} \qquad\qquad Ans.$$

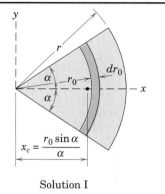

Solution I

Helpful Hints

① Note carefully that we must distinguish between the variable r_0 and the constant r.

② Be careful not to use r_0 as the centroidal coordinate for the element.

Solution II.   The area may also be covered by swinging a triangle of differential area about the vertex and through the total angle of the sector. This triangle, shown in the illustration, has an area $dA = (r/2)(r \, d\theta)$, where higher-order terms are neglected. From Sample Problem 5/2 the centroid of the triangular element of area is two-thirds of its altitude from its vertex, so that the x-coordinate to the centroid of the element is $x_c = \frac{2}{3}r \cos \theta$. Applying the first of Eqs. 5/5a gives

$$[A\bar{x} = \int x_c \, dA] \qquad (r^2\alpha)\bar{x} = \int_{-\alpha}^{\alpha} (\tfrac{2}{3}r \cos \theta)(\tfrac{1}{2}r^2 \, d\theta)$$

$$r^2\alpha\bar{x} = \frac{2}{3}r^3 \sin \alpha$$

and as before $\qquad\qquad \bar{x} = \frac{2}{3}\frac{r \sin \alpha}{\alpha} \qquad\qquad Ans.$

For a semicircular area $2\alpha = \pi$, which gives $\bar{x} = 4r/3\pi$. By symmetry we see immediately that this result also applies to the quarter-circular area where the measurement is made as shown.

It should be noted that, if we had chosen a second-order element $r_0 \, dr_0 \, d\theta$, one integration with respect to θ would yield the ring with which *Solution I* began. On the other hand, integration with respect to r_0 initially would give the triangular element with which *Solution II* began.

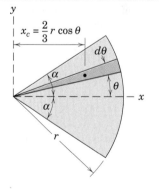

Solution II

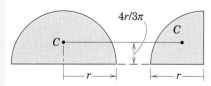

Sample Problem 5/4

Locate the centroid of the area under the curve $x = ky^3$ from $x = 0$ to $x = a$.

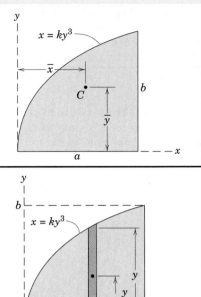

Solution I. A vertical element of area $dA = y\,dx$ is chosen as shown in the figure. The x-coordinate of the centroid is found from the first of Eqs. 5/5a. Thus,

① $[A\bar{x} = \int x_c\,dA]$ $\qquad \bar{x}\int_0^a y\,dx = \int_0^a xy\,dx$

Substituting $y = (x/k)^{1/3}$ and $k = a/b^3$ and integrating give

$$\frac{3ab}{4}\bar{x} = \frac{3a^2b}{7} \qquad \bar{x} = \tfrac{4}{7}a \qquad\qquad Ans.$$

In the solution for $\bar{y}$ from the second of Eqs. 5/5a, the coordinate to the centroid of the rectangular element is $y_c = y/2$, where y is the height of the strip governed by the equation of the curve $x = ky^3$. Thus, the moment principle becomes

$[A\bar{y} = \int y_c\,dA]$ $\qquad \dfrac{3ab}{4}\bar{y} = \displaystyle\int_0^a \left(\dfrac{y}{2}\right)y\,dx$

Substituting $y = b(x/a)^{1/3}$ and integrating give

$$\frac{3ab}{4}\bar{y} = \frac{3ab^2}{10} \qquad \bar{y} = \tfrac{2}{5}b \qquad\qquad Ans.$$

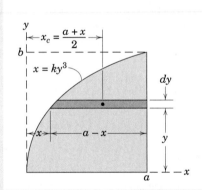

Solution II. The horizontal element of area shown in the lower figure may be employed in place of the vertical element. The x-coordinate to the centroid of the rectangular element is seen to be $x_c = x + \frac{1}{2}(a - x) = (a + x)/2$, which is simply the average of the coordinates a and x of the ends of the strip. Hence,

$[A\bar{x} = \int x_c\,dA]$ $\qquad \bar{x}\displaystyle\int_0^b (a - x)\,dy = \int_0^b \left(\dfrac{a + x}{2}\right)(a - x)\,dy$

The value of $\bar{y}$ is found from

$[A\bar{y} = \int y_c\,dA]$ $\qquad \bar{y}\displaystyle\int_0^b (a - x)\,dy = \int_0^b y(a - x)\,dy$

where $y_c = y$ for the horizontal strip. The evaluation of these integrals will check the previous results for $\bar{x}$ and $\bar{y}$.

Helpful Hint

① Note that $x_c = x$ for the vertical element.

Sample Problem 5/5

Hemispherical volume. Locate the centroid of the volume of a hemisphere of radius r with respect to its base.

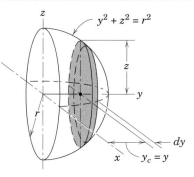

Solution I

Solution I. With the axes chosen as shown in the figure, $\bar{x} = \bar{z} = 0$ by symmetry. The most convenient element is a circular slice of thickness dy parallel to the x-z plane. Since the hemisphere intersects the y-z plane in the circle $y^2 + z^2 = r^2$, the radius of the circular slice is $z = +\sqrt{r^2 - y^2}$. The volume of the elemental slice becomes

①

$$dV = \pi(r^2 - y^2)\, dy$$

The second of Eqs. 5/6a requires

$$[V\bar{y} = \int y_c\, dV] \qquad \bar{y} \int_0^r \pi(r^2 - y^2)\, dy = \int_0^r y\pi(r^2 - y^2)\, dy$$

where $y_c = y$. Integrating gives

$$\tfrac{2}{3}\pi r^3 \bar{y} = \tfrac{1}{4}\pi r^4 \qquad \bar{y} = \tfrac{3}{8}r \qquad\qquad Ans.$$

Solution II. Alternatively we may use for our differential element a cylindrical shell of length y, radius z, and thickness dz, as shown in the lower figure. By expanding the radius of the shell from zero to r, we cover the entire volume. By symmetry the centroid of the elemental shell lies at its center, so that $y_c = y/2$. The volume of the element is $dV = (2\pi z\, dz)(y)$. Expressing y in terms of z from the equation of the circle gives $y = +\sqrt{r^2 - z^2}$. Using the value of $\tfrac{2}{3}\pi r^3$ computed in *Solution I* for the volume of the hemisphere and substituting in the second of Eqs. 5/6a give us

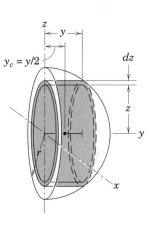

Solution II

$$[V\bar{y} = \int y_c\, dV] \qquad (\tfrac{2}{3}\pi r^3)\bar{y} = \int_0^r \frac{\sqrt{r^2 - z^2}}{2}\,(2\pi z\sqrt{r^2 - z^2})\, dz$$

$$= \int_0^r \pi(r^2 z - z^3)\, dz = \frac{\pi r^4}{4}$$

$$\bar{y} = \tfrac{3}{8}r \qquad\qquad Ans.$$

Solutions I and II are of comparable use since each involves an element of simple shape and requires integration with respect to one variable only.

Solution III. As an alternative, we could use the angle θ as our variable with limits of 0 and $\pi/2$. The radius of either element would become $r \sin\theta$, whereas the thickness of the slice in *Solution I* would be $dy = (r\, d\theta)\sin\theta$ and that of the shell in *Solution II* would be $dz = (r\, d\theta)\cos\theta$. The length of the shell would be $y = r\cos\theta$.

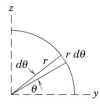

Solution III

Helpful Hint

① Can you identify the higher-order element of volume which is omitted from the expression for dV?

PROBLEMS

Introductory Problems

5/1 Place your pencil on the position of your best visual estimate of the centroid of the triangular area. Check the horizontal position of your estimate by referring to the results of Sample Problem 5/2.

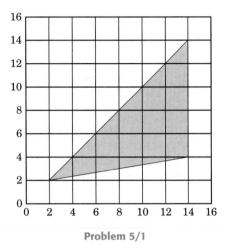

Problem 5/1

5/2 With your pencil make a dot on the position of your best visual estimate of the centroid of the area of the circular sector. Check your estimate by using the results of Sample Problem 5/3.

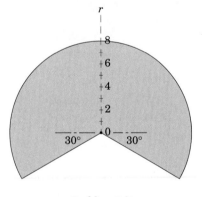

Problem 5/2

5/3 Specify the x- and z-coordinates of the center of mass of the semicylindrical shell.

Ans. $\bar{x} = -120$ mm, $\bar{z} = 43.6$ mm

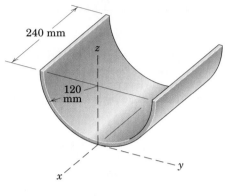

Problem 5/3

5/4 Specify the x-, y-, and z-coordinates of the mass center of the quadrant of the homogeneous solid cylinder.

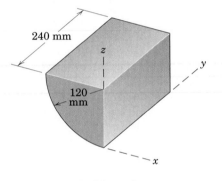

Problem 5/4

5/5 Determine the y-coordinate of the centroid of the area by direct integration.

Ans. $\bar{y} = \dfrac{14R}{9\pi}$

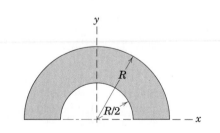

Problem 5/5

5/6 Determine the coordinates of the centroid of the shaded area.

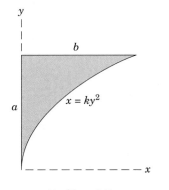

Problem 5/6

5/7 Determine the y-coordinate of the centroid of the area under the sine curve shown.

Ans. $\bar{y} = \dfrac{\pi a}{8}$

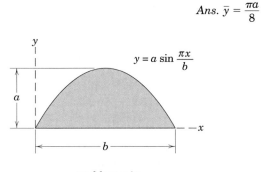

Problem 5/7

5/8 Determine the coordinates of the centroid of the shaded area.

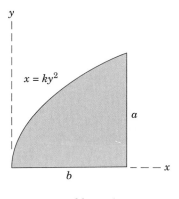

Problem 5/8

5/9 By direct integration, determine the coordinates of the centroid of the trapezoidal area.

Ans. $\bar{x} = 2.35,\ \bar{y} = 3.56$

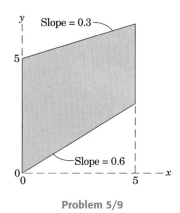

Problem 5/9

5/10 Find the distance $\bar{z}$ from the vertex of the right-circular cone to the centroid of its volume.

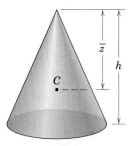

Problem 5/10

5/11 Determine the x- and y-coordinates of the centroid of the trapezoidal area.

Ans. $\bar{x} = \dfrac{a^2 + b^2 + ab}{3(a + b)},\ \bar{y} = \dfrac{h(2a + b)}{3(a + b)}$

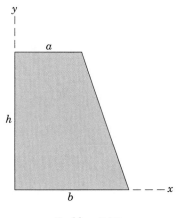

Problem 5/11

5/12 Determine the x- and y-coordinates of the centroid of the shaded area.

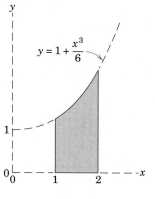

Problem 5/12

Representative Problems

5/13 Locate the centroid of the shaded area.

Ans. $\bar{x} = 2a/5, \ \bar{y} = 3b/8$

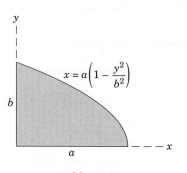

Problem 5/13

5/14 Locate the centroid of the shaded area shown.

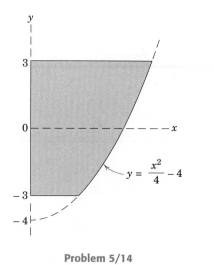

Problem 5/14

5/15 The mass per unit length of the slender rod varies with position according to $\rho = \rho_0(1 - x/2)$, where x is in feet. Determine the location of the center of mass of the rod.

Ans. $\bar{x} = \frac{4}{9}$ ft

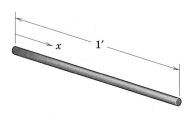

Problem 5/15

5/16 Determine the coordinates of the centroid of the shaded area.

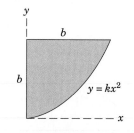

Problem 5/16

5/17 Calculate the coordinates of the centroid of the segment of the circular area.

Ans. $\bar{x} = \bar{y} = \dfrac{2a}{3(\pi - 2)}$

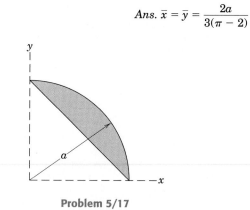

Problem 5/17

5/18 Determine the x-coordinate of the mass center of the tapered steel rod of length L where the diameter at the large end is twice the diameter at the small end.

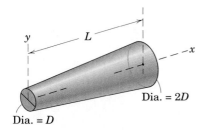

Problem 5/18

5/19 Determine the y-coordinate of the centroid of the shaded area.

$$Ans. \; \bar{y} = \frac{b}{2}$$

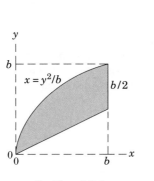

Problem 5/19

5/20 Let $c \to \infty$ and determine the x- and y-coordinates of the centroid of the shaded area.

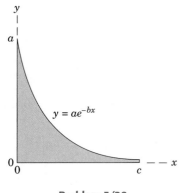

Problem 5/20

5/21 Determine the x- and y-coordinates of the centroid of the shaded area shown.

$$Ans. \; \bar{x} = \frac{a}{3\left(\dfrac{\pi}{2} - 1\right)}, \; \bar{y} = \frac{b}{3\left(\dfrac{\pi}{2} - 1\right)}$$

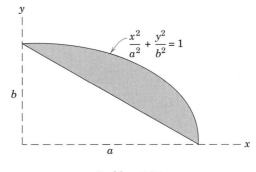

Problem 5/21

5/22 Determine the y-coordinate of the centroid of the shaded area.

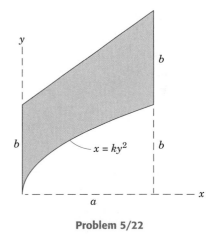

Problem 5/22

5/23 Use the results of Sample Problem 5/5 to compute the coordinates of the mass center of the portion of the solid homogeneous cylinder shown.

$$Ans. \; \bar{x} = \bar{y} = -8/(3\pi) \text{ in.}, \; \bar{z} = 5 \text{ in.}$$

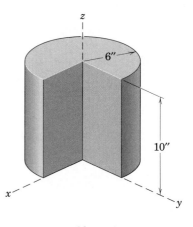

Problem 5/23

5/24 Determine the coordinates of the centroid of the shaded area.

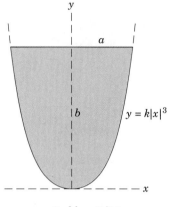

Problem 5/24

5/25 Determine the x- and y-coordinates of the centroid of the shaded area.

$$Ans. \ \bar{x} = \frac{a}{\pi - 1}, \ \bar{y} = \frac{7b}{6(\pi - 1)}$$

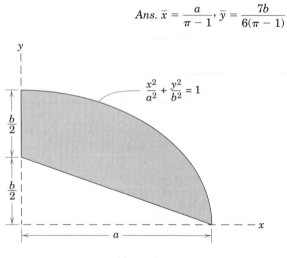

Problem 5/25

5/26 Determine the x- and y-coordinates of the shaded area.

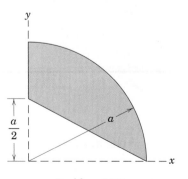

Problem 5/26

5/27 Determine the y-coordinate of the centroid of the shaded area.

$$Ans. \ \bar{y} = \frac{14\sqrt{2}}{9\pi} a$$

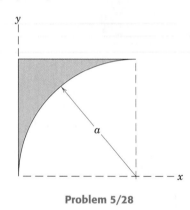

Problem 5/27

5/28 Locate the centroid of the area shown in the figure by direct integration. (*Caution:* Carefully observe the proper sign of the radical involved.)

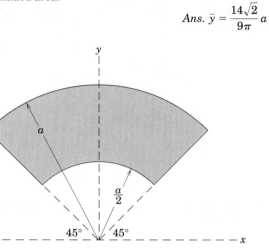

Problem 5/28

5/29 Locate the centroid of the shaded area between the two curves.

$$Ans. \ \bar{x} = \frac{24}{25}, \ \bar{y} = \frac{6}{7}$$

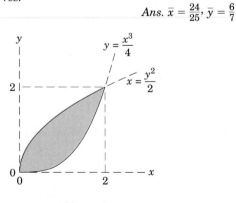

Problem 5/29

5/30 The figure represents a flat piece of sheet metal symmetrical about axis *A-A* and having a parabolic upper boundary. Choose your own coordinates and calculate the distance $\bar{h}$ from the base to the center of gravity of the piece.

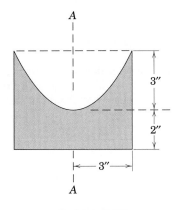

Problem 5/30

5/31 Determine the *z*-coordinate of the centroid of the volume obtained by revolving the shaded area under the parabola about the *z*-axis through 180°.

Ans. $\bar{z} = 2a/3$

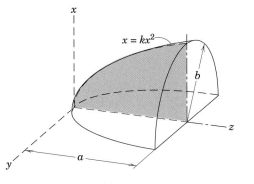

Problem 5/31

5/32 Determine the *x*-coordinate of the centroid of the solid spherical segment.

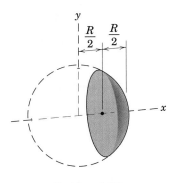

Problem 5/32

5/33 Determine the *y*-coordinate of the centroid of the shaded area shown. (Observe the caution cited with Prob. 5/28.)

Ans. $\bar{y} = 0.339a$

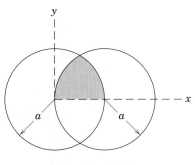

Problem 5/33

5/34 The thickness of the triangular plate varies linearly with *y* from a value t_0 along its base $y = 0$ to $2t_0$ at $y = h$. Determine the *y*-coordinate of the center of mass of the plate.

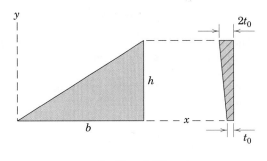

Problem 5/34

5/35 Calculate the distance $\bar{h}$ measured from the base to the centroid of the volume of the frustum of the right-circular cone.

Ans. $\bar{h} = \frac{11}{56}h$

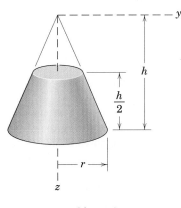

Problem 5/35

5/36 Determine the x-coordinate of the mass center of the portion of the spherical shell of uniform but small thickness.

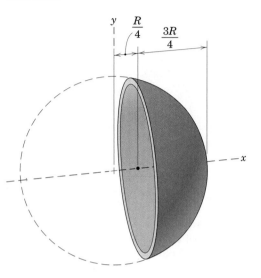

Problem 5/36

5/37 The homogeneous slender rod has a uniform cross section and is bent into the shape shown. Calculate the y-coordinate of the mass center of the rod. (*Reminder:* A differential arc length is $dL = \sqrt{(dx)^2 + (dy)^2} = \sqrt{1 + (dx/dy)^2}\, dy$.)

Ans. $\bar{y} = 57.4$ mm

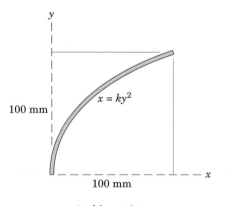

Problem 5/37

5/38 Determine the z-coordinate of the centroid of the volume obtained by revolving the shaded triangular area about the z-axis through 360°.

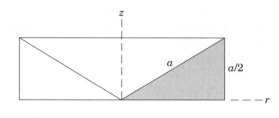

Problem 5/38

5/39 Locate the mass center of the homogeneous solid body whose volume is determined by revolving the shaded area through 360° about the z-axis.

Ans. $\bar{z} = 263$ mm

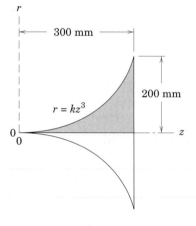

Problem 5/39

▶ **5/40** Determine the z-coordinate of the mass center of the homogeneous quarter-spherical shell which has a radius r.

Ans. $\bar{z} = \dfrac{r}{2}$

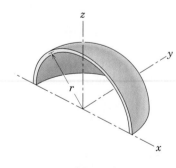

Problem 5/40

▶ **5/41** Determine the y-coordinate of the centroid of the plane area shown. Set $h = 0$ in your result and compare with the result $\bar{y} = \dfrac{4a}{3\pi}$ for a full semicircular area (see Sample Problem 5/3 and Table D/3). Also evaluate your result for the conditions $h = \dfrac{a}{4}$ and $h = \dfrac{a}{2}$.

$$Ans.\ \bar{y} = \frac{\dfrac{2}{3}[a^2 - h^2]^{3/2}}{a^2\left[\dfrac{\pi}{2} - \sin^{-1}\dfrac{h}{a}\right] - h\sqrt{a^2 - h^2}}$$

$$h = \frac{a}{4}: \bar{y} = 0.562a, \quad h = \frac{a}{2}: \bar{y} = 0.705a$$

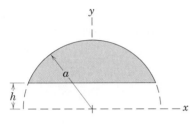

Problem 5/41

▶ **5/42** Determine the x-coordinate of the mass center of the homogeneous hemisphere with the smaller hemispherical portion removed.

$$Ans.\ \bar{x} = \frac{45}{112}R$$

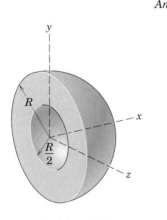

Problem 5/42

▶ **5/43** Determine the x-coordinate of the mass center of the solid homogeneous body shown.

$$Ans.\ \bar{x} = 1.542R$$

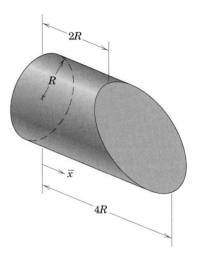

Problem 5/43

▶ **5/44** Determine the x-coordinate of the mass center of the cylindrical shell of small uniform thickness.

$$Ans.\ \bar{x} = 1.583R$$

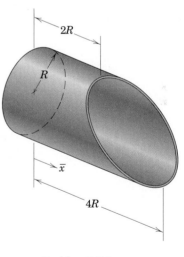

Problem 5/44

5/4   COMPOSITE BODIES AND FIGURES; APPROXIMATIONS

When a body or figure can be conveniently divided into several parts whose mass centers are easily determined, we use the principle of moments and treat each part as a finite element of the whole. Such a body is illustrated schematically in Fig. 5/13. Its parts have masses m_1, m_2, m_3 with the respective mass-center coordinates $\bar{x}_1$, $\bar{x}_2$, $\bar{x}_3$ in the x-direction. The moment principle gives

$$(m_1 + m_2 + m_3)\overline{X} = m_1\bar{x}_1 + m_2\bar{x}_2 + m_3\bar{x}_3$$

where $\overline{X}$ is the x-coordinate of the center of mass of the whole. Similar relations hold for the other two coordinate directions.

We generalize, then, for a body of any number of parts and express the sums in condensed form to obtain the mass-center coordinates

$$\overline{X} = \frac{\Sigma m\bar{x}}{\Sigma m} \qquad \overline{Y} = \frac{\Sigma m\bar{y}}{\Sigma m} \qquad \overline{Z} = \frac{\Sigma m\bar{z}}{\Sigma m} \qquad \text{(5/7)}$$

Analogous relations hold for composite lines, areas, and volumes, where the m's are replaced by L's, A's, and V's, respectively. Note that if a hole or cavity is considered one of the component parts of a composite body or figure, the corresponding mass represented by the cavity or hole is treated as a negative quantity.

An Approximation Method

In practice the boundaries of an area or volume might not be expressible in terms of simple geometrical shapes or as shapes which can be represented mathematically. For such cases we must resort to a method of approximation. As an example, consider the problem of locat-

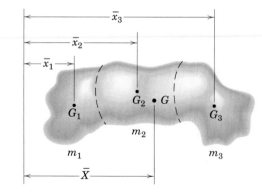

Figure 5/13

ing the centroid C of the irregular area shown in Fig. 5/14. The area is divided into strips of width Δx and variable height h. The area A of each strip, such as the one shown in red, is $h\,\Delta x$ and is multiplied by the coordinates x_c and y_c of its *centroid* to obtain the moments of the element of area. The sum of the moments for all strips divided by the total area of the strips will give the corresponding centroidal coordinate. A systematic tabulation of the results will permit an orderly evaluation of the total area ΣA, the sums $\Sigma A x_c$ and $\Sigma A y_c$, and the centroidal coordinates

$$\bar{x} = \frac{\Sigma A x_c}{\Sigma A} \qquad \bar{y} = \frac{\Sigma A y_c}{\Sigma A}$$

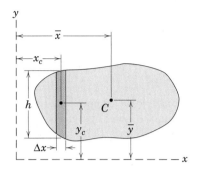

Figure 5/14

We can increase the accuracy of the approximation by decreasing the widths of the strips. In all cases the average height of the strip should be estimated in approximating the areas. Although it is usually advantageous to use elements of constant width, it is not necessary. In fact, we may use elements of any size and shape which approximate the given area to satisfactory accuracy.

Irregular Volumes

To locate the centroid of an irregular volume, we may reduce the problem to one of locating the centroid of an area. Consider the volume shown in Fig. 5/15, where the magnitudes A of the cross-sectional areas normal to the x-direction are plotted against x as shown. A vertical strip of area under the curve is $A\,\Delta x$, which equals the corresponding element of volume ΔV. Thus, the area under the plotted curve represents the volume of the body, and the x-coordinate of the centroid of the area under the curve is given by

$$\bar{x} = \frac{\Sigma(A\,\Delta x)x_c}{\Sigma A\,\Delta x} \qquad \text{which equals} \qquad \bar{x} = \frac{\Sigma V x_c}{\Sigma V}$$

for the centroid of the actual volume.

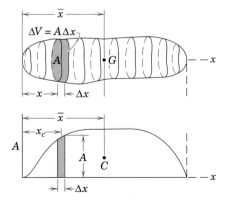

Figure 5/15

Sample Problem 5/6

Locate the centroid of the shaded area.

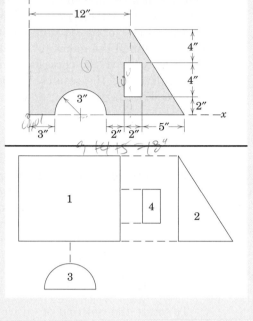

Solution. The composite area is divided into the four elementary shapes shown in the lower figure. The centroid locations of all these shapes may be obtained from Table D/3. Note that the areas of the "holes" (parts 3 and 4) are taken as negative in the following table:

PART	A in.^{2}	$\bar{x}$ in.	$\bar{y}$ in.	$\bar{x}A$ in.^{3}	$\bar{y}A$ in.^{3}
1	120	6	5	720	600
2	30	14	10/3	420	100
3	−14.14	6	1.273	−84.8	−18
4	−8	12	4	−96	−32
TOTALS	127.9			959	650

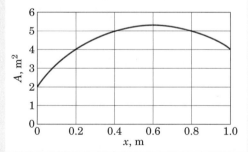

The area counterparts to Eqs. 5/7 are now applied and yield

$$\left[\bar{X} = \frac{\Sigma A\bar{x}}{\Sigma A}\right] \qquad \bar{X} = \frac{959}{127.9} = 7.50 \text{ in.} \qquad Ans.$$

$$\left[\bar{Y} = \frac{\Sigma A\bar{y}}{\Sigma A}\right] \qquad \bar{Y} = \frac{650}{127.9} = 5.08 \text{ in.} \qquad Ans.$$

Sample Problem 5/7

Approximate the x-coordinate of the volume centroid of a body whose length is 1 m and whose cross-sectional area varies with x as shown in the figure.

Solution. The body is divided into five sections. For each section, the average area, volume, and centroid location are determined and entered in the following table:

INTERVAL	A_{av} m^2	Volume V m^3	$\bar{x}$ m	$V\bar{x}$ m^4
0–0.2	3	0.6	0.1	0.060
0.2–0.4	4.5	0.90	0.3	0.270
0.4–0.6	5.2	1.04	0.5	0.520
0.6–0.8	5.2	1.04	0.7	0.728
0.8–1.0	4.5	0.90	0.9	0.810
TOTALS		4.48		2.388

$$\textcircled{1} \quad \left[\bar{X} = \frac{\Sigma V\bar{x}}{\Sigma V}\right] \qquad \bar{X} = \frac{2.388}{4.48} = 0.533 \text{ m} \qquad Ans.$$

Helpful Hint

① Note that the shape of the body as a function of y and z does not affect $\bar{X}$.

Sample Problem 5/8

Locate the center of mass of the bracket-and-shaft combination. The vertical face is made from sheet metal which has a mass of 25 kg/m^2. The material of the horizontal base has a mass of 40 kg/m^2, and the steel shaft has a density of 7.83 Mg/m^3.

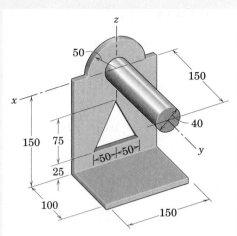

Dimensions in millimeters

Solution. The composite body may be considered to be composed of the five elements shown in the lower portion of the illustration. The triangular part will be taken as a negative mass. For the reference axes indicated it is clear by symmetry that the x-coordinate of the center of mass is zero.

The mass m of each part is easily calculated and should need no further explanation. For Part 1 we have from Sample Problem 5/3

$$\bar{z} = \frac{4r}{3\pi} = \frac{4(50)}{3\pi} = 21.2 \text{ mm}$$

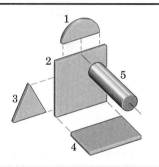

For Part 3 we see from Sample Problem 5/2 that the centroid of the triangular mass is one-third of its altitude above its base. Measurement from the coordinate axes becomes

$$\bar{z} = -[150 - 25 - \tfrac{1}{3}(75)] = -100 \text{ mm}$$

The y- and z-coordinates to the mass centers of the remaining parts should be evident by inspection. The terms involved in applying Eqs. 5/7 are best handled in the form of a table as follows:

PART	m kg	$\bar{y}$ mm	$\bar{z}$ mm	$m\bar{y}$ kg·m	$m\bar{z}$ kg·mm
1	0.098	0	21.2	0	2.08
2	0.562	0	−75.0	0	−42.19
3	−0.094	0	−100.0	0	9.38
4	0.600	50.0	−150.0	30.0	−90.00
5	1.476	75.0	0	110.7	0
TOTALS	2.642			140.7	−120.73

Equations 5/7 are now applied and the results are

$$\left[\bar{Y} = \frac{\Sigma m\bar{y}}{\Sigma m}\right] \qquad \bar{Y} = \frac{140.7}{2.642} = 53.3 \text{ mm} \qquad\qquad Ans.$$

$$\left[\bar{Z} = \frac{\Sigma m\bar{z}}{\Sigma m}\right] \qquad \bar{Z} = \frac{-120.73}{2.642} = -45.7 \text{ mm} \qquad\qquad Ans.$$

PROBLEMS

Introductory Problems

5/45 Determine the *y*-coordinate of the centroid of the shaded area.

Ans. $\overline{Y} = 37.1$ mm

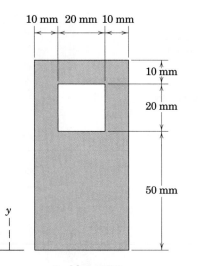

Problem 5/45

5/46 Calculate the *y*-coordinate of the centroid of the shaded area.

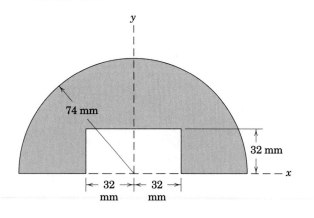

Problem 5/46

5/47 Determine the coordinates of the centroid of the trapezoidal area shown.

Ans. $\overline{X} = 25.3$ in., $\overline{Y} = 28.0$ in.

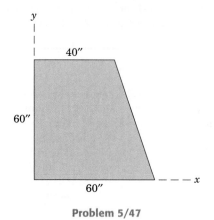

Problem 5/47

5/48 Determine the *x*- and *y*-coordinates of the centroid of the shaded area.

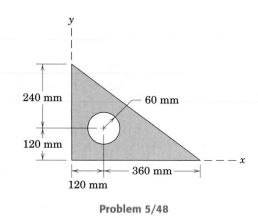

Problem 5/48

5/49 Determine the *y*-coordinate of the centroid of the shaded area in terms of *h*.

Ans. $\overline{Y} = 0.412h$

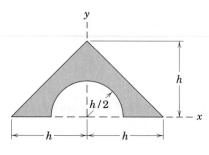

Problem 5/49

5/50 Determine the *y*-coordinate of the centroid of the shaded area. The triangle is equilateral.

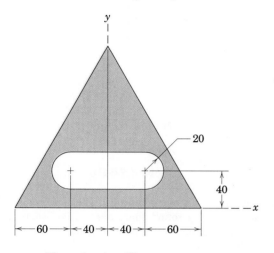

Dimensions in millimeters

Problem 5/50

5/51 Determine the *x*- and *y*-coordinates of the centroid of the shaded area.

Ans. $\overline{X} = \overline{Y} = 4.32$ in.

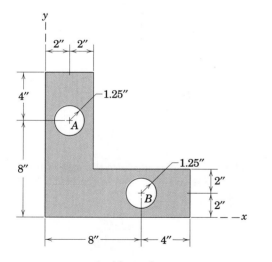

Problem 5/51

5/52 Determine the *x*- and *y*-coordinates of the centroid of the shaded area.

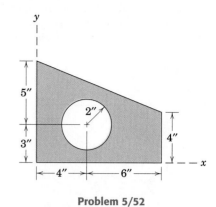

Problem 5/52

5/53 Determine the *x*- and *y*-coordinates of the centroid of the shaded area.

Ans. $\overline{X} = 4.02b, \overline{Y} = 1.628b$

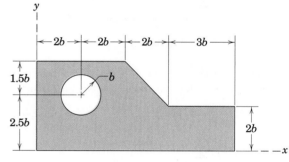

Problem 5/53

Representative Problems

5/54 By inspection, state the quadrant in which the centroid of the shaded area is located. Then determine the coordinates of the centroid. The plate center is M.

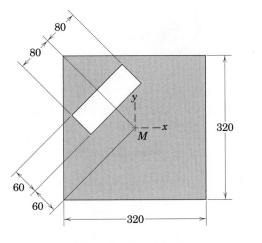

Dimensions in millimeters

Problem 5/54

5/55 Determine the distance $\overline{H}$ from the bottom of the base plate to the centroid of the built-up structural section shown.

Ans. $\overline{H} = 39.3$ mm

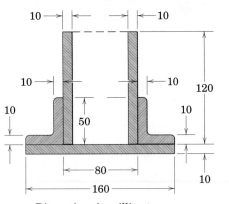

Dimensions in millimeters

Problem 5/55

5/56 Locate the mass center of the slender rod bent into the shape shown.

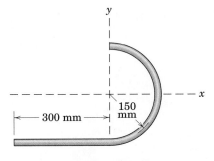

Problem 5/56

5/57 The rigidly connected unit consists of a 2-kg circular disk, a 1.5-kg round shaft, and a 1-kg square plate. Determine the z-coordinate of the mass center of the unit.

Ans. $\overline{Z} = 70$ mm

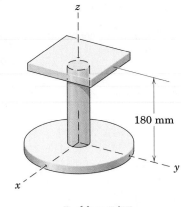

Problem 5/57

5/58 Determine the height above the base of the centroid of the cross-sectional area of the beam. Neglect the fillets.

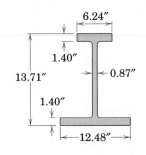

Problem 5/58

5/59 Determine the x-coordinate of the mass center of the bracket constructed of uniform steel plate.

Ans. $\overline{X} = 0.1975$ m

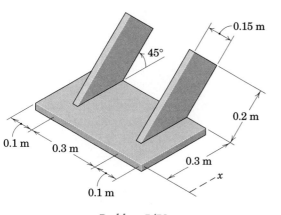

Problem 5/59

5/60 Determine the x- and y-coordinates of the centroid of the area of Prob. 5/26 by the method of this article.

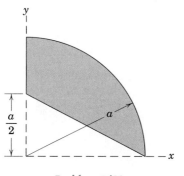

Problem 5/60

5/61 The homogeneous hemisphere with the smaller hemispherical portion removed is repeated here from Prob. 5/42. By the method of this article, determine the x-coordinate of the mass center.

Ans. $\overline{X} = \dfrac{45}{112}R$

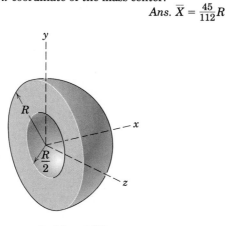

Problem 5/61

5/62 The two upper lengths of the welded Y-shaped assembly of uniform slender rods have a mass per unit length of 0.3 kg/m, while the lower length has a mass of 0.5 kg/m. Locate the mass center of the assembly.

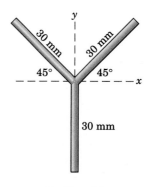

Problem 5/62

5/63 Determine the coordinates of the mass center of the welded assembly of uniform slender rods made from the same bar stock.

Ans. $\overline{X} = \dfrac{3a}{6 + \pi}, \overline{Y} = -\dfrac{2a}{6 + \pi}, \overline{Z} = \dfrac{\pi a}{6 + \pi}$

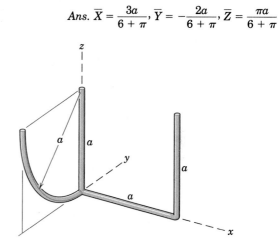

Problem 5/63

5/64 Determine the x-, y-, and z-coordinates of the mass center of the sheet-metal bracket whose thickness is small in comparison with the other dimensions.

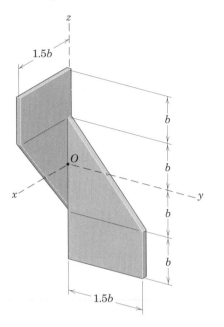

Problem 5/64

5/65 Determine the x- and y-coordinates of the centroid of the shaded area.

Ans. $\overline{X} = 5.56$ in., $\overline{Y} = 4.04$ in.

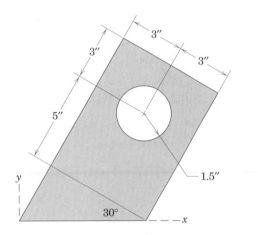

Problem 5/65

5/66 Determine the distance $\overline{H}$ from the bottom of the base to the mass center of the bracket casting.

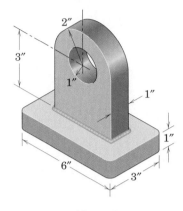

Problem 5/66

5/67 An underwater instrument is modeled as shown in the figure. Determine the coordinates of the centroid of this composite volume.

Ans. $\overline{X} = 38.5$ mm, $\overline{Y} = 13.52$ mm, $\overline{Z} = 0$

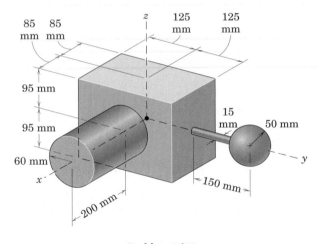

Problem 5/67

5/68 Calculate the coordinates of the mass center of the metal die casting shown.

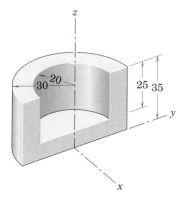

Dimensions in millimeters

Problem 5/68

5/69 Determine the dimension h of the rectangular opening in the square plate which will result in the mass center of the remaining plate being as close to the upper edge as possible.

Ans. $h = 0.586a$

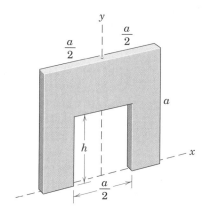

Problem 5/69

5/70 Determine the depth h of the circular hole in the cube for which the z-coordinate of the mass center will have the maximum possible value.

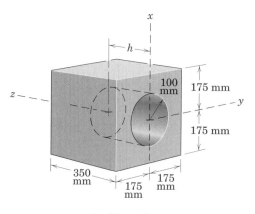

Problem 5/70

▶**5/71** An opening is formed in the thin cylindrical shell. Determine the x-, y-, and z-coordinates of the mass center of the homogeneous body.

Ans. $\overline{X} = -0.509L,\ \overline{Y} = 0.0443R,\ \overline{Z} = -0.01834R$

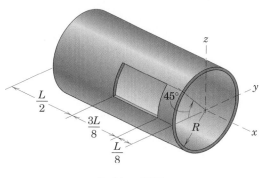

Problem 5/71

▶**5/72** Determine the y-coordinate of the centroid of the shaded area. Use the result of Prob. 5/41.

Ans. $\overline{Y} = 0.353$ m

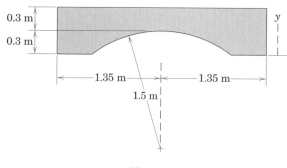

Problem 5/72

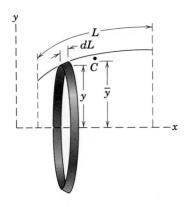

Figure 5/16

5/5   THEOREMS OF PAPPUS*

A very simple method exists for calculating the surface area generated by revolving a plane curve about a nonintersecting axis in the plane of the curve. In Fig. 5/16 the line segment of length L in the x-y plane generates a surface when revolved about the x-axis. An element of this surface is the ring generated by dL. The area of this ring is its circumference times its slant height or $dA = 2\pi y\, dL$. The total area is then

$$A = 2\pi \int y\, dL$$

Because $\bar{y}L = \int y\, dL$, the area becomes

$$\boxed{A = 2\pi\bar{y}L} \tag{5/8}$$

where $\bar{y}$ is the y-coordinate of the centroid C for the line of length L. Thus, the generated area is the same as the lateral area of a right-circular cylinder of length L and radius $\bar{y}$.

In the case of a volume generated by revolving an area about a nonintersecting line in its plane, an equally simple relation exists for finding the volume. An element of the volume generated by revolving the area A about the x-axis, Fig. 5/17, is the elemental ring of cross-section dA and radius y. The volume of the element is its circumference times dA or $dV = 2\pi y\, dA$, and the total volume is

$$V = 2\pi \int y\, dA$$

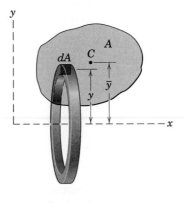

Figure 5/17

*Attributed to Pappus of Alexandria, a Greek geometer who lived in the third century A.D. The theorems often bear the name of Guldinus (Paul Guldin, 1577–1643), who claimed original authorship, although the works of Pappus were apparently known to him.

Because $\bar{y}A = \int y\,dA$, the volume becomes

$$V = 2\pi\bar{y}A \qquad (5/9)$$

where $\bar{y}$ is the y-coordinate of the centroid C of the revolved area A. Thus, we obtain the generated volume by multiplying the generating area by the circumference of the circular path described by its centroid.

The two theorems of Pappus, expressed by Eqs. 5/8 and 5/9, are useful for determining areas and volumes of revolution. They are also used to find the centroids of plane curves and plane areas when we know the corresponding areas and volumes created by revolving these figures about a nonintersecting axis. Dividing the area or volume by 2π times the corresponding line segment length or plane area gives the distance from the centroid to the axis.

If a line or an area is revolved through an angle θ less than 2π, we can determine the generated surface or volume by replacing 2π by θ in Eqs. 5/8 and 5/9. Thus, the more general relations are

$$A = \theta\bar{y}L \qquad (5/8a)$$

and

$$V = \theta\bar{y}A \qquad (5/9a)$$

where θ is expressed in radians.

Sample Problem 5/9

Determine the volume V and surface area A of the complete torus of circular cross section.

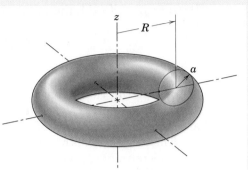

Solution. The torus can be generated by revolving the circular area of radius a through 360° about the z-axis. With the use of Eq. 5/9a, we have

① $$V = \theta \bar{r} A = 2\pi(R)(\pi a^2) = 2\pi^2 R a^2 \qquad \textit{Ans.}$$

Similarly, using Eq. 5/8a gives

$$A = \theta \bar{r} L = 2\pi(R)(2\pi a) = 4\pi^2 R a \qquad \textit{Ans.}$$

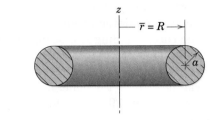

Helpful Hint

① We note that the angle θ of revolution is 2π for the complete ring. This common but special-case result is given by Eq. 5/9.

Sample Problem 5/10

Calculate the volume V of the solid generated by revolving the 60-mm right-triangular area through 180° about the z-axis. If this body were constructed of steel, what would be its mass m?

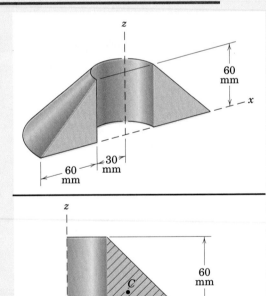

Solution. With the angle of revolution $\theta = 180°$, Eq. 5/9a gives

① $$V = \theta \bar{r} A = \pi[30 + \tfrac{1}{3}(60)][\tfrac{1}{2}(60)(60)] = 2.83(10^5) \text{ mm}^3 \qquad \textit{Ans.}$$

The mass of the body is then

$$m = \rho V = \left[7830 \, \frac{\text{kg}}{\text{m}^3}\right][2.83(10^5) \text{ mm}^3]\left[\frac{1 \text{ m}}{1000 \text{ mm}}\right]^3$$

$$= 2.21 \text{ kg} \qquad \textit{Ans.}$$

Helpful Hint

① Note that θ must be in radians.

PROBLEMS

Introductory Problems

5/73 Using the methods of this article, determine the surface area A and volume V of the body formed by revolving the rectangular area through 360° about the z-axis.

Ans. $A = 2640$ mm^2, $V = 3170$ mm^3

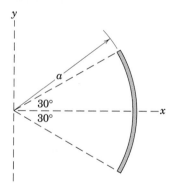

Problem 5/73

5/74 The circular arc is rotated through 360° about the y-axis. Determine the outer surface area S of the resulting body, which is a portion of a sphere.

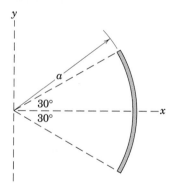

Problem 5/74

5/75 The area of the circular sector is rotated through 180° about the y-axis. Determine the volume of the resulting body, which is a portion of a sphere.

Ans. $V = \dfrac{\pi a^3}{3}$

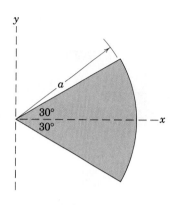

Problem 5/75

5/76 Compute the volume V of the solid generated by revolving the right triangle about the z-axis through 180°.

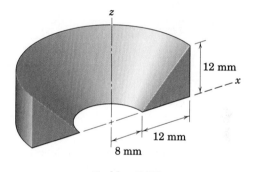

Problem 5/76

5/77 The body shown in cross section is a half-circular ring formed by revolving the cross-hatched area 180° about the z-axis. Determine the surface area A of the body.

Ans. $A = 90\ 000\ \text{mm}^2$

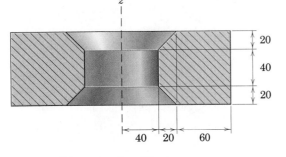

Dimensions in millimeters

Problem 5/77

5/78 Calculate the volume V of the complete ring of cross section shown.

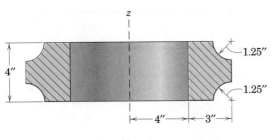

Problem 5/78

5/79 Determine the volume V generated by revolving the quarter-circular area about the z-axis through an angle of 90°.

Ans. $V = \left(\dfrac{\pi a^3}{24}\right)(3\pi + 4)$

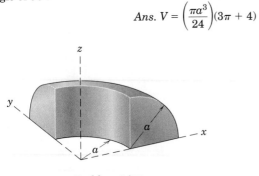

Problem 5/79

Representative Problems

5/80 The area shown is rotated through 360° about the y-axis. Determine the volume of the resulting body, which is a sphere with a significant portion removed.

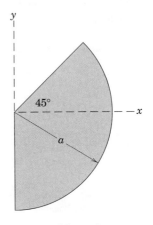

Problem 5/80

5/81 The water storage tank is a shell of revolution and is to be sprayed with two coats of paint which has a coverage of 500 ft² per gallon. The engineer (who remembers mechanics) consults a scale drawing of the tank and determines that the curved line ABC has a length of 34 ft and that its centroid is 8.2 ft from the centerline of the tank. How many gallons of paint will be used for the tank including the vertical cylindrical column?

Ans. 8.82 gal

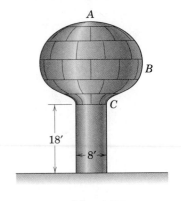

Problem 5/81

5/82 Determine the total surface area A and volume V of the complete solid shown in cross section. Determine the mass of the body if it is constructed of steel.

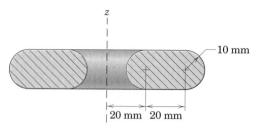

Problem 5/82

5/83 Calculate the volume V of the rubber gasket formed by the complete ring of the semicircular cross section shown. Also compute the surface area A of the outside of the ring.

Ans. $V = 10.09$ in.^{3}, $A = 29.3$ in.^{2}

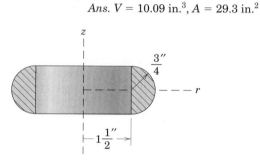

Problem 5/83

5/84 The body shown in cross section is a complete circular ring formed by revolving the cross-hatched area about the z-axis. Determine the surface area A and volume V of the body.

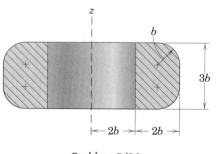

Problem 5/84

5/85 The body shown in cross section is a large neoprene washer. Compute its suface area A and volume V.

Ans. $A = 7290$ mm^2, $V = 24\ 400$ mm^3

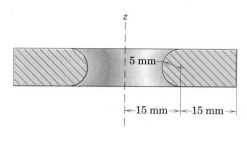

Problem 5/85

5/86 The two circular arcs AB and BC are revolved about the vertical axis to obtain the surface of revolution shown. Compute the area A of the outside of this surface.

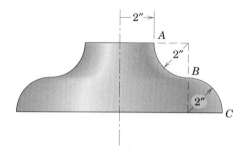

Problem 5/86

5/87 A thin shell, shown in section, has the form generated by revolving the arc about the z-axis through $360°$. Determine the surface area A of one of the two sides of the shell.

Ans. $A = 4\pi r(R\alpha - r \sin \alpha)$

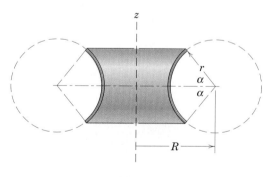

Problem 5/87

5/88 Calculate the volume formed by completely revolving the cross-sectional area shown about the z-axis of symmetry.

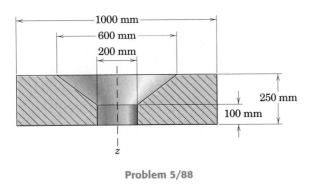

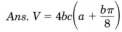

Problem 5/88

5/89 Calculate the weight W of the aluminum casting shown. The solid is generated by revolving the trapezoidal area shown about the z-axis through $180°$.

Ans. $W = 10.08$ lb

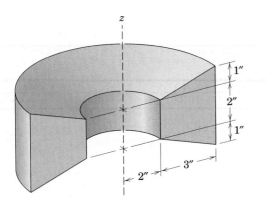

Problem 5/89

5/90 Determine the volume V and total surface area A of the solid generated by revolving the area shown through $180°$ about the z-axis.

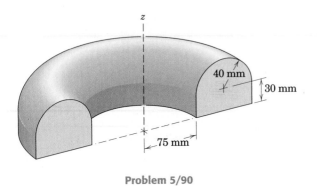

Problem 5/90

5/91 The shaded area is bounded by one half-cycle of a sine wave and the axis of the sine wave. Determine the volume generated by completely revolving the area about the x-axis.

$$Ans.\ V = 4bc\left(a + \frac{b\pi}{8}\right)$$

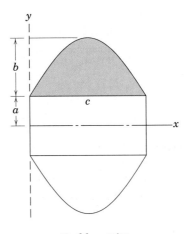

Problem 5/91

5/92 Find the volume V of the solid generated by revolving the shaded area about the z-axis through $90°$.

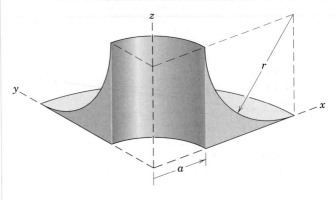

Problem 5/92

5/93 Calculate the mass m of concrete required to construct the arched dam shown. Concrete has a density of 2.40 Mg/m³.

Ans. $m = 1.126(10^6)$ Mg

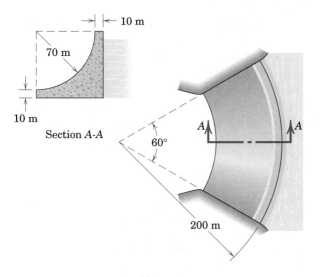

Section *A-A*

Problem 5/93

5/94 In order to provide sufficient support for the stone masonry arch designed as shown, it is necessary to know its total weight W. Use the results of Prob. 5/11 and determine W. The density of stone masonry is 2.40 Mg/m³.

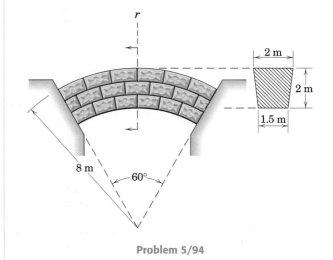

Problem 5/94

SECTION B SPECIAL TOPICS

5/6 BEAMS—EXTERNAL EFFECTS

Beams are structural members which offer resistance to bending due to applied loads. Most beams are long prismatic bars, and the loads are usually applied normal to the axes of the bars.

Beams are undoubtedly the most important of all structural members, so it is important to understand the basic theory underlying their design. To analyze the load-carrying capacities of a beam we must first establish the equilibrium requirements of the beam as a whole and any portion of it considered separately. Second, we must establish the relations between the resulting forces and the accompanying internal resistance of the beam to support these forces. The first part of this analysis requires the application of the principles of statics. The second part involves the strength characteristics of the material and is usually treated in studies of the mechanics of solids or the mechanics of materials.

This article is concerned with the *external* loading and reactions acting on a beam. In Art. 5/7 we calculate the distribution along the beam of the *internal* force and moment.

Types of Beams

Beams supported so that their external support reactions can be calculated by the methods of statics alone are called *statically determinate beams*. A beam which has more supports than needed to provide equilibrium is *statically indeterminate*. To determine the support reactions for such a beam we must consider its load-deformation properties in addition to the equations of static equilibrium. Figure 5/18 shows examples

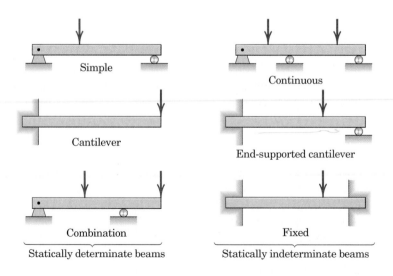

Figure 5/18

of both types of beams. In this article we will analyze statically determinate beams only.

Beams may also be identified by the type of external loading they support. The beams in Fig. 5/18 are supporting concentrated loads, whereas the beam in Fig. 5/19 is supporting a distributed load. The intensity w of a distributed load may be expressed as force per unit length of beam. The intensity may be constant or variable, continuous or discontinuous. The intensity of the loading in Fig. 5/19 is constant from C to D and variable from A to C and from D to B. The intensity is discontinuous at D, where it changes magnitude abruptly. Although the intensity itself is not discontinuous at C, the rate of change of intensity dw/dx is discontinuous.

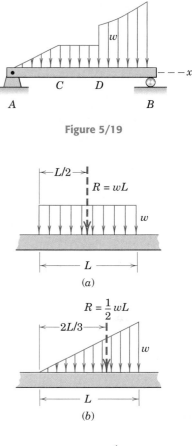

Figure 5/19

Distributed Loads

Loading intensities which are constant or which vary linearly are easily handled. Figure 5/20 illustrates the three most common cases and the resultants of the distributed loads in each case.

In cases a and b of Fig. 5/20, we see that the resultant load R is represented by the area formed by the intensity w (force per unit length of beam) and the length L over which the force is distributed. The resultant passes through the centroid of this area.

In part c of Fig. 5/20, the trapezoidal area is broken into a rectangular and a triangular area, and the corresponding resultants R_1 and R_2 of these subareas are determined separately. Note that a single resultant could be determined by using the composite technique for finding centroids, which was discussed in Art. 5/4. Usually, however, the determination of a single resultant is unnecessary.

For a more general load distribution, Fig. 5/21, we must start with a differential increment of force $dR = w\,dx$. The total load R is then the sum of the differential forces, or

$$R = \int w\,dx$$

As before, the resultant R is located at the centroid of the area under consideration. The x-coordinate of this centroid is found by the principle of moments $R\bar{x} = \int xw\,dx$, or

$$\bar{x} = \frac{\displaystyle\int xw\,dx}{R}$$

For the distribution of Fig. 5/21, the vertical coordinate of the centroid need not be found.

Once the distributed loads have been reduced to their equivalent concentrated loads, the external reactions acting on the beam may be found by a straightforward static analysis as developed in Chapter 3.

Figure 5/20

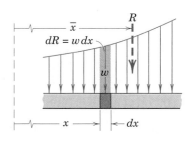

Figure 5/21

Sample Problem 5/11

Determine the equivalent concentrated load(s) and external reactions for the simply supported beam which is subjected to the distributed load shown.

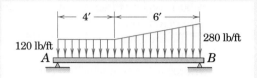

Solution. The area associated with the load distribution is divided into the rectangular and triangular areas shown. The concentrated-load values are determined by computing the areas, and these loads are located at the centroids of the respective areas. ①

Once the concentrated loads are determined, they are placed on the free-body diagram of the beam along with the external reactions at A and B. Using principles of equilibrium, we have

$[\Sigma M_A = 0]$ $\qquad$ $1200(5) + 480(8) - R_B(10) = 0$

$\qquad\qquad\qquad\qquad$ $R_B = 984 \text{ lb}$ $\qquad\qquad$ *Ans.*

$[\Sigma M_B = 0]$ $\qquad$ $R_A(10) - 1200(5) - 480(2) = 0$

$\qquad\qquad\qquad\qquad$ $R_A = 696 \text{ lb}$ $\qquad\qquad$ *Ans.*

Helpful Hint

① Note that it is usually unnecessary to reduce a given distributed load to a *single* concentrated load.

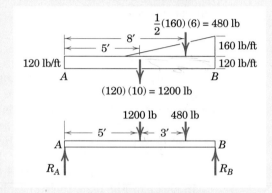

Sample Problem 5/12

Determine the reaction at the support A of the loaded cantilever beam.

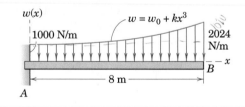

Solution. The constants in the load distribution are found to be $w_0 = 1000$ N/m and $k = 2$ N/m⁴. The load R is then ①

$$R = \int w \, dx = \int_0^8 (1000 + 2x^3) \, dx = \left(1000x + \frac{x^4}{2}\right)\Big|_0^8 = 10\,050 \text{ N}$$

② The x-coordinate of the centroid of the area is found by

$$\bar{x} = \frac{\int xw \, dx}{R} = \frac{1}{10\,050} \int_0^8 x(1000 + 2x^3) \, dx$$

$$= \frac{1}{10\,050} \left(500x^2 + \tfrac{2}{5}x^5\right)\Big|_0^8 = 4.49 \text{ m}$$

From the free-body diagram of the beam, we have

$[\Sigma M_A = 0]$ $\qquad\qquad$ $M_A - (10\,050)(4.49) = 0$

$\qquad\qquad\qquad\qquad$ $M_A = 45\,100 \text{ N·m}$ $\qquad\qquad$ *Ans.*

$[\Sigma F_y = 0]$ $\qquad\qquad$ $A_y = 10\,050 \text{ N}$ $\qquad\qquad$ *Ans.*

Note that $A_x = 0$ by inspection.

Helpful Hints

① Use caution with the units of the constants w_0 and k.

② The student should recognize that the calculation of R and its location $\bar{x}$ is simply an application of centroids as treated in Art. 5/3.

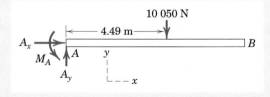

PROBLEMS

Introductory Problems

5/95 Calculate the supporting force R_A and moment M_A at A for the loaded cantilever beam.

Ans. $R_A = 2.4$ kN, $M_A = 14.4$ kN·m CCW

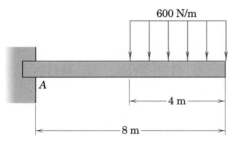

600 N/m

A

4 m

8 m

Problem 5/95

5/96 Calculate the reactions at A and B for the beam subjected to the triangular load distribution.

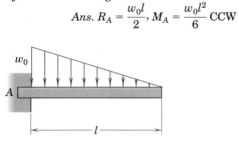

700 lb/ft

A B

3′   3′

Problem 5/96

5/97 Determine the reactions at the built-in end of the beam subjected to the triangular load distribution.

Ans. $R_A = \dfrac{w_0 l}{2}$, $M_A = \dfrac{w_0 l^2}{6}$ CCW

w_0

A

l

Problem 5/97

5/98 Determine the reactions at A and B for the loaded beam.

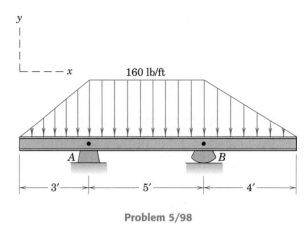

y

x

160 lb/ft

A B

3′   5′   4′

Problem 5/98

5/99 Find the reaction at A due to the uniform loading and the applied couple.

Ans. $R_A = 1600$ lb, $M_A = 800$ lb-ft CW

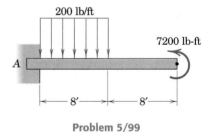

200 lb/ft

7200 lb-ft

A

8′   8′

Problem 5/99

5/100 Determine the reactions at A for the cantilever beam subjected to the distributed and concentrated loads.

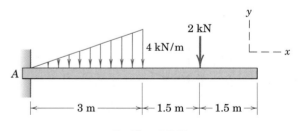

y

2 kN

4 kN/m

A

x

3 m   1.5 m   1.5 m

Problem 5/100

5/101 Determine the reactions at A and B for the beam subjected to a combination of distributed and point loads.

Ans. $A_x = 750$ N, $A_y = 3.07$ kN, $B_y = 1.224$ kN

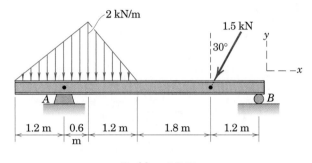

Problem 5/101

5/102 Calculate the supporting reactions at A and B for the beam subjected to the two linearly distributed loads.

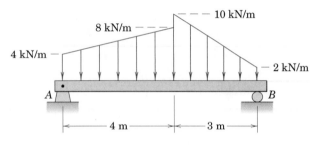

Problem 5/102

5/103 Determine the reactions at the supports of the beam which is loaded as shown.

Ans. $R_A = 2230$ N, $R_B = 2170$ N

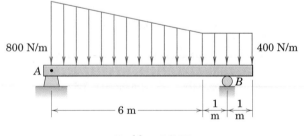

Problem 5/103

5/104 The beam is subjected to the distributed load and the couple shown. If M is slowly increased starting from zero, at what value M_0 will contact at B change from the lower surface to the upper surface?

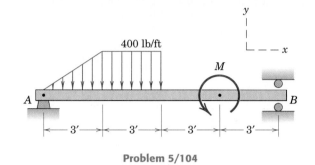

Problem 5/104

Representative Problems

5/105 Determine the force and moment reactions at A for the beam which is subjected to the load combination shown.

Ans. $R_A = 5.5$ kips, $M_A = 71.5$ kip-ft CW

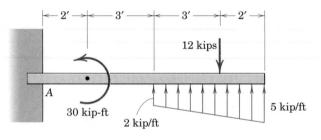

Problem 5/105

5/106 Determine the force and moment reactions at the support A of the built-in beam which is subjected to the sine-wave load distribution.

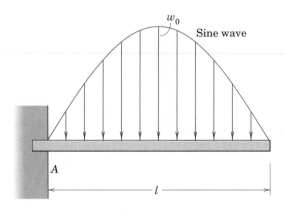

Problem 5/106

5/107 Determine the reactions at points A and B of the beam subjected to the elliptical and uniform load distributions. At which surface, upper or lower, is the reaction at A exerted?

Ans. $A = 5.15$ kN, $B = 5.37$ kN, upper

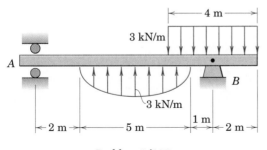

Problem 5/107

5/108 The cantilever beam is subjected to a parabolic distribution of load symmetrical about the middle of the beam. Determine the supporting force R_A and moment M_A acting on the beam at A.

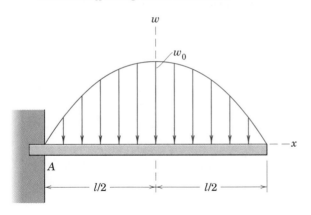

Problem 5/108

5/109 Determine the force and moment reactions at the support A of the cantilever beam subjected to the load distribution shown.

Ans. $R_A = \frac{2}{3}w_0 b$, $M_A = \frac{14}{15}w_0 b^2$ CW

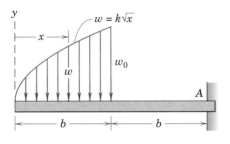

Problem 5/109

5/110 A cantilever beam supports the variable load shown. Calculate the supporting force R_A and moment M_A at A.

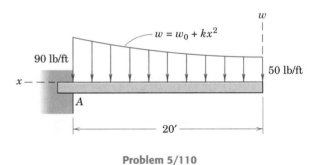

Problem 5/110

5/111 Determine the reactions at points A and B of the inclined beam subjected to the vertical load distribution shown. The value of the load distribution at the right end of the beam is 5 kN per *horizontal* meter.

Ans. $A = 2.5$ kN, $B = 6.61$ kN

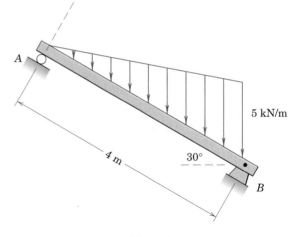

Problem 5/111

5/112 Determine the reactions at the support for the beam which is subjected to the combination of uniform and parabolic loading distributions.

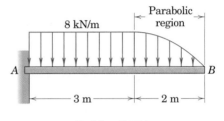

Problem 5/112

5/113 A beam is subjected to the variable loading shown. Calculate the support reactions at A and B.

$$Ans.\ R_A = 1900\ \text{lb},\ R_B = 1600\ \text{lb}$$

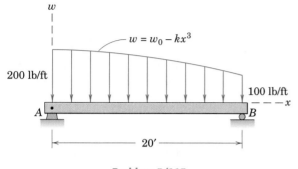

Problem 5/113

5/114 Determine the reactions at A and B for the beam subjected to the distributed and concentrated loads.

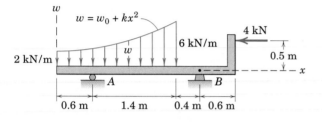

Problem 5/114

5/115 Determine the reactions at the supports of the beam which is acted on by the combination of uniform and parabolic loading distributions.

$$Ans.\ R_A = R_B = 7\ \text{kN}$$

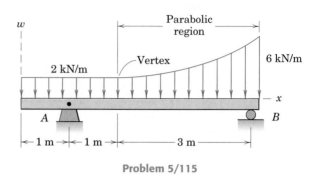

Problem 5/115

▶**5/116** The transition between the loads of 10 kN/m and 37 kN/m is accomplished by means of a cubic function of form $w = k_0 + k_1 x + k_2 x^2 + k_3 x^3$, the slope of which is zero at its end points $x = 1$ m and $x = 4$ m. Determine the reactions at A and B.

$$Ans.\ R_A = 43.1\ \text{kN},\ R_B = 74.4\ \text{kN}$$

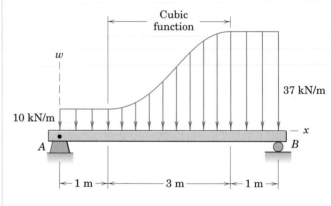

Problem 5/116

5/7   BEAMS—INTERNAL EFFECTS

The previous article treated the reduction of a distributed force to one or more equivalent concentrated forces and the subsequent determination of the external reactions acting on the beam. In this article we introduce internal beam effects and apply principles of statics to calculate the internal shear force and bending moment as functions of location along the beam.

Shear, Bending, and Torsion

In addition to supporting tension or compression, a beam can resist shear, bending, and torsion. These three effects are illustrated in Fig. 5/22. The force V is called the *shear force*, the couple M is called the *bending moment*, and the couple T is called a *torsional moment*. These effects represent the vector components of the resultant of the forces acting on a transverse section of the beam as shown in the lower part of the figure.

Consider the shear force V and bending moment M caused by forces applied to the beam in a single plane. The conventions for positive values of shear V and bending moment M shown in Fig. 5/23 are the ones generally used. From the principle of action and reaction we can see that the directions of V and M are reversed on the two sections. It is frequently impossible to tell without calculation whether the shear and moment at a particular section are positive or negative. For this reason it is advisable to represent V and M in their positive directions on the free-body diagrams and let the algebraic signs of the calculated values indicate the proper directions.

As an aid to the physical interpretation of the bending couple M, consider the beam shown in Fig. 5/24 bent by the two equal and opposite positive moments applied at the ends. The cross section of the beam is treated as an H-section with a very narrow center web and heavy top and bottom flanges. For this beam we may neglect the load carried by the small web compared with that carried by the two flanges. The upper flange of the beam clearly is shortened and is under compression, whereas the lower flange is lengthened and is under tension. The resultant of the two forces, one tensile and the other compressive, acting on any section is a couple and has the value of the bending moment on the section. If a beam having some other cross-sectional shape were loaded

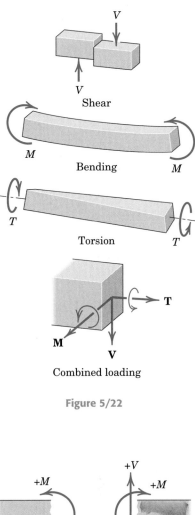

Figure 5/22

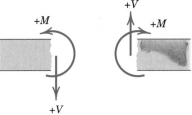

Figure 5/23

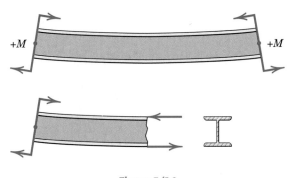

Figure 5/24

in the same way, the distribution of force over the cross section would be different, but the resultant would be the same couple.

Shear-Force and Bending-Moment Diagrams

The variation of shear force V and bending moment M over the length of a beam provides information necessary for the design analysis of the beam. In particular, the maximum magnitude of the bending moment is usually the primary consideration in the design or selection of a beam, and its value and position should be determined. The variations in shear and moment are best shown graphically, and the expressions for V and M when plotted against distance along the beam give the *shear-force* and *bending-moment diagrams* for the beam.

The first step in the determination of the shear and moment relations is to establish the values of all external reactions on the beam by applying the equations of equilibrium to a free-body diagram of the beam as a whole. Next, we isolate a portion of the beam, either to the right or to the left of an arbitrary transverse section, with a free-body diagram, and apply the equations of equilibrium to this isolated portion of the beam. These equations will yield expressions for the shear force V and bending moment M acting at the cut section on the part of the beam isolated. The part of the beam which involves the smaller number of forces, either to the right or to the left of the arbitrary section, usually yields the simpler solution.

We should avoid using a transverse section which coincides with the location of a concentrated load or couple, as such a position represents a point of discontinuity in the variation of shear or bending moment. Finally, it is important to note that the calculations for V and M on each section chosen should be consistent with the positive convention illustrated in Fig. 5/23.

General Loading, Shear, and Moment Relationships

For any beam with distributed loads we can establish certain general relationships which will aid greatly in the determination of the shear and moment distributions along the beam. Figure 5/25 represents a portion of a loaded beam, where an element dx of the beam is isolated. The loading w represents the force per unit length of beam. At the location x the shear V and moment M acting on the element are drawn in their positive directions. On the opposite side of the element where the

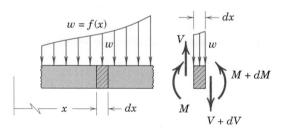

Figure 5/25

coordinate is $x + dx$, these quantities are also shown in their positive directions. They must, however, be labeled $V + dV$ and $M + dM$, since V and M change with x. The applied loading w may be considered constant over the length of the element, since this length is a differential quantity and the effect of any change in w disappears in the limit compared with the effect of w itself.

Equilibrium of the element requires that the sum of the vertical forces be zero. Thus, we have

$$V - w\, dx - (V + dV) = 0$$

or

$$\boxed{w = -\frac{dV}{dx}} \qquad \textbf{(5/10)}$$

We see from Eq. 5/10 that the slope of the shear diagram must everywhere be equal to the negative of the value of the applied loading. Equation 5/10 holds on either side of a concentrated load but not at the concentrated load because of the discontinuity produced by the abrupt change in shear.

We may now express the shear force V in terms of the loading w by integrating Eq. 5/10. Thus,

$$\int_{V_0}^{V} dV = -\int_{x_0}^{x} w\, dx$$

or

$$V = V_0 + \text{(the negative of the area under the loading curve from } x_0 \text{ to } x)$$

In this expression V_0 is the shear force at x_0 and V is the shear force at x. Summing the area under the loading curve is usually a simple way to construct the shear-force diagram.

Equilibrium of the element in Fig. 5/25 also requires that the moment sum be zero. Summing moments about the left side of the element gives

$$M + w\, dx\, \frac{dx}{2} + (V + dV)\, dx - (M + dM) = 0$$

The two M's cancel, and the terms $w(dx)^2/2$ and $dV\, dx$ may be dropped, since they are differentials of higher order than those which remain. This leaves

$$\boxed{V = \frac{dM}{dx}} \qquad \textbf{(5/11)}$$

Because of its economical use of material in achieving bending stiffness, the I-beam is a very common structural element.

© Jake Wyman/Photonica/Getty Images

which expresses the fact that the shear everywhere is equal to the slope of the moment curve. Equation 5/11 holds on either side of a concentrated couple but not at the concentrated couple because of the discontinuity caused by the abrupt change in moment.

We may now express the moment M in terms of the shear V by integrating Eq. 5/11. Thus,

$$\int_{M_0}^{M} dM = \int_{x_0}^{x} V \, dx$$

or

$$M = M_0 + \text{(area under the shear diagram from } x_0 \text{ to } x)$$

In this expression M_0 is the bending moment at x_0 and M is the bending moment at x. For beams where there is no externally applied moment M_0 at $x_0 = 0$, the total moment at any section equals the area under the shear diagram up to that section. Summing the area under the shear diagram is usually the simplest way to construct the moment diagram.

When V passes through zero and is a continuous function of x with $dV/dx \neq 0$, the bending moment M will be a maximum or a minimum, since $dM/dx = 0$ at such a point. Critical values of M also occur when V crosses the zero axis discontinuously, which occurs for beams under concentrated loads.

We observe from Eqs. 5/10 and 5/11 that the degree of V in x is one higher than that of w. Also M is of one higher degree in x than is V. Consequently, M is two degrees higher in x than w. Thus for a beam loaded by $w = kx$, which is of the first degree in x, the shear V is of the second degree in x and the bending moment M is of the third degree in x.

Equations 5/10 and 5/11 may be combined to yield

$$\boxed{\frac{d^2M}{dx^2} = -w} \tag{5/12}$$

Thus, if w is a known function of x, the moment M can be obtained by two integrations, provided that the limits of integration are properly evaluated each time. This method is usable only if w is a continuous function of x.*

When bending in a beam occurs in more than a single plane, we may perform a separate analysis in each plane and combine the results vectorially.

---

*When w is a discontinuous function of x, it is possible to introduce a special set of expressions called *singularity functions* which permit writing analytical expressions for shear V and moment M over an interval which includes discontinuities. These functions are not discussed in this book.

Sample Problem 5/13

Determine the shear and moment distributions produced in the simple beam by the 4-kN concentrated load.

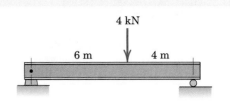

Solution. From the free-body diagram of the entire beam we find the support reactions, which are

$$R_1 = 1.6 \text{ kN} \qquad R_2 = 2.4 \text{ kN}$$

A section of the beam of length x is next isolated with its free-body diagram on which we show the shear V and the bending moment M in their positive directions. Equilibrium gives

$$[\Sigma F_y = 0] \qquad\qquad 1.6 - V = 0 \qquad V = 1.6 \text{ kN}$$

$$[\Sigma M_{R_1} = 0] \qquad\qquad M - 1.6x = 0 \qquad M = 1.6x$$

① These values of V and M apply to all sections of the beam to the left of the 4-kN load.

A section of the beam to the right of the 4-kN load is next isolated with its free-body diagram on which V and M are shown in their positive directions. Equilibrium requires

$$[\Sigma F_y = 0] \qquad\qquad V + 2.4 = 0 \qquad V = -2.4 \text{ kN}$$

$$[\Sigma M_{R_2} = 0] \qquad -(2.4)(10 - x) + M = 0 \qquad M = 2.4(10 - x)$$

These results apply only to sections of the beam to the right of the 4-kN load.

The values of V and M are plotted as shown. The maximum bending moment occurs where the shear changes direction. As we move in the positive x-direction starting with $x = 0$, we see that the moment M is merely the accumulated area under the shear diagram.

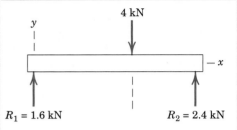

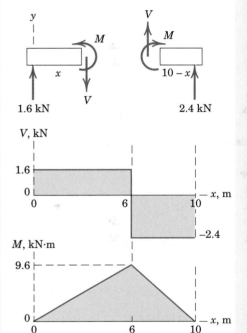

Helpful Hint

① We must be careful not to take our section at a concentrated load (such as $x = 6$ m) since the shear and moment relations involve discontinuities at such positions.

Sample Problem 5/14

The cantilever beam is subjected to the load intensity (force per unit length) which varies as $w = w_0 \sin (\pi x/l)$. Determine the shear force V and bending moment M as functions of the ratio x/l.

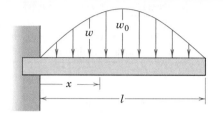

Solution. The free-body diagram of the entire beam is drawn first so that the shear force V_0 and bending moment M_0 which act at the supported end at $x = 0$ can be computed. By convention V_0 and M_0 are shown in their positive mathematical senses. A summation of vertical forces for equilibrium gives

$$[\Sigma F_y = 0] \qquad V_0 - \int_0^l w \, dx = 0 \qquad V_0 = \int_0^l w_0 \sin \frac{\pi x}{l} \, dx = \frac{2w_0 l}{\pi}$$

① A summation of moments about the left end at $x = 0$ for equilibrium gives

$$[\Sigma M = 0] \qquad -M_0 - \int_0^l x(w \, dx) = 0 \qquad M_0 = -\int_0^l w_0 x \sin \frac{\pi x}{l} \, dx$$

$$M_0 = \frac{-w_0 l^2}{\pi^2} \left[\sin \frac{\pi x}{l} - \frac{\pi x}{l} \cos \frac{\pi x}{l} \right]_0^l = -\frac{w_0 l^2}{\pi}$$

From a free-body diagram of an arbitrary section of length x, integration of Eq. 5/10 permits us to find the shear force internal to the beam. Thus,

② $$[dV = -w \, dx] \qquad \int_{V_0}^V dV = -\int_0^x w_0 \sin \frac{\pi x}{l} \, dx$$

$$V - V_0 = \left[\frac{w_0 l}{\pi} \cos \frac{\pi x}{l} \right]_0^x \qquad V - \frac{2w_0 l}{\pi} = \frac{w_0 l}{\pi} \left(\cos \frac{\pi x}{l} - 1 \right)$$

or in dimensionless form

$$\frac{V}{w_0 l} = \frac{1}{\pi} \left(1 + \cos \frac{\pi x}{l} \right) \qquad\qquad Ans.$$

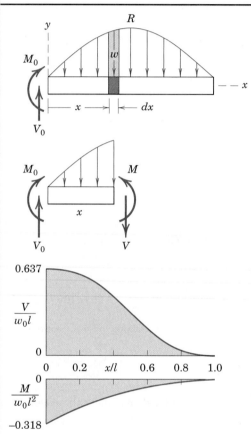

The bending moment is obtained by integration of Eq. 5/11, which gives

$$[dM = V \, dx] \qquad \int_{M_0}^M dM = \int_0^x \frac{w_0 l}{\pi} \left(1 + \cos \frac{\pi x}{l} \right) dx$$

$$M - M_0 = \frac{w_0 l}{\pi} \left[x + \frac{l}{\pi} \sin \frac{\pi x}{l} \right]_0^x$$

$$M = -\frac{w_0 l^2}{\pi} + \frac{w_0 l}{\pi} \left[x + \frac{l}{\pi} \sin \frac{\pi x}{l} - 0 \right]$$

or in dimensionless form

$$\frac{M}{w_0 l^2} = \frac{1}{\pi} \left(\frac{x}{l} - 1 + \frac{1}{\pi} \sin \frac{\pi x}{l} \right) \qquad\qquad Ans.$$

The variations of $V/w_0 l$ and $M/w_0 l^2$ with x/l are shown in the bottom figures. The negative values of $M/w_0 l^2$ indicate that physically the bending moment is in the direction opposite to that shown.

Helpful Hints

① In this case of symmetry it is clear that the resultant $R = V_0 = 2w_0 l/\pi$ of the load distribution acts at midspan, so that the moment requirement is simply $M_0 = -Rl/2 = -w_0 l^2/\pi$. The minus sign tells us that physically the bending moment at $x = 0$ is opposite to that represented on the free-body diagram.

② The free-body diagram serves to remind us that the integration limits for V as well as for x must be accounted for. We see that the expression for V is positive, so that the shear force is as represented on the free-body diagram.

Sample Problem 5/15

Draw the shear-force and bending-moment diagrams for the loaded beam and determine the maximum moment M and its location x from the left end.

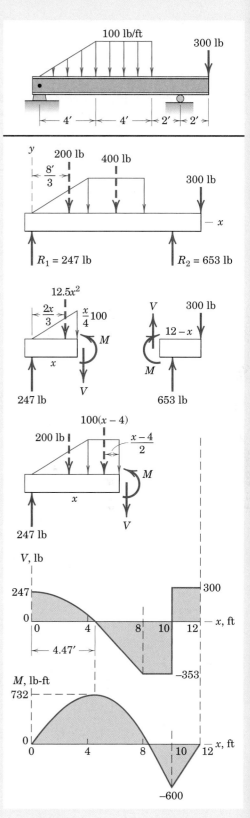

Solution. The support reactions are most easily obtained by considering the resultants of the distributed loads as shown on the free-body diagram of the beam as a whole. The first interval of the beam is analyzed from the free-body diagram of the section for $0 < x < 4$ ft. A summation of vertical forces and a moment summation about the cut section yield

$[\Sigma F_y = 0]$ $\qquad V = 247 - 12.5x^2$

$[\Sigma M = 0]$ $\qquad M + (12.5x^2)\dfrac{x}{3} - 247x = 0$ $\qquad M = 247x - 4.17x^3$

These values of V and M hold for $0 < x < 4$ ft and are plotted for that interval in the shear and moment diagrams shown.

From the free-body diagram of the section for which $4 < x < 8$ ft, equilibrium in the vertical direction and a moment sum about the cut section give

$[\Sigma F_y = 0]$ $\quad V + 100(x-4) + 200 - 247 = 0$ $\quad V = 447 - 100x$

$[\Sigma M = 0]$ $\quad M + 100(x-4)\dfrac{x-4}{2} + 200[x - \frac{2}{3}(4)] - 247x = 0$

$\qquad\qquad M = -267 + 447x - 50x^2$

These values of V and M are plotted on the shear and moment diagrams for the interval $4 < x < 8$ ft.

The analysis of the remainder of the beam is continued from the free-body diagram of the portion of the beam to the right of a section in the next interval. It should be noted that V and M are represented in their positive directions. A vertical-force summation and a moment summation about the section yield

$$V = -353 \text{ lb} \qquad \text{and} \qquad M = 2930 - 353x$$

These values of V and M are plotted on the shear and moment diagrams for the interval $8 < x < 10$ ft.

The last interval may be analyzed by inspection. The shear is constant at $+300$ lb, and the moment follows a straight-line relation beginning with zero at the right end of the beam.

The maximum moment occurs at $x = 4.47$ ft, where the shear curve crosses the zero axis, and the magnitude of M is obtained for this value of x by substitution into the expression for M for the second interval. The maximum moment is

$$M = 732 \text{ lb-ft} \qquad\qquad \textit{Ans.}$$

As before, note that the change in moment M up to any section equals the area under the shear diagram up to that section. For instance, for $x < 4$ ft,

$[\Delta M = \int V\,dx]$ $\qquad M - 0 = \displaystyle\int_0^x (247 - 12.5x^2)\,dx$

and, as above, $\qquad M = 247x - 4.17x^3$

PROBLEMS

Introductory Problems

5/117 Determine the shear-force and bending-moment distributions produced in the beam by the concentrated load. What are the values of the shear and moment at $x = l/2$?

$$Ans. \ V = P, M = -\frac{Pl}{6}$$

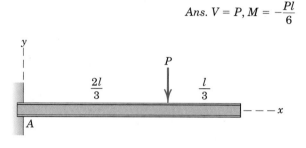

Problem 5/117

5/118 Determine the shear V at a section B between A and C and the moment M at the support A.

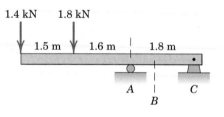

Problem 5/118

5/119 Draw the shear and moment diagrams for the diving board, which supports the 175-lb man poised to dive. Specify the bending moment with the maximum magnitude.

$$Ans. \ M_B = -1488 \ \text{lb-ft}$$

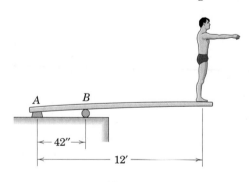

Problem 5/119

5/120 Draw the shear and moment diagrams for the beam loaded at its center by the couple C.

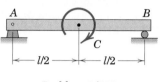

Problem 5/120

5/121 Draw the shear and moment diagrams for the beam subjected to the end couple. What is the moment M at a section 0.5 m to the right of B?

$$Ans. \ M = -120 \ \text{N} \cdot \text{m}$$

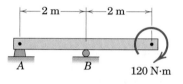

Problem 5/121

5/122 Draw the shear and moment diagrams for the loaded beam. What are the values of the shear and moment at the middle of the beam?

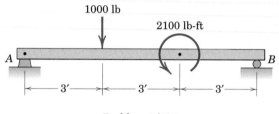

Problem 5/122

5/123 Draw the shear and moment diagrams for the beam shown and find the bending moment M at section C.

$$Ans. \ M_C = -1667 \ \text{lb-ft}$$

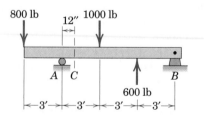

Problem 5/123

Representative Problems

5/124 Construct the bending-moment diagram for the cantilevered shaft AB of the rigid unit shown.

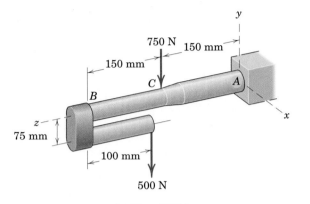

Problem 5/124

5/125 Determine the shear V and moment M at a section of the loaded beam 200 mm to the right of A.
Ans. $V = 0.15$ kN, $M = 0.15$ kN·m

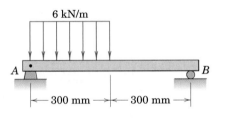

Problem 5/125

5/126 Determine the shear V and moment M in the beam at a section 2 ft to the right of end A.

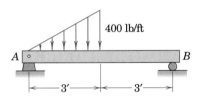

Problem 5/126

5/127 Draw the shear and moment diagrams for the cantilever beam with the linear loading, repeated here from Prob. 5/97. Find the maximum magnitude of the bending moment M.
$$\text{Ans. } |M|_{\max} = \frac{w_0 l^2}{6}$$

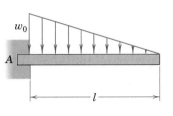

Problem 5/127

5/128 Draw the shear and moment diagrams for the beam shown. Determine the distance b, measured from the left end, to the point where the bending moment is zero between the supports.

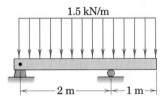

Problem 5/128

5/129 Draw the shear and moment diagrams for the loaded beam and find the maximum magnitude M of the bending moment.
$$\text{Ans. } M = \tfrac{5}{6} Pl$$

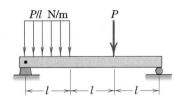

Problem 5/129

5/130 Draw the shear and moment diagrams for the loaded cantilever beam where the end couple M_1 is adjusted so as to produce zero moment at the fixed end of the beam. Find the bending moment M at $x = 4$ ft.

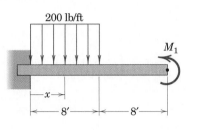

200 lb/ft

M_1

x

8′ 8′

Problem 5/130

5/131 Draw the shear and moment diagrams for the linearly loaded simple beam shown. Determine the maximum magnitude of the bending moment M.

$$Ans. \ |M|_{max} = \frac{w_0 l^2}{12}$$

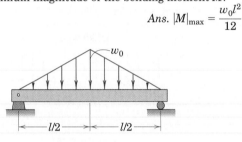

w_0

$l/2$ $l/2$

Problem 5/131

5/132 The shear force in pounds in a certain beam is given by $V = 2200x - 40x^3$ where x is the distance in feet measured along the beam. Determine the corresponding variation with x of the normal loading w in pounds per foot of length. Also determine the bending moment M at $x = 3$ ft if the bending moment at $x = 1$ ft is -1600 lb-ft.

5/133 The resistance of a beam of uniform width to bending is found to be proportional to the square of the beam depth y. For the cantilever beam shown the depth is h at the support. Find the required depth y as a function of the length x in order for all sections to be equally effective in their resistance to bending.

$$Ans. \ y = h\sqrt{x/l}$$

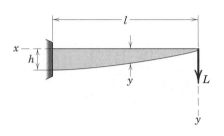

l

x

h

y

L

y

Problem 5/133

5/134 Determine the maximum bending moment M and the corresponding value of x in the crane beam and indicate the section where this moment acts.

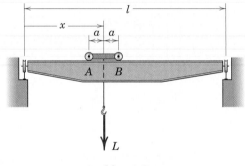

l

x

a a

A B

L

Problem 5/134

5/135 The angle strut is welded to the end C of the I-beam and supports the 1.6-kN vertical force. Determine the bending moment at B and the distance x to the left of C at which the bending moment is zero. Also construct the moment diagram for the beam.

Ans. $M_B = -0.40$ kN·m, $x = 0.2$ m

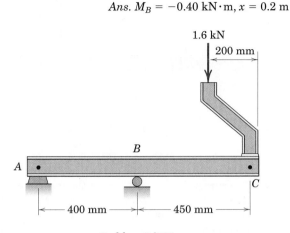

Problem 5/135

5/136 Plot the shear and moment diagrams for the beam loaded with both the distributed and point loads. What are the values of the shear and moment at $x = 6$ m? Determine the maximum bending moment M_{max}.

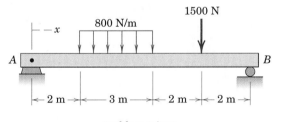

Problem 5/136

5/137 Repeat Prob. 5/136, where the 1500-N load has been replaced by the 4.2-kN·m couple.

Ans. $V = -1400$ N, $M = 0$, $M_{max} = 2800$ N·m

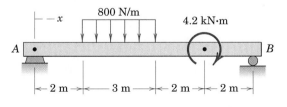

Problem 5/137

5/138 The beam is subjected to the two similar loadings shown where the maximum intensity of loading, in force per unit length, is w_0. Derive expressions for the shear V and moment M in the beam in terms of the distance x measured from the center of the beam.

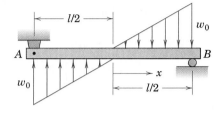

Problem 5/138

5/139 The distributed load decreases linearly from 4 to 2 kN/m in a distance of 2 m along a certain beam in equilibrium. If the shear force and bending moment at section A are +3 kN and +2 kN·m, respectively, calculate the shear force and bending moment at section B.

Ans. $V_B = -3$ kN, $M_B = 4/3$ kN·m

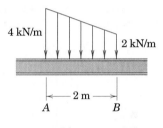

Problem 5/139

5/140 Derive expressions for the shear force V and bending moment M as functions of x in the cantilever beam loaded as shown.

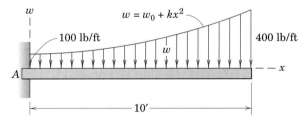

Problem 5/140

5/141 Derive expressions for the shear V and moment M in terms of x for the cantilever beam of Prob. 5/108 shown again here.

$$Ans. \; V = w_0\left(\frac{l}{3} - x + \frac{4x^3}{3l^2}\right)$$

$$M = w_0\left(-\frac{l^2}{16} + \frac{xl}{3} - \frac{x^2}{2} + \frac{x^4}{3l^2}\right)$$

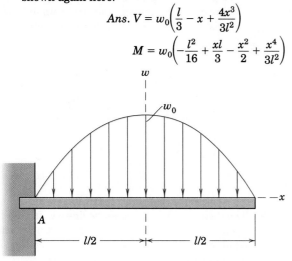

Problem 5/141

5/142 A curved cantilever beam has the form of a quarter circular arc. Determine the expressions for the shear V and the bending moment M as functions of θ.

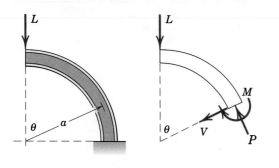

Problem 5/142

▶**5/143** The beam supports a uniform unit load w. Determine the location x of the two supports so as to minimize the maximum bending moment M_{max} in the beam. Specify M_{max}.

$$Ans. \; x = 0.207L, \; M_{max} = 0.0214wL^2$$

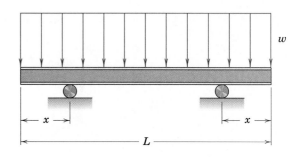

Problem 5/143

▶**5/144** The curved cantilever beam in the form of a quarter-circular arc supports a load of w N/m applied along the curve of the beam on its upper surface. Determine the magnitudes of the torsional moment T and bending moment M in the beam as functions of θ.

$$Ans. \; T = wr^2\left(\frac{\pi}{2} - \theta - \cos\theta\right)$$

$$M = wr^2(1 - \sin\theta)$$

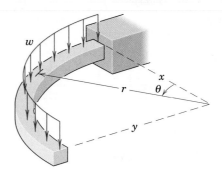

Problem 5/144

5/8   FLEXIBLE CABLES

One important type of structural member is the flexible cable which is used in suspension bridges, transmission lines, messenger cables for supporting heavy trolley or telephone lines, and many other applications. To design these structures we must know the relations involving the tension, span, sag, and length of the cables. We determine these quantities by examining the cable as a body in equilibrium. In the analysis of flexible cables we assume that any resistance offered to bending is negligible. This assumption means that the force in the cable is always in the direction of the cable.

Flexible cables may support a series of distinct concentrated loads, as shown in Fig. 5/26a, or they may support loads continuously distributed over the length of the cable, as indicated by the variable-intensity loading w in 5/26b. In some instances the weight of the cable is negligible compared with the loads it supports. In other cases the weight of the cable may be an appreciable load or the sole load and cannot be neglected. Regardless of which of these conditions is present, the equilibrium requirements of the cable may be formulated in the same manner.

General Relationships

If the intensity of the variable and continuous load applied to the cable of Fig. 5/26b is expressed as w units of force per unit of horizontal length x, then the resultant R of the vertical loading is

$$R = \int dR = \int w \, dx$$

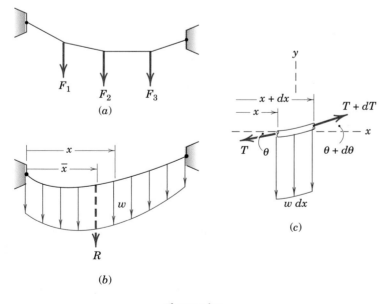

Figure 5/26

where the integration is taken over the desired interval. We find the position of R from the moment principle, so that

$$R\bar{x} = \int x \, dR \qquad \bar{x} = \frac{\int x \, dR}{R}$$

The elemental load $dR = w \, dx$ is represented by an elemental strip of vertical length w and width dx of the shaded area of the loading diagram, and R is represented by the total area. It follows from the foregoing expressions that R passes through the *centroid* of the shaded area.

The equilibrium condition of the cable is satisfied if each infinitesimal element of the cable is in equilibrium. The free-body diagram of a differential element is shown in Fig. 5/26c. At the general position x the tension in the cable is T, and the cable makes an angle θ with the horizontal x-direction. At the section $x + dx$ the tension is $T + dT$, and the angle is $\theta + d\theta$. Note that the changes in both T and θ are taken to be positive with a positive change in x. The vertical load $w \, dx$ completes the free-body diagram. The equilibrium of vertical and horizontal forces requires, respectively, that

$$(T + dT) \sin (\theta + d\theta) = T \sin \theta + w \, dx$$

$$(T + dT) \cos (\theta + d\theta) = T \cos \theta$$

The trigonometric expansion for the sine and cosine of the sum of two angles and the substitutions $\sin d\theta = d\theta$ and $\cos d\theta = 1$, which hold in the limit as $d\theta$ approaches zero, yield

$$(T + dT)(\sin \theta + \cos \theta \, d\theta) = T \sin \theta + w \, dx$$

$$(T + dT)(\cos \theta - \sin \theta \, d\theta) = T \cos \theta$$

Dropping the second-order terms and simplifying give us

$$T \cos \theta \, d\theta + dT \sin \theta = w \, dx$$

$$-T \sin \theta \, d\theta + dT \cos \theta = 0$$

which we write as

$$d(T \sin \theta) = w \, dx \qquad \text{and} \qquad d(T \cos \theta) = 0$$

The second relation expresses the fact that the horizontal component of T remains unchanged, which is clear from the free-body diagram. If we introduce the symbol $T_0 = T \cos \theta$ for this constant horizontal force, we may then substitute $T = T_0/\cos \theta$ into the first of the two equations just derived and obtain $d(T_0 \tan \theta) = w \, dx$. Because $\tan \theta = dy/dx$, the equilibrium equation may be written in the form

$$\boxed{\frac{d^2y}{dx^2} = \frac{w}{T_0}} \qquad\qquad \textbf{(5/13)}$$

Equation 5/13 is the differential equation for the flexible cable. The solution to the equation is that functional relation $y = f(x)$ which satisfies the equation and also satisfies the conditions at the fixed ends of the cable, called *boundary conditions*. This relationship defines the shape of the cable, and we will use it to solve two important and limiting cases of cable loading.

Parabolic Cable

When the intensity of vertical loading w is constant, the condition closely approximates that of a suspension bridge where the uniform weight of the roadway may be expressed by the constant w. The mass of the cable itself is not distributed uniformly with the horizontal but is relatively small, and thus we neglect its weight. For this limiting case we will prove that the cable hangs in a *parabolic arc*.

We start with a cable suspended from two points A and B which are not on the same horizontal line, Fig. 5/27a. We place the coordinate origin at the lowest point of the cable, where the tension is horizontal and is T_0. Integration of Eq. 5/13 once with respect to x gives

$$\frac{dy}{dx} = \frac{wx}{T_0} + C$$

where C is a constant of integration. For the coordinate axes chosen, $dy/dx = 0$ when $x = 0$, so that $C = 0$. Thus,

$$\frac{dy}{dx} = \frac{wx}{T_0}$$

which defines the slope of the curve as a function of x. One further integration yields

$$\int_0^y dy = \int_0^x \frac{wx}{T_0}\, dx \qquad \text{or} \qquad \boxed{y = \frac{wx^2}{2T_0}} \qquad \textbf{(5/14)}$$

Alternatively, you should be able to obtain the identical results with the indefinite integral together with the evaluation of the constant of in-

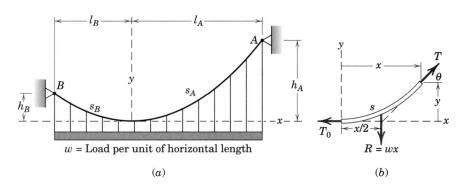

w = Load per unit of horizontal length

(a)   (b)

Figure 5/27

tegration. Equation 5/14 gives the shape of the cable, which we see is a vertical parabola. The constant horizontal component of cable tension becomes the cable tension at the origin.

Inserting the corresponding values $x = l_A$ and $y = h_A$ in Eq. 5/14 gives

$$T_0 = \frac{w l_A{}^2}{2 h_A} \qquad \text{so that} \qquad y = h_A (x/l_A)^2$$

The tension T is found from a free-body diagram of a finite portion of the cable, shown in Fig. 5/27b. From the Pythagorean theorem

$$T = \sqrt{T_0{}^2 + w^2 x^2}$$

Elimination of T_0 gives

$$T = w\sqrt{x^2 + (l_A{}^2/2h_A)^2} \tag{5/15}$$

The maximum tension occurs where $x = l_A$ and is

$$T_{\max} = w l_A \sqrt{1 + (l_A/2h_A)^2} \tag{5/15a}$$

We obtain the length s_A of the cable from the origin to point A by integrating the expression for a differential length $ds = \sqrt{(dx)^2 + (dy)^2}$. Thus,

$$\int_0^{s_A} ds = \int_0^{l_A} \sqrt{1 + (dy/dx)^2}\, dx = \int_0^{l_A} \sqrt{1 + (wx/T_0)^2}\, dx$$

Although we can integrate this expression in closed form, for computational purposes it is more convenient to express the radical as a convergent series and then integrate it term by term. For this purpose we use the binomial expansion

$$(1 + x)^n = 1 + nx + \frac{n(n-1)}{2!} x^2 + \frac{n(n-1)(n-2)}{3!} x^3 + \cdots$$

which converges for $x^2 < 1$. Replacing x in the series by $(wx/T_0)^2$ and setting $n = \frac{1}{2}$ give the expression

$$s_A = \int_0^{l_A} \left(1 + \frac{w^2 x^2}{2 T_0{}^2} - \frac{w^4 x^4}{8 T_0{}^4} + \cdots\right) dx$$

$$= l_A \left[1 + \frac{2}{3}\left(\frac{h_A}{l_A}\right)^2 - \frac{2}{5}\left(\frac{h_A}{l_A}\right)^4 + \cdots\right] \tag{5/16}$$

This series is convergent for values of $h_A/l_A < \frac{1}{2}$, which holds for most practical cases.

The relationships which apply to the cable section from the origin to point B can be easily obtained by replacing h_A, l_A, and s_A by h_B, l_B, and s_B, respectively.

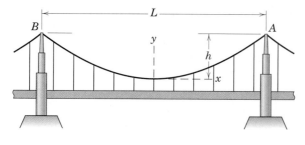

Figure 5/28

For a suspension bridge where the supporting towers are on the same horizontal line, Fig. 5/28, the total span is $L = 2l_A$, the sag is $h = h_A$, and the total length of the cable is $S = 2s_A$. With these substitutions, the maximum tension and the total length become

$$T_{max} = \frac{wL}{2} \sqrt{1 + (L/4h)^2} \qquad\qquad \textbf{(5/15b)}$$

$$S = L \left[1 + \frac{8}{3} \left(\frac{h}{L} \right)^2 - \frac{32}{5} \left(\frac{h}{L} \right)^4 + \cdots \right] \qquad \textbf{(5/16a)}$$

This series converges for all values of $h/L < \frac{1}{4}$. In most cases h is much smaller than $L/4$, so that the three terms of Eq. 5/16a give a sufficiently accurate approximation.

Catenary Cable

Consider now a uniform cable, Fig. 5/29a, suspended from two points A and B and hanging under the action of its own weight only. We will show in this limiting case that the cable assumes a curved shape known as a *catenary*.

The free-body diagram of a finite portion of the cable of length s measured from the origin is shown in part b of the figure. This free-body diagram differs from the one in Fig. 5/27b in that the total vertical force supported is equal to the weight of the cable section of length s rather

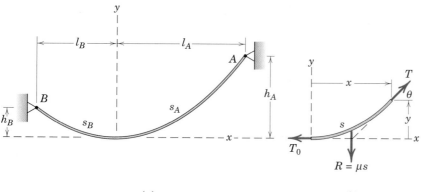

(a)   (b)

Figure 5/29

than the load distributed uniformly with respect to the horizontal. If the cable has a weight μ per unit of its length, the resultant R of the load is $R = \mu s$, and the incremental vertical load $w\,dx$ of Fig. 5/26c is replaced by $\mu\,ds$. With this replacement the differential relation, Eq. 5/13, for the cable becomes

$$\frac{d^2y}{dx^2} = \frac{\mu}{T_0}\frac{ds}{dx} \tag{5/17}$$

Because $s = f(x, y)$, we must change this equation to one containing only the two variables.

We may substitute the identity $(ds)^2 = (dx)^2 + (dy)^2$ to obtain

$$\frac{d^2y}{dx^2} = \frac{\mu}{T_0}\sqrt{1 + \left(\frac{dy}{dx}\right)^2} \tag{5/18}$$

Equation 5/18 is the differential equation of the curve (catenary) formed by the cable. This equation is easier to solve if we substitute $p = dy/dx$ to obtain

$$\frac{dp}{\sqrt{1 + p^2}} = \frac{\mu}{T_0}\,dx$$

Integrating this equation gives us

$$\ln\left(p + \sqrt{1 + p^2}\right) = \frac{\mu}{T_0}x + C$$

The constant C is zero because $dy/dx = p = 0$ when $x = 0$. Substituting $p = dy/dx$, changing to exponential form, and clearing the equation of the radical give

$$\frac{dy}{dx} = \frac{e^{\mu x/T_0} - e^{-\mu x/T_0}}{2} = \sinh\frac{\mu x}{T_0}$$

where the hyperbolic function* is introduced for convenience. The slope may be integrated to obtain

$$y = \frac{T_0}{\mu}\cosh\frac{\mu x}{T_0} + K$$

The integration constant K is evaluated from the boundary condition $x = 0$ when $y = 0$. This substitution requires that $K = -T_0/\mu$, and hence,

$$y = \frac{T_0}{\mu}\left(\cosh\frac{\mu x}{T_0} - 1\right) \tag{5/19}$$

---

*See Arts. C/8 and C/10, Appendix C, for the definition and integral of hyperbolic functions.

Equation 5/19 is the equation of the curve (catenary) formed by the cable hanging under the action of its weight only.

From the free-body diagram in Fig. 5/29*b* we see that $dy/dx = \tan \theta = \mu s/T_0$. Thus, from the previous expression for the slope,

$$s = \frac{T_0}{\mu} \sinh \frac{\mu x}{T_0} \tag{5/20}$$

We obtain the tension T in the cable from the equilibrium triangle of the forces in Fig. 5/29*b*. Thus,

$$T^2 = \mu^2 s^2 + T_0{}^2$$

which, when combined with Eq. 5/20, becomes

$$T^2 = T_0{}^2 \left(1 + \sinh^2 \frac{\mu x}{T_0} \right) = T_0{}^2 \cosh^2 \frac{\mu x}{T_0}$$

or

$$T = T_0 \cosh \frac{\mu x}{T_0} \tag{5/21}$$

We may also express the tension in terms of y with the aid of Eq. 5/19, which, when substituted into Eq. 5/21, gives

$$T = T_0 + \mu y \tag{5/22}$$

Equation 5/22 shows that the change in cable tension from that at the lowest position depends only on μy.

Most problems dealing with the catenary involve solutions of Eqs. 5/19 through 5/22, which can be handled by a graphical approximation or solved by computer. The procedure for a graphical or computer solution is illustrated in Sample Problem 5/17 following this article.

The solution of catenary problems where the sag-to-span ratio is small may be approximated by the relations developed for the parabolic cable. A small sag-to-span ratio means a tight cable, and the uniform distribution of weight along the cable is not very different from the same load intensity distributed uniformly along the horizontal.

Many problems dealing with both the catenary and parabolic cables involve suspension points which are not on the same level. In such cases we may apply the relations just developed to the part of the cable on each side of the lowest point.

Tramway in Juneau, Alaska

Sample Problem 5/16

The light cable supports a mass of 12 kg per meter of horizontal length and is suspended between the two points on the same level 300 m apart. If the sag is 60 m, find the tension at midlength, the maximum tension, and the total length of the cable.

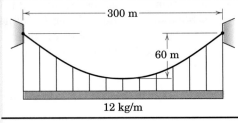

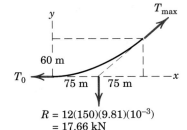

$R = 12(150)(9.81)(10^{-3})$
$= 17.66$ kN

Solution. With a uniform horizontal distribution of load, the solution of part (b) of Art. 5/8 applies, and we have a parabolic shape for the cable. For $h = 60$ m, $L = 300$ m, and $w = 12(9.81)(10^{-3})$ kN/m the relation following Eq. 5/14 with $l_A = L/2$ gives for the midlength tension

$$\left[T_0 = \frac{wL^2}{8h} \right] \qquad T_0 = \frac{0.1177(300)^2}{8(60)} = 22.1 \text{ kN} \qquad Ans.$$

The maximum tension occurs at the supports and is given by Eq. 5/15b. Thus,

① $$\left[T_{max} = \frac{wL}{2} \sqrt{1 + \left(\frac{L}{4h} \right)^2} \right]$$

$$T_{max} = \frac{12(9.81)(10^{-3})(300)}{2} \sqrt{1 + \left(\frac{300}{4(60)} \right)^2} = 28.3 \text{ kN} \qquad Ans.$$

The sag-to-span ratio is 60/300 = 1/5 < 1/4. Therefore, the series expression developed in Eq. 5/16a is convergent, and we may write for the total length

$$S = 300 \left[1 + \frac{8}{3} \left(\frac{1}{5} \right)^2 - \frac{32}{5} \left(\frac{1}{5} \right)^4 + \cdots \right]$$

$$= 300[1 + 0.1067 - 0.01024 + \cdots]$$

$$= 329 \text{ m} \qquad Ans.$$

Helpful Hint

① *Suggestion:* Check the value of T_{max} directly from the free-body diagram of the right-hand half of the cable, from which a force polygon may be drawn.

Sample Problem 5/17

Replace the cable of Sample Problem 5/16, which is loaded uniformly along the horizontal, by a cable which has a mass of 12 kg per meter of its own length and supports its own weight only. The cable is suspended between two points on the same level 300 m apart and has a sag of 60 m. Find the tension at midlength, the maximum tension, and the total length of the cable.

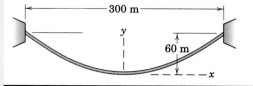

Solution. With a load distributed uniformly along the length of the cable, the solution of part (c) of Art. 5/8 applies, and we have a catenary shape of the cable. Equations 5/20 and 5/21 for the cable length and tension both involve the minimum tension T_0 at midlength, which must be found from Eq. 5/19. Thus, for $x = 150$ m, $y = 60$ m, and $\mu = 12(9.81)(10^{-3}) = 0.1177$ kN/m, we have

$$60 = \frac{T_0}{0.1177}\left[\cosh\frac{(0.1177)(150)}{T_0} - 1\right]$$

or

$$\frac{7.06}{T_0} = \cosh\frac{17.66}{T_0} - 1$$

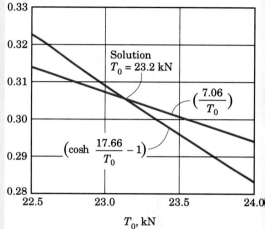

This equation can be solved graphically. We compute the expression on each side of the equals sign and plot it as a function of T_0. The intersection of the two curves establishes the equality and determines the correct value of T_0. This plot is shown in the figure accompanying this problem and yields the solution

$$T_0 = 23.2 \text{ kN}$$

Alternatively, we may write the equation as

$$f(T_0) = \cosh\frac{17.66}{T_0} - \frac{7.06}{T_0} - 1 = 0$$

and set up a computer program to calculate the value(s) of T_0 which renders $f(T_0) = 0$. See Art. C/11 of Appendix C for an explanation of one applicable numerical method.

The maximum tension occurs for maximum y and from Eq. 5/22 is

$$T_{\max} = 23.2 + (0.1177)(60) = 30.2 \text{ kN} \qquad\qquad Ans.$$

① From Eq. 5/20 the total length of the cable becomes

$$2s = 2\frac{23.2}{0.1177}\sinh\frac{(0.1177)(150)}{23.2} = 330 \text{ m} \qquad\qquad Ans.$$

Helpful Hint

① Note that the solution of Sample Problem 5/16 for the parabolic cable gives a very close approximation to the values for the catenary even though we have a fairly large sag. The approximation is even better for smaller sag-to-span ratios.

PROBLEMS

(The problems marked with an asterisk (*) involve transcendental equations which may be solved with a computer or by graphical methods.)

Introductory Problems

5/145 A coil of surveyor's tape 100 ft in length weighs 0.624 lb. When the tape is stretched between two points on the same level by a tension of 10 lb at each end, calculate the sag h of the tape in the middle.

Ans. $h = 9.36$ in.

5/146 The Golden Gate Bridge in San Francisco has a main span of 4200 ft, a sag of 470 ft, and a total static loading of 21,300 lb per lineal foot of horizontal measurement. The weight of both of the main cables is included in this figure and is assumed to be uniformly distributed along the horizontal. The angle made by the cable with the horizontal at the top of the tower is the same on each side of each tower. Calculate the midspan tension T_0 in each of the main cables and the compressive force C exerted by each cable on the top of each tower.

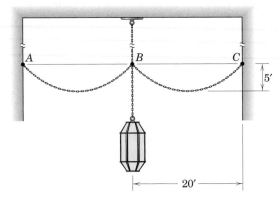

Problem 5/146

5/147 Calculate the tension T_0 in the cable at A necessary to support the load distributed uniformly with respect to the horizontal. Also find the angle θ made by the cable with the horizontal at the attachment point B.

Ans. $T_0 = 18.03$ kN, $\theta = 48.8°$

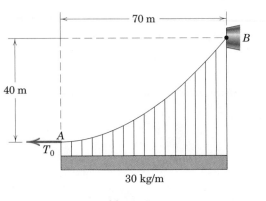

Problem 5/147

5/148 A cable supports a load of 40 kg/m uniformly distributed along the horizontal and is suspended from two fixed points A and B located as shown. Calculate the cable tensions at A and B and the minimum tension T_0.

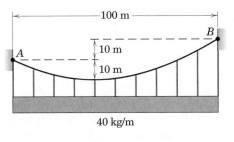

Problem 5/148

*5/149 A light fixture is suspended from the ceiling of an outside portico. Four chains, two of which are shown, prevent excessive motion of the fixture during windy conditions. If the chains weigh 15 lb per foot of length, determine the chain tension at C and the length L of chain BC.

Ans. $T_C = 236$ lb, $L = 23.0$ ft

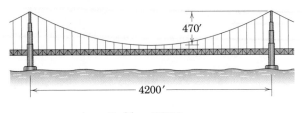

Problem 5/149

5/150 An advertising balloon is moored to a post with a cable which has a mass of 0.12 kg/m. In a wind the cable tensions at A and B are 110 N and 230 N, respectively. Determine the height h of the balloon.

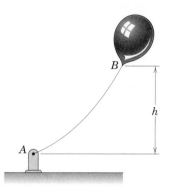

Problem 5/150

5/151 A horizontal 350-mm-diameter water pipe is supported over a ravine by the cable shown. The pipe and the water within it have a combined mass of 1400 kg per meter of its length. Calculate the compression C exerted by the cable on each support. The angles made by the cable with the horizontal are the same on both sides of each support.

Ans. $C = 549$ kN

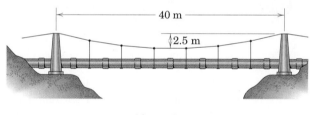

Problem 5/151

5/152 Strain-gage measurements made on the cables of the suspension bridge at position A indicate an increase of 480,000 lb of tension in *each* of the two main cables because the bridge has been repaved. Determine the total weight w' of added paving material used per foot of roadway.

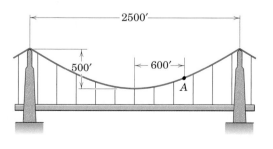

Problem 5/152

*5/153 The glider A is being towed in level flight and is 400 ft behind and 100 ft below the tow plane B. The tangent to the cable at the glider is horizontal. The cable weighs 0.5 lb per foot of length. Calculate the horizontal tension T_0 in the cable at the glider. Neglect air resistance and compare your result with that obtained by approximating the cable shape by a parabola.

Ans. $T_0 = 408$ lb, $(T_0)_{\text{par}} = 400$ lb

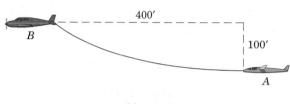

Problem 5/153

*5/154 Find the total length L of chain which will have a sag of 6 ft when suspended from two points on the same horizontal line 30 ft apart.

Problem 5/154

5/155 A cable weighing 40 newtons per meter of length is suspended from point A and passes over the small pulley at B. Determine the mass m of the attached cylinder which will produce a sag of 10 m. With the small sag-to-span ratio, approximation as a parabolic cable may be used.

Ans. $m = 480$ kg

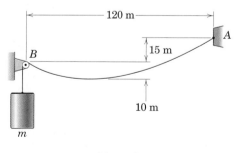

Problem 5/155

*5/156 Repeat Prob. 5/155, but do not use the approximation of a parabolic cable. Compare your results with the printed answer for Prob. 5/155.

5/157 A floating dredge is anchored in position with a single stern cable which has a horizontal direction at the attachment A and extends a horizontal distance of 250 m to an anchorage B on shore. A tension of 300 kN is required in the cable at A. If the cable has a mass of 22 kg per meter of its length, compute the required height H of the anchorage above water level and find the length s of cable between A and B.

Ans. $H = 24.5$ m, $s = 251$ m

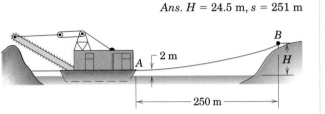

Problem 5/157

*5/158 A series of spherical floats are equally spaced and securely fastened to a flexible cable of length 20 m. Ends A and B are anchored 16 m apart to the bottom of a fresh-water lake at a depth of 8 m. The floats and cable have a combined weight of 100 N per meter of cable length, and the buoyancy of the water produces an upward force of 560 N per meter of cable length. Calculate the depth h below the surface to the top of the line of floats. Also find the angle θ made by the line of floats with the horizontal at A.

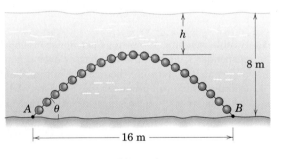

Problem 5/158

*5/159 Determine the length L of chain required from B to A and the corresponding tension at A if the slope of the chain is to be horizontal as it enters the guide at A. The weight of the chain is 140 N per meter of its length.

Ans. $L = 8.71$ m, $T_A = 1559$ N

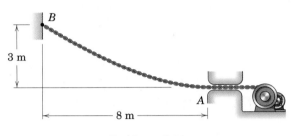

Problem 5/159

*5/160 Numerous small flotation devices are attached to the cable, and the difference between buoyancy and weight results in a net upward force of 30 newtons per meter of cable length. Determine the force T which must be applied to cause the cable configuration shown.

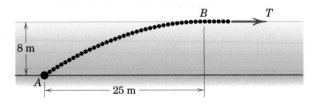

Problem 5/160

*5/161 A rope 40 m in length is suspended between two points which are separated by a horizontal distance of 10 m. Compute the distance h to the lowest part of the loop.

Ans. $h = 18.53$ m

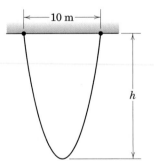

Problem 5/161

5/162 The blimp is moored to the ground winch in a gentle wind with 100 m of 12-mm cable which has a mass of 0.51 kg/m. A torque of 400 N·m on the drum is required to start winding in the cable. At this condition the cable makes an angle of 30° with the vertical as it approaches the winch. Calculate the height H of the blimp. The diameter of the drum is 0.5 m.

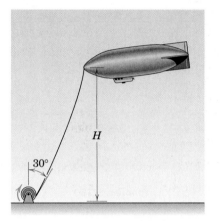

Problem 5/162

*5/163 A cable installation crew wishes to establish the dependence of the tension T on the cable sag h. Plot both the minimum tension T_0 and the tension T at the supports A and B as functions of h for $1 \le h \le 10$ m. The cable mass per unit length is 3 kg/m. State the values of T_0 and T for $h = 2$ m.

Ans. $T_0 = 6630$ N, $T = 6690$ N

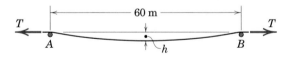

Problem 5/163

*5/164 The cable of Prob. 5/163 is now placed on supports A and B whose elevation differs by 9 m as shown. Plot the minimum tension T_0, the tension T_A at support A, and the tension T_B at support B as functions of h for $1 \le h \le 10$ m, where h is the sag below point A. State all three tensions for $h = 2$ m and compare with the results of Prob. 5/163. The cable mass per unit length is 3 kg/m.

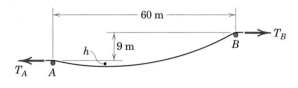

Problem 5/164

*5/165 In preparing to spray-clean a wall, a person arranges a hose as shown in the figure. The hose is horizontal at A and has a mass of 0.75 kg/m when empty and 1.25 kg/m when full of water. Determine the necessary tension T and angle θ for both the empty and full hose.

Ans. $T = 63.0$ N (empty), $T = 105.0$ N (full)
$\theta = 40.0°$ in both cases

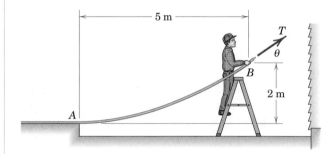

Problem 5/165

*5/166 A length of cable which has a mass of 1.2 kg/m is to have a sag of 2.4 m when suspended from the two points A and B on the same horizontal line 10 m apart. For comparison purposes, determine the length L of cable required and plot its configuration for the two cases of (a) assuming a parabolic shape and (b) using the proper catenary model. In order to more clearly distinguish between the two cases, also plot the difference ($y_C - y_P$) as a function of x, where C and P refer to catenary and parabola, respectively.

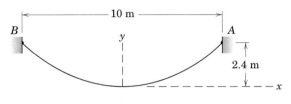

Problem 5/166

*5/167 A power line is suspended from two towers 200 m apart on the same horizontal line. The cable has a mass of 18.2 kg per meter of length and has a sag of 32 m at midspan. If the cable can support a maximum tension of 60 kN, determine the mass ρ of ice per meter which can form on the cable without exceeding the maximum cable tension.

Ans. $\rho = 13.44$ kg of ice per meter

*5/168 A cable which has a mass of 0.5 kg per meter of length is attached to point A. A tension T_B is applied to point B, causing the angle θ_A to be 15°. Determine T_B and θ_B.

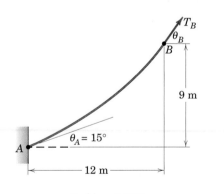

Problem 5/168

*5/169 The moving cable for a ski lift has a mass of 10 kg/m and carries equally spaced chairs and passengers, whose added mass is 20 kg/m when averaged over the length of the cable. The cable leads horizontally from the supporting guide wheel at A. Calculate the tensions in the cable at A and B and the length s of the cable between A and B.

Ans. $T_A = 27.4$ kN, $T_B = 33.3$ kN, $s = 64.2$ m

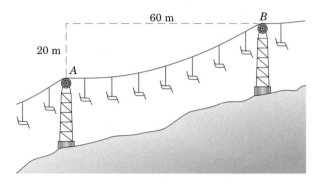

Problem 5/169

*5/170 A cable ship tows a plow A during a survey of the ocean floor for later burial of a telephone cable. The ship maintains a constant low speed with the plow at a depth of 600 ft and with a sufficient length of cable so that it leads horizontally from the plow, which is 1600 ft astern of the ship. The tow cable has an effective weight of 3.10 lb/ft when the buoyancy of the water is accounted for. Also, the forces on the cable due to movement through the water are neglected at the low speed. Compute the horizontal force T_0 applied to the plow and the maximum tension in the cable. Also find the length of the tow cable from point A to point B.

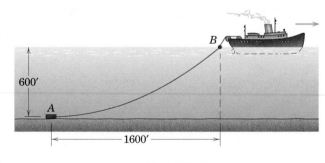

Problem 5/170

*5/171 For aesthetic reasons, chains are sometimes used instead of downspouts on small buildings in order to direct roof runoff water from the gutter down to ground level. The architect of the illustrated building specified a 6-m vertical chain from A to B, but the builder decided to use a 6.1-m chain from A to C as shown in order to place the water farther from the structure. By what percentage n did the builder increase the magnitude of the force exerted on the gutter at A over that figured by the architect? The chain weighs 100 N per meter of its length.

Ans. $n = 29.0\%$

*5/172 A 50-kg traffic signal is suspended by two 21-m cables which have a mass of 1.2 kg per meter of length. Determine the vertical deflection δ of the junction ring A relative to its position before the signal is added.

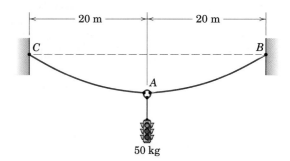

Problem 5/172

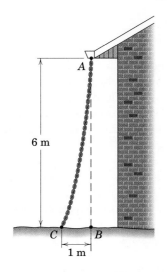

Problem 5/171

5/9   FLUID STATICS

So far in this chapter we have treated the action of forces on and between *solid* bodies. In this article we consider the equilibrium of bodies subjected to forces due to fluid pressures. A *fluid* is any continuous substance which, when at rest, is unable to support shear force. A shear force is one tangent to the surface on which it acts and is developed when differential velocities exist between adjacent layers of fluids. Thus, a fluid at rest can exert only normal forces on a bounding surface. Fluids may be either gaseous or liquid. The statics of fluids is generally called *hydrostatics* when the fluid is a liquid and *aerostatics* when the fluid is a gas.

Fluid Pressure

The pressure at any given point in a fluid is the same in all directions (Pascal's law). We may prove this by considering the equilibrium of an infinitesimal triangular prism of fluid as shown in Fig. 5/30. The fluid pressures normal to the faces of the element are p_1, p_2, p_3, and p_4 as shown. With force equal to pressure times area, the equilibrium of forces in the x- and y-directions gives

$$p_1 \, dy \, dz = p_3 \, ds \, dz \sin \theta \qquad p_2 \, dx \, dz = p_3 \, ds \, dz \cos \theta$$

Since $ds \sin \theta = dy$ and $ds \cos \theta = dx$, these questions require that

$$p_1 = p_2 = p_3 = p$$

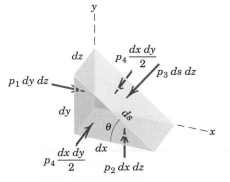

Figure 5/30

By rotating the element through 90°, we see that p_4 is also equal to the other pressures. Thus, the pressure at any point in a fluid at rest is the same in all directions. In this analysis we need not account for the weight of the fluid element because, when the weight per unit volume (density ρ times g) is multiplied by the volume of the element, a differential quantity of third order results which disappears in the limit compared with the second-order pressure-force terms.

In all fluids at rest the pressure is a function of the vertical dimension. To determine this function, we consider the forces acting on a differential element of a vertical column of fluid of cross-sectional area dA, as shown in Fig. 5/31. The positive direction of vertical measurement h is taken downward. The pressure on the upper face is p, and that on the lower face is p plus the change in p, or $p + dp$. The weight of the element equals ρg multiplied by its volume. The normal forces on the lateral surface, which are horizontal and do not affect the balance of forces in the vertical direction, are not shown. Equilibrium of the fluid element in the h-direction requires

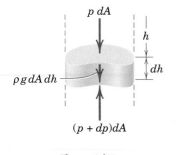

Figure 5/31

$$p \, dA + \rho g \, dA \, dh - (p + dp) \, dA = 0$$

$$dp = \rho g \, dh \qquad \qquad \textbf{(5/23)}$$

This differential relation shows us that the pressure in a fluid increases with depth or decreases with increased elevation. Equation 5/23 holds for both liquids and gases, and agrees with our common observations of air and water pressures.

Fluids which are essentially incompressible are called *liquids*, and for most practical purposes we may consider their density ρ constant for every part of the liquid.* With ρ a constant, integration of Eq. 5/23 gives

$$p = p_0 + \rho gh \qquad\qquad (5/24)$$

The pressure p_0 is the pressure on the surface of the liquid where $h = 0$. If p_0 is due to atmospheric pressure and the measuring instrument records only the increment above atmospheric pressure,[†] the measurement gives what is called *gage pressure*. It is computed from $p = \rho gh$.

The common unit for pressure in SI units is the kilopascal (kPa), which is the same as a kilonewton per square meter (10^3 N/m^2). In computing pressure, if we use Mg/m^3 for ρ, m/s^2 for g, and m for h, then the product ρgh gives us pressure in kPa directly. For example, the pressure at a depth of 10 m in fresh water is

$$p = \rho gh = \left(1.0\, \frac{\text{Mg}}{\text{m}^3}\right)\left(9.81\, \frac{\text{m}}{\text{s}^2}\right)(10 \text{ m}) = 98.1\left(10^3\, \frac{\text{kg}\cdot\text{m}}{\text{s}^2}\, \frac{1}{\text{m}^2}\right)$$

$$= 98.1 \text{ kN/m}^2 = 98.1 \text{ kPa}$$

In the U.S. customary system, fluid pressure is generally expressed in pounds per square inch (lb/in.^{2}) or occasionally in pounds per square foot (lb/ft^2). Thus, at a depth of 10 ft in fresh water the pressure is

$$p = \rho gh = \left(62.4\, \frac{\text{lb}}{\text{ft}^3}\right)\left(\frac{1}{1728}\, \frac{\text{ft}^3}{\text{in.}^3}\right)(120 \text{ in.}) = 4.33 \text{ lb/in.}^2$$

Hydrostatic Pressure on Submerged Rectangular Surfaces

A body submerged in a liquid, such as a gate valve in a dam or the wall of a tank, is subjected to fluid pressure acting normal to its surface and distributed over its area. In problems where fluid forces are appreciable, we must determine the resultant force due to the distribution of pressure on the surface and the position at which this resultant acts. For systems open to the atmosphere, the atmospheric pressure p_0 acts over all surfaces and thus yields a zero resultant. In such cases, then, we need to consider only the gage pressure $p = \rho gh$, which is the increment above atmospheric pressure.

Consider the special but common case of the action of hydrostatic pressure on the surface of a rectangular plate submerged in a liquid. Figure 5/32a shows such a plate 1-2-3-4 with its top edge horizontal and with the plane of the plate making an arbitrary angle θ with the vertical plane. The horizontal surface of the liquid is represented by the x-y' plane. The fluid pressure (gage) acting normal to the plate at point 2 is

---

*See Table D/1, Appendix D, for table of densities.

[†]Atmospheric pressure at sea level may be taken to be 101.3 kPa or 14.7 lb/in.^{2}

represented by the arrow 6-2 and equals ρg times the vertical distance from the liquid surface to point 2. This same pressure acts at all points along the edge 2-3. At point 1 on the lower edge, the fluid pressure equals ρg times the depth of point 1, and this pressure is the same at all points along edge 1-4. The variation of pressure p over the area of the plate is governed by the linear depth relationship and therefore it is represented by the arrow p, shown in Fig. 5/32b, which varies linearly from the value 6-2 to the value 5-1. The resultant force produced by this pressure distribution is represented by R, which acts at some point P called the *center of pressure*.

The conditions which prevail at the vertical section 1-2-6-5 in Fig. 5/32a are identical to those at section 4-3-7-8 and at every other vertical section normal to the plate. Thus, we may analyze the problem from the two-dimensional view of a vertical section as shown in Fig. 5/32b for section 1-2-6-5. For this section the pressure distribution is trapezoidal. If b is the horizontal width of the plate measured normal to the plane of the figure (dimension 2-3 in Fig. 5/32a), an element of plate area over which

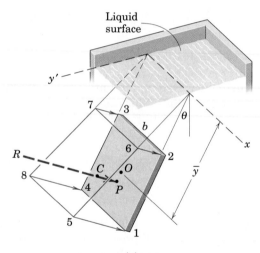

(a)

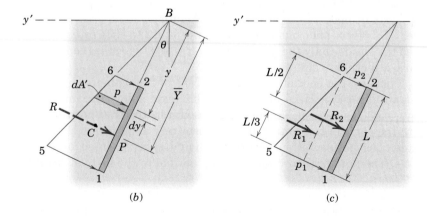

(b)   (c)

Figure 5/32

the pressure $p = \rho g h$ acts is $dA = b \, dy$, and an increment of the resultant force is $dR = p \, dA = bp \, dy$. But $p \, dy$ is merely the shaded increment of trapezoidal area dA', so that $dR = b \, dA'$. We may therefore express the resultant force acting on the entire plate as the trapezoidal area 1-2-6-5 times the width b of the plate,

$$R = b \int dA' = bA'$$

Be careful not to confuse the physical area A of the plate with the geometrical area A' defined by the trapezoidal distribution of pressure.

The trapezoidal area representing the pressure distribution is easily expressed by using its average altitude. The resultant force R may therefore be written in terms of the average pressure $p_{av} = \frac{1}{2}(p_1 + p_2)$ times the plate area A. The average pressure is also the pressure which exists at the average depth, measured to the centroid O of the plate. An alternative expression for R is therefore

$$R = p_{av}A = \rho g \bar{h} A$$

where $\bar{h} = \bar{y} \cos \theta$.

We obtain the line of action of the resultant force R from the principle of moments. Using the x-axis (point B in Fig. 5/32b) as the moment axis yields $R\bar{Y} = \int y(pb \, dy)$. Substituting $p \, dy = dA'$ and $R = bA'$ and canceling b give

$$\bar{Y} = \frac{\int y \, dA'}{\int dA'}$$

which is simply the expression for the centroidal coordinate of the trapezoidal area A'. In the two-dimensional view, therefore, the resultant R passes through the centroid C of the trapezoidal area defined by the pressure distribution in the vertical section. Clearly $\bar{Y}$ also locates the centroid C of the truncated prism 1-2-3-4-5-6-7-8 in Fig. 5/32a through which the resultant passes.

For a trapezoidal distribution of pressure, we may simplify the calculation by dividing the trapezoid into a rectangle and a triangle, Fig. 5/32c, and separately considering the force represented by each part. The force represented by the rectangular portion acts at the center O of the plate and is $R_2 = p_2 A$, where A is the area 1-2-3-4 of the plate. The force R_1 represented by the triangular increment of pressure distribution is $\frac{1}{2}(p_1 - p_2)A$ and acts through the centroid of the triangular portion shown.

Hydrostatic Pressure on Cylindrical Surfaces

The determination of the resultant R due to distributed pressure on a submerged curved surface involves more calculation than for a flat surface. For example, consider the submerged cylindrical surface shown in Fig. 5/33a where the elements of the curved surface are parallel to the horizontal surface x-y' of the liquid. Vertical sections perpendicular to the surface all disclose the same curve AB and the same pressure distri-

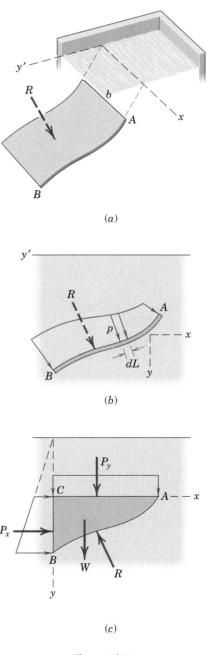

(a)

(b)

(c)

Figure 5/33

bution. Thus, the two-dimensional representation in Fig. 5/33b may be used. To find R by a direct integration, we need to integrate the x- and y-components of dR along the curve AB, since dR continuously changes direction. Thus,

$$R_x = b \int (p \, dL)_x = b \int p \, dy \quad \text{and} \quad R_y = b \int (p \, dL)_y = b \int p \, dx$$

A moment equation would now be required if we wished to establish the position of R.

A second method for finding R is usually much simpler. Consider the equilibrium of the block of liquid ABC directly above the surface, shown in Fig. 5/33c. The resultant R then appears as the equal and opposite reaction of the surface on the block of liquid. The resultants of the pressures along AC and CB are P_y and P_x, respectively, and are easily obtained. The weight W of the liquid block is calculated from the area ABC of its section multiplied by the constant dimension b and by ρg. The weight W passes through the centroid of area ABC. The equilibrant R is then determined completely from the equilibrium equations which we apply to the free-body diagram of the fluid block.

Hydrostatic Pressure on Flat Surfaces of Any Shape

Figure 5/34a shows a flat plate of any shape submerged in a liquid. The horizontal surface of the liquid is the plane x-y', and the plane of the plate makes an angle θ with the vertical. The force acting on a differential strip of area dA parallel to the surface of the liquid is $dR = p \, dA = \rho g h \, dA$. The pressure p has the same magnitude throughout the length of the strip, because there is no change of

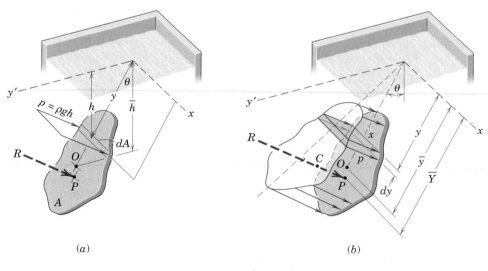

(a) (b)

Figure 5/34

depth along the strip. We obtain the total force acting on the exposed area A by integration, which gives

$$R = \int dR = \int p \, dA = \rho g \int h \, dA$$

Substituting the centroidal relation $\bar{h}A = \int h \, dA$ gives us

$$\boxed{R = \rho g \bar{h} A} \qquad\qquad \textbf{(5/25)}$$

The quantity $\rho g \bar{h}$ is the pressure which exists at the depth of the centroid O of the area and is the average pressure over the area.

We may also represent the resultant R geometrically by the volume V' of the figure shown in Fig. 5/34b. Here the fluid pressure p is represented as a dimension normal to the plate regarded as a base. We see that the resulting volume is a truncated right cylinder. The force dR acting on the differential area $dA = x \, dy$ is represented by the elemental volume $p \, dA$ shown by the shaded slice, and the total force is represented by the total volume of the cylinder. We see from Eq. 5/25 that the average altitude of the truncated cylinder is the average pressure $\rho g \bar{h}$ which exists at a depth corresponding to the centroid O of the area exposed to pressure.

For problems where the centroid O or the volume V' is not readily apparent, a direct integration may be performed to obtain R. Thus,

$$R = \int dR = \int p \, dA = \int \rho g h x \, dy$$

where the depth h and the length x of the horizontal strip of differential area must be expressed in terms of y to carry out the integration.

After the resultant is obtained, we must determine its location. Using the principle of moments with the x-axis of Fig. 5/34b as the moment axis, we obtain

$$R\bar{Y} = \int y \, dR \qquad \text{or} \qquad \bar{Y} = \frac{\int y(px \, dy)}{\int px \, dy} \qquad \textbf{(5/26)}$$

This second relation satisfies the definition of the coordinate $\bar{Y}$ to the centroid of the volume V' of the pressure-area truncated cylinder. We conclude, therefore, that the resultant R passes through the centroid C of the volume described by the plate area as base and the linearly varying pressure as the perpendicular coordinate. The point P at which R is applied to the plate is the center of pressure. Note that the center of pressure P and the centroid O of the plate area are *not* the same.

Buoyancy

Archimedes is credited with discovering the *principle of buoyancy*. This principle is easily explained for any fluid, gaseous or liquid, in equilibrium. Consider a portion of the fluid defined by an imaginary closed

Glen Canyon Dam at Lake Powell, Arizona

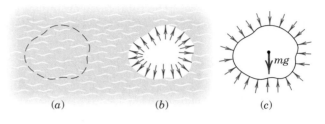

(a)   (b)   (c)

Figure 5/35

surface, as illustrated by the irregular dashed boundary in Fig. 5/35a. If the body of the fluid could be sucked out from within the closed cavity and replaced simultaneously by the forces which it exerted on the boundary of the cavity, Fig. 5/35b, the equilibrium of the surrounding fluid would not be disturbed. Furthermore, a free-body diagram of the fluid portion before removal, Fig. 5/35c, shows that the resultant of the pressure forces distributed over its surface must be equal and opposite to its weight *mg* and must pass through the center of mass of the fluid element. If we replace the fluid element by a body of the same dimensions, the surface forces acting on the body held in this position will be identical to those acting on the fluid element. Thus, the resultant force exerted on the surface of an object immersed in a fluid is equal and opposite to the weight of fluid displaced and passes through the center of mass of the displaced fluid. This resultant force is called the force of *buoyancy*

$$F = \rho g V \qquad\qquad (5/27)$$

where ρ is the density of the fluid, g is the acceleration due to gravity, and V is the volume of the fluid displaced. In the case of a liquid whose density is constant, the center of mass of the displaced liquid coincides with the centroid of the displaced volume.

Thus when the density of an object is less than the density of the fluid in which it is fully immersed, there is an imbalance of force in the vertical direction, and the object rises. When the immersing fluid is a liquid, the object continues to rise until it comes to the surface of the liquid and then comes to rest in an equilibrium position, assuming that the density of the new fluid above the surface is less than the density of the object. In the case of the surface boundary between a liquid and a gas, such as water and air, the effect of the gas pressure on that portion of the floating object above the liquid is balanced by the added pressure in the liquid due to the action of the gas on its surface.

An important problem involving buoyancy is the determination of the stability of a floating object, such as a ship hull shown in cross section in an upright position in Fig. 5/36a. Point *B* is the centroid of the displaced volume and is called the *center of buoyancy*. The resultant of the forces exerted on the hull by the water pressure is the buoyancy force *F* which passes through *B* and is equal and opposite to the weight *W* of the ship. If the ship is caused to list through an angle α, Fig. 5/36b,

The design of ship hulls must take into account fluid dynamics as well as fluid statics.

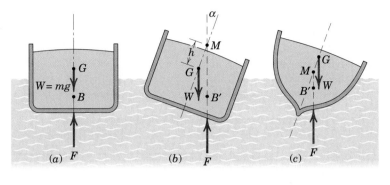

Figure 5/36

the shape of the displaced volume changes, and the center of buoyancy shifts to B'.

The point of intersection of the vertical line through B' with the centerline of the ship is called the *metacenter M*, and the distance h of M from the center of mass G is called the *metacentric height*. For most hull shapes h remains practically constant for angles of list up to about 20°. When M is above G, as in Fig. 5/36b, there is a righting moment which tends to bring the ship back to its upright position. If M is below G, as for the hull of Fig. 5/36c, the moment accompanying the list is in the direction to increase the list. This is clearly a condition of instability and must be avoided in the design of any ship.

© Cousteau Society/The Image Bank/Getty Images

Submersible vessels must be designed with extremely large external pressures in mind.

Sample Problem 5/18

A rectangular plate, shown in vertical section AB, is 4 m high and 6 m wide (normal to the plane of the paper) and blocks the end of a fresh-water channel 3 m deep. The plate is hinged about a horizontal axis along its upper edge through A and is restrained from opening by the fixed ridge B which bears horizontally against the lower edge of the plate. Find the force B exerted on the plate by the ridge.

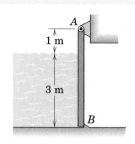

Solution. The free-body diagram of the plate is shown in section and includes the vertical and horizontal components of the force at A, the unspecified weight $W = mg$ of the plate, the unknown horizontal force B, and the resultant R of the triangular distribution of pressure against the vertical face.

The density of fresh water is $\rho = 1.000$ Mg/m^3 so that the average pressure is

① $[p_{av} = \rho g \bar{h}]$ $p_{av} = 1.000(9.81)(\frac{3}{2}) = 14.72$ kPa

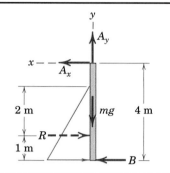

The resultant R of the pressure forces against the plate becomes

$[R = p_{av}A]$ $R = (14.72)(3)(6) = 265$ kN

This force acts through the centroid of the triangular distribution of pressure, which is 1 m above the bottom of the plate. A zero moment summation about A establishes the unknown force B. Thus,

$[\Sigma M_A = 0]$ $3(265) - 4B = 0$ $B = 198.7$ kN   *Ans.*

Helpful Hint

① Note that the units of pressure ρgh are

$$\left(10^3 \frac{\text{kg}}{\text{m}^3}\right)\left(\frac{\text{m}}{\text{s}^2}\right)(\text{m}) = \left(10^3 \frac{\text{kg} \cdot \text{m}}{\text{s}^2}\right)\left(\frac{1}{\text{m}^2}\right)$$
$$= \text{kN/m}^2 = \text{kPa}.$$

Sample Problem 5/19

The air space in the closed fresh-water tank is maintained at a pressure of 0.80 lb/in.^{2} (above atmospheric). Determine the resultant force R exerted by the air and water on the end of the tank.

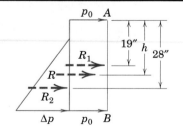

Side view   End view

Solution. The pressure distribution on the end surface is shown, where $p_0 = 0.80$ lb/in.^{2} The specific weight of fresh water is $\mu = \rho g = 62.4/1728 = 0.0361$ lb/in.^{3} so that the increment of pressure Δp due to the water is

$$\Delta p = \mu \, \Delta h = 0.0361(30) = 1.083 \text{ lb/in.}^2$$

① The resultant forces R_1 and R_2 due to the rectangular and triangular distributions of pressure, respectively, are

$$R_1 = p_0 A_1 = 0.80(38)(25) = 760 \text{ lb}$$

$$R_2 = \Delta p_{av} A_2 = \frac{1.083}{2}(30)(25) = 406 \text{ lb}$$

The resultant is then $R = R_1 + R_2 = 760 + 406 = 1166$ lb.   *Ans.*

We locate R by applying the moment principle about A noting that R_1 acts through the center of the 38-in. depth and that R_2 acts through the centroid of the triangular pressure distribution 20 in. below the surface of the water and $20 + 8 = 28$ in. below A. Thus,

$[Rh = \Sigma M_A]$ $1166h = 760(19) + 406(28)$ $h = 22.1$ in.   *Ans.*

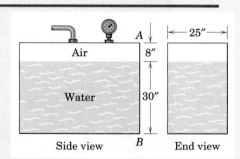

Helpful Hint

① Dividing the pressure distribution into these two parts is decidedly the simplest way in which to make the calculation.

Sample Problem 5/20

Determine completely the resultant force R exerted on the cylindrical dam surface by the water. The density of fresh water is 1.000 Mg/m³, and the dam has a length b, normal to the paper, of 30 m.

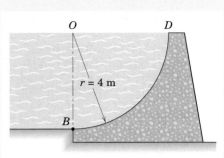

Solution. The circular block of water BDO is isolated and its free-body diagram is drawn. The force P_x is

① $$P_x = \rho g \bar{h} A = \frac{\rho g r}{2} \, br = \frac{(1.000)(9.81)(4)}{2} (30)(4) = 2350 \text{ kN}$$

The weight W of the water passes through the mass center G of the quarter-circular section and is

$$mg = \rho g V = (1.000)(9.81) \frac{\pi (4)^2}{4} (30) = 3700 \text{ kN}$$

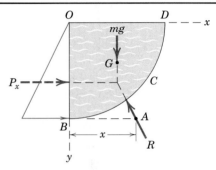

Equilibrium of the section of water requires

$[\Sigma F_x = 0]$ $R_x = P_x = 2350 \text{ kN}$

$[\Sigma F_y = 0]$ $R_y = mg = 3700 \text{ kN}$

The resultant force R exerted by the fluid on the dam is equal and opposite to that shown acting on the fluid and is

$[R = \sqrt{R_x^2 + R_y^2}]$ $R = \sqrt{(2350)^2 + (3700)^2} = 4380 \text{ kN}$ *Ans.*

The x-coordinate of the point A through which R passes may be found from the principle of moments. Using B as a moment center gives

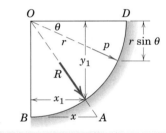

$$P_x \frac{r}{3} + mg \frac{4r}{3\pi} - R_y x = 0, \; x = \frac{2350 \left(\dfrac{4}{3}\right) + 3700 \left(\dfrac{16}{3\pi}\right)}{3700} = 2.55 \text{ m} \quad \textit{Ans.}$$

② **Alternative Solution.** The force acting on the dam surface may be obtained by a direct integration of the components

$$dR_x = p \, dA \cos \theta \quad \text{and} \quad dR_y = p \, dA \sin \theta$$

where $p = \rho g h = \rho g r \sin \theta$ and $dA = b(r \, d\theta)$. Thus,

$$R_x = \int_0^{\pi/2} \rho g r^2 b \sin \theta \cos \theta \, d\theta = -\rho g r^2 b \left[\frac{\cos 2\theta}{4} \right]_0^{\pi/2} = \frac{1}{2} \rho g r^2 b$$

$$R_y = \int_0^{\pi/2} \rho g r^2 b \sin^2 \theta \, d\theta = \rho g r^2 b \left[\frac{\theta}{2} - \frac{\sin 2\theta}{4} \right]_0^{\pi/2} = \frac{1}{4} \pi \rho g r^2 b$$

Thus, $R = \sqrt{R_x^2 + R_y^2} = \frac{1}{2} \rho g r^2 b \sqrt{1 + \pi^2/4}$. Substituting the numerical values gives

$$R = \frac{1}{2}(1.000)(9.81)(4^2)(30)\sqrt{1 + \pi^2/4} = 4380 \text{ kN} \quad \textit{Ans.}$$

Since dR always passes through point O, we see that R also passes through O and, therefore, the moments of R_x and R_y about O must cancel. So we write $R_x y_1 = R_y x_1$, which gives us

$$x_1/y_1 = R_x/R_y = (\tfrac{1}{2}\rho g r^2 b)/(\tfrac{1}{4}\pi \rho g r^2 b) = 2/\pi$$

By similar triangles we see that

$$x/r = x_1/y_1 = 2/\pi \quad \text{and} \quad x = 2r/\pi = 2(4)/\pi = 2.55 \text{ m} \quad \textit{Ans.}$$

Helpful Hints

① See note ① in Sample Problem 5/18 if there is any question about the units for $\rho g \bar{h}$.

② This approach by integration is feasible here mainly because of the simple geometry of the circular arc.

Sample Problem 5/21

Determine the resultant force R exerted on the semicircular end of the water tank shown in the figure if the tank is filled to capacity. Express the result in terms of the radius r and the water density ρ.

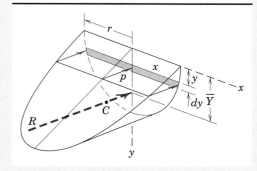

Solution I. We will obtain R first by a direct integration. With a horizontal strip of area $dA = 2x\,dy$ acted on by the pressure $p = \rho gy$, the increment of the resultant force is $dR = p\,dA$ so that

$$R = \int p\,dA = \int \rho gy(2x\,dy) = 2\rho g \int_0^r y\sqrt{r^2 - y^2}\,dy$$

Integrating gives $\qquad\qquad\qquad R = \frac{2}{3}\rho g r^3 \qquad\qquad\qquad$ *Ans.*

The location of R is determined by using the principle of moments. Taking moments about the x-axis gives

$$\left[R\overline{Y} = \int y\,dR\right] \qquad \frac{2}{3}\rho g r^3 \overline{Y} = 2\rho g \int_0^r y^2 \sqrt{r^2 - y^2}\,dy$$

Integrating gives $\quad \dfrac{2}{3}\rho g r^3 \overline{Y} = \dfrac{\rho g r^4}{4}\dfrac{\pi}{2} \quad$ and $\quad \overline{Y} = \dfrac{3\pi r}{16} \qquad$ *Ans.*

Solution II. We may use Eq. 5/25 directly to find R, where the average pressure is $\rho g \overline{h}$ and $\overline{h}$ is the coordinate to the centroid of the area over which the pressure acts. For a semicircular area $\overline{h} = 4r/(3\pi)$.

$$[R = \rho g \overline{h} A] \qquad R = \rho g \frac{4r}{3\pi}\frac{\pi r^2}{2} = \frac{2}{3}\rho g r^3 \qquad\qquad \textit{Ans.}$$

which is the volume of the pressure-area figure.

The resultant R acts through the centroid C of the volume defined by the
① pressure-area figure. Calculation of the centroidal distance $\overline{Y}$ involves the same integral obtained in *Solution I*.

Helpful Hint

① Be very careful not to make the mistake of assuming that R passes through the centroid of the area over which the pressure acts.

Sample Problem 5/22

A buoy in the form of a uniform 8-m pole 0.2 m in diameter has a mass of 200 kg and is secured at its lower end to the bottom of a fresh-water lake with 5 m of cable. If the depth of the water is 10 m, calculate the angle θ made by the pole with the horizontal.

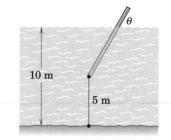

Solution. The free-body diagram of the buoy shows its weight acting through G, the vertical tension T in the anchor cable, and the buoyancy force B which passes through centroid C of the submerged portion of the buoy. Let x be the distance from G to the waterline. The density of fresh water is $\rho = 10^3$ kg/m³, so that the buoyancy force is

$$[B = \rho g V] \qquad B = 10^3 (9.81)\pi(0.1)^2 (4 + x)\ \text{N}$$

Moment equilibrium, $\Sigma M_A = 0$, about A gives

$$200(9.81)(4\cos\theta) - [10^3 (9.81)\pi(0.1)^2 (4 + x)]\frac{4 + x}{2}\cos\theta = 0$$

Thus, $\qquad x = 3.14$ m $\qquad$ and $\qquad \theta = \sin^{-1}\left(\dfrac{5}{4 + 3.14}\right) = 44.5°$ $\qquad$ *Ans.*

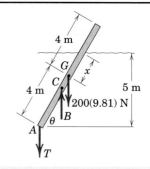

PROBLEMS

Introductory Problems

5/173 Determine the maximum height h to which a vacuum pump can cause the fresh water to rise. Assume standard atmospheric pressure of $1.0133(10^5)$ Pa. Repeat your calculations for mercury.

Ans. $h = 10.33$ m (water)
$h = 0.761$ m (mercury)

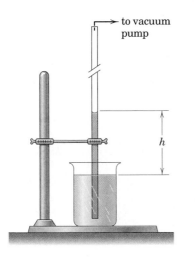

Problem 5/173

5/174 A beaker of fresh water is in place on the scale when the 1-lb stainless-steel weight is added to the beaker. What is the normal force which the weight exerts on the bottom of the beaker? By how much does the scale reading increase as the weight is added? Explain your answer.

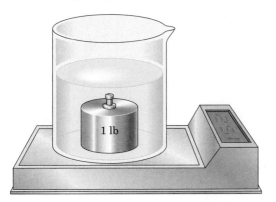

Problem 5/174

5/175 Specify the magnitude and location of the resultant force which acts on each side and the bottom of the aquarium due to the fresh water inside it.

Ans. Bottom force = 824 N
Side forces = 235 N, 549 N
All four side forces at $\frac{2}{3}$ depth

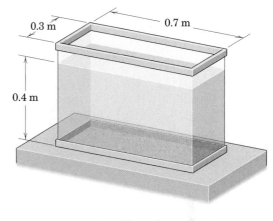

Problem 5/175

5/176 A rectangular block of density ρ_1 floats in a liquid of density ρ_2. Determine the ratio $r = h/c$, where h is the submerged depth of block. Evaluate r for an oak block floating in fresh water and for steel floating in mercury.

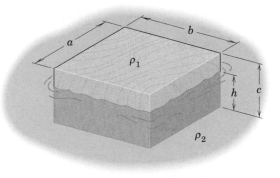

Problem 5/176

5/177 Calculate the depth h of mud for which the 3-m concrete retaining wall is on the verge of tipping about its forward edge A. The density of mud may be taken to be 1760 kg/m³ and that of concrete to be 2400 kg/m³.

Ans. $h = 1.99$ m

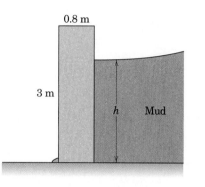

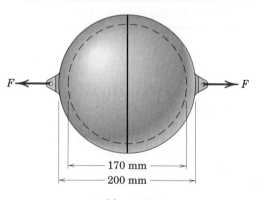

Problem 5/177

5/178 The two hemispherical shells are perfectly sealed together over their mating surfaces, and the air inside is partially evacuated to a pressure of 14 kPa. Atmospheric pressure is 101.3 kPa. Determine the force F required to separate the shells.

Problem 5/178

5/179 A deep-submersible diving chamber designed in the form of a spherical shell 1500 mm in diameter is ballasted with lead so that its weight slightly exceeds its buoyancy. Atmospheric pressure is maintained within the sphere during an ocean dive to a depth of 3 km. The thickness of the shell is 25 mm. For this depth calculate the compressive stress σ which acts on a diametral section of the shell, as indicated in the right-hand view.

Ans. $\sigma = 463$ MPa

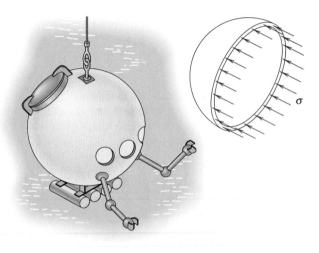

Problem 5/179

5/180 Fresh water in a channel is contained by the uniform 2.5-m plate freely hinged at A. If the gate is designed to open when the depth of the water reaches 0.8 m as shown in the figure, what must be the weight w (in newtons per meter of horizontal length into the paper) of the gate?

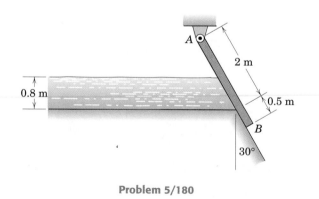

Problem 5/180

5/181 When the seawater level inside the hemispherical chamber reaches the 0.6-m level shown in the figure, the plunger is lifted, allowing a surge of sea water to enter the vertical pipe. For this fluid level (*a*) determine the average pressure σ supported by the seal area of the valve before force is applied to lift the plunger and (*b*) determine the force P (in addition to the force needed to support its weight) required to lift the plunger. Assume atmospheric pressure in all airspaces and in the seal area when contact ceases under the action of P.

Ans. $\sigma = 10.74$ kPa, $P = 1.687$ kN

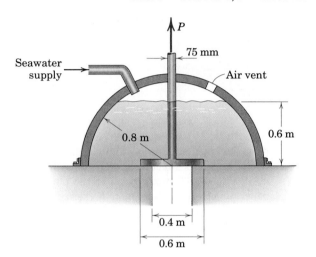

Problem 5/181

Representative Problems

5/182 The figure shows the end view of a long homogeneous solid cylinder which floats in a liquid and has a removed segment. Show that $\theta = 0$ and $\theta = 180°$ are the two values of the angle between its centerline and the vertical for which the cylinder floats in stable positions.

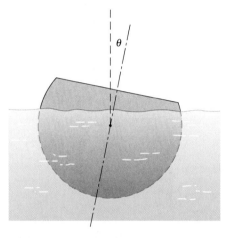

Problem 5/182

5/183 One of the critical problems in the design of deep-submergence vehicles is to provide viewing ports which will withstand tremendous hydrostatic pressures without fracture or leakage. The figure shows the cross section of an experimental acrylic window with spherical surfaces under test in a high-pressure liquid chamber. If the pressure p is raised to a level that simulates the effect of a dive to a depth of 1 km in sea water, calculate the average pressure σ supported by the gasket A.

Ans. $\sigma = 26.4$ MPa

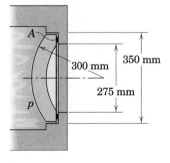

Problem 5/183

5/184 The solid concrete cylinder 6 ft long and 4 ft in diameter is supported in a half-submerged position in fresh water by a cable which passes over a fixed pulley at A. Compute the tension T in the cable. The cylinder is waterproofed by a plastic coating. (Consult Table D/1, Appendix D, as needed.)

Problem 5/184

5/185 A marker buoy consisting of a cylinder and cone has the dimensions shown and weighs 625 lb when out of the water. Determine the protrusion h when the buoy is floating in salt water. The buoy is weighted so that a low center of mass ensures stability.

Ans. h = 3.89 ft

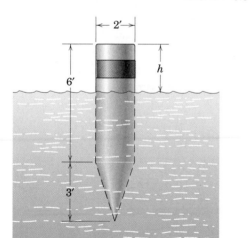

Problem 5/185

5/186 A fresh-water channel 10 ft wide (normal to the plane of the paper) is blocked at its end by a rectangular barrier, shown in section ABD. Supporting struts BC are spaced every 2 ft along the 10-ft width. Determine the compression C in each strut. Neglect the weights of the members.

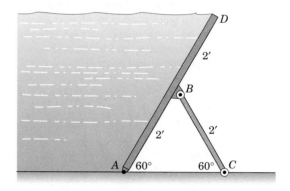

Problem 5/186

5/187 A vertical section of an oil sump is shown. The access plate covers a rectangular opening which has a dimension of 400 mm normal to the plane of the paper. Calculate the total force R exerted by the oil on the plate and the location x of R. The oil has a density of 900 kg/m³.

Ans. R = 1377 N, x = 323 mm

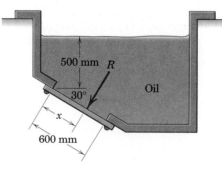

Problem 5/187

5/188 A homogeneous solid sphere of radius r is resting on the bottom of a tank containing a liquid of density ρ_l, which is greater than the density ρ_s of the sphere. As the tank is filled, a depth h is reached at which the sphere begins to float. Determine the expression for the density ρ_s of the sphere.

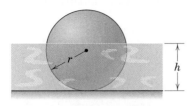

Problem 5/188

5/189 The rectangular gate shown in section is 10 ft long (perpendicular to the paper) and is hinged about its upper edge B. The gate divides a channel leading to a fresh-water lake on the left and a salt-water tidal basin on the right. Calculate the torque M on the shaft of the gate at B required to prevent the gate from opening when the salt-water level drops to $h = 3$ ft.

Ans. $M = 11.22(10^4)$ lb-ft

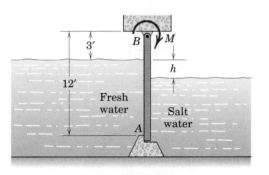

Problem 5/189

5/190 The hydraulic cylinder operates the toggle which closes the vertical gate against the pressure of fresh water on the opposite side. The gate is rectangular with a horizontal width of 2 m perpendicular to the paper. For a depth $h = 3$ m of water, calculate the required oil pressure p which acts on the 150-mm-diameter piston of the hydraulic cylinder.

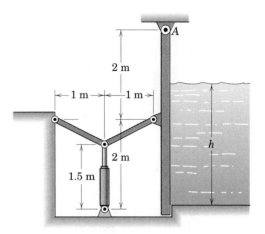

Problem 5/190

5/191 The design of a floating oil-drilling platform consists of two rectangular pontoons and six cylindrical columns which support the working platform. When ballasted, the entire structure has a displacement of 26,000 tons (expressed in long tons of 2240 lb). Calculate the total draft h of the structure when it is moored in the ocean. The specific weight of salt water is 64 lb/ft^3. Neglect the vertical components of the mooring forces.

Ans. $h = 74.5$ ft

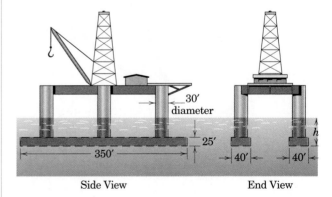

Problem 5/191

5/192 Determine the total force R exerted on the triangular window by the fresh water in the tank. The water level is even with the top of the window. Also determine the distance H from R to the water level.

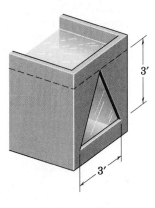

Problem 5/192

5/193 The barge crane of rectangular proportions has a 12-ft by 30-ft cross section over its entire length of 80 ft. If the maximum permissible submergence and list in sea water are represented by the position shown, determine the corresponding maximum safe load w which the barge can handle at the 20-ft extended position of the boom. Also find the total displacement W in long tons of the unloaded barge (1 long ton equals 2240 lb). The distribution of machinery and ballast places the center of gravity G of the barge, minus the load w, at the center of the hull.

Ans. $w = 100{,}800$ lb, $W = 366$ long tons

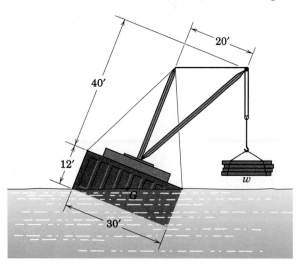

Problem 5/193

5/194 The cast-iron plug seals the drainpipe of an open fresh-water tank which is filled to a depth of 20 ft. Determine the tension T required to remove the plug from its tapered hole. Atmospheric pressure exists in the drainpipe and in the seal area as the plug is being removed. Neglect mechanical friction between the plug and its supporting surface.

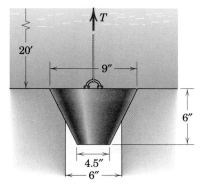

Problem 5/194

5/195 The Quonset hut is subjected to a horizontal wind, and the pressure p against the circular roof is approximated by $p_0 \cos \theta$. The pressure is positive on the windward side of the hut and is negative on the leeward side. Determine the total horizontal shear Q on the foundation per unit length of roof measured normal to the paper.

Ans. $Q = \frac{1}{2}\pi r p_0$

Problem 5/195

5/196 The upstream side of an arched dam has the form of a vertical cylindrical surface of 500-ft radius and subtends an angle of 60°. If the fresh water is 100 ft deep, determine the total force R exerted by the water on the dam face.

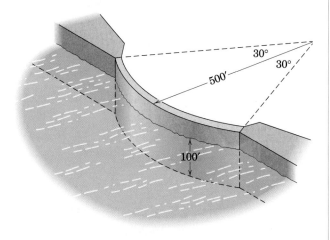

Problem 5/196

5/197 The fresh-water side of a concrete dam has the shape of a vertical parabola with vertex at A. Determine the position b of the base point B through which acts the resultant force of the water against the dam face C.

Ans. b = 28.1 m

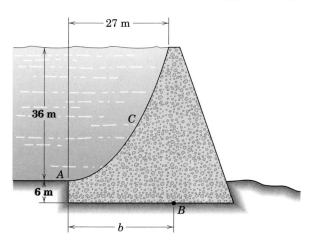

Problem 5/197

5/198 The small access hole A allows maintenance workers to enter the storage tank at ground level when it is empty. Two designs, (a) and (b), are shown for the hole cover. If the tank is full of fresh water, estimate the average pressure σ in the seal area of design (a) and the average increase ΔT in the initial tension in each of the 16 bolts of design (b). You may take the pressure over the hole area to be constant, and the pressure in the seal area of design (b) may be assumed to be atmospheric.

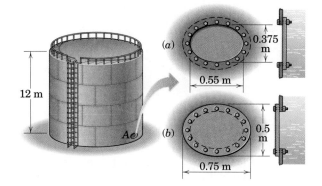

Problem 5/198

5/199 The 3-m plank shown in section has a density of 800 kg/m^3 and is hinged about a horizontal axis through its upper edge O. Calculate the angle θ assumed by the plank with the horizontal for the level of fresh water shown.

Ans. θ = 48.2°

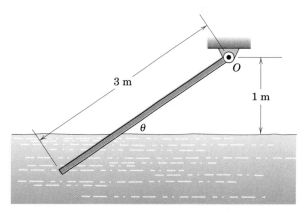

Problem 5/199

5/200 The deep-submersible research vessel has a passenger compartment in the form of a spherical steel shell with a mean radius of 1.000 m and a thickness of 35 mm. Calculate the mass of lead ballast which the vessel must carry so that the combined weight of the steel shell and lead ballast exactly cancels the combined buoyancy of these two parts alone. (Consult Table D/1, Appendix D, as needed.)

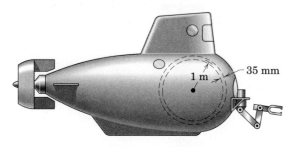

1 m — 35 mm

Problem 5/200

5/201 The elements of a new method for constructing concrete foundation walls for new houses are shown in the figure. Once the footing F is in place, polystyrene forms A are erected and a thin concrete mixture B is poured between the forms. Ties T prevent the forms from separating. After the concrete cures, the forms are left in place for insulation. As a design exercise, make a conservative estimate for the uniform tie spacing d if the tension in each tie is not to exceed 6.5 kN. The horizontal tie spacing is the same as the vertical spacing. State any assumptions. The density of wet concrete is 2400 kg/m³.

Ans. $d = 0.300$ m

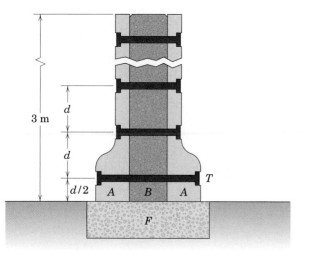

3 m
d
d
$d/2$
T
A B A
F

Problem 5/201

5/202 The trapezoidal viewing window in a sea-life aquarium has the dimensions shown. With the aid of appropriate diagrams and coordinates, describe two methods by which the resultant force R on the glass due to water pressure, and the vertical location of R, could be found if numerical values were supplied.

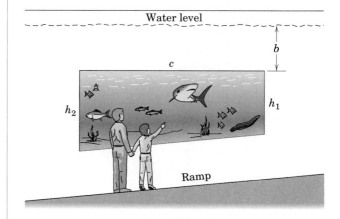

Water level
b
c
h_2
h_1
Ramp

Problem 5/202

▶**5/203** The sphere is used as a valve to close the hole in the fresh-water tank. As the depth h decreases, the tension T required to open the valve decreases because the downward force on the sphere decreases with less pressure. Determine the depth h for which T equals the weight of the sphere.

Ans. $h = 9.33$ in.

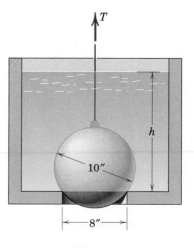

T
h
10″
8″

Problem 5/203

▶5/204 The accurate determination of the vertical position of the center of mass G of a ship is difficult to achieve by calculation. It is more easily obtained by a simple inclining experiment on the loaded ship. With reference to the figure, a known external mass m_0 is placed a distance d from the centerline, and the angle of list θ is measured by means of the deflection of a plumb bob. The displacement of the ship and the location of the metacenter M are known. Calculate the metacentric height $\overline{GM}$ for a 12 000-t ship inclined by a 27-t mass placed 7.8 m from the centerline if a 6-m plumb line is deflected a distance $a = 0.2$ m. The mass m_0 is at a distance b = 1.8 m above M. [Note that the metric ton (t) equals 1000 kg and is the same as the megagram (Mg).]

Ans. $\overline{GM} = 0.530$ m

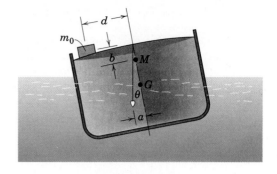

Problem 5/204

5/10   CHAPTER REVIEW

In Chapter 5 we have studied various common examples of forces distributed throughout volumes, over areas, and along lines. In all these problems we often need to determine the resultant of the distributed forces and the location of the resultant.

Finding Resultants of Distributed Forces

To find the resultant and line of action of a distributed force:

1. Begin by multiplying the intensity of the force by the appropriate element of volume, area, or length in terms of which the intensity is expressed. Then sum (integrate) the incremental forces over the region involved to obtain their resultant.

2. To locate the line of action of the resultant, use the principle of moments. Evaluate the sum of the moments, about a convenient axis, of all of the increments of force. Equate this sum to the moment of the resultant about the same axis. Then solve for the unknown moment arm of the resultant.

Gravitational Forces

When force is distributed throughout a mass, as in the case of gravitational attraction, the intensity is the force of attraction ρg per unit of volume, where ρ is the density and g is the gravitational acceleration. For bodies whose density is constant, we saw in Section A that ρg cancels when the moment principle is applied. This leaves us with a strictly geometric problem of finding the centroid of the figure, which coincides with the mass center of the physical body whose boundary defines the figure.

1. For flat plates and shells which are homogeneous and have constant thickness, the problem becomes one of finding the properties of an area.

2. For slender rods and wires of uniform density and constant cross section, the problem becomes one of finding the properties of a line segment.

Integration of Differential Relationships

For problems which require the integration of differential relationships, keep in mind the following considerations.

1. Select a coordinate system which provides the simplest description of the boundaries of the region of integration.

2. Eliminate higher-order differential quantities whenever lower-order differential quantities will remain.

3. Choose a first-order differential element in preference to a second-order element and a second-order element in preference to a third-order element.

4. Whenever possible, choose a differential element which avoids discontinuities within the region of integration.

Distributed Forces in Beams, Cables, and Fluids

In Section B we used these guidelines along with the principles of equilibrium to solve for the effects of distributed forces in beams, cables, and fluids. Remember that:

1. For beams and cables the force intensity is expressed as force per unit length.
2. For fluids the force intensity is expressed as force per unit area, or pressure.

Although beams, cables, and fluids are physically quite different applications, their problem formulations share the common elements cited above.

REVIEW PROBLEMS

5/205 Determine the x- and y-coordinates of the centroid of the shaded area.

$$Ans. \; \bar{x} = \frac{37}{84}, \; \bar{y} = \frac{13}{30}$$

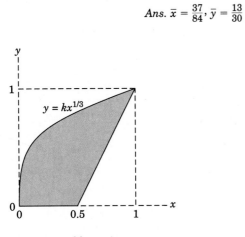

Problem 5/205

5/206 Determine the x- and y-coordinates of the centroid of the shaded area.

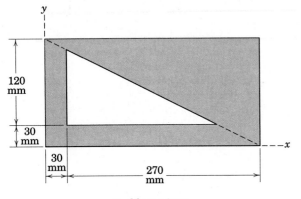

Problem 5/206

5/207 The cross section shown is for a complete cast-iron body of revolution about the z-axis. Compute its mass m.

$$Ans. \; m = 11.55 \text{ kg}$$

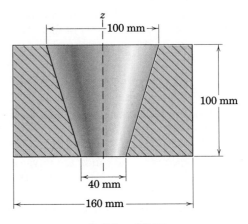

Problem 5/207

5/208 The assembly consists of four rods cut from the same bar stock. The curved member is a circular arc of radius b. Determine the y- and z-coordinates of the mass center of the assembly.

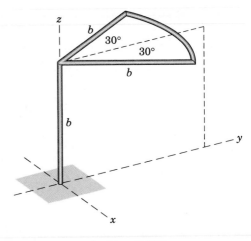

Problem 5/208

5/209 Determine the area A of the curved surface $ABCD$ of the solid of revolution shown.

$$Ans.\ A = \frac{\pi a^2}{2}(\pi - 1)$$

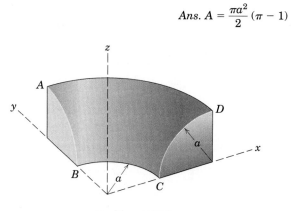

Problem 5/209

5/210 The uniform bar of mass m is bent into the quarter-circular arc in the vertical plane and is hinged freely at A. The bar is held in position by the horizontal wire from O to B. Determine the magnitude R of the force on the pin at A.

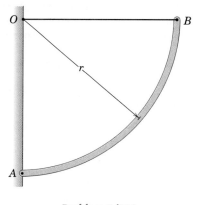

Problem 5/210

5/211 The tapered body has a horizontal cross section which is circular. Determine the height $\bar{h}$ of its mass center above the base of the homogeneous body.

$$Ans.\ \bar{h} = \frac{11}{28}H$$

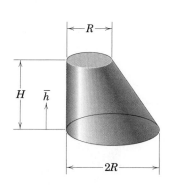

Problem 5/211

5/212 Shown in the figure is a body similar to that of Prob. 5/211, only now the body is a thin shell with open top and bottom. Determine the height $\bar{h}$ of its mass center above the base of the homogeneous body and compare your result with the printed answer for Prob. 5/211.

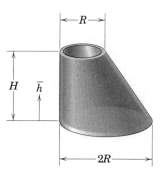

Problem 5/212

5/213 The homogeneous body shown is constructed by beginning with the solid body of Prob. 5/211 and then removing a cylindrical volume of diameter just less than R as shown, leaving a thin wall at the left portion of the body as well as at the very top. Determine the height $\bar{h}$ of the mass center above the base of the body.

$$Ans. \ \bar{h} = \frac{5}{16}H$$

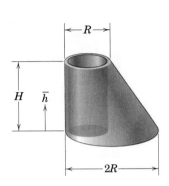

Problem 5/213

5/214 Draw the shear and moment diagrams for the beam, which supports the uniform load of 50 lb per foot of beam length distributed over its midsection. Determine the maximum bending moment and its location.

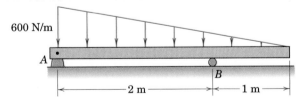

Problem 5/214

5/215 Determine the maximum bending moment M_{max} for the loaded beam and specify the distance x to the right of end A where M_{max} exists.

$$Ans. \ M_{max} = 186.4 \ N \cdot m \ at \ x = 0.879 \ m$$

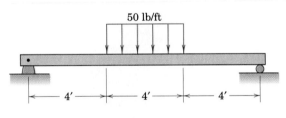

Problem 5/215

5/216 Sketch the shear and moment diagrams for each of

the four beams loaded and supported as shown.

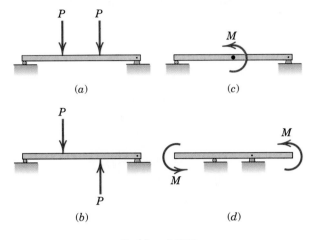

Problem 5/216

5/217 The triangular sign is attached to the post embedded in the concrete base at B. Calculate the shear force V, the bending moment M, and the torsional moment T in the post at B during a storm where the wind velocity normal to the sign reaches 100 km/h. The air pressure (called stagnation pressure) against the vertical surface corresponding to this velocity is 1.4 kPa.

$$Ans. \ V = 2.8 \ kN, \ M = 6.53 \ kN \cdot m, \ T = 1.867 \ kN \cdot m$$

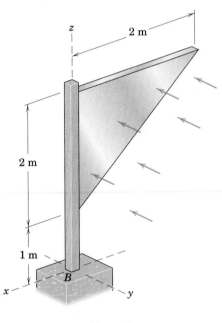

Problem 5/217

5/218 The circular disk rotates about an axis through its center O and has three holes of diameter d positioned as shown. A fourth hole is to be drilled in the disk at the same radius r so that the disk will be in balance (mass center at O). Determine the required diameter D of the new hole and its angular position.

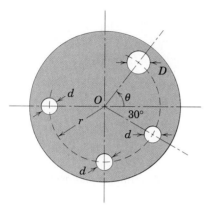

Problem 5/218

5/219 Determine the depth h of the square hole in the solid circular cylinder for which the z-coordinate of the mass center will have the maximum possible value.

Ans. $h = 5.10$ in.

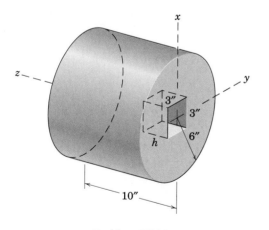

Problem 5/219

5/220 A flat plate seals a triangular opening in the vertical wall of a tank of liquid of density ρ. The plate is hinged about the upper edge O of the triangle. Determine the force P required to hold the gate in a closed position against the pressure of the liquid.

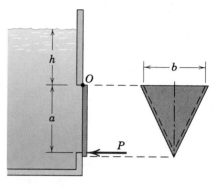

Problem 5/220

5/221 A prismatic structure of height h and base b is subjected to a horizontal wind load whose pressure p increases from zero at the base to p_0 at the top according to $p = k\sqrt{y}$. Determine the resisting moment M at the base of the structure.

Ans. $M = \frac{4}{35} p_0 b h^2$

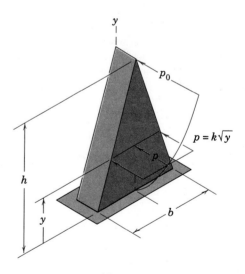

Problem 5/221

5/222 As part of a preliminary design study, the effects of wind loads on a 900-ft building are investigated. For the parabolic distribution of wind pressure shown in the figure, compute the force and moment reactions at the base A of the building due to the wind load. The depth of the building (perpendicular to the paper) is 200 ft.

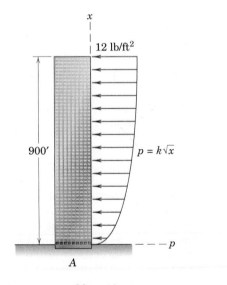

12 lb/ft^2

900′

$p = k\sqrt{x}$

A

Problem 5/222

▶**5/223** Regard the tall building of Prob. 5/222 as a uniform upright beam. Determine and plot the shear force and bending moment in the structure as functions of the height x above the ground. Evaluate your expressions at $x = 450$ ft.

$$\text{Ans. } V = 1.440(10^6) - \frac{160}{3}x^{3/2} \text{ lb}$$

$$M = 7.78(10^8) - 1.440(10^6)x + \frac{64}{3}x^{5/2} \text{ lb-ft}$$

$$V|_{x=450'} = 0.931(10^6) \text{ lb}$$

$$M|_{x=450'} = 2.21(10^8) \text{ lb-ft}$$

▶**5/224** Locate the center of mass of the thin spherical shell which is formed by cutting out one-eighth of a complete shell.

$$\text{Ans. } \bar{x} = \bar{y} = \bar{z} = \frac{r}{2}$$

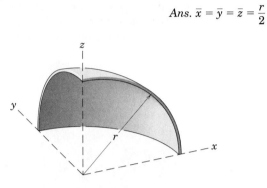

Problem 5/224

▶**5/225** Locate the center of mass of the homogeneous bell-shaped shell of uniform but negligible thickness.

$$\text{Ans. } \bar{z} = \frac{a}{\pi - 2}$$

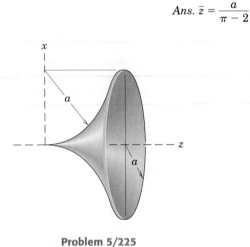

Problem 5/225

Computer-Oriented Problems*

*5/226 Construct the shear and moment diagrams for the loaded beam of Prob. 5/113, repeated here. Determine the maximum values of the shear and moment and their locations on the beam.

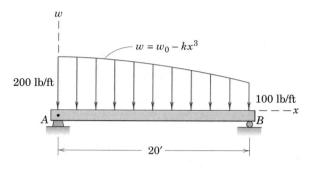

Problem 5/226

*5/227 Construct the shear and moment diagrams for the loaded beam shown. Determine the maximum bending moment and its location.

Ans. $M_{max} = 6.23$ kN·m at $x = 2.13$ m

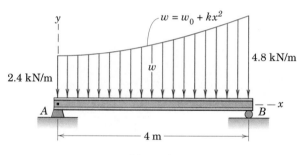

Problem 5/227

*5/228 Find the angle θ which will place the mass center of the thin ring a distance $r/10$ from the center of the arc.

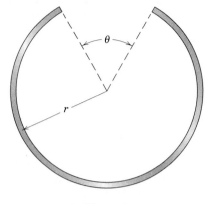

Problem 5/228

*5/229 A homogeneous charge of solid propellant for a rocket is in the shape of the circular cylinder formed with a concentric hole of depth x. For the dimensions shown, plot $\overline{X}$, the x-coordinate of the mass center of the propellant, as a function of the depth x of the hole from $x = 0$ to $x = 600$ mm. Determine the maximum value of $\overline{X}$ and show that it is equal to the corresponding value of x.

Ans. $\overline{X}_{max} = 322$ mm

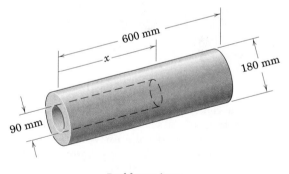

Problem 5/229

*5/230 A cable weighing 10 lb/ft is attached to point A and passes over the small pulley at B on the same horizontal line with A. Determine the sag h and length S of the cable between A and B if a tension of 2500 lb is applied to the cable over the pulley.

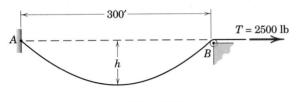

Problem 5/230

*5/231 An underwater detection instrument A is attached to the midpoint of a 100-m cable suspended between two ships 50 m apart. Determine the depth h of the instrument, which has negligible mass. Does the result depend on the mass of the cable or on the density of the water?

Ans. $h = 39.8$ m

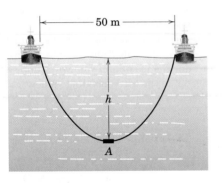

Problem 5/231

*5/232 The center of buoyancy B of a ship's hull is the centroid of its submerged volume. The underwater cross-sectional areas A of the transverse sections of the tugboat hull shown are tabulated for every five meters of waterline length. With an appropriate computer program, determine to the nearest 0.5 m the distance $\bar{x}$ of B aft of point A.

x, m	A, m^2	x, m	A, m^2
0	0	30	23.8
5	7.1	35	19.5
10	15.8	40	12.5
15	22.1	45	5.1
20	24.7	50	0
25	25.1		

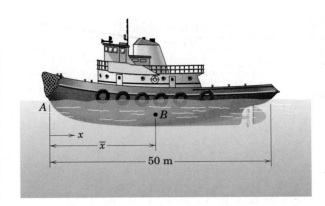

Problem 5/232

*5/233 A right circular cylinder of density ρ_1 floats in a liquid of density ρ_2. If d is the diameter of the cylinder and h is the submerged depth, plot the ratio $r = \dfrac{h}{d}$ as a function of $\dfrac{\rho_1}{\rho_2}$ over the range $0 \le \dfrac{\rho_1}{\rho_2} \le 1$. Evaluate r for a pine cylinder floating in sea water.

Ans. $r = 0.473$

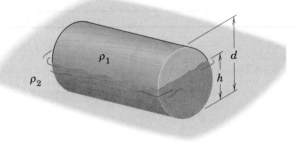

Problem 5/233

*5/234 A length of power cable is suspended from the two towers as shown. The cable has a mass of 20 kg per meter of cable. If the maximum allowable cable tension is 75 kN, determine the mass ρ of ice per meter which can form on the cable without the maximum allowable tension being exceeded. If additional stretch in the cable is neglected, does the addition of the ice change the cable configuration?

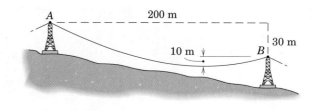

Problem 5/234

This test of a brake disk dramatically shows the transformation of mechanical energy to heat energy via kinetic friction between the stationary brake pads and the rotating disk.

FRICTION

6/1   INTRODUCTION

In the preceding chapters we have usually assumed that the forces of action and reaction between contacting surfaces act normal to the surfaces. This assumption characterizes the interaction between smooth surfaces and was illustrated in Example 2 of Fig. 3/1. Although this ideal assumption often involves only a relatively small error, there are many problems in which we must consider the ability of contacting surfaces to support tangential as well as normal forces. Tangential forces generated between contacting surfaces are called *friction forces* and occur to some degree in the interaction between all real surfaces. Whenever a tendency exists for one contacting surface to slide along another surface, the friction forces developed are always in a direction to oppose this tendency.

In some types of machines and processes we want to minimize the retarding effect of friction forces. Examples are bearings of all types, power screws, gears, the flow of fluids in pipes, and the propulsion of aircraft and missiles through the atmosphere. In other situations we wish to maximize the effects of friction, as in brakes, clutches, belt drives, and wedges. Wheeled vehicles depend on friction for both starting and stopping, and ordinary walking depends on friction between the shoe and the ground.

Friction forces are present throughout nature and exist in all machines no matter how accurately constructed or carefully lubricated. A machine or process in which friction is small enough to be neglected is said to be *ideal*. When friction must be taken into account, the machine or process is termed *real*. In all cases where there is sliding motion between parts, the friction forces result in a loss of energy which is dissipated in the form of heat. Wear is another effect of friction.

SECTION A FRICTIONAL PHENOMENA

6/2 TYPES OF FRICTION

In this article we briefly discuss the types of frictional resistance encountered in mechanics. The next article contains a more detailed account of the most common type of friction, dry friction.

(a) Dry Friction. Dry friction occurs when the unlubricated surfaces of two solids are in contact under a condition of sliding or a tendency to slide. A friction force tangent to the surfaces of contact occurs both during the interval leading up to impending slippage and while slippage takes place. The direction of this friction force always opposes the motion or impending motion. This type of friction is also called *Coulomb friction*. The principles of dry or Coulomb friction were developed largely from the experiments of Coulomb in 1781 and from the work of Morin from 1831 to 1834. Although we do not yet have a comprehensive theory of dry friction, in Art. 6/3 we describe an analytical model sufficient to handle the vast majority of problems involving dry friction. This model forms the basis for most of this chapter.

(b) Fluid Friction. Fluid friction occurs when adjacent layers in a fluid (liquid or gas) are moving at different velocities. This motion causes frictional forces between fluid elements, and these forces depend on the relative velocity between layers. When there is no relative velocity, there is no fluid friction. Fluid friction depends not only on the velocity gradients within the fluid but also on the viscosity of the fluid, which is a measure of its resistance to shearing action between fluid layers. Fluid friction is treated in the study of fluid mechanics and will not be discussed further in this book.

(c) Internal Friction. Internal friction occurs in all solid materials which are subjected to cyclical loading. For highly elastic materials the recovery from deformation occurs with very little loss of energy due to internal friction. For materials which have low limits of elasticity and which undergo appreciable plastic deformation during loading, a considerable amount of internal friction may accompany this deformation. The mechanism of internal friction is associated with the action of shear deformation, which is discussed in references on materials science. Because this book deals primarily with the external effects of forces, we will not discuss internal friction further.

6/3   DRY FRICTION

The remainder of this chapter describes the effects of dry friction acting on the exterior surfaces of rigid bodies. We will now explain the mechanism of dry friction with the aid of a very simple experiment.

Mechanism of Dry Friction

Consider a solid block of mass m resting on a horizontal surface, as shown in Fig. 6/1a. We assume that the contacting surfaces have some roughness. The experiment involves the application of a horizontal force P which continuously increases from zero to a value sufficient to move the block and give it an appreciable velocity. The free-body diagram of the block for any value of P is shown in Fig. 6/1b, where the tangential friction force exerted by the plane on the block is labeled F. This friction force acting on the body will *always* be in a direction to oppose motion or the tendency toward motion of the body. There is also a normal force N which in this case equals mg, and the total force R exerted by the supporting surface on the block is the resultant of N and F.

A magnified view of the irregularities of the mating surfaces, Fig. 6/1c, helps us to visualize the mechanical action of friction. Support is necessarily intermittent and exists at the mating humps. The direction of each of the reactions on the block, R_1, R_2, R_3, etc. depends not only on the geometric profile of the irregularities but also on the extent of local deformation at each contact point. The total normal force N is the sum

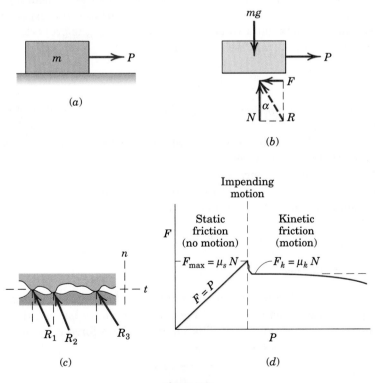

Figure 6/1

of the n-components of the R's, and the total frictional force F is the sum of the t-components of the R's. When the surfaces are in relative motion, the contacts are more nearly along the tops of the humps, and the t-components of the R's are smaller than when the surfaces are at rest relative to one another. This observation helps to explain the well-known fact that the force P necessary to maintain motion is generally less than that required to start the block when the irregularities are more nearly in mesh.

If we perform the experiment and record the friction force F as a function of P, we obtain the relation shown in Fig. 6/1d. When P is zero, equilibrium requires that there be no friction force. As P is increased, the friction force must be equal and opposite to P as long as the block does not slip. During this period the block is in equilibrium, and all forces acting on the block must satisfy the equilibrium equations. Finally, we reach a value of P which causes the block to slip and to move in the direction of the applied force. At this same time the friction force decreases slightly and abruptly. It then remains essentially constant for a time but then decreases still more as the velocity increases.

Static Friction

The region in Fig. 6/1d up to the point of slippage or impending motion is called the range of *static friction*, and in this range the value of the friction force is determined by the *equations of equilibrium*. This friction force may have any value from zero up to and including the maximum value. For a given pair of mating surfaces the experiment shows that this maximum value of static friction F_{max} is proportional to the normal force N. Thus, we may write

$$\boxed{F_{max} = \mu_s N} \tag{6/1}$$

where μ_s is the proportionality constant, called the *coefficient of static friction*.

Be aware that Eq. 6/1 describes only the *limiting* or *maximum* value of the static friction force and *not* any lesser value. Thus, the equation applies only to cases where motion is impending with the friction force at its peak value. For a condition of static equilibrium when motion is *not* impending, the static friction force is

$$F < \mu_s N$$

Kinetic Friction

After slippage occurs, a condition of *kinetic friction* accompanies the ensuing motion. Kinetic friction force is usually somewhat less than the maximum static friction force. The kinetic friction force F_k is also proportional to the normal force. Thus,

$$\boxed{F_k = \mu_k N} \tag{6/2}$$

where μ_k is the *coefficient of kinetic friction*. It follows that μ_k is generally less than μ_s. As the velocity of the block increases, the kinetic friction decreases somewhat, and at high velocities, this decrease may be significant. Coefficients of friction depend greatly on the exact condition of the surfaces, as well as on the relative velocity, and are subject to considerable uncertainty.

Because of the variability of the conditions governing the action of friction, in engineering practice it is frequently difficult to distinguish between a static and a kinetic coefficient, especially in the region of transition between impending motion and motion. Well-greased screw threads under mild loads, for example, often exhibit comparable frictional resistance whether they are on the verge of turning or whether they are in motion.

In the engineering literature we frequently find expressions for maximum static friction and for kinetic friction written simply as $F = \mu N$. It is understood from the problem at hand whether maximum static friction or kinetic friction is described. Although we will frequently distinguish between the static and kinetic coefficients, in other cases no distinction will be made, and the friction coefficient will be written simply as μ. In those cases you must decide which of the friction conditions, maximum static friction for impending motion or kinetic friction, is involved. We emphasize again that many problems involve a static friction force which is less than the maximum value at impending motion, and therefore under these conditions the friction relation Eq. 6/1 cannot be used.

Figure 6/1c shows that rough surfaces are more likely to have larger angles between the reactions and the n-direction than do smoother surfaces. Thus, for a pair of mating surfaces, a friction coefficient reflects the roughness, which is a geometric property of the surfaces. With this geometric model of friction, we describe mating surfaces as "smooth" when the friction forces they can support are negligibly small. It is meaningless to speak of a coefficient of friction for a single surface.

Friction Angles

The direction of the resultant R in Fig. 6/1b measured from the direction of N is specified by $\tan \alpha = F/N$. When the friction force reaches its limiting static value $F_{\max}$, the angle α reaches a maximum value ϕ_s. Thus,

$$\tan \phi_s = \mu_s$$

When slippage is occurring, the angle α has a value ϕ_k corresponding to the kinetic friction force. In like manner,

$$\tan \phi_k = \mu_k$$

In practice we often see the expression $\tan \phi = \mu$, in which the coefficient of friction may refer to either the static or the kinetic case,

This mountain climber depends on friction between his body and the rocks as well as friction between the rope and mechanical devices through which the rope can slip.

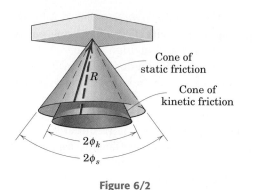

Figure 6/2

depending on the particular problem. The angle ϕ_s is called the *angle of static friction*, and the angle ϕ_k is called the *angle of kinetic friction*. The friction angle for each case clearly defines the limiting direction of the total reaction R between two contacting surfaces. If motion is impending, R must be one element of a right-circular cone of vertex angle $2\phi_s$, as shown in Fig. 6/2. If motion is not impending, R is within the cone. This cone of vertex angle $2\phi_s$ is called the *cone of static friction* and represents the locus of possible directions for the reaction R at impending motion. If motion occurs, the angle of kinetic friction applies, and the reaction must lie on the surface of a slightly different cone of vertex angle $2\phi_k$. This cone is the *cone of kinetic friction*.

Factors Affecting Friction

Further experiment shows that the friction force is essentially independent of the apparent or projected area of contact. The true contact area is much smaller than the projected value, since only the peaks of the contacting surface irregularities support the load. Even relatively small normal loads result in high stresses at these contact points. As the normal force increases, the true contact area also increases as the material undergoes yielding, crushing, or tearing at the points of contact.

A comprehensive theory of dry friction must go beyond the mechanical explanation presented here. For example, there is evidence that molecular attraction may be an important cause of friction under conditions where the mating surfaces are in very close contact. Other factors which influence dry friction are the generation of high local temperatures and adhesion at contact points, relative hardness of mating surfaces, and the presence of thin surface films of oxide, oil, dirt, or other substances.

Some typical values of coefficients of friction are given in Table D/1, Appendix D. These values are only approximate and are subject to considerable variation, depending on the exact conditions prevailing. They may be used, however, as typical examples of the magnitudes of frictional effects. To make a reliable calculation involving friction, the appropriate friction coefficient should be determined by experiments which duplicate the surface conditions of the application as closely as possible.

Types of Friction Problems

We can now recognize the following three types of problems encountered in applications involving dry friction. The first step in solving a friction problem is to identify its type.

1. In the *first* type of problem, the condition of impending motion is known to exist. Here a body which is in equilibrium is on the verge of slipping, and the friction force equals the limiting static friction $F_{max} = \mu_s N$. The equations of equilibrium will, of course, also hold.

2. In the *second* type of problem, neither the condition of impending motion nor the condition of motion is known to exist. To determine the actual friction conditions, we first assume static equilibrium and then solve for the friction force F necessary for equilibrium. Three outcomes are possible:

   (a) $F < (F_{max} = \mu_s N)$: Here the friction force necessary for equilibrium can be supported, and therefore the body is in static equilibrium as assumed. We emphasize that the *actual* friction force F is *less than* the limiting value F_{max} given by Eq. 6/1 and that F is determined *solely* by the equations of equilibrium.

   (b) $F = (F_{max} = \mu_s N)$: Since the friction force F is at its maximum value F_{max}, motion impends, as discussed in problem type (1). The assumption of static equilibrium is valid.

   (c) $F > (F_{max} = \mu_s N)$: Clearly this condition is impossible, because the surfaces cannot support more force than the maximum $\mu_s N$. The assumption of equilibrium is therefore invalid, and motion occurs. The friction force F is equal to $\mu_k N$ from Eq. 6/2.

3. In the *third* type of problem, relative motion is known to exist between the contacting surfaces, and thus the kinetic coefficient of friction clearly applies. For this problem type, Eq. 6/2 always gives the kinetic friction force directly.

The foregoing discussion applies to all dry contacting surfaces and, to a limited extent, to moving surfaces which are partially lubricated.

Sample Problem 6/1

Determine the maximum angle θ which the adjustable incline may have with the horizontal before the block of mass m begins to slip. The coefficient of static friction between the block and the inclined surface is μ_s.

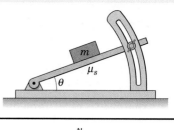

Solution. The free-body diagram of the block shows its weight $W = mg$, the normal force N, and the friction force F exerted by the incline on the block. The friction force acts in the direction to oppose the slipping which would occur if no friction were present.

① Equilibrium in the x- and y-directions requires

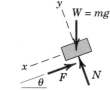

$$[\Sigma F_x = 0] \qquad mg \sin \theta - F = 0 \qquad F = mg \sin \theta$$

$$[\Sigma F_y = 0] \qquad -mg \cos \theta + N = 0 \qquad N = mg \cos \theta$$

Dividing the first equation by the second gives $F/N = \tan \theta$. Since the maximum angle occurs when $F = F_{max} = \mu_s N$, for impending motion we have

② $$\mu_s = \tan \theta_{max} \qquad \text{or} \qquad \theta_{max} = \tan^{-1} \mu_s \qquad \qquad Ans.$$

Helpful Hints

① We choose reference axes along and normal to the direction of F to avoid resolving both F and N into components.

② This problem describes a very simple way to determine a static coefficient of friction. The maximum value of θ is known as the *angle of repose*.

Sample Problem 6/2

Determine the range of values which the mass m_0 may have so that the 100-kg block shown in the figure will neither start moving up the plane nor slip down the plane. The coefficient of static friction for the contact surfaces is 0.30.

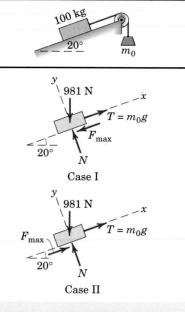
Case I

Case II

Solution. The maximum value of m_0 will be given by the requirement for motion impending up the plane. The friction force on the block therefore acts down the plane, as shown in the free-body diagram of the block for Case I in the figure. With the weight $mg = 100(9.81) = 981$ N, the equations of equilibrium give

$$[\Sigma F_y = 0] \qquad N - 981 \cos 20° = 0 \qquad N = 922 \text{ N}$$

$$[F_{max} = \mu_s N] \qquad F_{max} = 0.30(922) = 277 \text{ N}$$

$$[\Sigma F_x = 0] \qquad m_0(9.81) - 277 - 981 \sin 20° = 0 \qquad m_0 = 62.4 \text{ kg} \qquad Ans.$$

The minimum value of m_0 is determined when motion is impending down the
① plane. The friction force on the block will act up the plane to oppose the tendency to move, as shown in the free-body diagram for Case II. Equilibrium in the x-direction requires

$$[\Sigma F_x = 0] \qquad m_0(9.81) + 277 - 981 \sin 20° = 0 \qquad m_0 = 6.01 \text{ kg} \qquad Ans.$$

Thus, m_0 may have any value from 6.01 to 62.4 kg, and the block will remain at rest.

In both cases equilibrium requires that the resultant of F_{max} and N be concurrent with the 981-N weight and the tension T.

Helpful Hint

① We see from the results of Sample Problem 6/1 that the block would slide down the incline without the restraint of attachment to m_0 since tan 20° > 0.30. Thus, a value of m_0 will be required to maintain equilibrium.

Sample Problem 6/3

Determine the magnitude and direction of the friction force acting on the 100-kg block shown if, first, $P = 500$ N and, second, $P = 100$ N. The coefficient of static friction is 0.20, and the coefficient of kinetic friction is 0.17. The forces are applied with the block initially at rest.

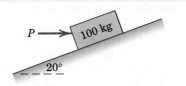

Solution.   There is no way of telling from the statement of the problem whether the block will remain in equilibrium or whether it will begin to slip following the application of P. It is therefore necessary that we make an assumption, so we will take the friction force to be up the plane, as shown by the solid arrow. From the free-body diagram a balance of forces in both x- and y-directions gives

$[\Sigma F_x = 0]$ $\qquad\qquad$ $P \cos 20° + F - 981 \sin 20° = 0$

$[\Sigma F_y = 0]$ $\qquad\qquad$ $N - P \sin 20° - 981 \cos 20° = 0$

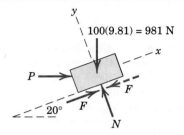

Case I. $P = 500$ N

Substitution into the first of the two equations gives

$$F = -134.3 \text{ N}$$

The negative sign tells us that *if* the block is in equilibrium, the friction force acting on it is in the direction opposite to that assumed and therefore is down the plane, as represented by the dashed arrow. We cannot reach a conclusion on the magnitude of F, however, until we verify that the surfaces are capable of supporting 134.3 N of friction force. This may be done by substituting $P = 500$ N into the second equation, which gives

$$N = 1093 \text{ N}$$

The maximum static friction force which the surfaces can support is then

$[F_{\max} = \mu_s N]$ $\qquad\qquad$ $F_{\max} = 0.20(1093) = 219$ N

Since this force is greater than that required for equilibrium, we conclude that the assumption of equilibrium was correct. The answer is, then,

$$F = 134.3 \text{ N down the plane} \qquad\qquad Ans.$$

Case II. $P = 100$ N

Substitution into the two equilibrium equations gives

$$F = 242 \text{ N} \qquad N = 956 \text{ N}$$

But the maximum possible static friction force is

$[F_{\max} = \mu_s N]$ $\qquad\qquad$ $F_{\max} = 0.20(956) = 191.2$ N

It follows that 242 N of friction cannot be supported. Therefore, equilibrium cannot exist, and we obtain the correct value of the friction force by using the kinetic coefficient of friction accompanying the motion down the plane. Hence, the answer is

$[F_k = \mu_k N]$ $\qquad\qquad$ $F = 0.17(956) = 162.5$ N up the plane $\qquad\qquad$ *Ans.*

Helpful Hint

① We should note that even though ΣF_x is no longer equal to zero, equilibrium does exist in the y-direction, so that $\Sigma F_y = 0$. Therefore, the normal force N is 956 N whether or not the block is in equilibrium.

Sample Problem 6/4

The homogeneous rectangular block of mass m, width b, and height H is placed on the horizontal surface and subjected to a horizontal force P which moves the block along the surface with a constant velocity. The coefficient of kinetic friction between the block and the surface is μ_k. Determine (a) the greatest value which h may have so that the block will slide without tipping over and (b) the location of a point C on the bottom face of the block through which the resultant of the friction and normal forces acts if $h = H/2$.

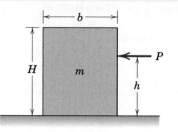

Solution. (a) With the block on the verge of tipping, we see that the entire reaction between the plane and the block will necessarily be at A. The free-body diagram of the block shows this condition. Since slipping occurs, the friction force is the limiting value $\mu_k N$, and the angle θ becomes $\theta = \tan^{-1} \mu_k$. The resultant of F_k and N passes through a point B through which P must also pass, since ① three coplanar forces in equilibrium are concurrent. Hence, from the geometry of the block

$$\tan \theta = \mu_k = \frac{b/2}{h} \qquad h = \frac{b}{2\mu_k} \qquad \textit{Ans.}$$

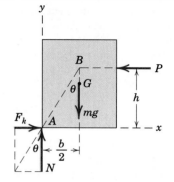

If h were greater than this value, moment equilibrium about A would not be satisfied, and the block would tip over.

Alternatively, we may find h by combining the equilibrium requirements for the x- and y-directions with the moment-equilibrium equation about A. Thus,

$$[\Sigma F_y = 0] \qquad N - mg = 0 \qquad N = mg$$

$$[\Sigma F_x = 0] \qquad F_k - P = 0 \qquad P = F_k = \mu_k N = \mu_k mg$$

$$[\Sigma M_A = 0] \qquad Ph - mg\frac{b}{2} = 0 \qquad h = \frac{mgb}{2P} = \frac{mgb}{2\mu_k mg} = \frac{b}{2\mu_k} \qquad \textit{Ans.}$$

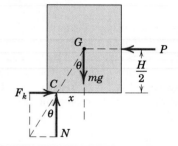

(b) With $h = H/2$ we see from the free-body diagram for case (b) that the resultant of F_k and N passes through a point C which is a distance x to the left of the vertical centerline through G. The angle θ is still $\theta = \phi = \tan^{-1} \mu_k$ as long as the block is slipping. Thus, from the geometry of the figure we have

②
$$\frac{x}{H/2} = \tan \theta = \mu_k \qquad \text{so} \qquad x = \mu_k H/2 \qquad \textit{Ans.}$$

If we were to replace μ_k by the static coefficient μ_s, then our solutions would describe the conditions under which the block is (a) on the verge of tipping and (b) on the verge of slipping, both from a rest position.

Helpful Hints

① Recall that the equilibrium equations apply to a body moving with a constant velocity (zero acceleration) just as well as to a body at rest.

② Alternatively, we could equate the moments about G to zero, which would give us $F(H/2) - Nx = 0$. Thus, with $F_k = \mu_k N$ we get $x = \mu_x H/2$.

Sample Problem 6/5

The three flat blocks are positioned on the 30° incline as shown, and a force P parallel to the incline is applied to the middle block. The upper block is prevented from moving by a wire which attaches it to the fixed support. The coefficient of static friction for each of the three pairs of mating surfaces is shown. Determine the maximum value which P may have before any slipping takes place.

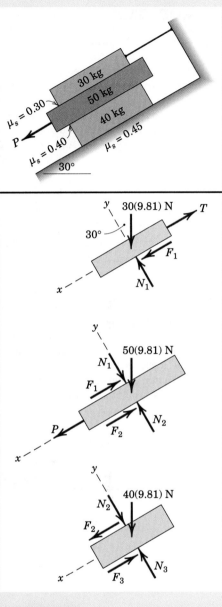

Solution. The free-body diagram of each block is drawn. The friction forces are assigned in the directions to oppose the relative motion which would occur if ① no friction were present. There are two possible conditions for impending motion. Either the 50-kg block slips and the 40-kg block remains in place, or the 50- and 40-kg blocks move together with slipping occurring between the 40-kg block and the incline.

The normal forces, which are in the y-direction, may be determined without reference to the friction forces, which are all in the x-direction. Thus,

$[\Sigma F_y = 0]$ (30-kg) $N_1 - 30(9.81) \cos 30° = 0$ $N_1 = 255$ N

(50-kg) $N_2 - 50(9.81) \cos 30° - 255 = 0$ $N_2 = 680$ N

(40-kg) $N_3 - 40(9.81) \cos 30° - 680 = 0$ $N_3 = 1019$ N

We will assume arbitrarily that only the 50-kg block slips, so that the 40-kg block remains in place. Thus, for impending slippage at both surfaces of the 50-kg block, we have

$[F_{max} = \mu_s N]$ $F_1 = 0.30(255) = 76.5$ N $F_2 = 0.40(680) = 272$ N

The assumed equilibrium of forces at impending motion for the 50-kg block gives

$[\Sigma F_x = 0]$ $P - 76.5 - 272 + 50(9.81) \sin 30° = 0$ $P = 103.1$ N

We now check on the validity of our initial assumption. For the 40-kg block with $F_2 = 272$ N the friction force F_3 would be given by

$[\Sigma F_x = 0]$ $272 + 40(9.81) \sin 30° - F_3 = 0$ $F_3 = 468$ N

But the maximum possible value of F_3 is $F_3 = \mu_s N_3 = 0.45(1019) = 459$ N. Thus, 468 N cannot be supported and our initial assumption was wrong. We conclude, therefore, that slipping occurs first between the 40-kg block and the incline. With the corrected value $F_3 = 459$ N, equilibrium of the 40-kg block for its impending motion requires

② $[\Sigma F_x = 0]$ $F_2 + 40(9.81) \sin 30° - 459 = 0$ $F_2 = 263$ N

Equilibrium of the 50-kg block gives, finally,

$[\Sigma F_x = 0]$ $P + 50(9.81) \sin 30° - 263 - 76.5 = 0$

$P = 93.8$ N   *Ans.*

Thus, with $P = 93.8$ N, motion impends for the 50-kg and 40-kg blocks as a unit.

Helpful Hints

① In the absence of friction the middle block, under the influence of P, would have a greater movement than the 40-kg block, and the friction force F_2 will be in the direction to oppose this motion as shown.

② We see now that F_2 is less than $\mu_s N_2 = 272$ N.

PROBLEMS

Introductory Problems

6/1 The force P is applied to the 90-kg crate, which is stationary before the force is applied. Determine the magnitude and direction of the friction force F exerted by the horizontal surface on the crate if (a) $P = 300$ N, (b) $P = 400$ N, and (c) $P = 500$ N.

Ans. (a) $F = 300$ N, (b) $F = 400$ N
(c) $F = 353$ N (all to the left)

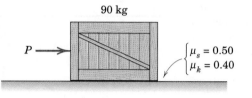

Problem 6/1

6/2 The 50-kg block rests on the horizontal surface, and a force $P = 200$ N, whose direction can be varied, is applied to the block. (a) If the block begins to slip when θ is reduced to 30°, calculate the coefficient of static friction μ_s between the block and the surface. (b) If P is applied with $\theta = 45°$, calculate the friction force F.

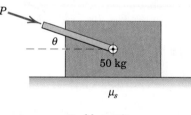

Problem 6/2

6/3 The force P is applied to the 100-lb block when it is at rest. Determine the magnitude and direction of the friction force exerted by the surface on the block if (a) $P = 0$, (b) $P = 40$ lb, and (c) $P = 60$ lb. (d) What value of P is required to initiate motion up the incline? The static and kinetic coefficients of friction between the block and the incline are $\mu_s = 0.25$ and $\mu_k = 0.20$, respectively.

Ans. (a) $F = 19.32$ lb (up the incline)
(b) $F = 11.71$ lb (down the incline)
(c) $F = 15.21$ lb (down the incline)
(d) $P = 48.8$ lb

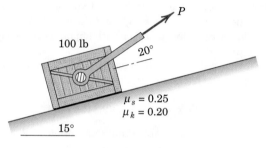

Problem 6/3

6/4 The designer of a ski resort wishes to have a portion of a beginner's slope on which the skier's speed will remain fairly constant. Tests indicate the average coefficients of friction between skis and snow to be $\mu_s = 0.10$ and $\mu_k = 0.08$. What should be the slope angle θ of the constant-speed section?

Problem 6/4

6/5 Determine the magnitude and direction of the friction force which the vertical wall exerts on the 100-lb block if (a) $\theta = 15°$ and (b) $\theta = 30°$.

Ans. (a) $F = 46.4$ lb up, (b) $F = 40$ lb up

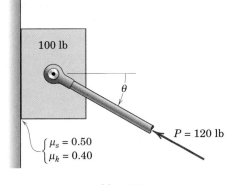

100 lb

θ

$P = 120$ lb

$\begin{cases} \mu_s = 0.50 \\ \mu_k = 0.40 \end{cases}$

Problem 6/5

6/6 The light bar is used to support the 50-kg block in its vertical guides. If the coefficient of static friction is 0.30 at the upper end of the bar and 0.40 at the lower end of the bar, find the friction force acting at each end for $x = 75$ mm. Also find the maximum value of x for which the bar will not slip.

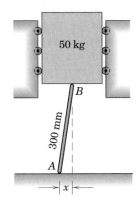

50 kg

B

300 mm

A

x

Problem 6/6

6/7 Two men are sliding a 100-kg crate up an incline. If the lower man pushes horizontally with a force of 500 N and if the coefficient of kinetic friction is 0.40, determine the tension T which the upper man must exert in the rope to maintain motion of the crate.

Ans. $T = 465$ N

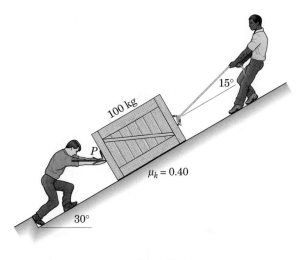

15°

100 kg

P

$\mu_k = 0.40$

30°

Problem 6/7

6/8 The force P is applied to the 200-lb block A which rests atop the 100-lb crate. The system is at rest when P is first applied. Determine what happens to each body if (a) $P = 60$ lb, (b) $P = 80$ lb, and (c) $P = 120$ lb.

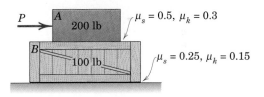

P 　A
　　　200 lb 　　$\mu_s = 0.5,\ \mu_k = 0.3$

B
　　100 lb 　　$\mu_s = 0.25,\ \mu_k = 0.15$

Problem 6/8

6/9 The 30-kg homogeneous cylinder of 400-mm diameter rests against the vertical and inclined surfaces as shown. If the coefficient of static friction between the cylinder and the surfaces is 0.30, calculate the applied clockwise couple M which would cause the cylinder to slip.

Ans. M = 32.9 N·m

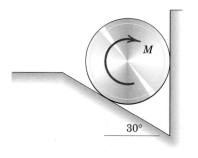

Problem 6/9

6/10 Calculate the force P required to initiate motion of the 60-lb block up the 10° incline. The coefficient of static friction for each pair of surfaces is 0.30.

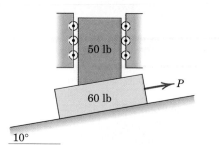

Problem 6/10

Representative Problems

6/11 The uniform pole of length l and mass m is placed against the supporting surfaces shown. If the coefficient of static friction is $\mu_s = 0.25$ at both A and B, determine the maximum angle θ at which the pole can be placed before it begins to slip.

Ans. θ = 59.9°

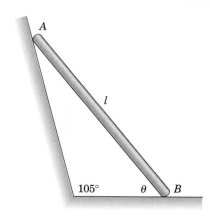

Problem 6/11

6/12 The 100-lb wheel rolls on its hub up the circular incline under the action of the 25-lb weight attached to a cord around the rim. Determine the angle θ at which the wheel comes to rest, assuming that friction is sufficient to prevent slippage. What is the minimum coefficient of friction which will permit this position to be reached with no slipping?

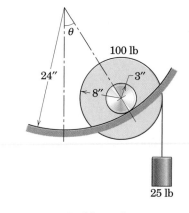

Problem 6/12

6/13 A person (mass m_1) who is kneeling on a plywood panel (mass m_2) wishes to push himself and the panel away from the right-hand wall. Determine a condition on μ_2 which would allow him to do this, as opposed to his slipping relative to a stationary panel. The two applicable coefficients of static friction are μ_1 and μ_2. Evaluate μ_2 for $m_1 = 80$ kg, $m_2 = 10$ kg, and $\mu_1 = 0.60$.

Ans. $\mu_2 = 0.533$

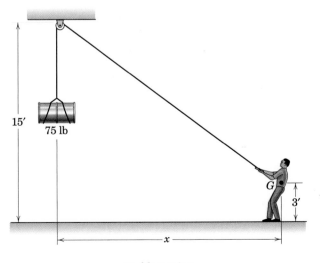

Problem 6/13

6/14 The 180-lb man with center of gravity G supports the 75-lb drum as shown. Find the greatest distance x at which the man can position himself without slipping if the coefficient of static friction between his shoes and the ground is 0.40.

6/15 The center of mass of the vertical 800-kg panel is at G. The panel is mounted on wheels which permit ease of horizontal movement along the fixed rail. If the bearing of the wheel at A becomes "frozen" so that the wheel cannot turn, compute the force P required to slide the panel. The coefficient of kinetic friction between the wheel and the rail is 0.30.

Ans. $P = 932$ N

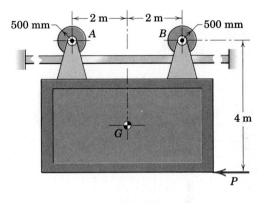

Problem 6/15

6/16 Determine the horizontal force P required to cause slippage to occur. The friction coefficients for the three pairs of mating surfaces are indicated. The top block is free to move vertically.

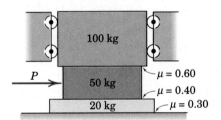

Problem 6/16

Problem 6/14

6/17 A 180-lb man pulls the 100-lb cart up the incline at steady speed. Determine the minimum coefficient μ_s of static friction for which the man's shoes will not slip. Also determine the distance s required for equilibrium of the man's body.

Ans. $\mu_s = 0.409$, $s = 5.11$ in.

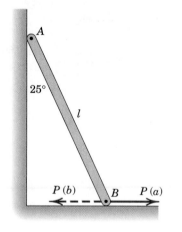

Problem 6/17

6/18 The uniform bar AB of mass m and length l is leaning against the vertical wall as shown. The coefficient of static friction between the supporting surfaces and the ends of the pole is $\mu_s = 0.30$. Determine the force P which causes impending slip if P (a) acts to the right and (b) acts to the left.

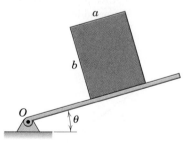

Problem 6/18

6/19 The wheel shown will roll to the left when the angle α of the cord is small. When α is large, the wheel rolls to the right. Determine by inspection from the geometry of the free-body diagram the angle α for which the wheel will not roll in either direction. If the coefficient of static friction is μ_s and the mass of the wheel is m, determine the value of P for which the wheel will slip for the critical value of α.

Ans. $\alpha = \sin^{-1}\dfrac{r_1}{r_2}$, $P = \dfrac{\mu_s m g r_2}{r_1 + \mu_s\sqrt{r_2^2 - r_1^2}}$

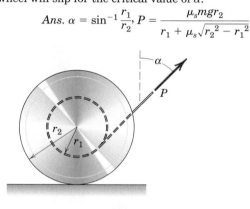

Problem 6/19

6/20 The homogeneous rectangular block of mass m rests on the inclined plane which is hinged about a horizontal axis through O. If the coefficient of static friction between the block and the plane is μ, specify the conditions which determine whether the block tips before it slips or slips before it tips as the angle θ is gradually increased.

Problem 6/20

6/21 The inverted track T with freely floating cylinder C comprise a system which is designed to hold paper or other thin materials P in place. The coefficient of static friction is μ for all interfaces. What minimum value of μ ensures that the device will work no matter how heavy the supported material P is?

Ans. $\mu = 0.268$

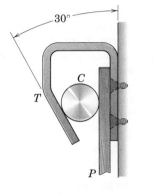

Problem 6/21

6/22 The system of two blocks, cable, and fixed pulley is initially at rest. Determine the horizontal force P necessary to cause motion when (*a*) P is applied to the 5-kg block and (*b*) P is applied to the 10-kg block. Determine the corresponding tension T in the cable for each case.

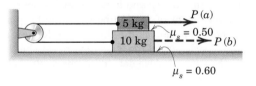

Problem 6/22

6/23 The figure shows a device, called a jam cleat, which secures a rope under tension by reason of large friction forces developed. For the position shown determine the minimum coefficient μ_s of static friction between the rope and the cam surfaces for which the cleat will be self-locking. Also find the magnitude of the total reaction R on each of the cam bearings.

Ans. $\mu_s = 0.208$, $R = 1471$ N

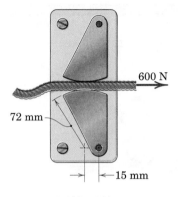

Problem 6/23

6/24 Determine the range of weights W for which the 100-lb block is in equilibrium. All wheels and pulleys have negligible friction.

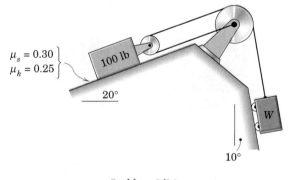

Problem 6/24

6/25 What force P must the two men exert on the rope to slide the uniform 20-ft plank on the overhead rack? The plank weighs 200 lb, and the coefficient of kinetic friction between the plank and each support is 0.50.

Ans. P = 162.3 lb

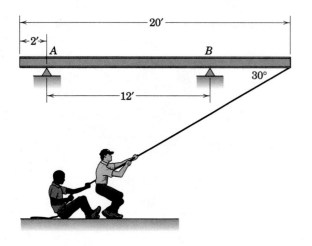

Problem 6/25

6/26 Calculate the force T required to rotate the 200-kg reel of telephone cable which rests on its hubs and bears against a vertical wall. The coefficient of kinetic friction for each pair of contacting surfaces is 0.60.

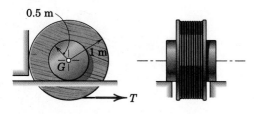

Problem 6/26

6/27 The 1600-kg car is just beginning to negotiate the 16° ramp. If the car has rear-wheel drive, determine the minimum coefficient of static friction required at B.

Ans. μ_s = 0.365

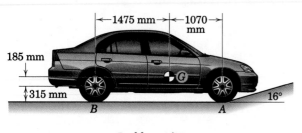

Problem 6/27

6/28 Repeat Prob. 6/27, but now the car has front-wheel drive.

6/29 Repeat Prob. 6/27, but now the car has all-wheel drive. Assume that slipping occurs at A and B simultaneously.

Ans. μ_s = 0.1575

6/30 The uniform rod with center of mass at G is supported by the pegs A and B, which are fixed in the wheel. If the coefficient of friction between the rod and pegs is μ, determine the angle θ through which the wheel may be slowly turned about its horizontal axis through O, starting from the position shown, before the rod begins to slip. Neglect the diameter of the rod compared with the other dimensions.

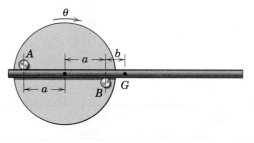

Problem 6/30

6/31 A metal hoop of negligible thickness has a mass m and mean radius r and hangs from a support at A. If the coefficient of friction between the hoop and the support is μ, determine the vertical force P applied to the periphery of the hoop required to slip the hoop on the support. Also determine the angle β at which slipping occurs.

Ans. $P = mg\ \dfrac{\mu}{\sqrt{1+\mu^2}-\mu}, \beta = \tan^{-1}\mu$

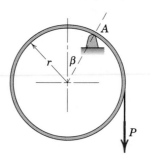

Problem 6/31

6/32 The roll of paper is to be rolled slowly up the incline by a tension T applied to the paper as it is pulled horizontally off the roll. The coefficient of friction between the roll and the incline is 0.20. Prove whether the paper rolls without slipping or whether it slips.

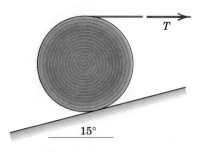

Problem 6/32

6/33 Assuming the worker is strong enough, can he maintain equilibrium of the uniform 7-m plank in the position shown by exerting a horizontal pull on the rope? The coefficient of static friction between the 60-kg plank and the top of the building is 0.30. If the answer is yes, then compute the friction force existing in the equilibrium state.

Ans. F = 117.7 N

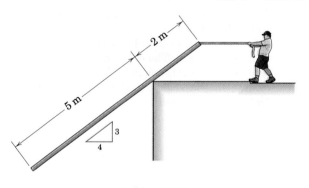

Problem 6/33

6/34 Two automobiles, both of which have the mass center located as shown midway between the front and rear axles, are identical except that one is front-wheel-drive and the other is rear-wheel-drive. The cars are driven at constant speed up ramps of various inclinations. From a theoretical design viewpoint, which car could climb the ramp of higher inclination angle θ? Justify your answer.

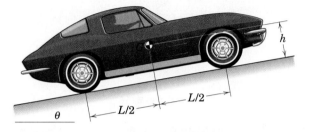

Problem 6/34

6/35 The 10-kg solid cylinder is resting in the inclined V-block. If the coefficient of static friction between the cylinder and the block is 0.50, determine (*a*) the friction force F acting on the cylinder at each side before force P is applied and (*b*) the value of P required to start sliding the cylinder up the incline.

Ans. (*a*) *F* = 24.5 N, (*b*) *P* = 109.1 N

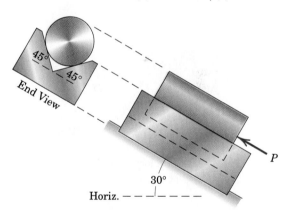

Problem 6/35

6/36 The semicylindrical shell of mass m and radius r is rolled through an angle θ by the horizontal force P applied to its rim. If the coefficient of friction is μ_s, determine the angle θ at which the shell slips on the horizontal surface as P is gradually increased. What value of μ_s will permit θ to reach 90°?

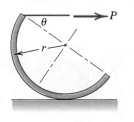

Problem 6/36

6/37 The homogeneous semicylinder rests on a horizontal surface and is subjected to the force P applied to a cord firmly attached to its periphery. The force P is slowly increased and kept normal to the flat surface of the semicylinder. If slipping is observed just as θ reaches 40°, determine the coefficient of static friction μ_s and the value of P when slipping occurs.

Ans. $\mu_s = 0.1223$, $P = 0.1661mg$

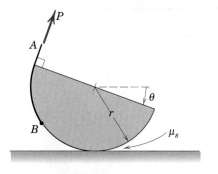

Problem 6/37

6/38 If the coefficient of friction between the solid semicircular cylinder and the incline is 0.30, determine the maximum angle θ with the horizontal at which the incline can be tilted before the cylinder slips. For this condition what is the corresponding angle α through which the cylinder has rolled with respect to the incline?

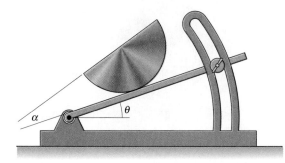

Problem 6/38

6/39 The head of the simple woodworker's clamp is welded to its steel shaft of diameter d and is designed to hold the workpiece A in place. The clamp is set by first tapping it on the head with a mallet. The shaft is prevented from slipping up in the slightly oversized hole in the benchtop by friction forces at the contact points B and C. If the coefficient of static friction between the shaft and the bench is μ_s, determine the minimum throat distance b for which the clamp will not slip. The clamping force between the head and the workbench at D may be assumed to be vertical, and the slight tilt of the shaft from the vertical may be neglected.

Ans. $b = \dfrac{1}{2}\left(\dfrac{h}{\mu_s} - d\right)$

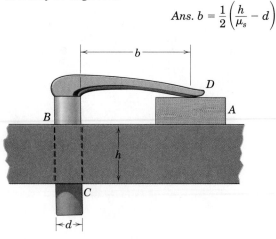

Problem 6/39

6/40 The movable left-hand jaw of the C-clamp can be slid along the frame to increase the capacity of the clamp. To prevent slipping of the jaw on the frame when the clamp is under load, the dimension x must exceed a certain minimum value. For given values of a and b and a static friction coefficient μ_s, specify this design minimum value of x to prevent slipping of the jaw.

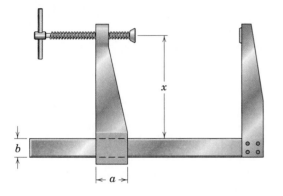

Problem 6/40

6/41 A uniform block of mass m is at rest on an incline θ. Determine the maximum force P which can be applied to the block in the direction shown before slipping begins. The coefficient of static friction between the block and the incline is μ_s. Also determine the angle β between the horizontal direction of P and the direction of initial movement of the block.

$$Ans.\ P = mg\sqrt{\mu_s^2 \cos^2\theta - \sin^2\theta}$$
$$\beta = \tan^{-1}\left(\frac{mg\sin\theta}{P}\right)$$

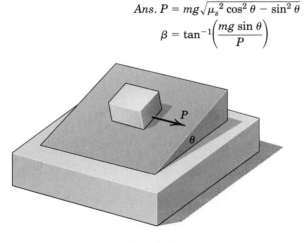

Problem 6/41

6/42 The uniform I-beam of mass m is supported at its ends on two fixed horizontal rails as shown. Determine the maximum horizontal force P which can be applied without causing the beam to slip, and find the corresponding value of the friction force at A. The coefficient of static friction between the beam and the rails is μ_s. Also, take $b < l/2$.

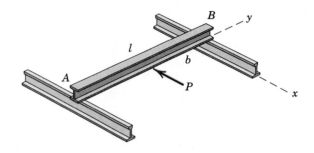

Problem 6/42

6/43 The two bars AC and BD are constructed from the same stock. Determine and plot the angle θ at which the bar AC will slip as a function of the coefficient of static friction μ_s over the range $0 \leq \mu_s \leq 1$. Friction at the two pins is negligible. State the values of θ for $\mu_s = 0$, 0.50, and 1.

$$Ans.\ \theta \rightarrow 90° \text{ as } \mu_s \rightarrow 0,\ \theta = 51.1° \text{ for } \mu_s = 0.50$$
$$\theta = 31.8° \text{ for } \mu_s = 1$$

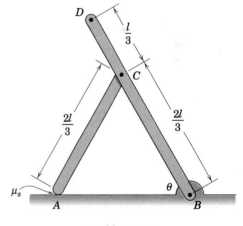

Problem 6/43

6/44 Determine the distance s to which the 90-kg painter can climb without causing the 4-m ladder to slip at its lower end A. The top of the 15-kg ladder has a small roller, and at the ground the coefficient of static friction is 0.25. The mass center of the painter is directly above her feet.

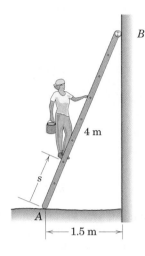

Problem 6/44

6/45 The uniform slender rod is slowly lowered from the upright position ($\theta = 90°$) by means of the cord attached to its upper end and passing under the small fixed pulley. If the rod is observed to slip at its lower end when $\theta = 40°$, determine the coefficient of static friction at the horizontal surface.

Ans. $\mu_s = 0.761$

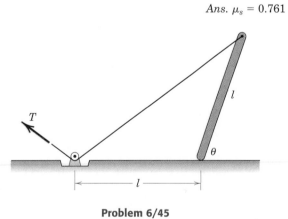

Problem 6/45

6/46 A woman pedals her bicycle up a 5-percent grade on a slippery road at a steady speed. The woman and bicycle have a combined mass of 82 kg with mass center at G. If the rear wheel is on the verge of slipping, determine the coefficient of friction μ_s between the rear tire and the road. If the coefficient of friction were doubled, what would be the friction force F acting on the rear wheel? (Why may we neglect friction under the front wheel?)

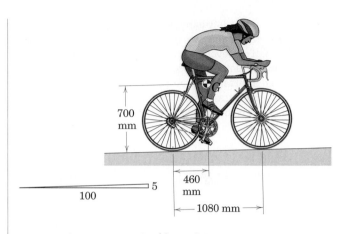

Problem 6/46

6/47 A block of mass m_0 is placed between the vertical wall and the upper end A of the uniform slender bar of mass m. If the coefficient of static friction is μ_s between the block and the wall and also between the block and the bar, determine a general expression for the minimum value $\theta_{\min}$ of the angle θ for which the block will remain in equilibrium. Evaluate your expression for the conditions $\mu_s = 0.5$ and

(a) $\dfrac{m}{m_0} = 0.1$,

(b) $\dfrac{m}{m_0} = 1$, and

(c) $\dfrac{m}{m_0} = 10$.

For each case, state the minimum coefficient of static friction $(\mu_s)_B$ necessary to prevent slippage at B.

Ans. (a) $\theta_{\min} = 61.2°$, $(\mu_s)_B = 1.667$ (unusual!)
(b) $\theta_{\min} = 45°$, $(\mu_s)_B = 0.667$
(c) $\theta_{\min} = 10.30°$, $(\mu_s)_B = 0.0952$

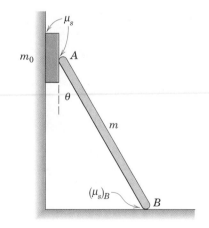

Problem 6/47

6/48 A block of mass m_0 is placed between the vertical wall and the small ideal roller at the upper end A of the uniform slender bar of mass m. The lower end B of the bar rests on the horizontal surface. If the coefficient of static friction is μ_s at B and also between the block and the wall, determine a general expression for the minimum value θ_{min} of θ for which the block will remain in equilibrium. Evaluate your expression for $\mu_s = 0.5$ and $m/m_0 = 10$. For these conditions, check for possible slipping at B.

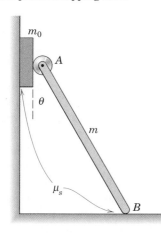

Problem 6/48

6/49 The double-block brake shown is applied to the flywheel by means of the action of the spring. To release the brake, a force P is applied to the control rod. In the operating position with $P = 0$, the spring is compressed 30 mm. Select a spring with an appropriate constant (stiffness) k which will provide sufficient force to brake the flywheel under the torque $M = 100$ N·m if the applicable coefficient of friction for both brake shoes is 0.20. Neglect the dimensions of the shoes.

Ans. $k = 20.8(10^3)$ N/m

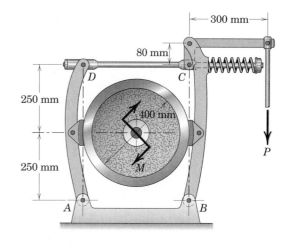

Problem 6/49

▶**6/50** The miniature roller clutch is designed to enable the outer ring to rotate clockwise relative to the shaft (cross-hatched) but to prevent counterclockwise rotation. This is accomplished by the frictional engagement of the rockers, called sprags (only three of many are shown). Calculate the minimum coefficient of friction μ_{min} for the contacting surfaces necessary to prevent counterclockwise rotation of the outer ring.

Ans. $\mu_{min} = 0.309$

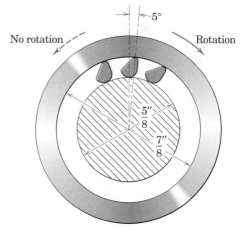

Problem 6/50

SECTION B   APPLICATIONS OF FRICTION IN MACHINES

In Section B we investigate the action of friction in various machine applications. Because the conditions in these applications are normally either limiting static or kinetic friction, we will use the variable μ (rather than μ_s or μ_k) in general. Depending on whether motion is impending or actually occurring, μ can be interpreted as either the static or kinetic coefficient of friction.

6/4   WEDGES

A wedge is one of the simplest and most useful machines. A wedge is used to produce small adjustments in the position of a body or to apply large forces. Wedges largely depend on friction to function. When sliding of a wedge is impending, the resultant force on each sliding surface of the wedge will be inclined from the normal to the surface by an amount equal to the friction angle. The component of the resultant along the surface is the friction force, which is always in the direction to oppose the motion of the wedge relative to the mating surfaces.

Figure 6/3a shows a wedge used to position or lift a large mass m, where the vertical loading is mg. The coefficient of friction for each pair of surfaces is $\mu = \tan \phi$. The force P required to start the wedge is found from the equilibrium triangles of the forces on the load and on the wedge. The free-body diagrams are shown in Fig. 6/3b, where the reactions are inclined at an angle ϕ from their respective normals and are in the direction to oppose the motion. We neglect the mass of the wedge. From the free-body diagrams we write the force equilibrium conditions by equating to zero the sum of the force vectors acting on each body. The solutions of these equations are shown in part c of the figure, where R_2 is found first in the upper diagram using the known value of mg. The force P is then found from the lower triangle once the value of R_2 has been established.

If P is removed and the wedge remains in place, equilibrium of the wedge requires that the equal reactions R_1 and R_2 be collinear as shown in Fig. 6/4, where the wedge angle α is taken to be less than ϕ. Part a of the figure represents impending slippage at the upper surface, and part c of the figure represents impending slippage at the lower surface. In order for the wedge to slide out of its space, slippage must occur at *both* surfaces simultaneously; otherwise, the wedge is *self-locking*, and there is a finite range of possible intermediate angular positions of R_1 and R_2 for which the wedge will remain in place. Figure 6/4b illustrates this range and shows that simultaneous slippage is not possible if $\alpha < 2\phi$. You are encouraged to construct additional diagrams for the case where $\alpha > \phi$ and verify that the wedge is self-locking as long as $\alpha < 2\phi$.

If the wedge is self-locking and is to be withdrawn, a pull P on the wedge will be required. To oppose the new impending motion, the reactions R_1 and R_2 must act on the opposite sides of their normals from those when the wedge was inserted. The solution can be obtained as

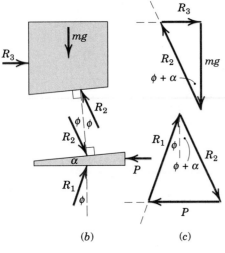

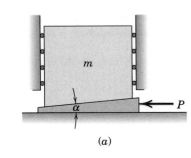

(a)

(b)   (c)

Forces to raise load

Figure 6/3

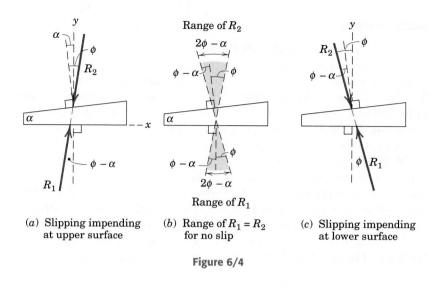

(a) Slipping impending
at upper surface

(b) Range of $R_1 = R_2$
for no slip

(c) Slipping impending
at lower surface

Figure 6/4

with the case of raising the load. The free-body diagrams and vector polygons for this condition are shown in Fig. 6/5.

Wedge problems lend themselves to graphical solutions as indicated in the three figures. The accuracy of a graphical solution is easily held within tolerances consistent with the uncertainty of friction coefficients. Algebraic solutions may also be obtained from the trigonometry of the equilibrium polygons.

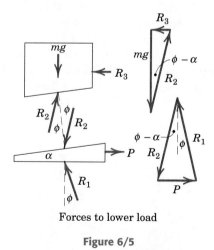

Forces to lower load

Figure 6/5

6/5   SCREWS

Screws are used for fastening and for transmitting power or motion. In each case the friction developed in the threads largely determines the action of the screw. For transmitting power or motion the square thread is more efficient than the V-thread, and the analysis here is confined to the square thread.

Force Analysis

Consider the square-threaded jack, Fig. 6/6, under the action of the axial load W and a moment M applied about the axis of the screw. The screw has a lead L (advancement per revolution) and a mean radius r. The force R exerted by the thread of the jack frame on a small representative portion of the screw thread is shown on the free-body diagram of the screw. Similar reactions exist on all segments of the screw thread where contact occurs with the thread of the base.

If M is just sufficient to turn the screw, the thread of the screw will slide around and up on the fixed thread of the frame. The angle ϕ made by R with the normal to the thread is the angle of friction, so that $\tan \phi = \mu$. The moment of R about the vertical axis of the screw is $Rr \sin (\alpha + \phi)$, and the total moment due to all reactions on the threads is $\Sigma Rr \sin (\alpha + \phi)$. Since $r \sin (\alpha + \phi)$ appears in each term, we may factor it out. The moment equilibrium equation for the screw becomes

$$M = [r \sin (\alpha + \phi)] \, \Sigma R$$

Equilibrium of forces in the axial direction further requires that

$$W = \Sigma R \cos (\alpha + \phi) = [\cos (\alpha + \phi)] \, \Sigma R$$

Combining the expressions for M and W gives

$$M = Wr \tan (\alpha + \phi) \qquad\qquad \textbf{(6/3)}$$

To determine the helix angle α, unwrap the thread of the screw for one complete turn and note that $\alpha = \tan^{-1} (L/2\pi r)$.

We may use the unwrapped thread of the screw as an alternative model to simulate the action of the entire screw, as shown in Fig. 6/7a. The equivalent force required to push the movable thread up the fixed incline is $P = M/r$, and the triangle of force vectors gives Eq. 6/3 immediately.

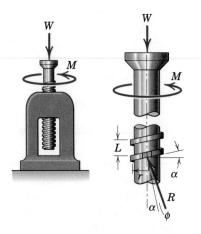

Figure 6/6

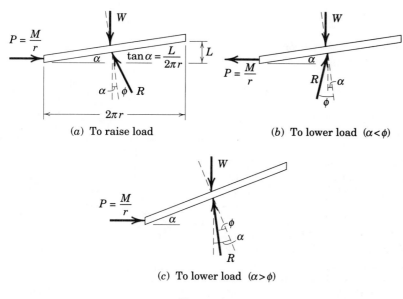

(a) To raise load

(b) To lower load ($\alpha < \phi$)

(c) To lower load ($\alpha > \phi$)

Figure 6/7

Conditions for Unwinding

If the moment M is removed, the friction force changes direction so that ϕ is measured to the other side of the normal to the thread. The screw will remain in place and be self-locking provided that $\alpha < \phi$, and will be on the verge of unwinding if $\alpha = \phi$.

To lower the load by unwinding the screw, we must reverse the direction of M as long as $\alpha < \phi$. This condition is illustrated in Fig. 6/7b for our simulated thread on the fixed incline. An equivalent force $P = M/r$ must be applied to the thread to pull it down the incline. From the triangle of vectors we therefore obtain the moment required to lower the screw, which is

$$M = Wr \tan (\phi - \alpha) \tag{6/3a}$$

If $\alpha > \phi$, the screw will unwind by itself, and Fig. 6/7c shows that the moment required to prevent unwinding is

$$M = Wr \tan (\alpha - \phi) \tag{6/3b}$$

Sample Problem 6/6

The horizontal position of the 500-kg rectangular block of concrete is adjusted by the 5° wedge under the action of the force **P**. If the coefficient of static friction for both pairs of wedge surfaces is 0.30 and if the coefficient of static friction between the block and the horizontal surface is 0.60, determine the least force P required to move the block.

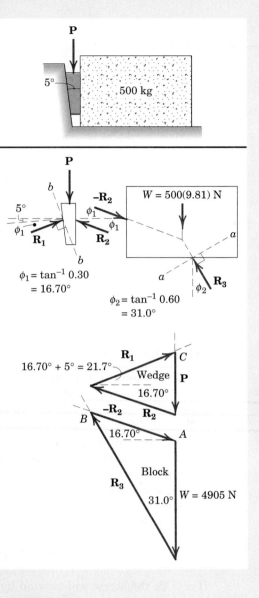

Solution.   The free-body diagrams of the wedge and the block are drawn with the reactions $\mathbf{R}_1$, $\mathbf{R}_2$, and $\mathbf{R}_3$ inclined with respect to their normals by the

① amount of the friction angles for impending motion. The friction angle for limiting static friction is given by $\phi = \tan^{-1}\mu$. Each of the two friction angles is computed and shown on the diagram.

We start our vector diagram expressing the equilibrium of the block at a convenient point A and draw the only known vector, the weight **W** of the block. Next we add $\mathbf{R}_3$, whose 31.0° inclination from the vertical is now known. The vector $-\mathbf{R}_2$, whose 16.70° inclination from the horizontal is also known, must close the polygon for equilibrium. Thus, point B on the lower polygon is determined by the intersection of the known directions of $\mathbf{R}_3$ and $-\mathbf{R}_2$, and their magnitudes become known.

For the wedge we draw $\mathbf{R}_2$, which is now known, and add $\mathbf{R}_1$, whose direction is known. The directions of $\mathbf{R}_1$ and **P** intersect at C, thus giving us the solution for the magnitude of **P**.

Algebraic solution.   The simplest choice of reference axes for calculation purposes is, for the block, in the direction a-a normal to $\mathbf{R}_3$ and, for the wedge, in

② the direction b-b normal to $\mathbf{R}_1$. The angle between $\mathbf{R}_2$ and the a-direction is $16.70° + 31.0° = 47.7°$. Thus, for the block

$$[\Sigma F_a = 0] \qquad 500(9.81)\sin 31.0° - R_2 \cos 47.7° = 0$$

$$R_2 = 3750 \text{ N}$$

For the wedge the angle between $\mathbf{R}_2$ and the b-direction is $90° - (2\phi_1 + 5°) = 51.6°$, and the angle between **P** and the b-direction is $\phi_1 + 5° = 21.7°$. Thus,

$$[\Sigma F_b = 0] \qquad 3750 \cos 51.6° - P \cos 21.7° = 0$$

$$P = 2500 \text{ N} \qquad\qquad Ans.$$

Graphical solution.   The accuracy of a graphical solution is well within the uncertainty of the friction coefficients and provides a simple and direct result. By laying off the vectors to a reasonable scale following the sequence described, we obtain the magnitudes of **P** and the **R**'s easily by scaling them directly from the diagrams.

Helpful Hints

① Be certain to note that the reactions are inclined from their normals in the direction to oppose the motion. Also, we note the equal and opposite reactions $\mathbf{R}_2$ and $-\mathbf{R}_2$.

② It should be evident that we avoid simultaneous equations by eliminating reference to $\mathbf{R}_3$ for the block and $\mathbf{R}_1$ for the wedge.

Sample Problem 6/7

The single-threaded screw of the vise has a mean diameter of 1 in. and has 5 square threads per inch. The coefficient of static friction in the threads is 0.20. A 60-lb pull applied normal to the handle at A produces a clamping force of 1000 lb between the jaws of the vice. (a) Determine the frictional moment M_B, developed at B, due to the thrust of the screw against the body of the jaw. (b) Determine the force Q applied normal to the handle at A required to loosen the vise.

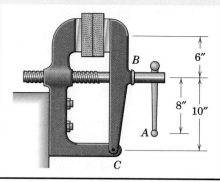

Solution.   From the free-body diagram of the jaw we first obtain the tension T in the screw.

$$[\Sigma M_C = 0] \qquad 1000(16) - 10T = 0 \qquad T = 1600 \text{ lb}$$

The helix angle α and the friction angle ϕ for the thread are given by

① $$\alpha = \tan^{-1}\frac{L}{2\pi r} = \tan^{-1}\frac{1/5}{2\pi(0.5)} = 3.64°$$

$$\phi = \tan^{-1}\mu = \tan^{-1}0.20 = 11.31°$$

where the mean radius of the thread is $r = 0.5$ in.

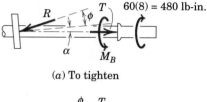

(a) To tighten

(a) To tighten.   The isolated screw is simulated by the free-body diagram shown where all of the forces acting on the threads of the screw are represented by a single force R inclined at the friction angle ϕ from the normal to the thread. The moment applied about the screw axis is $60(8) = 480$ lb-in. in the clockwise direction as seen from the front of the vise. The frictional moment M_B due to the friction forces acting on the collar at B is in the counterclockwise direction to oppose the impending motion. From Eq. 6/3 with T substituted for W the net moment acting on the screw is

$$M = Tr \tan (\alpha + \phi)$$

$$480 - M_B = 1600(0.5) \tan (3.64° + 11.31°)$$

$$M_B = 266 \text{ lb-in.} \qquad\qquad Ans.$$

(b) To loosen

(b) To loosen.   The free-body diagram of the screw on the verge of being loosened is shown with R acting at the friction angle from the normal in the direction to counteract the impending motion. Also shown is the frictional moment $M_B = 266$ lb-in. acting in the clockwise direction to oppose the motion. The angle between R and the screw axis is now $\phi - \alpha$, and we use Eq. 6/3a with the net moment equal to the applied moment M' minus M_B. Thus

$$M = Tr \tan (\phi - \alpha)$$

$$M' - 266 = 1600(0.5) \tan (11.31° - 3.64°)$$

$$M' = 374 \text{ lb-in.}$$

Thus, the force on the handle required to loosen the vise is

$$Q = M'/d = 374/8 = 46.8 \text{ lb} \qquad\qquad Ans.$$

Helpful Hints

① Be careful to calculate the helix angle correctly. Its tangent is the lead L (advancement per revolution) divided by the mean circumference $2\pi r$ and not by the diameter $2r$.

② Note that R swings to the opposite side of the normal as the impending motion reverses direction.

PROBLEMS

(Unless otherwise instructed, neglect the weights of the wedges and screws in the problems which follow.)

Introductory Problems

6/51 If the coefficient of friction between the steel wedge and the moist fibers of the newly cut stump is 0.20, determine the maximum angle α which the wedge may have and not pop out of the wood after being driven by the sledge.

Ans. $\alpha = 22.6°$

Problem 6/51

6/52 The elements of a deodorant dispenser are shown in the figure. To advance the deodorant D, the knob C is turned. The threaded shaft engages the threads in the movable deodorant base A; the shaft has no threads at the fixed support B. The 6-mm-diameter double square thread has a lead of 5 mm. Determine the coefficient of static friction μ_s for which the deodorant will not retract under the action of the force P. Neglect friction at B.

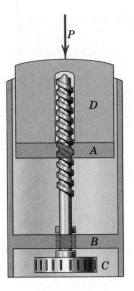

Problem 6/52

6/53 A 1600-kg rear-wheel-drive car is being driven up the ramp at a slow steady speed. Determine the minimum coefficient of static friction μ_s for which the portable ramp will not slip forward. Also determine the required friction force F_A at each rear drive wheel.

Ans. $\mu_s = 0.3, F_A = 1294$ N

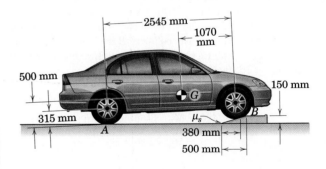

Problem 6/53

6/54 The 100-kg industrial door with mass center at G is being positioned for repair by insertion of the 5° wedge under corner B. Horizontal movement is prevented by the small ledge at corner A. If the coefficients of static friction at both the top and bottom wedge surfaces are 0.6, determine the force P required to lift the door at B.

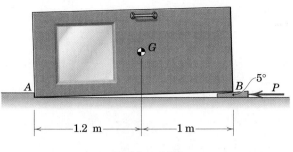

Problem 6/54

6/55 Calculate the rightward force P' which would re-move the wedge from under the door of Prob. 6/54. Assume that corner A does not slip for your calcula-tion of P', but then check this assumption; the coeffi-cient of static friction at A is 0.6.

Ans. $P' = 582$ N

6/56 The bar clamp is being used to clamp two boards to-gether while the glue between them cures. What torque M must be applied to the handle of the screw in order to produce an 80-lb compression between the boards? The $\frac{3}{8}$-in.-diameter single-thread screw has 12 square threads per inch, and the effective co-efficient of friction is 0.2. Neglect any friction in the pivot contact at C. What torque M' is required to loosen the clamp?

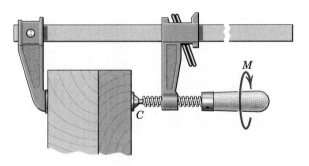

Problem 6/56

6/57 Specify the required torque M on the power screw necessary to overcome a resistance of 450 N to mo-tion of the carriage along its horizontal ways. The screw has a mean diameter of 25 mm and has two separate square threads which move the carriage 20 mm per revolution of the screw. The coefficient of friction may be taken as 0.20.

Ans. $M = 2.69$ N·m

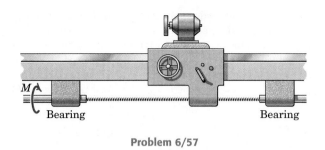

Problem 6/57

Representative Problems

6/58 The coefficient of static friction μ_s between the 100-lb body and the 15° wedge is 0.20. Determine the magnitude of the force P required to raise the 100-lb body if (*a*) rollers of negligible friction are present under the wedge, as illustrated, and (*b*) the rollers are removed and the coefficient of static friction $\mu_s = 0.20$ applies at this surface as well.

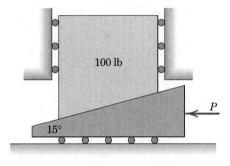

Problem 6/58

6/59 For both conditions (*a*) and (*b*) as stated in Prob. 6/58, determine the magnitude and direction of the force P' required to lower the 100-lb body.

Ans. (*a*) $P' = 6.45$ lb left, (*b*) $P' = 13.55$ lb right

6/60 The large turnbuckle supports a cable tension of 10,000 lb. The $1\frac{1}{4}$-in. screws have a mean diameter of 1.150 in. and have five square threads per inch. The coefficient of friction for the greased threads does not exceed 0.25. Determine the moment M ap-plied to the body of the turnbuckle (*a*) to tighten it and (*b*) to loosen it. Both screws have single threads and are prevented from turning.

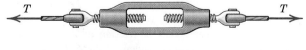

Problem 6/60

6/61 The two 4° wedges are used to position the vertical column under a load L. What is the least value of the coefficient of friction μ_2 for the bottom pair of surfaces for which the column may be raised by applying a single horizontal force P to the upper wedge?

Ans. $(\mu_2)_{\text{min}} = 0.378$

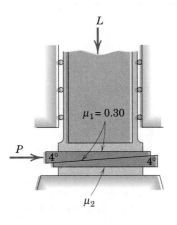

Problem 6/61

6/62 The collar A has a force fit on shaft B and is to be removed from the shaft by the wheel-puller mechanism shown. The screw has a single square thread with a mean diameter of 20 mm and a lead L of 6 mm. If a torque of 24 N·m is required to turn wheel C to slip the collar off the shaft, determine the average pressure p (compressive stress) between the collar and the shaft. The coefficient of friction for the screw at E is 0.25, and that for the shaft and collar is 0.30. Friction at the ball end D of the shaft is negligible.

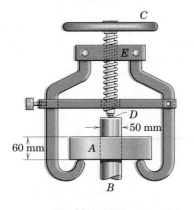

Problem 6/62

6/63 Compute the force P required to move the 20-kg wheel. The coefficient of friction at A is 0.25 and that for both pairs of wedge surfaces is 0.30. Also, the spring S is under a compression of 100 N, and the rod offers negligible support to the wheel.

Ans. $P = 114.7$ N

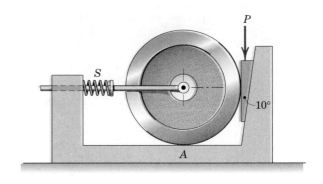

Problem 6/63

6/64 Work Prob. 6/63 if the compression in the spring is 200 N. All other conditions remain unchanged.

6/65 The vertical position of the 100-kg block is adjusted by the screw-activated wedge. Calculate the moment M which must be applied to the handle of the screw to raise the block. The single-thread screw has square threads with a mean diameter of 30 mm and advances 10 mm for each complete turn. The coefficient of friction for the screw threads is 0.25, and the coefficient of friction for all mating surfaces of the block and wedge is 0.40. Neglect friction at the ball joint A.

Ans. $M = 7.30$ N·m

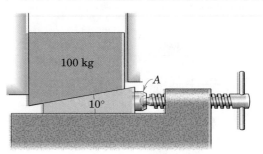

Problem 6/65

6/66 Calculate the moment M' which must be applied to the handle of the screw of Prob. 6/65 to withdraw the wedge and lower the 100-kg load.

6/67 The bench hold-down clamp is being used to clamp two boards together while they are being glued. What torque M must be applied to the screw in order to produce a 200-lb compression between the boards? The $\frac{1}{2}$-in.-diameter single-thread screw has 12 square threads per inch, and the coefficient of friction in the threads may be taken to be 0.20. Neglect any friction in the small ball contact at A and assume that the contact force at A is directed along the axis of the screw. What torque M' is required to loosen the clamp?

Ans. $M = 48.2$ lb-in., $M' = 27.4$ lb-in.

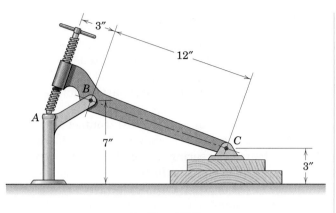

Problem 6/67

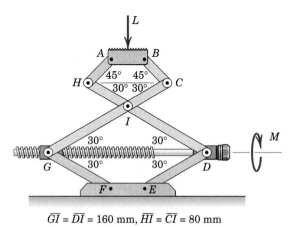

$$\overline{GI} = \overline{DI} = 160 \text{ mm}, \ \overline{HI} = \overline{CI} = 80 \text{ mm}$$

Problem 6/69

6/68 The design of a joint to connect two shafts by a flat 5° tapered cotter is shown by the two views in the figure. If the shafts are under a constant tension T of 200 lb, find the force P required to move the cotter and take up any slack in the joint. The coefficient of friction between the cotter and the sides of the slots is 0.20. Neglect any horizontal friction between the shafts.

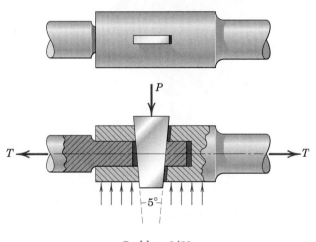

Problem 6/68

▶**6/69** A scissors-type jack with a single square thread which engages the threaded collar G and turns in a ball thrust bearing at D is being designed. The thread is to have a mean diameter of 10 mm and a lead (advancement per revolution) of 3 mm. With a coefficient of friction of 0.20 for the greased threads, (*a*) calculate the torque M on the screw required to raise a load of 1000 kg from the position shown and (*b*) calculate the torque M required to lower the load from the same position. Assume that platform AB and line DG will remain horizontal under load. Neglect friction in the bearing at D.

Ans. (*a*) $M = 35.7 \text{ N·m}$, (*b*) $M = 12.15 \text{ N·m}$

▶**6/70** Replace the square thread of the screw jack in Fig. 6/6 by a V-thread as indicated in the figure accompanying this problem and determine the moment M on the screw required to raise the load W. The force R acting on a representative small section of the thread is shown with its relevant projections. The vector R_1 is the projection of R in the plane of the figure containing the axis of the screw. The analysis is begun with an axial force and a moment summation and includes substitutions for the angles γ and β in terms of θ, α, and the friction angle $\phi = \tan^{-1} \mu$. The helix angle of the single thread is exaggerated for clarity.

$$\text{Ans. } M = Wr \ \frac{\tan \alpha + \mu \sqrt{1 + \tan^2 \frac{\theta}{2} \cos^2 \alpha}}{1 - \mu \tan \alpha \sqrt{1 + \tan^2 \frac{\theta}{2} \cos^2 \alpha}}$$

$$\text{where } \alpha = \tan^{-1} \frac{L}{2\pi r}$$

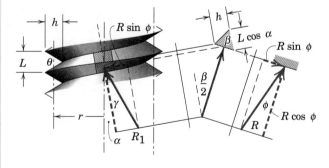

Problem 6/70

6/6 JOURNAL BEARINGS

A *journal bearing* is one which gives lateral support to a shaft in contrast to axial or thrust support. For dry bearings and for many partially lubricated bearings we may apply the principles of dry friction. These principles provide a satisfactory approximation for design purposes.

A dry or partially lubricated journal bearing with contact or near contact between the shaft and the bearing is shown in Fig. 6/8, where the clearance between the shaft and bearing is greatly exaggerated to clarify the action. As the shaft begins to turn in the direction shown, it will roll up the inner surface of the bearing until it slips. Here it will remain in a more or less fixed position during rotation. The torque M required to maintain rotation and the radial load L on the shaft will cause

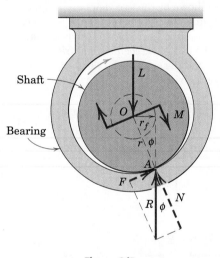

Figure 6/8

a reaction R at the contact point A. For vertical equilibrium R must equal L but will not be collinear with it. Thus, R will be tangent to a small circle of radius r_f called the *friction circle*. The angle between R and its normal component N is the friction angle ϕ. Equating the sum of the moments about A to zero gives

$$M = Lr_f = Lr \sin \phi \qquad\qquad (6/4)$$

For a small coefficient of friction, the angle ϕ is small, and the sine and tangent may be interchanged with only small error. Since $\mu = \tan \phi$, a good approximation to the torque is

$$M = \mu Lr \qquad\qquad (6/4a)$$

This relation gives the amount of torque or moment which must be applied to the shaft to overcome friction for a dry or partially lubricated journal bearing.

6/7 THRUST BEARINGS; DISK FRICTION

Friction between circular surfaces under distributed normal pressure occurs in pivot bearings, clutch plates, and disk brakes. To examine these applications, we consider the two flat circular disks shown in Fig. 6/9. Their shafts are mounted in bearings (not shown) so that they can be brought into contact under the axial force P. The maximum torque which this clutch can transmit is equal to the torque M required to slip one disk against the other. If p is the normal pressure at any location between the plates, the frictional force acting on an elemental area is $\mu p\, dA$, where μ is the friction coefficient and dA is the area $r\, dr\, d\theta$ of the element. The moment of this elemental friction force about the shaft axis is $\mu p r\, dA$, and the total moment becomes

$$M = \int \mu p r\, dA$$

where we evaluate the integral over the area of the disk. To carry out this integration, we must know the variation of μ and p with r.

In the following examples we will assume that μ is constant. Furthermore, if the surfaces are new, flat, and well supported, it is reasonable to assume that the pressure p is uniform over the entire surface so that $\pi R^2 p = P$. Substituting the constant value of p in the expression for M gives

$$M = \frac{\mu P}{\pi R^2} \int_0^{2\pi} \int_0^R r^2\, dr\, d\theta = \tfrac{2}{3}\mu P R \qquad \textbf{(6/5)}$$

We may interpret this result as equivalent to the moment due to a friction force μP acting at a distance $\tfrac{2}{3}R$ from the shaft center.

If the friction disks are rings, as in the collar bearing shown in Fig. 6/10, the limits of integration are the inside and outside radii R_i and R_o, respectively, and the frictional torque becomes

$$M = \tfrac{2}{3}\mu P\, \frac{R_o{}^3 - R_i{}^3}{R_o{}^2 - R_i{}^2} \qquad \textbf{(6/5a)}$$

After the initial wearing-in period is over, the surfaces retain their new relative shape and further wear is therefore constant over the surface. This wear depends on both the circumferential distance traveled

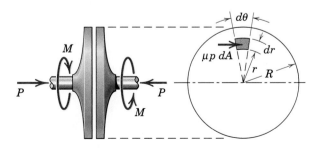

Figure 6/9

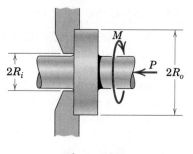

Figure 6/10

and the pressure p. Since the distance traveled is proportional to r, the expression $rp = K$ may be written, where K is a constant. The value of K is determined from the equilibrium condition for the axial forces, which gives

$$P = \int p \, dA = K \int_0^{2\pi} \int_0^R dr \, d\theta = 2\pi KR$$

With $pr = K = P/(2\pi R)$, we may write the expression for M as

$$M = \int \mu pr \, dA = \frac{\mu P}{2\pi R} \int_0^{2\pi} \int_0^R r \, dr \, d\theta$$

which becomes

$$M = \tfrac{1}{2}\mu PR \tag{6/6}$$

The frictional moment for worn-in plates is, therefore, only $(\tfrac{1}{2})/(\tfrac{2}{3})$, or $\tfrac{3}{4}$ as much as for new surfaces. If the friction disks are rings of inside radius R_i and outside radius R_o, substitution of these limits gives for the frictional torque for worn-in surfaces

$$M = \tfrac{1}{2}\mu P(R_o + R_i) \tag{6/6a}$$

You should be prepared to deal with other disk-friction problems where the pressure p is some other function of r.

Sample Problem 6/8

The bell crank fits over a 100-mm-diameter shaft which is fixed and cannot rotate. The horizontal force T is applied to maintain equilibrium of the crank under the action of the vertical force $P = 100$ N. Determine the maximum and minimum values which T may have without causing the crank to rotate in either direction. The coefficient of static friction μ between the shaft and the bearing surface of the crank is 0.20.

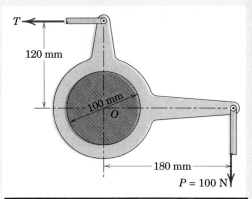

Solution.   Impending rotation occurs when the reaction R of the fixed shaft on the bell crank makes an angle $\phi = \tan^{-1} \mu$ with the normal to the bearing surface and is, therefore, tangent to the friction circle. Also, equilibrium requires that the three forces acting on the crank be concurrent at point C. These facts are shown in the free-body diagrams for the two cases of impending motion.

The following calculations are needed:

Friction angle $\phi = \tan^{-1} \mu = \tan^{-1} 0.20 = 11.31°$

Radius of friction circle $r_f = r \sin \phi = 50 \sin 11.31° = 9.81$ mm

Angle $\theta = \tan^{-1} \dfrac{120}{180} = 33.7°$

Angle $\beta = \sin^{-1} \dfrac{r_f}{OC} = \sin^{-1} \dfrac{9.81}{\sqrt{(120)^2 + (180)^2}} = 2.60°$

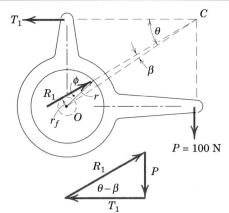

(*a*) Counterclockwise motion impends

(a) Impending counterclockwise motion.   The equilibrium triangle of forces is drawn and gives

$$T_1 = P \cot (\theta - \beta) = 100 \cot (33.7° - 2.60°)$$

$$T_1 = T_{max} = 165.8 \text{ N} \qquad\qquad Ans.$$

(b) Impending clockwise motion.   The equilibrium triangle of forces for this case gives

$$T_2 = P \cot (\theta + \beta) = 100 \cot (33.7° + 2.60°)$$

$$T_2 = T_{min} = 136.2 \text{ N} \qquad\qquad Ans.$$

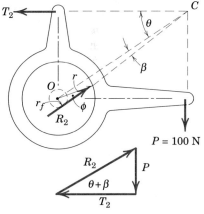

(*b*) Clockwise motion impends

PROBLEMS

Introductory Problems

6/71 The two flywheels are mounted on a common shaft which is supported by a journal bearing between them. Each flywheel has a mass of 40 kg, and the diameter of the shaft is 40 mm. If a 3-N·m couple M on the shaft is required to maintain rotation of the flywheels and shaft at a constant low speed, compute (a) the coefficient of friction in the bearing and (b) the radius r_f of the friction circle.

Ans. (a) $\mu = 0.1947$, (b) $r_f = 3.82$ mm

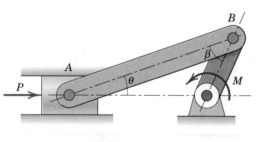

Problem 6/71

6/72 The figure shows a slider-crank mechanism in equilibrium with force P applied to the slider and couple M applied to the crank. If P is increased, the slider will start to move to the right, and both angles θ and β will decrease for the position shown. Draw the free-body diagram of link AB showing the line of action of the compressive forces on AB in relation to the friction circle at each end.

Problem 6/72

6/73 Circular disk A is placed on top of disk B and is subjected to a compressive force of 80 lb. The diameters of A and B are 9 in. and 12 in., respectively, and the pressure under each disk is constant over its surface. If the coefficient of friction between A and B is 0.40, determine the couple M which will cause A to slip on B. Also, what is the minimum coefficient of friction μ between B and the supporting surface C which will prevent B from rotating?

Ans. $M = 96$ lb-in.
$\mu = 0.3$

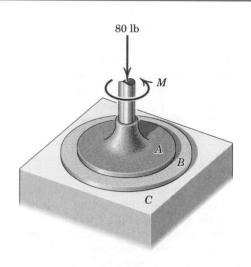

Problem 6/73

6/74 Determine the tension T in the cable to raise the 800-kg load if the coefficient of friction for the 30-mm bearing is 0.25. Also find the tension T_0 in the stationary section of the cable. The mass of the cable and pulley is small and may be neglected.

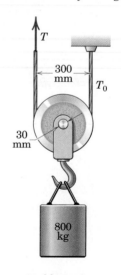

Problem 6/74

6/75 Calculate the tension T required to lower the 800-kg load described in Prob. 6/74. Also find T_0.

Ans. $T = 3830$ N, $T_0 = 4020$ N

6/76 Calculate the torque M required to rotate the 540-lb reel of telephone cable clockwise against the 400-lb tension in the cable. The diameter of the bearing is 2.50 in., and the coefficient of friction for the bearing is 0.30.

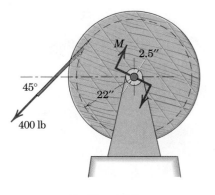

Problem 6/76

6/77 Calculate the torque M' on the 540-lb telephone-cable reel of Prob. 6/76 which will permit the reel to turn slowly counterclockwise under the action of the 400-lb tension.

Ans. $M' = 707$ lb-ft

6/78 The operator of the 6-lb disk sander applies a torque M_z and an additional downward force of 10 lb to the sander. After the sanding disk has been used for a while, the coefficient of friction between the disk and the surface being sanded decreases linearly with radial distance—from 0.80 at the center to 0.50 at the outside of the 6-in.-diameter disk. Assume that the pressure p under the disk is constant and determine M_z.

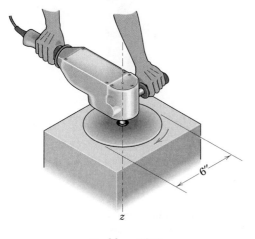

Problem 6/78

Representative Problems

6/79 The mass of the drum D and its cable is 45 kg, and the coefficient of friction μ for the bearing is 0.20. Determine the force P required to raise the 40-kg cylinder if the bearing friction is (*a*) neglected and (*b*) included in the analysis. The weight of the shaft is negligible.

Ans. (*a*) $P = 245$ N, (*b*) $P = 259$ N

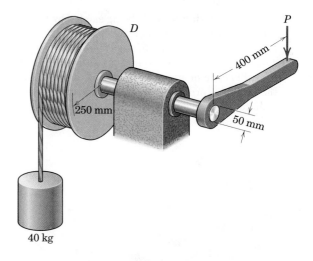

40 kg

Problem 6/79

6/80 Determine the force P required to lower the 40-kg cylinder of Prob. 6/79. Compare your answer with the stated results of that problem. Is the no-friction value of P equal to the average of the forces required to raise and lower the cylinder?

6/81 The axial section of the two mating circular disks is shown. Derive the expression for the torque M required to turn the upper disk on the fixed lower one if the pressure p between the disks follows the relation $p = k/r^2$, where k is a constant to be determined. The coefficient of friction μ is constant over the entire surface.

Ans. $M = \mu L \dfrac{r_o - r_i}{\ln(r_o/r_i)}$

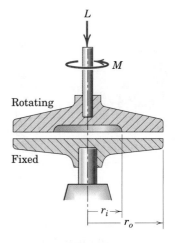

Problem 6/81

6/82 An end of the thin board is being sanded by the disk sander under application of the force P. If the effective coefficient of kinetic friction is μ and if the pressure is essentially constant over the board end, determine the moment M which must be applied by the motor in order to rotate the disk at a constant angular speed. The board end is centered along the radius of the disk.

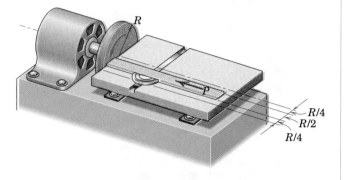

Problem 6/82

6/83 The front wheels of an experimental rear-drive vehicle have a radius of 300 mm and are designed with disk-type brakes consisting of a ring A with outside and inside radii of 150 mm and 75 mm, respectively. The ring, which does not turn with the wheel, is forced against the wheel disk with a force P. If the pressure between the ring and the wheel disk is uniform over the mating surfaces, compute the friction force F between each front tire and the horizontal road for an axial force $P = 1$ kN when the vehicle is powered at constant speed with the wheels turning. The coefficient of friction between the disk and ring is 0.35.

Ans. $F = 136.1$ N

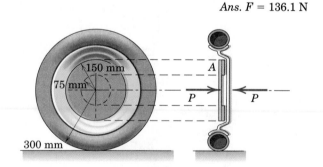

Problem 6/83

6/84 An automobile disk brake consists of a flat-faced rotor and a caliper which contains a disk pad on each side of the rotor. For equal forces P behind the two pads with the pressure p uniform over the pad, show that the moment applied to the hub is independent of the angular span β of the pads. Would pressure variation with θ change the moment?

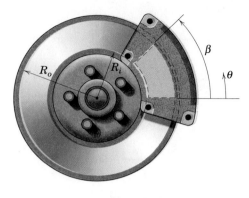

Problem 6/84

6/85 In a design test on friction, shaft A is fitted loosely in the wrist-pin bearing of the connecting rod with center of gravity at G as shown. With the rod initially in the vertical position, the shaft is rotated slowly until the rod slips at the angle α. Write an exact expression for the coefficient of friction μ.

$$Ans. \ \mu = \frac{1}{\sqrt{\left(\dfrac{d/2}{\bar{r}\sin\alpha}\right)^2 - 1}}$$

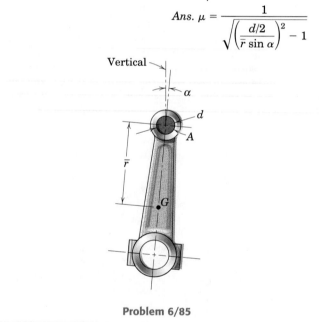

Problem 6/85

6/86 For the flat sanding disk of radius a, the pressure p developed between the disk and the sanded surface decreases linearly with r from a value p_0 at the center to $p_0/2$ at $r = a$. If the coefficient of friction is μ, derive the expression for the torque M required to turn the shaft under an axial force L.

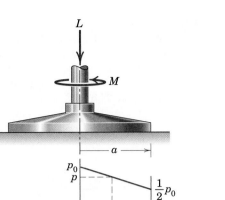

Problem 6/86

6/87 Each of the four wheels of the vehicle weighs 40 lb and is mounted on a 4-in.-diameter journal (shaft). The total weight of the vehicle is 960 lb, including wheels, and is distributed equally on all four wheels. If a force $P = 16$ lb is required to keep the vehicle rolling at a constant low speed on a horizontal surface, calculate the coefficient of friction which exists in the wheel bearings. (*Hint:* Draw a complete free-body diagram of one wheel.)

Ans. $\mu = 0.204$

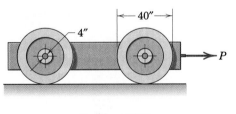

Problem 6/87

6/88 Determine the angle θ with the horizontal made by the steepest slope on which the vehicle of Prob. 6/87 can stand without rolling by itself in the absence of a force P. Take the coefficient of friction in the wheel bearings to be 0.20.

6/89 The 10-Mg crate is lowered into an underground storage facility on a two-screw elevator designed as shown. Each screw has a mass of 0.9 Mg, is 120 mm in mean diameter, and has a single square thread with a lead of 11 mm. The screws are turned in synchronism by a motor unit in the base of the facility. The entire mass of the crate, screws, and 3-Mg elevator platform is supported equally by flat collar bearings at A, each of which has an outside diameter of 250 mm and an inside diameter of 125 mm. The pressure on the bearings is assumed to be uniform over the bearing surface. If the coefficient of friction for the collar bearing and the screws at B is 0.15, calculate the torque M which must be applied to each screw (*a*) to raise the elevator and (*b*) to lower the elevator.

Ans. (a) $M = 1747$ N·m, (b) $M = 1519$ N·m

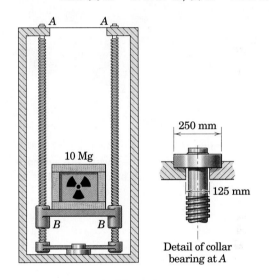

Problem 6/89

6/90 In the figure is shown a multiple-disk clutch designed for marine use. The driving disks A are splined to the driving shaft B so that they are free to slip along the shaft but must rotate with it. The disks C drive the housing D by means of the bolts E, along which they are free to slide. In the clutch shown there are five pairs of friction surfaces. Assume the pressure is uniformly distributed over the area of the disks and determine the maximum torque M which can be transmitted if the coefficient of friction is 0.15 and $P = 500$ N.

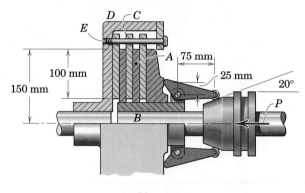

Problem 6/90

▶**6/91** Determine the expression for the torque M required to turn the shaft whose thrust L is supported by a conical pivot bearing. The coefficient of friction is μ, and the bearing pressure is constant.

$$Ans.\ M = \frac{\mu L}{3 \sin \frac{\alpha}{2}} \frac{d_2^{\,3} - d_1^{\,3}}{d_2^{\,2} - d_1^{\,2}}$$

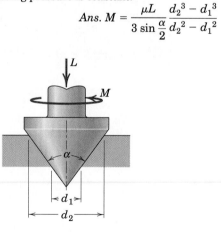

Problem 6/91

▶**6/92** A thrust bearing for a shaft under an axial load L is designed as a partial spherical cup of radius r. If the pressure between the bearing surfaces at any point varies according to $p = p_0 \cos \theta$, derive the expression for the torque M required to maintain constant rotational speed. The coefficient of friction is μ.

$$Ans.\ M = \frac{3\sqrt{3}}{7} \mu r L$$

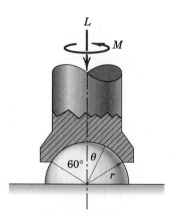

Problem 6/92

6/8   Flexible Belts

The impending slippage of flexible cables, belts, and ropes over sheaves and drums is important in the design of belt drives of all types, band brakes, and hoisting rigs.

Figure 6/11a shows a drum subjected to the two belt tensions T_1 and T_2, the torque M necessary to prevent rotation, and a bearing reaction R. With M in the direction shown, T_2 is greater than T_1. The free-body diagram of an element of the belt of length $r\,d\theta$ is shown in part b of the figure. We analyze the forces acting on this differential element by establishing the equilibrium of the element, in a manner similar to that used for other variable-force problems. The tension increases from T at the angle θ to $T + dT$ at the angle $\theta + d\theta$. The normal force is a differential dN, since it acts on a differential element of area. Likewise the friction force, which must act on the belt in a direction to oppose slipping, is a differential and is $\mu\,dN$ for impending motion.

Equilibrium in the t-direction gives

$$T \cos \frac{d\theta}{2} + \mu\,dN = (T + dT) \cos \frac{d\theta}{2}$$

or

$$\mu\,dN = dT$$

since the cosine of a differential quantity is unity in the limit. Equilibrium in the n-direction requires that

$$dN = (T + dT) \sin \frac{d\theta}{2} + T \sin \frac{d\theta}{2}$$

or

$$dN = T\,d\theta$$

where we have used the facts that the sine of a differential angle in the limit equals the angle and that the product of two differentials must be neglected in the limit compared with the first-order differentials remaining.

Combining the two equilibrium relations gives

$$\frac{dT}{T} = \mu\,d\theta$$

Integrating between corresponding limits yields

$$\int_{T_1}^{T_2} \frac{dT}{T} = \int_0^\beta \mu\,d\theta$$

or

$$\ln \frac{T_2}{T_1} = \mu\beta$$

where the ln (T_2/T_1) is a natural logarithm (base e). Solving for T_2 gives

$$T_2 = T_1 e^{\mu\beta} \qquad\qquad \textbf{(6/7)}$$

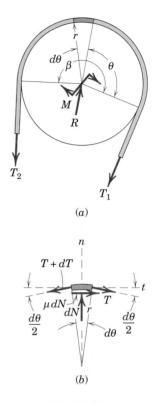

(a)

(b)

Figure 6/11

© Image Source/Media Bakery

Just one turn of a line around a fixed cylinder can produce a large change in tension.

Note that β is the total angle of belt contact and must be expressed in radians. If a rope were wrapped around a drum n times, the angle β would be $2\pi n$ radians. Equation 6/7 holds equally well for a noncircular section where the total angle of contact is β. This conclusion is evident from the fact that the radius r of the circular drum in Fig. 6/11 does not enter into the equations for the equilibrium of the differential element of the belt.

The relation expressed by Eq. 6/7 also applies to belt drives where both the belt and the pulley are rotating at constant speed. In this case the equation describes the ratio of belt tensions for slippage or impending slippage. When the speed of rotation becomes large, the belt tends to leave the rim, so Eq. 6/7 involves some error in this case.

6/9   ROLLING RESISTANCE

Deformation at the point of contact between a rolling wheel and its supporting surface introduces a resistance to rolling, which we mention only briefly. This resistance is not due to tangential friction forces and therefore is an entirely different phenomenon from that of dry friction.

To describe rolling resistance, we consider the wheel shown in Fig. 6/12 under the action of a load L on the axle and a force P applied at its center to produce rolling. The deformation of the wheel and supporting surfaces as shown is greatly exaggerated. The distribution of pressure p over the area of contact is similar to the distribution shown. The resultant R of this distribution acts at some point A and must pass through the wheel center for the wheel to be in equilibrium. We find the force P necessary to maintain rolling at constant speed by equating the moments of all forces about A to zero. This gives us

$$P = \frac{a}{r} L = \mu_r L$$

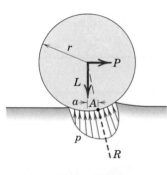

Figure 6/12

where the moment arm of P is taken to be r. The ratio $\mu_r = a/r$ is called the *coefficient of rolling resistance*. This coefficient is the ratio of resisting force to normal force and thus is analogous to the coefficient of static or kinetic friction. On the other hand, there is no slippage or impending slippage in the interpretation of μ_r.

Because the dimension a depends on many factors which are difficult to quantify, a comprehensive theory of rolling resistance is not available. The distance a is a function of the elastic and plastic properties of the mating materials, the radius of the wheel, the speed of travel, and the roughness of the surfaces. Some tests indicate that a varies only slightly with wheel radius, and thus a is often taken to be independent of the rolling radius. Unfortunately, the quantity a has also been called the coefficient of rolling friction in some references. However, a has the dimension of length and therefore is not a dimensionless coefficient in the usual sense.

Sample Problem 6/9

A flexible cable which supports the 100-kg load is passed over a fixed circular drum and subjected to a force P to maintain equilibrium. The coefficient of static friction μ between the cable and the fixed drum is 0.30. (*a*) For $\alpha = 0$, determine the maximum and minimum values which P may have in order not to raise or lower the load. (*b*) For $P = 500$ N, determine the minimum value which the angle α may have before the load begins to slip.

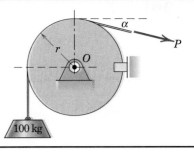

Solution. Impending slipping of the cable over the fixed drum is given by Eq. 6/7, which is $T_2/T_1 = e^{\mu\beta}$.

① (*a*) With $\alpha = 0$ the angle of contact is $\beta = \pi/2$ rad. For impending upward motion of the load, $T_2 = P_{max}$, $T_1 = 981$ N, and we have

② $$P_{max}/981 = e^{0.30(\pi/2)} \qquad P_{max} = 981(1.602) = 1572 \text{ N} \qquad Ans.$$

For impending downward motion of the load, $T_2 = 981$ N and $T_1 = P_{min}$. Thus,

$$981/P_{min} = e^{0.30(\pi/2)} \qquad P_{min} = 981/1.602 = 612 \text{ N} \qquad Ans.$$

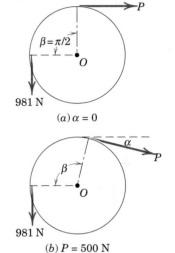

(*a*) $\alpha = 0$

(*b*) $P = 500$ N

(*b*) With $T_2 = 981$ N and $T_1 = P = 500$ N, Eq. 6/7 gives us

$$981/500 = e^{0.30\beta} \qquad 0.30\beta = \ln(981/500) = 0.674$$

$$\beta = 2.25 \text{ rad} \qquad \text{or} \qquad \beta = 2.25\left(\frac{360}{2\pi}\right) = 128.7°$$

③ $$\alpha = 128.7° - 90° = 38.7° \qquad\qquad Ans.$$

Helpful Hints

① We are careful to note that β must be expressed in radians.

② In our derivation of Eq. 6/7 be certain to note that $T_2 > T_1$.

③ As was noted in the derivation of Eq. 6/7, the radius of the drum does not enter into the calculations. It is only the angle of contact and the coefficient of friction which determine the limiting conditions for impending motion of the flexible cable over the curved surface.

PROBLEMS

Introductory Problems

6/93 It is observed that the two cylinders will remain in slow steady motion as indicated in the drawing. Determine the coefficient of friction μ between the cord and the fixed shaft.

Ans. $\mu = 0.244$

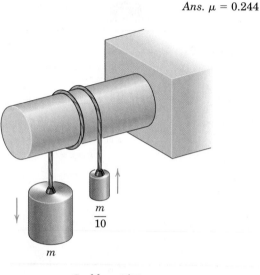

Problem 6/93

6/94 Determine the force P required to (*a*) raise and (*b*) lower the cylinder of weight W at a slow steady speed. The coefficient of friction between the cord and the fixed shaft is 0.4.

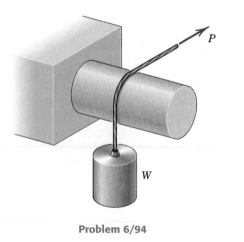

Problem 6/94

6/95 A dockworker adjusts a spring line (rope) which keeps a ship from drifting alongside a wharf. If he exerts a pull of 200 N on the rope, which has $1\frac{1}{4}$ turns around the mooring bit, what force T can he support? The coefficient of friction between the rope and the cast-steel mooring bit is 0.30.

Ans. $T = 2.11$ kN

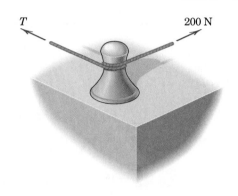

Problem 6/95

6/96 If the dockworker of Prob. 6/95 is to support a tension of 18 kN in the rope leading to the ship, how many turns around the mooring bit are necessary if he exerts a pull of 240 N on the free end? The coefficient of friction between the rope and the bit is 0.30.

6/97 The 20-lb block and the 50-lb block are connected by a rope hung over the fixed curved surface. If the system is on the verge of slipping, calculate the coefficient of friction between the rope and the surface.

Ans. $\mu = 0.292$

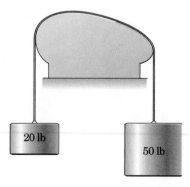

Problem 6/97

6/98 The 180-lb rock climber is lowered over the edge of the cliff by his two companions, who together exert a horizontal pull T of 75 lb on the rope. Compute the coefficient of friction μ between the rope and the rock.

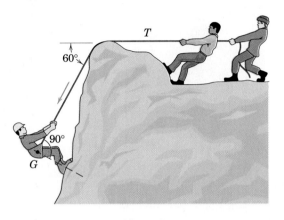

Problem 6/98

6/99 With what horizontal force P must the worker pull in order to move the 35-kg log up the 15° incline? The coefficient of friction between the log and the incline is 0.6 and that between the rope and the boulder is 0.4.

Ans. P = 320 N

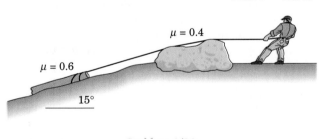

Problem 6/99

6/100 In western movies, cowboys are frequently observed hitching their horses by casually winding a few turns of the reins around a horizontal pole and letting the end hang free as shown—no knots! If the freely hanging length of rein weighs 2 oz and the number of turns is as shown, what tension T does the horse have to produce in the direction shown in order to gain freedom? The coefficient of friction between the reins and wooden pole is 0.70.

Problem 6/100

6/101 For a certain coefficient of friction μ and a certain angle α, the force P required to raise m is 4 kN and that required to lower m at a constant slow speed is 1.6 kN. Calculate the mass m.

Ans. m = 258 kg

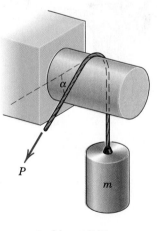

Problem 6/101

Representative Problems

6/102 The tape slides around the two fixed pegs as shown and is under the action of the horizontal tensions $T_1 = 40$ N and $T_2 = 160$ N. Determine the coefficient of friction μ between the tape and the pegs.

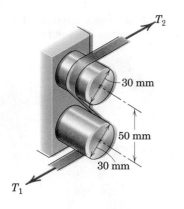

Problem 6/102

6/103 The 180-lb tree surgeon lowers himself with the rope over a horizontal limb of the tree. If the coefficient of friction between the rope and the limb is 0.60, compute the force which the man must exert on the rope to let himself down slowly.

Ans. P = 23.7 lb

Problem 6/103

6/104 A counterclockwise moment $M = 1500$ lb-in. is applied to the flywheel. If the coefficient of friction between the band and the wheel is 0.20, compute the minimum force P necessary to prevent the wheel from rotating.

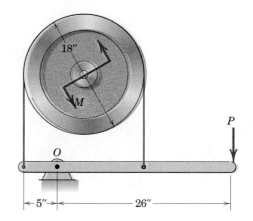

Problem 6/104

6/105 Magnetic tape passes around the light idler pulleys B and over the fixed circular recording head A with a constant speed. The tape tension is unchanged as it passes around the idler pulleys. Calculate the minimum spacing a in the design of this unit for which the ratio of the tensions T_1 and T_2 will not exceed 1.15. The coefficient of friction between the tape and the head is 0.10.

Ans. a = 62.2 mm

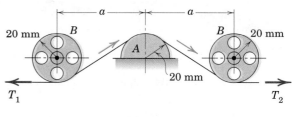

Problem 6/105

6/106 The uniform drum *A* with center of mass at mid-length is suspended by a rope which passes over the fixed cylindrical surface *B*. The coefficient of static friction between the rope and the surface over which it passes is μ. Determine the maximum value which the dimension *a* may have before the drum tips out of its horizontal position.

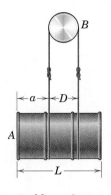

Problem 6/106

6/107 A device designed for lowering a person in a sling down a rope at a constant controlled rate is shown in the figure. The rope passes around a central shaft fixed to the frame and leads freely out of the lower collar. The number of turns is adjusted by turning the lower collar, which winds or unwinds the rope around the shaft. Entrance of the rope into the upper collar at *A* is equivalent to $\frac{1}{4}$ of a turn, and passage around the corner at *B* is also equivalent to $\frac{1}{4}$ of a turn. Friction of the rope through the straight portions of the collars averages 10 N for each collar. If three complete turns around the shaft, in addition to the corner turns, are required for a 75-kg man to lower himself at a constant rate without exerting a pull on the free end of the rope, calculate the coefficient of friction μ between the rope and the contact surfaces of the device. Neglect the small helix angle of the rope around the shaft.

Ans. $\mu = 0.1948$

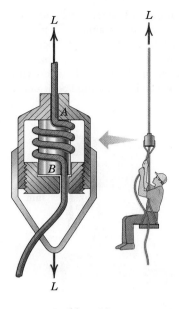

Problem 6/107

6/108 The endless belt of an escalator passes around idler drum *A* and is driven by a torque *M* applied to drum *B*. Belt tension is adjusted by a turnbuckle at *C*, which produces an initial tension of 1000 lb in each side of the belt when the escalator is unloaded. For the design of this system calculate the minimum coefficient of friction μ between drum *B* and the belt to prevent slipping if the escalator handles 30 people uniformly distributed along the belt and averaging 150 lb each. (*Note:* It can be shown that the increase in belt tension on the upper side of drum *B* and the decrease in belt tension at the lower drum *A* are each equal to half the component of the combined passenger weight along the incline.)

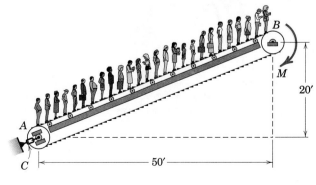

Problem 6/108

6/109 The uniform I-beam has a mass of 74 kg per meter of length and is supported by the rope over the fixed 300-mm drum. If the coefficient of friction between the rope and the drum is 0.5, calculate the least value of the force P which will cause the beam to tip from its horizontal position.

Ans. P = 160.3 N

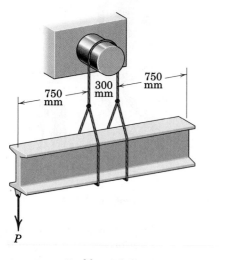

750 mm   300 mm   750 mm

P

Problem 6/109

6/110 For the design of the band brake shown, find the couple M required to turn the pipe in the V-block against the action of the flexible band. A force $P = 25$ lb is applied to the lever, which is pivoted about O. The coefficient of friction between the band and the pipe is 0.30, and that between the pipe and the block is 0.40. The weights of the parts are negligible.

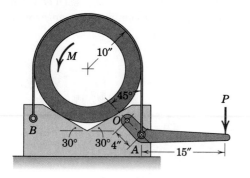

M 10″   45° P

B 30°   30°   4″ O A 15″

Problem 6/110

6/111 The cylinder of mass m is attached to the ring A, which is suspended by the cable which passes over the pulley, as shown in part a of the figure. A couple M applied to the pulley turns it until slipping of the cable on the pulley occurs at the position $\theta = 20°$, shown in part b of the figure. Calculate the coefficient of friction μ between the cable and the pulley.

Ans. μ = 0.214

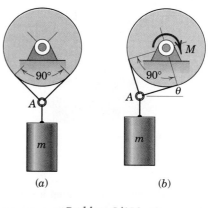

90° A m 90° M A θ m

(*a*)   (*b*)

Problem 6/111

6/112 Shown in the figure is the design of a band-type oil-filter wrench. If the coefficient of friction between the band and the fixed filter is 0.25, determine the minimum value of h which ensures that the wrench will not slip on the filter, regardless of the magnitude of the force P. Neglect the mass of the wrench and assume that the effect of the small part at A is equivalent to that of a band wrap which begins at the three-o'clock position and runs clockwise.

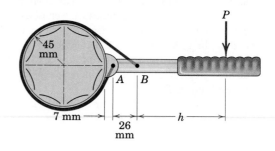

45 mm P A B 7 mm   26 mm h

Problem 6/112

6/113 Determine the range of cylinder mass m over which the system is in equilibrium. Consider the two cases in which the coefficient of friction between the cord and its supporting fixed surface is (*a*) 0 and (*b*) 0.3. In both cases, include the indicated friction between the 100-kg body and the incline.

Ans. (*a*) 6.01 kg $\leq m \leq$ 62.4 kg
(*b*) 3.38 kg $\leq m \leq$ 111.0 kg

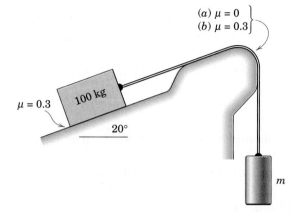

Problem 6/113

6/114 Replace the flat belt and pulley of Fig. 6/11 by a V-belt and matching grooved pulley as indicated by the cross-sectional view accompanying this problem. Derive the relation among the belt tensions, the angle of contact, and the coefficient of friction for the V-belt when slipping impends. A V-belt design with $\alpha = 35°$ would be equivalent to increasing the coefficient of friction for a flat belt of the same material by what factor n?

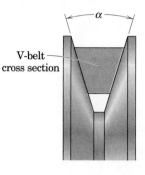

Problem 6/114

6/10 CHAPTER REVIEW

In our study of friction we have concentrated on dry or Coulomb friction where a simple mechanical model of surface irregularities between the contacting bodies, Fig. 6/1, explains the phenomenon adequately for most engineering purposes. This model helps to visualize the three types of dry-friction problems which are encountered in practice. These problem types are:

1. Static friction of less than the maximum possible value and determined by the equations of equilibrium. (This usually requires a check to see that $F < \mu_s N$.)
2. Limiting static friction with impending motion ($F = \mu_s N$).
3. Kinetic friction where sliding motion occurs between contacting surfaces ($F = \mu_k N$).

Keep in mind the following when solving dry-friction problems:

1. A coefficient of friction applies to a given pair of mating surfaces. It is meaningless to speak of a coefficient of friction for a single surface.
2. The coefficient of static friction μ_s for a given pair of surfaces is usually slightly greater than the kinetic coefficient μ_k.
3. The friction force which acts on a body is always in the direction to oppose the slipping of the body which takes place or the slipping which would take place in the absence of friction.
4. When friction forces are distributed over a surface or along a line, we select a representative element of the surface or line and evaluate the force and moment effects of the elemental friction force acting on the element. We then integrate these effects over the entire surface or line.
5. Friction coefficients vary considerably, depending on the exact condition of the mating surfaces. Computing coefficients of friction to three significant figures represents an accuracy which cannot easily be duplicated by experiment. When cited, such values are included for purposes of computational check only. For design computations in engineering practice, any handbook value for a coefficient of static or kinetic friction must be viewed as an approximation.

Other forms of friction mentioned in the introductory article of the chapter are important in engineering. Problems which involve fluid friction, for example, are among the most important of the friction problems encountered in engineering, and are studied in the subject of fluid mechanics.

REVIEW PROBLEMS

6/115 The 80-kg block is placed on the 20° incline against the spring and released from rest. The coefficient of static friction between the block and the incline is 0.25. (*a*) Determine the maximum and minimum values of the initial compression force in the spring for which the block will not slip upon release. (*b*) Calculate the magnitude and direction of the friction force acting on the block if the spring compression force C is 200 N.

$$Ans.\ C_{max} = 453\ N,\ C_{min} = 84.0\ N$$
$$F = 68.4\ N\ up\ the\ incline$$

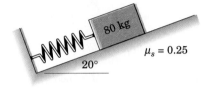

Problem 6/115

6/116 Before running, a jogger stretches his calf muscles by leaning against a wall as shown in the figure. This position is achieved with no vertical (friction) force acting between his hands and the wall. What is the minimum coefficient μ_s of static friction for which his feet will not slip?

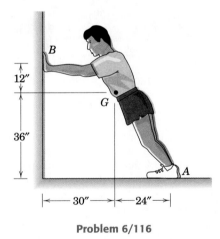

Problem 6/116

6/117 The homogeneous disk of mass m is resting on the right-angled supporting surfaces shown. The tension P in the cord is very gradually increased from zero. If the friction at both A and B is characterized by $\mu_s = 0.25$, what happens first—does the homogeneous disk slip in place or does it begin to roll up the incline? Determine the value of P at which this first movement occurs.

$$Ans.\ P = 0.232mg$$

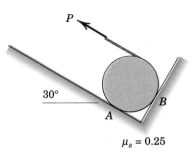

Problem 6/117

6/118 A frictional locking device allows bar A to move to the left but prevents movement to the right. If the coefficient of friction between the shoe B and the bar A is 0.40, specify the maximum length b of the link which will permit the device to work as described.

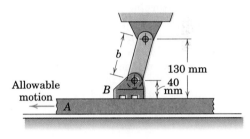

Problem 6/118

6/119 The toggle-wedge is an effective device to close the gap between two planks during construction of a wooden boat. For the combination shown, if a force P of 300 lb is required to move the wedge, determine the friction force F acting on the upper end A of the toggle. The coefficients of static and kinetic friction for all pairs of mating surfaces are taken to be 0.40.

$$Ans.\ F = 120.3\ lb$$

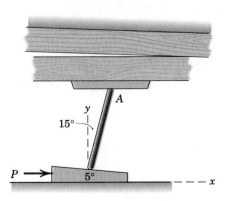

Problem 6/119

6/120 Three boxes are placed on the incline in contact with each other and released from rest. The coefficients of static friction under boxes A, B, and C are 0.30, 0.20, and 0.35, respectively. Describe what happens.

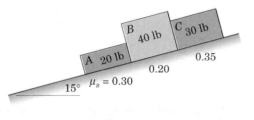

Problem 6/120

6/121 The device shown is used for coarse adjustment of the height of an experimental apparatus without a change in its horizontal position. Because of the slipjoint at A, turning the screw does not rotate the cylindrical leg above A. The mean diameter of the thread is $\frac{3}{8}$ in. and the coefficient of friction is 0.15. For a conservative design which neglects friction at the slipjoint, what should be the minimum number N of threads per inch to ensure that the single-threaded screw does not turn by itself under the weight of the apparatus?

Ans. N = 5.66 threads per inch

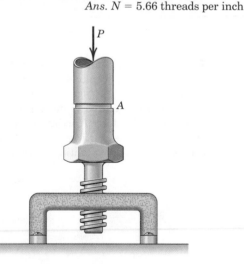

Problem 6/121

6/122 The coefficient of static friction between the collar of the drill-press table and the vertical column is 0.30. Will the collar and table slide down the column under the action of the drill thrust if the operator forgets to secure the clamp, or will friction be sufficient to hold it in place? Neglect the weight of the table and collar compared with the drill thrust and assume that contact occurs at the points A and B.

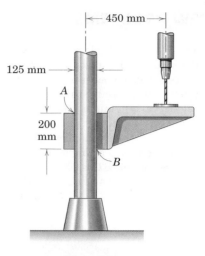

Problem 6/122

6/123 Calculate the torque M which the engine of the pickup truck must supply to the rear axle to roll the front wheels over the curbing from a rest position if the rear wheels do not slip. Determine the minimum effective coefficient of friction at the rear wheels to prevent slipping. The mass of the loaded truck with mass center at G is 1900 kg.

Ans. M = 3.00 kN·m, μ_{min} = 0.787

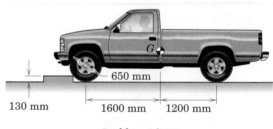

Problem 6/123

6/124 A compressive force of 600 N is to be applied to the two boards in the grip of the C-clamp. The threaded screw has a mean diameter of 10 mm and advances 2.5 mm per turn. The coefficient of static friction is 0.20. Determine the force F which must be applied normal to the handle at C in order to (a) tighten and (b) loosen the clamp. Neglect friction at point A.

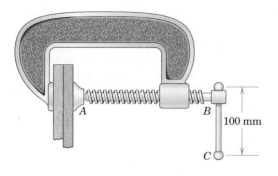

Problem 6/124

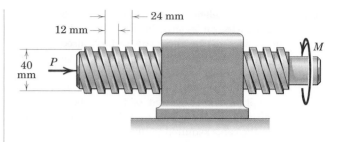

Problem 6/126

6/125 The screw of the small press has a mean diameter of 25 mm and has a single square thread with a lead of 8 mm. The flat thrust bearing at A is shown in the enlarged view and has surfaces which are well worn. If the coefficient of friction for both the threads and the bearing at A is 0.25, calculate the design torque M on the handwheel required (a) to produce a compressive force of 4 kN and (b) to loosen the press from the 4-kN compression.

Ans. (a) $M = 24.1$ N·m, (b) $M = 13.22$ N·m

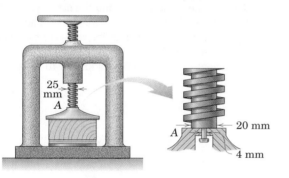

Problem 6/125

6/126 The 40-mm-diameter screw has a double square thread with a pitch of 12 mm and a lead of 24 mm. The screw and its mating threads in the fixed block are graphite-lubricated and have a friction coefficient of 0.15. If a torque $M = 60$ N·m is applied to the right-hand portion of the shaft, determine (a) the force P required to advance the shaft to the right and (b) the force P which would allow the shaft to move to the left at a constant speed.

6/127 The movable head of a universal testing machine has a mass of 2.2 Mg and is elevated into testing position by two 78-mm-diameter lead screws, each with a single thread and a lead of 13 mm. If the coefficient of friction in the threads is 0.25, how much torque M must be supplied to each screw (a) to raise the head and (b) to lower the head? The inner loading columns are not attached to the head during positioning.

Ans. (a) $M = 129.3$ N·m, (b) $M = 81.8$ N·m

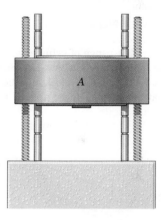

Problem 6/127

6/128 The reel of telephone cable has a mass of 3 Mg and is supported on its shaft in the V-notched blocks on both sides of the reel. The reel is raised off the ground by jacking up the supports so that cable may be pulled off in the horizontal direction as shown. The shaft is fastened to the reel and turns with it. If the coefficient of friction between the shaft and the V-surfaces is 0.30, calculate the pull P in the cable required to turn the reel.

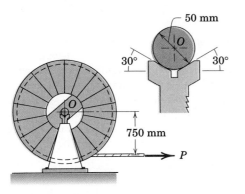

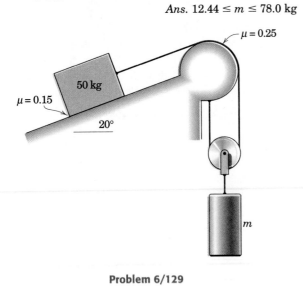

Problem 6/128

6/129 Determine the range of cylinder mass m for which the system is in equilibrium. The coefficient of friction between the 50-kg block and the incline is 0.15 and that between the cord and cylindrical support is 0.25.

Ans. $12.44 \leq m \leq 78.0$ kg

Problem 6/129

6/130 The lower end A of the uniform 100-lb plank rests on rollers which are free to move on the horizontal surface. If the coefficients of static and kinetic friction between the plank and the corner B are 0.80 and 0.70, respectively, compute the friction force F acting at B if the plank is released from rest in the position shown.

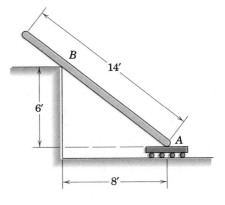

Problem 6/130

6/131 The cylinder weighs 80 lb and the attached uniform slender bar has an unknown weight W. The unit remains in static equilibrium for values of the angle θ ranging up to 45° but slips if θ exceeds 45°. If the coefficient of static friction is known to be 0.30, determine W.

Ans. $W = 70.1$ lb

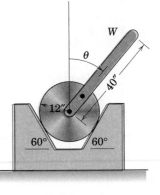

Problem 6/131

6/132 The truck unloads its cargo box by sliding it off the elevated rack, as the truck rolls slowly forward with its brakes applied for control. The box has a total weight of 10,000 lb with center of mass at G in the center of the box. The coefficient of static friction between the box and the rack is 0.30. Calculate the braking force F between the tires and the level road as the box is on the verge of slipping down the rack from the position shown and the truck is on the verge of rolling forward. No slipping occurs at the lower corner of the box.

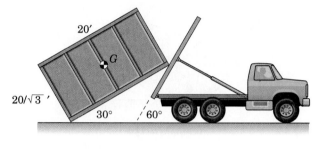

Problem 6/132

6/133 The figure shows a friction silent ratchet for turning the wheel counterclockwise under the action of a force P applied to the handle. The floating link AB engages the wheel so that $\alpha = 20°$. If a force $P = 150$ N is required to turn the wheel about its bearing at O, determine (a) the minimum coefficient of static friction between the link and the wheel which will ensure no slipping at B and (b) the magnitude R of the force on the pin at A.

Ans. (a) $(\mu_s)_{\text{min}} = 0.364$, (b) $R = 1.754$ kN

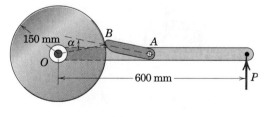

Problem 6/133

6/134 The jack shown is designed to lift small unit-body cars. The screw is threaded into the collar pivoted at B, and the shaft turns in a ball thrust bearing at A. The thread has a mean diameter of 10 mm and a lead (advancement per revolution) of 2 mm. The coefficient of friction for the threads is 0.20. Determine the force P normal to the handle at D required (a) to raise a mass of 500 kg from the position shown and (b) to lower the load from the same position. Neglect friction in the pivot and bearing at A.

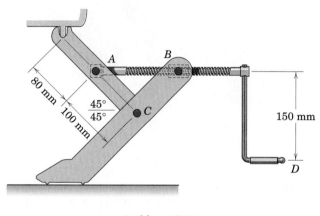

Problem 6/134

6/135 Calculate the couple M applied to the lower of the two 20-kg cylinders which will allow them to roll slowly *down* the incline. The coefficients of static and kinetic friction for all contacting surfaces are $\mu_s = 0.60$ and $\mu_k = 0.50$.

Ans. $M = 10.16$ N·m

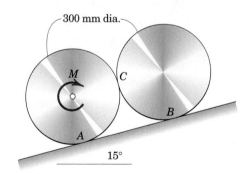

Problem 6/135

6/136 Flexible tape fastened at A makes a half turn around each of the two 100-mm-diameter drums and ends at the spring S, which is adjusted to produce a tension of 40 N in the tape. Calculate the torques M_1 and M_2 applied to the shafts of the drums to turn them at the same time in the direction shown. The coefficient of friction between drum 1 and the tape is 0.30, and that between drum 2 and the tape is 0.20.

Problem 6/136

6/137 The 8-kg block is resting on the 20° inclined plane with a coefficient of static friction $\mu_s = 0.50$. Determine the minimum horizontal force P which will cause the block to slip.

Ans. $P = 25.3$ N

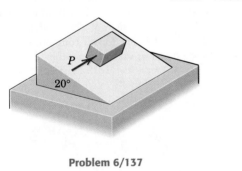

Problem 6/137

*Computer-Oriented Problems**

***6/138** A semicylinder of uniform density rests on a horizontal surface and is subjected to a force P applied as shown. If P is slowly increased and kept normal to the flat surface, plot the tilt angle θ as a function of P up to the point of slipping. Determine the tilt angle θ_{max} and the corresponding value P_{max} for which slipping occurs. The coefficient of static friction is 0.35.

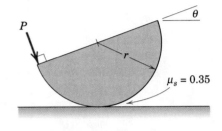

Problem 6/138

***6/139** Plot the force P required to move the 100-kg crate up the 10° incline starting from rest at various values of x from zero to 10 m. Determine the least possible value of P and the corresponding value of x.

Ans. $P_{min} = 517$ N at $x = 7.5$ m

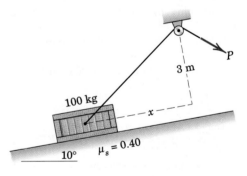

Problem 6/139

*6/140 The small roller on the upper end of the uniform rod rests against the vertical surface at A while the rounded end B rests on the platform which is slowly pivoted downward beginning at the horizontal position shown. For a coefficient of static friction $\mu_s = 0.40$ at B, determine the angle θ of the platform at which slipping will occur. Neglect the size and friction of the roller and the small thickness of the platform.

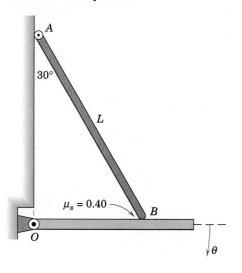

Problem 6/140

*6/141 A heavy cable with a mass of 12 kg per meter of length passes over the two fixed pipes 300 m apart on the same level. One end supports a 1600-kg cylinder. By experiment it is found that a downward force P of 60 kN is required to induce slipping of the cable over both pipes at a constant rate. Determine the coefficient of kinetic friction μ_k between the cable and the pipes, the maximum tension T in the cable between the pipes, and the sag h in the cable.

Ans. $\mu_k = 0.298$, $T = 30.7$ kN, $h = 58.1$ m

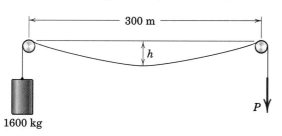

Problem 6/141

*6/142 The uniform slender rod of mass m and length L is initially at rest in a centered horizontal position on the fixed circular surface of radius $R = 0.6L$. If a force P normal to the bar is gradually applied to its end, determine the maximum equilibrium angle θ which the rod can reach before slipping takes place. The coefficient of static friction between the rod and its support is 0.15.

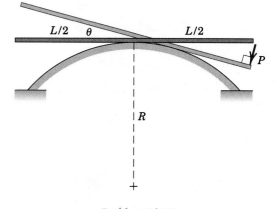

Problem 6/142

*6/143 Refer to Prob. 6/142, where the mass of the uniform rod is 3 kg, $L = 2$ m, and $R = 1.2$ m. As the force P is increased, the equilibrium angle θ also increases, provided no slippage occurs. The limiting value of θ is reached when $R\theta = L/2$, where θ is in radians. For values of θ between zero and $L/(2R)$, plot the corresponding equilibrium value of P, the normal contact force N, and the minimum value μ_s of the static friction coefficient which prevents slipping at that angle. Identify the angle θ_{cr} at which μ_s reaches a maximum value.

$$Ans.\ P = \frac{35.3\theta \cos \theta}{1 - 1.2\theta},\ N = \frac{29.4 \cos \theta}{1 - 1.2\theta}$$
$$\mu_s = (1 - 1.2\theta) \tan \theta$$
$$(\mu_s)_{\max} = 0.222 \text{ at } \theta = 25.5°$$

*6/144 The device is designed to permit an adjustment to the horizontal tension T in the cable passing around the two fixed wheels in order to lower the mass m. If the coefficient of friction between the cable and the wheel surfaces is 0.40, determine and plot the ratio T/mg as a function of θ in the range $0 < \theta < 90°$. Also find the value of the shear force V in the adjusting pin at D terms of mg for $\theta = 60°$.

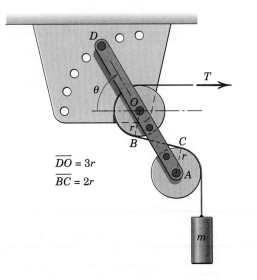

$$\overline{DO} = 3r$$
$$\overline{BC} = 2r$$

Problem 6/144

*6/145 The flat belt has a width $b = 50$ mm and is on the verge of slipping as it drives the pulley of radius $r = 75$ mm through a contact angle of 180°. If the tension in the slack side of the belt is $T_1 = 1000$ N, plot the normal pressure p between the belt and the pulley as a function of θ between $\theta = 0$ and $\theta = \pi$. Specify the maximum and minimum pressures p_2 and p_1, the tension T_2, and the resisting torque M on the pulley shaft. The coefficient of friction between the belt and the pulley is 0.30. (The pulley bearings are not shown.)

Ans. $p_1 = 267$ kPa, $p_2 = 684$ kPa
$M = 117.5$ N·m

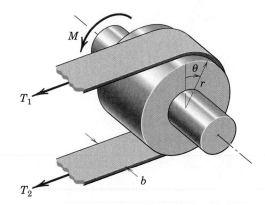

Problem 6/145

The analysis of multi-link structures which change configuration is generally best handled by a virtual-work approach. This construction platform is a typical example.

7 VIRTUAL WORK

CHAPTER OUTLINE

7/1 INTRODUCTION

In the previous chapters we have analyzed the equilibrium of a body by isolating it with a free-body diagram and writing the zero-force and zero-moment summation equations. This approach is usually employed for a body whose equilibrium position is known or specified and where one or more of the external forces is an unknown to be determined.

There is a separate class of problems in which bodies are composed of interconnected members which can move relative to each other. Thus various equilibrium configurations are possible and must be examined. For problems of this type, the force- and moment-equilibrium equations, although valid and adequate, are often not the most direct and convenient approach.

A method based on the concept of the work done by a force is more direct. Also, the method provides a deeper insight into the behavior of mechanical systems and enables us to examine the stability of systems in equilibrium. This method is called the *method of virtual work*.

7/2 WORK

We must first define the term *work* in its quantitative sense, in contrast to its common nontechnical usage.

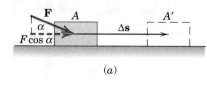

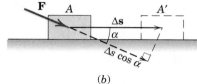

(a)

(b)

Figure 7/1

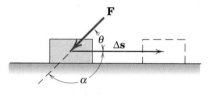

Figure 7/2

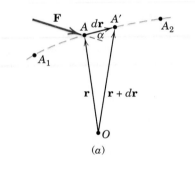

(a)

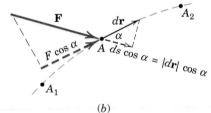

(b)

Figure 7/3

Work of a Force

Consider the constant force **F** acting on the body shown in Fig. 7/1a, whose movement along the plane from A to A' is represented by the vector Δ**s**, called the *displacement* of the body. By definition the work U done by the force **F** on the body during this displacement is the component of the force in the direction of the displacement times the displacement, or

$$U = (F \cos \alpha) \, \Delta s$$

From Fig. 7/1b we see that the same result is obtained if we multiply the magnitude of the force by the component of the displacement in the direction of the force. This gives

$$U = F(\Delta s \cos \alpha)$$

Because we obtain the same result regardless of the direction in which we resolve the vectors, we conclude that work U is a *scalar* quantity.

Work is positive when the working component of the force is in the same direction as the displacement. When the working component is in the direction opposite to the displacement, Fig. 7/2, the work done is negative. Thus,

$$U = (F \cos \alpha) \, \Delta s = -(F \cos \theta) \, \Delta s$$

We now generalize the definition of work to account for conditions under which the direction of the displacement and the magnitude and direction of the force are variable.

Figure 7/3a shows a force **F** acting on a body at a point A which moves along the path shown from A_1 to A_2. Point A is located by its position vector **r** measured from some arbitrary but convenient origin O. The infinitesimal displacement in the motion from A to A' is given by the differential change d**r** of the position vector. The work done by the force **F** during the displacement d**r** is defined as

$$\boxed{dU = \mathbf{F} \cdot d\mathbf{r}} \qquad (7/1)$$

If F denotes the magnitude of the force **F** and ds denotes the magnitude of the differential displacement d**r**, we use the definition of the dot product to obtain

$$dU = F \, ds \cos \alpha$$

We may again interpret this expression as the force component $F \cos \alpha$ in the direction of the displacement times the displacement, or as the displacement component $ds \cos \alpha$ in the direction of the force times the force, as represented in Fig. 7/3*b*. If we express **F** and *d***r** in terms of their rectangular components, we have

$$dU = (\mathbf{i}F_x + \mathbf{j}F_y + \mathbf{k}F_z) \cdot (\mathbf{i}\,dx + \mathbf{j}\,dy + \mathbf{k}\,dz)$$
$$= F_x\,dx + F_y\,dy + F_z\,dz$$

To obtain the total work U done by **F** during a finite movement of point A from A_1 to A_2, Fig. 7/3*a*, we integrate dU between these positions. Thus,

$$U = \int \mathbf{F} \cdot d\mathbf{r} = \int (F_x\,dx + F_y\,dy + F_z\,dz)$$

or

$$U = \int F \cos \alpha \, ds$$

To carry out this integration, we must know the relation between the force components and their respective coordinates, or the relations between F and s and between $\cos \alpha$ and s.

In the case of concurrent forces which are applied at any particular point on a body, the work done by their resultant equals the total work done by the several forces. This is because the component of the resultant in the direction of the displacement equals the sum of the components of the several forces in the same direction.

Work of a Couple

In addition to the work done by forces, couples also can do work. In Fig. 7/4*a* the couple M acts on the body and changes its angular position by an amount $d\theta$. The work done by the couple is easily determined from the combined work of the two forces which constitute the couple. In part *b* of the figure we represent the couple by two equal and opposite forces **F** and $-\mathbf{F}$ acting at two arbitrary points A and B such that $F = M/b$. During the infinitesimal movement in the plane of the figure, line AB moves to $A''B'$. We now take the displacement of A in two steps, first, a displacement $d\mathbf{r}_B$ equal to that of B and, second, a displacement $d\mathbf{r}_{A/B}$ (read as the displacement of A with respect to B) due to the rotation about B. Thus the work done by **F** during the displacement from A to A' is equal and opposite in sign to that due to $-\mathbf{F}$ acting through the equal displacement from B to B'. We therefore conclude that no work is done by a couple during a translation (movement without rotation).

During the rotation, however, **F** does work equal to $\mathbf{F} \cdot d\mathbf{r}_{A/B} = Fb\,d\theta$, where $dr_{A/B} = b\,d\theta$ and where $d\theta$ is the infinitesimal angle of rotation in radians. Since $M = Fb$, we have

$$\boxed{dU = M\,d\theta} \qquad (7/2)$$

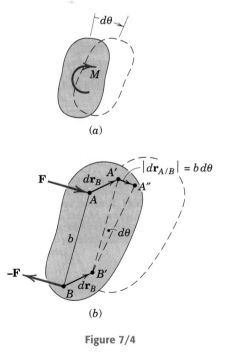

(a)

(b)

Figure 7/4

The work of the couple is positive if M has the same sense as $d\theta$ (clockwise in this illustration), and negative if M has a sense opposite to that of the rotation. The total work of a couple during a finite rotation in its plane becomes

$$U = \int M \, d\theta$$

Dimensions of Work

Work has the dimensions of (force) × (distance). In SI units the unit of work is the joule (J), which is the work done by a force of one newton moving through a distance of one meter in the direction of the force (J = N·m). In the U.S. customary system the unit of work is the foot-pound (ft-lb), which is the work done by a one-pound force moving through a distance of one foot in the direction of the force.

The dimensions of the work of a force and the moment of a force are the same although they are entirely different physical quantities. Note that work is a *scalar* given by the dot product and thus involves the product of a force and a distance, both measured along the same line. Moment, on the other hand, is a *vector* given by the cross product and involves the product of force and distance measured at right angles to the force. To distinguish between these two quantities when we write their units, in SI units we use the joule (J) for work and reserve the combined units newton-meter (N·m) for moment. In the U.S. customary system we normally use the sequence foot-pound (ft-lb) for work and pound-foot (lb-ft) for moment.

Virtual Work

We consider now a particle whose static equilibrium position is determined by the forces which act on it. Any assumed and arbitrary small displacement $\delta\mathbf{r}$ away from this natural position and consistent with the system constraints is called a *virtual displacement*. The term *virtual* is used to indicate that the displacement does not really exist but only is assumed to exist so that we may compare various possible equilibrium positions to determine the correct one.

The work done by any force $\mathbf{F}$ acting on the particle during the virtual displacement $\delta\mathbf{r}$ is called *virtual work* and is

$$\delta U = \mathbf{F} \cdot \delta\mathbf{r} \qquad \text{or} \qquad \delta U = F \, \delta s \cos \alpha$$

where α is the angle between $\mathbf{F}$ and $\delta\mathbf{r}$, and δs is the magnitude of $\delta\mathbf{r}$. The difference between $d\mathbf{r}$ and $\delta\mathbf{r}$ is that $d\mathbf{r}$ refers to an actual infinitesimal change in position and can be integrated, whereas $\delta\mathbf{r}$ refers to an infinitesimal virtual or assumed movement and cannot be integrated. Mathematically both quantities are first-order differentials.

A virtual displacement may also be a rotation $\delta\theta$ of a body. According to Eq. 7/2 the virtual work done by a couple M during a virtual angular displacement $\delta\theta$ is $\delta U = M \, \delta\theta$.

We may regard the force $\mathbf{F}$ or couple M as remaining constant during any infinitesimal virtual displacement. If we account for any change in **F** or *M* during the infinitesimal motion, higher-order terms will result which disappear in the limit. This consideration is the same mathematically as that which permits us to neglect the product $dx\,dy$ when writing $dA = y\,dx$ for the element of area under the curve $y = f(x)$.

7/3   EQUILIBRIUM

We now express the equilibrium conditions in terms of virtual work, first for a particle, then for a single rigid body, and then for a system of connected rigid bodies.

Equilibrium of a Particle

Consider the particle or small body in Fig. 7/5 which attains an equilibrium position as a result of the forces in the attached springs. If the mass of the particle were significant, then the weight mg would also be included as one of the forces. For an assumed virtual displacement $\delta\mathbf{r}$ of the particle away from its equilibrium position, the total virtual work done on the particle is

$$\delta U = \mathbf{F}_1 \cdot \delta\mathbf{r} + \mathbf{F}_2 \cdot \delta\mathbf{r} + \mathbf{F}_3 \cdot \delta\mathbf{r} + \cdots = \Sigma\mathbf{F} \cdot \delta\mathbf{r}$$

We now express $\Sigma\mathbf{F}$ in terms of its scalar sums and $\delta\mathbf{r}$ in terms of its component virtual displacements in the coordinate directions, as follows:

$$\delta U = \Sigma\mathbf{F} \cdot \delta\mathbf{r} = (\mathbf{i}\,\Sigma F_x + \mathbf{j}\,\Sigma F_y + \mathbf{k}\,\Sigma F_z) \cdot (\mathbf{i}\,\delta x + \mathbf{j}\,\delta y + \mathbf{k}\,\delta z)$$

$$= \Sigma F_x\,\delta x + \Sigma F_y\,\delta y + \Sigma F_z\,\delta z = 0$$

The sum is zero, since $\Sigma\mathbf{F} = \mathbf{0}$, which gives $\Sigma F_x = 0$, $\Sigma F_y = 0$, and $\Sigma F_z = 0$. The equation $\delta U = 0$ is therefore an alternative statement of the equilibrium conditions for a particle. This condition of zero virtual work for equilibrium is both necessary and sufficient, since we may apply it to virtual displacements taken one at a time in each of the three mutually perpendicular directions, in which case it becomes equivalent to the three known scalar requirements for equilibrium.

The principle of zero virtual work for the equilibrium of a single particle usually does not simplify this already simple problem because $\delta U = 0$ and $\Sigma\mathbf{F} = \mathbf{0}$ provide the same information. However, we introduce the concept of virtual work for a particle so that we can later apply it to systems of particles.

Equilibrium of a Rigid Body

We can easily extend the principle of virtual work for a single particle to a rigid body treated as a system of small elements or particles rigidly attached to one another. Because the virtual work done on each particle of the body in equilibrium is zero, it follows that the virtual work done on the entire rigid body is zero. Only the virtual work done by *external* forces appears in the evaluation of $\delta U = 0$ for the entire

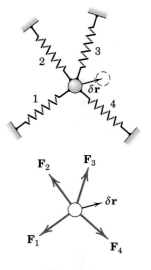

Figure 7/5

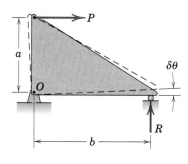

Figure 7/6

body, since all internal forces occur in pairs of equal, opposite, and collinear forces, and the net work done by these forces during any movement is zero.

As in the case of a particle, we again find that the principle of virtual work offers no particular advantage to the solution for a single rigid body in equilibrium. Any assumed virtual displacement defined by a linear or angular movement will appear in each term in $\delta U = 0$ and when canceled will leave us with the same expression we would have obtained by using one of the force or moment equations of equilibrium directly.

This condition is illustrated in Fig. 7/6, where we want to determine the reaction R under the roller for the hinged plate of negligible weight under the action of a given force P. A small assumed rotation $\delta\theta$ of the plate about O is consistent with the hinge constraint at O and is taken as the virtual displacement. The work done by P is $-Pa\ \delta\theta$, and the work done by R is $+Rb\ \delta\theta$. Therefore, the principle $\delta U = 0$ gives

$$-Pa\ \delta\theta + Rb\ \delta\theta = 0$$

Canceling $\delta\theta$ leaves

$$Pa - Rb = 0$$

which is simply the equation of moment equilibrium about O. Therefore, nothing is gained by using the virtual-work principle for a single rigid body. The principle is, however, decidedly advantageous for interconnected bodies, as discussed next.

Equilibrium of Ideal Systems of Rigid Bodies

We now extend the principle of virtual work to the equilibrium of an interconnected system of rigid bodies. Our treatment here will be limited to so-called *ideal systems*. These are systems composed of two or more rigid members linked together by mechanical connections which are incapable of absorbing energy through elongation or compression, and in which friction is small enough to be neglected.

Figure 7/7a shows a simple example of an ideal system where relative motion between its two parts is possible and where the equilibrium position is determined by the applied external forces **P** and **F**. We can identify three types of forces which act in such an interconnected system. They are as follows:

(1) *Active forces* are external forces capable of doing virtual work during possible virtual displacements. In Fig. 7/7a forces **P** and **F** are active forces because they would do work as the links move.

(2) *Reactive forces* are forces which act at fixed support positions where no virtual displacement takes place in the direction of the force. Reactive forces do no work during a virtual displacement. In Fig. 7/7b the horizontal force $\mathbf{F}_B$ exerted on the roller end of the member by the vertical guide can do no work because there can be no horizontal displacement of the roller. The reactive force $\mathbf{F}_O$ exerted on the system by the fixed support at O also does no work because O cannot move.

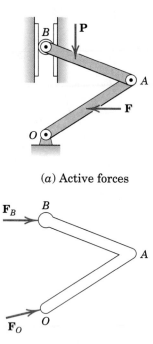

(a) Active forces

(b) Reactive forces

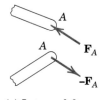

(c) Internal forces

Figure 7/7

(3) *Internal forces* are forces in the connections between members. During any possible movement of the system or its parts, the *net work done by the internal forces at the connections is zero*. This is so because the internal forces always exist in pairs of equal and opposite forces, as indicated for the internal forces $\mathbf{F}_A$ and $-\mathbf{F}_A$ at joint A in Fig. 7/7c. The work of one force therefore necessarily cancels the work of the other force during their identical displacements.

Principle of Virtual Work

Noting that only the external active forces do work during any possible movement of the system, we may now state the principle of virtual work as follows:

> The virtual work done by external active forces on an ideal mechanical system in equilibrium is zero for any and all virtual displacements consistent with the constraints.

By constraint we mean restriction of the motion by the supports. We state the principle mathematically by the equation

$$\delta U = 0 \tag{7/3}$$

where δU stands for the total virtual work done on the system by all active forces during a virtual displacement.

Only now can we see the real advantages of the method of virtual work. There are essentially two. First, it is not necessary for us to dismember ideal systems in order to establish the relations between the active forces, as is generally the case with the equilibrium method based on force and moment summations. Second, we may determine the relations between the active forces directly without reference to the reactive forces. These advantages make the method of virtual work particularly useful in determining the position of equilibrium of a system under known loads. This type of problem is in contrast to the problem of determining the forces acting on a body whose equilibrium position is known.

The method of virtual work is especially useful for the purposes mentioned but requires that the internal friction forces do negligible work during any virtual displacement. Consequently, if internal friction in a mechanical system is appreciable, the method of virtual work cannot be used for the system as a whole unless the work done by internal friction is included.

When using the method of virtual work, you should draw a diagram which isolates the system under consideration. Unlike the free-body diagram, where all forces are shown, the diagram for the method of virtual work need show only the *active forces*, since the reactive forces do not enter into the application of $\delta U = 0$. Such a drawing will be termed an *active-force diagram*. Figure 7/7a is an active-force diagram for the system shown.

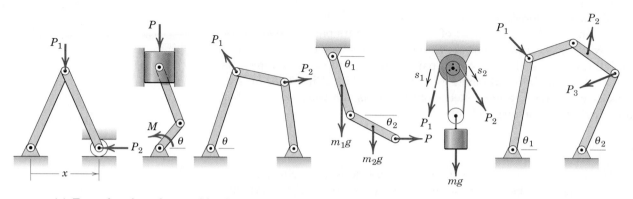

(a) Examples of one-degree-of-freedom systems

(b) Examples of two-degree-of-freedom systems

Figure 7/8

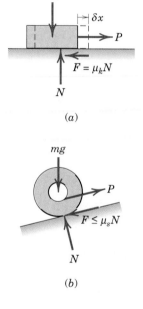

(a)

(b)

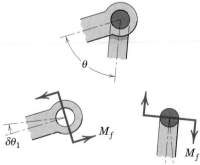

(c)

Figure 7/9

Degrees of Freedom

The number of *degrees of freedom* of a mechanical system is the number of independent coordinates needed to specify completely the configuration of the system. Figure 7/8a shows three examples of one-degree-of-freedom systems. Only one coordinate is needed to establish the position of every part of the system. The coordinate can be a distance or an angle. Figure 7/8b shows three examples of two-degree-of-freedom systems where two independent coordinates are needed to determine the configuration of the system. By the addition of more links to the mechanism in the right-hand figure, there is no limit to the number of degrees of freedom which can be introduced.

The principal of virtual work $\delta U = 0$ may be applied as many times as there are degrees of freedom. With each application, we allow only one independent coordinate to change at a time while holding the others constant. In our treatment of virtual work in this chapter, we consider only one-degree-of-freedom systems.*

Systems with Friction

When sliding friction is present to any appreciable degree in a mechanical system, the system is said to be "real." In real systems some of the positive work done on the system by external active forces (input work) is dissipated in the form of heat generated by the kinetic friction forces during movement of the system. When there is sliding between contacting surfaces, the friction force does negative work because its direction is always opposite to the movement of the body on which it acts. This negative work cannot be regained.

Thus, the kinetic friction force $\mu_k N$ acting on the sliding block in Fig. 7/9a does work on the block during the displacement x in the amount of $-\mu_k Nx$. During a virtual displacement δx, the friction force does work equal to $-\mu_k N\,\delta x$. The static friction force acting on the

---

*For examples of solutions to problems of two or more degrees of freedom, see Chapter 7 of the first author's *Statics, 2nd Edition*, 1971, or *SI Version*, 1975.

rolling wheel in Fig. 7/9b, on the other hand, does no work if the wheel does not slip as it rolls.

In Fig. 7/9c the moment M_f about the center of the pinned joint due to the friction forces which act at the contacting surfaces does negative work during any relative angular movement between the two parts. Thus, for a virtual displacement $\delta\theta$ between the two parts, which have the separate virtual displacements $\delta\theta_1$ and $\delta\theta_2$ as shown, the negative work done is $-M_f\,\delta\theta_1 - M_f\,\delta\theta_2 = -M_f(\delta\theta_1 + \delta\theta_2)$, or simply $-M_f\,\delta\theta$. For each part, M_f is in the sense to oppose the relative motion of rotation.

It was noted earlier in the article that a major advantage of the method of virtual work is in the analysis of an entire system of connected members without taking them apart. If there is appreciable kinetic friction internal to the system, it becomes necessary to dismember the system to determine the friction forces. In such cases the method of virtual work finds only limited use.

Mechanical Efficiency

Because of energy loss due to friction, the output work of a machine is always less than the input work. The ratio of the two amounts of work is the *mechanical efficiency e*. Thus,

$$e = \frac{\text{output work}}{\text{input work}}$$

The mechanical efficiency of simple machines which have a single degree of freedom and which operate in a uniform manner may be determined by the method of work by evaluating the numerator and denominator of the expression for e during a virtual displacement.

As an example, consider the block being moved up the inclined plane in Fig. 7/10. For the virtual displacement δs shown, the output work is that necessary to elevate the block, or $mg\,\delta s\,\sin\theta$. The input work is $T\,\delta s = (mg\sin\theta + \mu_k mg\cos\theta)\,\delta s$. The efficiency of the inclined plane is, therefore,

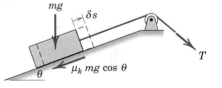

Figure 7/10

$$e = \frac{mg\,\delta s\,\sin\theta}{mg(\sin\theta + \mu_k\cos\theta)\,\delta s} = \frac{1}{1 + \mu_k\cot\theta}$$

As a second example, consider the screw jack described in Art. 6/5 and shown in Fig. 6/6. Equation 6/3 gives the moment M required to raise the load W, where the screw has a mean radius r and a helix angle α, and where the friction angle is $\phi = \tan^{-1}\mu_k$. During a small rotation $\delta\theta$ of the screw, the input work is $M\,\delta\theta = Wr\,\delta\theta\,\tan(\alpha + \phi)$. The output work is that required to elevate the load, or $Wr\,\delta\theta\,\tan\alpha$. Thus the efficiency of the jack can be expressed as

$$e = \frac{Wr\,\delta\theta\,\tan\alpha}{Wr\,\delta\theta\,\tan(\alpha + \phi)} = \frac{\tan\alpha}{\tan(\alpha + \phi)}$$

As friction is decreased, ϕ becomes smaller, and the efficiency approaches unity.

Sample Problem 7/1

Each of the two uniform hinged bars has a mass m and a length l, and is supported and loaded as shown. For a given force P determine the angle θ for equilibrium.

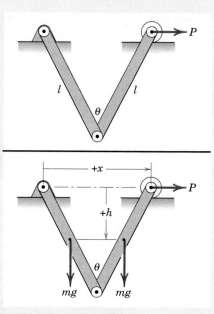

Solution. The active-force diagram for the system composed of the two members is shown separately and includes the weight mg of each bar in addition to the force P. All other forces acting externally on the system are reactive forces which do no work during a virtual movement δx and are therefore not shown.

The principle of virtual work requires that the total work of all external active forces be zero for any virtual displacement consistent with the constraints. Thus, for a movement δx the virtual work becomes

① $[\delta U = 0]$ $\qquad\qquad P\,\delta x + 2mg\,\delta h = 0$

We now express each of these virtual displacements in terms of the variable θ, the required quantity. Hence,

$$x = 2l \sin \frac{\theta}{2} \qquad \text{and} \qquad \delta x = l \cos \frac{\theta}{2} \delta\theta$$

② Similarly,

$$h = \frac{l}{2} \cos \frac{\theta}{2} \qquad \text{and} \qquad \delta h = -\frac{l}{4} \sin \frac{\theta}{2} \delta\theta$$

Substitution into the equation of virtual work gives us

$$Pl \cos \frac{\theta}{2} \delta\theta - 2mg \frac{l}{4} \sin \frac{\theta}{2} \delta\theta = 0$$

from which we get

$$\tan \frac{\theta}{2} = \frac{2P}{mg} \qquad \text{or} \qquad \theta = 2 \tan^{-1} \frac{2P}{mg} \qquad\qquad Ans.$$

To obtain this result by the principles of force and moment summation, it would be necessary to dismember the frame and take into account all forces acting on each member. Solution by the method of virtual work involves a simpler operation.

Helpful Hints

① Note carefully that with x positive to the right δx is also positive to the right in the direction of P, so that the virtual work is $P(+\delta x)$. With h positive down δh is also mathematically positive down in the direction of mg, so that the correct mathematical expression for the work is $mg(+\delta h)$. When we express δh in terms of $\delta\theta$ in the next step, δh will have a negative sign, thus bringing our mathematical expression into agreement with the physical observation that the weight mg does negative work as each center of mass moves upward with an increase in x and θ.

② We obtain δh and δx with the same mathematical rules of differentiation with which we may obtain dh and dx.

Sample Problem 7/2

The mass m is brought to an equilibrium position by the application of the couple M to the end of one of the two parallel links which are hinged as shown. The links have negligible mass, and all friction is assumed to be absent. Determine the expression for the equilibrium angle θ assumed by the links with the vertical for a given value of M. Consider the alternative of a solution by force and moment equilibrium.

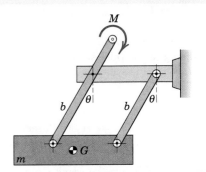

Solution. The active-force diagram shows the weight mg acting through the center of mass G and the couple M applied to the end of the link. There are no other external active forces or moments which do work on the system during a change in the angle θ.

The vertical position of the center of mass G is designated by the distance h below the fixed horizontal reference line and is $h = b \cos \theta + c$. The work done by mg during a movement δh in the direction of mg is

$$+mg\, \delta h = mg\, \delta(b \cos \theta + c)$$

$$= mg(-b \sin \theta\, \delta\theta + 0)$$

$$= -mgb \sin \theta\, \delta\theta$$

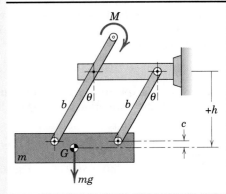

① The minus sign shows that the work is negative for a positive value of $\delta\theta$. The constant c drops out since its variation is zero.

With θ measured positive in the clockwise sense, $\delta\theta$ is also positive clockwise. Thus, the work done by the clockwise couple M is $+M\, \delta\theta$. Substitution into the virtual-work equation gives us

$[\delta U = 0]$ $M\, \delta\theta + mg\, \delta h = 0$

which yields

$$M\, \delta\theta = mgb \sin \theta\, \delta\theta$$

$$\theta = \sin^{-1} \frac{M}{mgb} \qquad\qquad \textit{Ans.}$$

Inasmuch as $\sin \theta$ cannot exceed unity, we see that for equilibrium, M is limited to values that do not exceed mgb.

The advantage of the virtual-work solution for this problem is readily seen when we observe what would be involved with a solution by force and moment equilibrium. For the latter approach, it would be necessary for us to draw separate free-body diagrams of all of the three moving parts and account for all of the internal reactions at the pin connections. To carry out these steps, it would be necessary for us to include in the analysis the horizontal position of G with respect to the attachment points of the two links, even though reference to this position would finally drop out of the equations when they were solved. We conclude, then, that the virtual-work method in this problem deals directly with cause and effect and avoids reference to irrelevant quantities.

Helpful Hint

① Again, as in Sample Problem 7/1, we are consistent mathematically with our definition of work, and we see that the algebraic sign of the resulting expression agrees with the physical change.

Sample Problem 7/3

For link *OA* in the horizontal position shown, determine the force *P* on the sliding collar which will prevent *OA* from rotating under the action of the couple *M*. Neglect the mass of the moving parts.

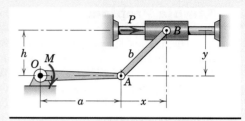

Solution. The given sketch serves as the active-force diagram for the system. All other forces are either internal or nonworking reactive forces due to the constraints.

We will give the crank *OA* a small clockwise angular movement $\delta\theta$ as our virtual displacement and determine the resulting virtual work done by *M* and *P*. From the horizontal position of the crank, the angular movement gives a downward displacement of *A* equal to

① $$\delta y = a\,\delta\theta$$

where $\delta\theta$ is, of course, expressed in radians.

From the right triangle for which link *AB* is the constant hypotenuse we may write

$$b^2 = x^2 + y^2$$

We now take the differential of the equation and get

② $$0 = 2x\,\delta x + 2y\,\delta y \qquad \text{or} \qquad \delta x = -\frac{y}{x}\,\delta y$$

Thus,

$$\delta x = -\frac{y}{x}\,a\,\delta\theta$$

and the virtual-work equation becomes

③ $[\delta U = 0]$ $\qquad M\,\delta\theta + P\,\delta x = 0 \qquad M\,\delta\theta + P\left(-\frac{y}{x}\,a\,\delta\theta\right) = 0$

$$P = \frac{Mx}{ya} = \frac{Mx}{ha} \qquad\qquad \textit{Ans.}$$

Again, we observe that the virtual-work method produces a direct relationship between the active force *P* and the couple *M* without involving other forces which are irrelevant to this relationship. Solution by the force and moment equations of equilibrium, although fairly simple in this problem, would require accounting for all forces initially and then eliminating the irrelevant ones.

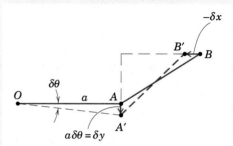

Helpful Hints

① Note that the displacement $a\,\delta\theta$ of point *A* would no longer equal δy if the crank *OA* were not in a horizontal position.

② The length *b* is constant so that $\delta b = 0$. Notice the negative sign, which merely tells us that if one change is positive, the other must be negative.

③ We could just as well use a counterclockwise virtual displacement for the crank, which would merely reverse the signs of all terms.

PROBLEMS

(Assume that the negative work of friction is negligible in the following problems unless otherwise indicated.)

Introductory Problems

7/1 Determine the moment M applied to the lower link through its shaft which is necessary to support the load P in terms of the angle θ. Neglect the weights of the parts.

Ans. $M = 2Pr \sin \theta$

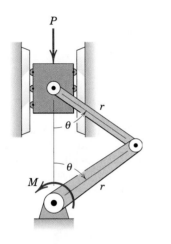

Problem 7/1

7/2 The arbor press works by a rack and pinion and is used to develop large forces, such as those required to produce force fits. If the mean radius of the pinion (gear) is r, determine the force R which can be developed by the press for a given force P on the handle.

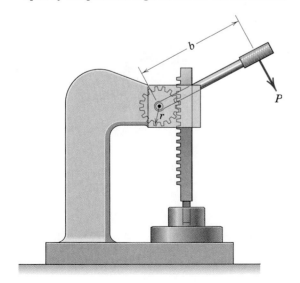

Problem 7/2

7/3 For a given force P determine the angle θ for equilibrium. Neglect the mass of the links.

Ans. $\theta = \cos^{-1} \dfrac{2P}{mg}$

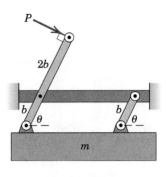

Problem 7/3

7/4 The foot-operated lift is used to raise a platform of mass m. Determine the necessary force P applied at the 10° angle to support the 80-kg load.

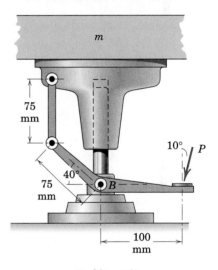

Problem 7/4

7/5 Each of the four uniform links has a mass m. Determine the horizontal force P required to hold them in place in the vertical plane as shown.

$$Ans.\ P = \tfrac{5}{2}mg \tan\theta$$

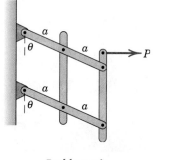

Problem 7/5

7/6 Each of the uniform links of the frame has a mass m and a length b. The equilibrium position of the frame in the vertical plane is determined by the couple M applied to the left-hand link. Express θ in terms of M.

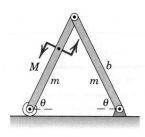

Problem 7/6

7/7 Replace the couple M of Prob. 7/6 by the horizontal force P as shown and determine the equilibrium angle θ.

$$Ans.\ \theta = \tan^{-1}\frac{2mg}{3P}$$

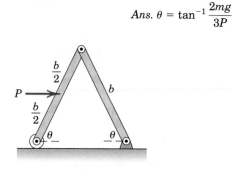

Problem 7/7

7/8 The spring of constant k is unstretched when $\theta = 0$. Derive an expression for the force P required to deflect the system to an angle θ. The mass of the bars is negligible.

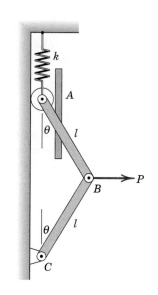

Problem 7/8

7/9 By means of a rack-and-pinion mechanism, large forces can be developed by the cork puller shown. If the mean radius of the pinion gears is 12 mm, determine the force R which is exerted on the cork for given forces P on the handles.

$$Ans.\ R = 11.67P$$

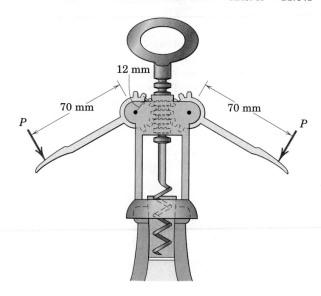

Problem 7/9

7/10 The hydraulic cylinder is used to spread the linkage and elevate the load m. For the position shown determine the compression C in the cylinder. Neglect the mass of all parts other than m.

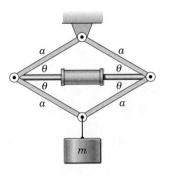

Problem 7/10

7/11 The upper jaw D of the toggle press slides with negligible frictional resistance along the fixed vertical column. Determine the required force F on the handle to produce a compression R on the roller for any given value of θ.

Ans. $F = 0.8R \cos \theta$

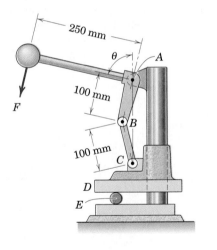

Problem 7/11

7/12 For each unit of movement of the free end of the rope in the direction of the applied force P, the 250-lb load moves one-fourth of a unit. If the mechanical efficiency e of the hoist is 0.75, calculate the force P required to raise the load and the force P' required to lower the load.

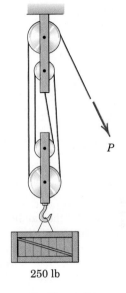

250 lb

Problem 7/12

7/13 The symmetrical linkage supports the cylinder of mass m in the vertical plane by the action of the two couples M applied to OA and OB as shown. Determine the couple value M in terms of θ. The mass of the links is negligible compared with m.

Ans. $M = mgb \cos \dfrac{\theta}{2}$

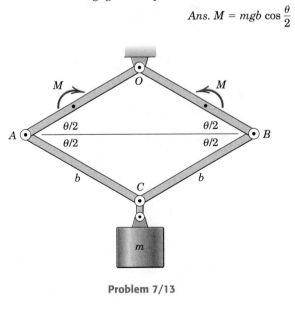

Problem 7/13

7/14 The portable car hoist is operated by the hydraulic cylinder which controls the horizontal movement of end A of the link in the horizontal slot. Determine the compression C in the piston rod of the cylinder to support the load P at a height h.

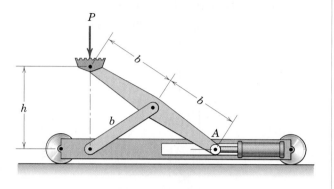

Problem 7/14

7/15 The hand-operated hoist is designed to lift a 100-kg load where 25 turns of the handle on the worm shaft produce one revolution of the drum. Assuming a 40-percent loss of energy due to friction in the mechanism, calculate the force F normal to the handle arm required to lift the load.

Ans. F = 61.3 N

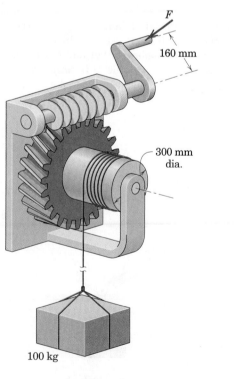

Problem 7/15

Representative Problems

7/16 The speed reducer shown is designed with a gear ratio of 40:1. With an input torque $M_1 = 30$ N·m, the measured output torque is $M_2 = 1180$ N·m. Determine the mechanical efficiency e of the unit.

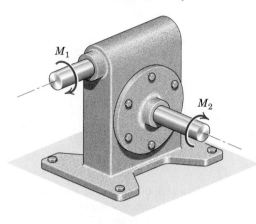

Problem 7/16

7/17 The folding linkage is composed of n identical sections, each of which consists of two identical bars of mass m each. Determine the horizontal force P necessary to maintain equilibrium in an arbitrary position characterized by the angle θ. Does P depend on the number n of sections present?

Ans. $P = mg \tan \dfrac{\theta}{2}$, no

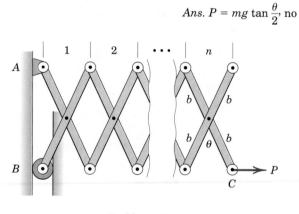

Problem 7/17

7/18 Replace the force P acting on the linkage of Prob. 7/17 by a couple M. Determine the moment M of the couple necessary to maintain equilibrium in an arbitrary position characterized by the angle θ. Does M depend on the number n of sections present?

7/19 Determine the couple M which must be applied at O in order to support the mechanism in the position $\theta = 30°$. The masses of the disk at C, bar OA, and bar BC are m_0, m, and $2m$, respectively.

$$Ans.\ M = (\tfrac{5}{4}m + m_0)gl\sqrt{3}$$

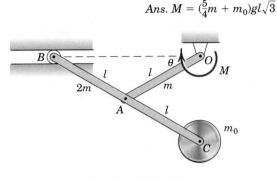

Problem 7/19

7/20 In testing the design of the screw-lift jack shown, 12 turns of the handle are required to elevate the lifting pad 1 in. If a force $F = 10$ lb applied normal to the crank is required to elevate a load $L = 2700$ lb, determine the efficiency e of the screw in raising the load.

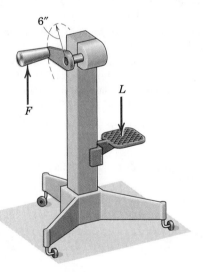

Problem 7/20

7/21 The torque M applied to the light link OA through its shaft at O rotates OA through an angle θ and raises end A of the uniform bar AB of mass m. End B is supported by a small roller on the horizontal surface. When $\theta = 0$, the bar AB is horizontal. Determine the equilibrium angle for a given value of M. What would happen if M were greater than $mgr/2$?

$$Ans.\ \theta = \sin^{-1}\left(\frac{2M}{mgr}\right)$$

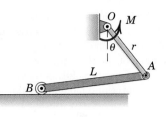

Problem 7/21

7/22 Calculate the efficiency with which the 5° wedge elevates the weight W under the action of the horizontal force P on the wedge. The coefficient of friction between the wedge and block is 0.30.

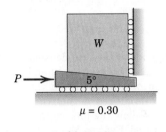

$\mu = 0.30$

Problem 7/22

7/23 Determine the force F which the person must apply tangent to the rim of the handwheel of a wheelchair in order to roll up the incline of angle θ. The combined mass of the chair and person is m. (If s is the displacement of the center of the wheel measured along the incline and β the corresponding angle in radians through which the wheel turns, it is easily shown that $s = R\beta$ if the wheel rolls without slipping.)

$$Ans.\ F = mg\frac{R}{r}\sin\theta$$

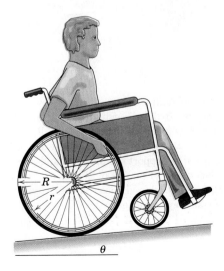

Problem 7/23

7/24 Specify the horizontal force F necessary to maintain equilibrium of the 80-kg platform in terms of the angle θ made by the supporting links with the horizontal. Each of the three uniform links has a mass of 10 kg. (Compare the solution by virtual work with a solution by force and moment equilibrium.)

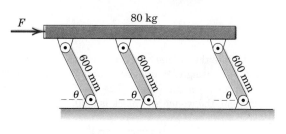

Problem 7/24

7/25 The cargo box of the food-delivery truck for aircraft servicing has a loaded mass m and is elevated by the application of a torque M on the lower end of the link which is hinged to the truck frame. The horizontal slots allow the linkage to unfold as the cargo box is elevated. Express M as a function of h.

$$\text{Ans. } M = 2mgb\sqrt{1 - \left(\frac{h}{2b}\right)^2}$$

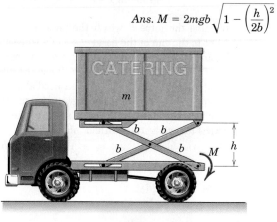

Problem 7/25

7/26 The portable work platform is elevated by means of the two hydraulic cylinders articulated at points C. Each cylinder is under a hydraulic pressure p and has a piston area A. Determine the pressure p required to support the platform and show that it is independent of θ. The platform, worker, and supplies have a combined mass m, and the masses of the links may be neglected.

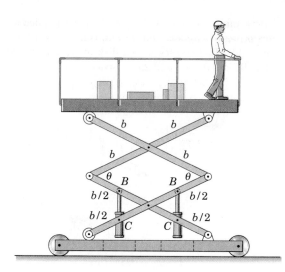

Problem 7/26

7/27 The existing design of an aircraft cargo loader is under review. The platform AD of the loader is elevated to the proper height by the mechanism shown. There are two sets of linkages and hydraulic lifts, one set on each side. Cables F which lift the platform are controlled by the hydraulic cylinders E whose piston rods elevate the pulleys G. If the total weight of the platform and containers is W, determine the compressive force P in each of the two piston rods. Does P depend on the height h? What force Q is supported by each link at its center joint when W is centered between A and D?

$$\text{Ans. } P = W, \text{ no; } Q = \frac{W}{2}\left(1 + \frac{2e}{\sqrt{l^2 - h^2}}\right)$$

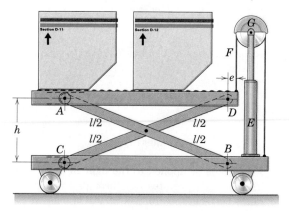

Problem 7/27

7/28 The postal scale consists of a sector of mass m_0 hinged at O and with center of mass at G. The pan and vertical link AB have a mass m_1 and are hinged to the sector at B. End A is hinged to the uniform link AC of mass m_2, which in turn is hinged to the fixed frame. The figure $OBAC$ forms a parallelogram, and the angle GOB is a right angle. Determine the relation between the mass m to be measured and the angle θ, assuming that $\theta = \theta_0$ when $m = 0$.

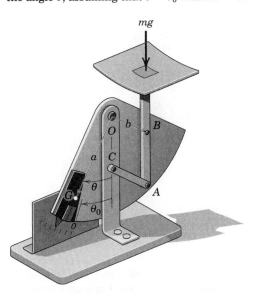

Problem 7/28

7/29 The elevation of the load of mass m is controlled by the adjusting screw which connects joints A and B. The change in the distance between A and B for one revolution of the screw equals the lead L of the screw (advancement per revolution). If a moment M_f is required to overcome friction in the threads and thrust bearing of the screw, determine the expression for the total moment M, applied to the adjusting screw, necessary to raise the load.

$$\text{Ans. } M = M_f + \frac{mgL}{\pi} \cot \theta$$

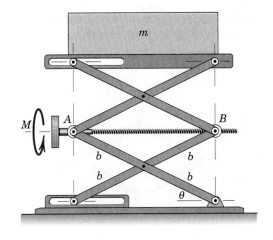

Problem 7/29

7/30 Express the compression C in the hydraulic cylinder of the car hoist in terms of the angle θ. The mass of the hoist is negligible compared with the mass m of the vehicle.

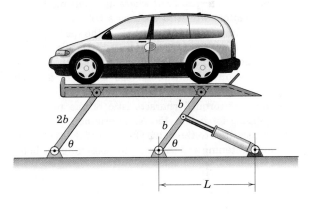

Problem 7/30

7/31 Determine the force N exerted on the log by each jaw of the fireplace tongs shown.

Ans. $N = 1.6P$

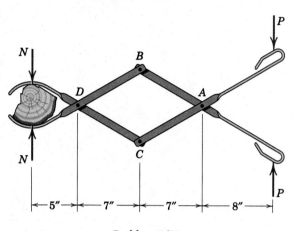

Problem 7/31

7/32 The antitorque wrench is designed for use by a crew member of a spacecraft where no stable platform exists against which to push as a bolt is turned. The pin A fits into an adjacent hole in the space structure which contains the bolt to be turned. Successive oscillations of the gear and handle unit turn the socket in one direction through the action of a ratchet mechanism. The reaction against the pin A provides the "antitorque" characteristic of the tool. For a gripping force $P = 150$ N, determine the torque M transmitted to the bolt. (One side of the tool is used for tightening and the other for loosening the bolt.)

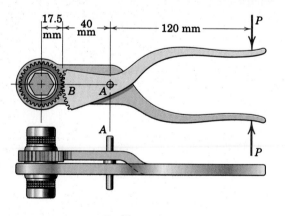

Problem 7/32

7/33 A power-operated loading platform designed for the back of a truck is shown in the figure. The position of the platform is controlled by the hydraulic cylinder, which applies force at C. The links are pivoted to the truck frame at A, B, and F. Determine the force P supplied by the cylinder in order to support the platform in the position shown. The mass of the platform and links may be neglected compared with that of the 250-kg crate with center of mass at G.

Ans. $P = 3.5$ kN

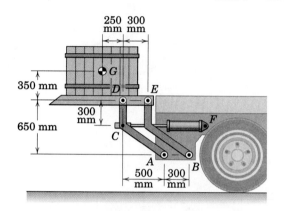

Problem 7/33

▶**7/34** In the screw-activated clamp, a torque M applied to the handle of the screw tightens the clamp by increasing the distance $\overline{BD}$, thereby producing a clamping force C. The screw is threaded through the pivoted collar at D with a single square thread having a lead (advancement per revolution) of $\frac{1}{6}$ in. Assume no friction and express the clamping force C in pounds in terms of the torque M measured in lb-in. for the position $\theta = 30°$.

Ans. $C = 26.9M$

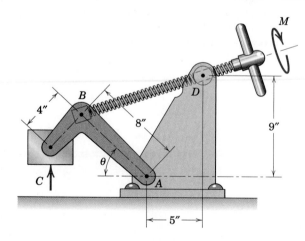

Problem 7/34

7/4   POTENTIAL ENERGY AND STABILITY

The previous article treated the equilibrium configuration of mechanical systems composed of individual members which we assumed to be perfectly rigid. We now extend our method to account for mechanical systems which include elastic elements in the form of springs. We introduce the concept of potential energy, which is useful for determining the stability of equilibrium.

Elastic Potential Energy

The work done on an elastic member is stored in the member in the form of *elastic potential energy* V_e. This energy is potentially available to do work on some other body during the relief of its compression or extension.

Consider a spring, Fig. 7/11, which is being compressed by a force F. We assume that the spring is elastic and linear, which means that the force F is directly proportional to the deflection x. We write this relation as $F = kx$, where k is the *spring constant* or *stiffness* of the spring. The work done on the spring by F during a movement dx is $dU = F\,dx$, so that the elastic potential energy of the spring for a compression x is the total work done on the spring

$$V_e = \int_0^x F\,dx = \int_0^x kx\,dx$$

or
$$\boxed{V_e = \tfrac{1}{2}kx^2} \qquad (7/4)$$

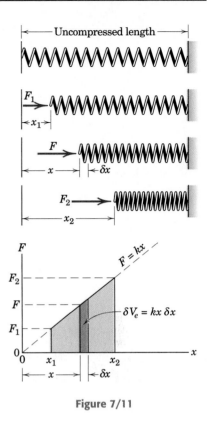

Figure 7/11

Thus, the potential energy of the spring equals the triangular area in the diagram of F versus x from 0 to x.

During an increase in the compression of the spring from x_1 to x_2, the work done on the spring equals its *change* in elastic potential energy or

$$\Delta V_e = \int_{x_1}^{x_2} kx\,dx = \tfrac{1}{2}k(x_2{}^2 - x_1{}^2)$$

which equals the trapezoidal area from x_1 to x_2.

During a virtual displacement δx of the spring, the virtual work done on the spring is the virtual change in elastic potential energy

$$\delta V_e = F\,\delta x = kx\,\delta x$$

During a decrease in the compression of the spring as it is relaxed from $x = x_2$ to $x = x_1$, the *change* (final minus initial) in the potential energy of the spring is negative. Consequently, if δx is negative, δV_e is also negative.

When we have a spring in tension rather than compression, the work and energy relations are the same as those for compression, where x now represents the elongation of the spring rather than its compression. While the spring is being stretched, the force again acts in the direction of the displacement, doing positive work on the spring and increasing its potential energy.

Because the force acting on the movable end of a spring is the negative of the force exerted by the spring on the body to which its movable end is attached, the *work done on the body is the negative of the potential energy change of the spring*.

A *torsional spring*, which resists the rotation of a shaft or another element, can also store and release potential energy. If the *torsional stiffness*, expressed as torque per radian of twist, is a constant K, and if θ is the angle of twist in radians, then the resisting torque is $M = K\theta$. The potential energy becomes $V_e = \int_0^\theta K\theta \, d\theta$ or

$$V_e = \tfrac{1}{2}K\theta^2 \qquad \textbf{(7/4a)}$$

which is analogous to the expression for the linear extension spring.

The units of elastic potential energy are the same as those of work and are expressed in joules (J) in SI units and in foot-pounds (ft-lb) in U.S. customary units.

Gravitational Potential Energy

In the previous article we treated the work of a gravitational force or weight acting on a body in the same way as the work of any other active force. Thus, for an upward displacement δh of the body in Fig. 7/12 the weight $W = mg$ does negative work $\delta U = -mg \, \delta h$. If, on the other hand, the body has a downward displacement δh, with h measured positive downward, the weight does positive work $\delta U = +mg \, \delta h$.

An alternative to the foregoing treatment expresses the work done by gravity in terms of a change in potential energy of the body. This alternative treatment is a useful representation when we describe a mechanical system in terms of its total energy. The *gravitational potential energy V_g* of a body is defined as the work done on the body by a force equal and opposite to the weight in bringing the body to the position under consideration from some arbitrary datum plane where the potential energy is defined to be zero. The potential energy, then, is the negative of the work done by the weight. When the body is raised, for example, the work done is converted into energy which is potentially available, since the body can do work on some other body as it returns to its original lower position. If we take V_g to be zero at $h = 0$, Fig. 7/12, then at a height h above the datum plane, the gravitational potential energy of the body is

$$V_g = mgh \qquad \textbf{(7/5)}$$

If the body is a distance h below the datum plane, its gravitational potential energy is $-mgh$.

Note that the datum plane for zero potential energy is arbitrary because only the *change* in potential energy matters, and this change is the same no matter where we place the datum plane. Note also that the gravitational potential energy is independent of the path followed in arriving at a particular level h. Thus, the body of mass m in Fig. 7/13 has

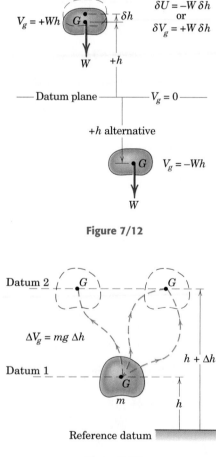

Figure 7/12

Figure 7/13

the same potential-energy change no matter which path it follows in going from datum plane 1 to datum plane 2 because Δh is the same for all three paths.

The virtual change in gravitational potential energy is simply

$$\delta V_g = mg\, \delta h$$

where δh is the upward virtual displacement of the mass center of the body. If the mass center has a downward virtual displacement, then δV_g is negative.

The units of gravitational potential energy are the same as those for work and elastic potential energy, joules (J) in SI units and foot-pounds (ft-lb) in U.S. customary units.

Energy Equation

We saw that the work done *by* a linear spring *on* the body to which its movable end is attached is the negative of the change in the elastic potential energy of the spring. Also, the work done by the gravitational force or weight mg is the negative of the change in gravitational potential energy. Therefore, when we apply the virtual-work equation to a system with springs and with changes in the vertical position of its members, we may replace the work of the springs and the work of the weights by the negative of the respective potential energy changes.

We can use these substitutions to write the total virtual work δU in Eq. 7/3 as the sum of the work $\delta U'$ done by all active forces, *other than spring forces and weight forces*, and the work $-(\delta V_e + \delta V_g)$ done by the spring and weight forces. Equation 7/3 then becomes

$$\boxed{\delta U' - (\delta V_e + \delta V_g) = 0} \quad \text{or} \quad \boxed{\delta U' = \delta V} \qquad \textbf{(7/6)}$$

where $V = V_e + V_g$ stands for the total potential energy of the system. With this formulation a spring becomes *internal* to the system, and the work of spring and gravitational forces is accounted for in the δV term.

Active-Force Diagrams

With the method of virtual work it is useful to construct the *active-force diagram* of the system you are analyzing. The boundary of the system must clearly distinguish those members which are part of the system from other bodies which are not part of the system. When we include an elastic member *within* the boundary of our system, the forces of interaction between it and the movable members to which it is attached are *internal* to the system. Thus these forces need not be shown because their effects are accounted for in the V_e term. Similarly, weight forces are not shown because their work is accounted for in the V_g term.

Figure 7/14 illustrates the difference between the use of Eqs. 7/3 and 7/6. We consider the body in part *a* of the figure to be a particle for simplicity, and we assume that the virtual displacement is along the fixed path. The particle is in equilibrium under the action of the applied forces F_1 and F_2, the gravitational force mg, the spring force kx, and a normal

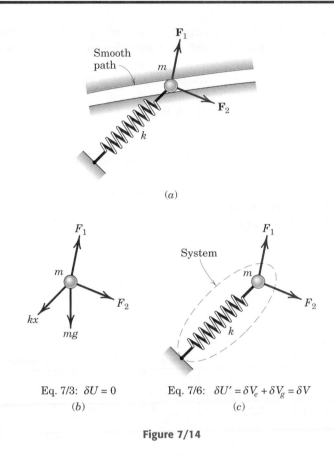

(a)

Eq. 7/3: $\delta U = 0$ $\qquad\qquad$ Eq. 7/6: $\delta U' = \delta V_e + \delta V_g = \delta V$

(b) $\qquad\qquad\qquad\qquad$ (c)

Figure 7/14

reaction force. In Fig 7/14b, where the particle alone is isolated, δU includes the virtual work of all forces shown on the active-force diagram of the particle. (The normal reaction exerted on the particle by the smooth guide does no work and is omitted.) In Fig. 7/14c the spring is *included* in the system, and $\delta U'$ is the virtual work of only F_1 and F_2, which are the only external forces whose work is not accounted for in the potential-energy terms. The work of the weight mg is accounted for in the δV_g term, and the work of the spring force is included in the δV_e term.

Principle of Virtual Work

Thus, for a mechanical system with elastic members and members which undergo changes in position, we may restate the principle of virtual work as follows:

> **The virtual work done by all external active forces (other than the gravitational and spring forces accounted for in the potential energy terms) on a mechanical system in equilibrium equals the corresponding change in the total elastic and gravitational potential energy of the system for any and all virtual displacements consistent with the constraints.**

Stability of Equilibrium

Consider now the case of a mechanical system where movement is accompanied by changes in gravitational and elastic potential energies and where no work is done on the system by nonpotential forces. The mechanism treated in Sample Problem 7/6 is an example of such a system. With $\delta U' = 0$ the virtual-work relation, Eq. 7/6, becomes

$$\delta(V_e + V_g) = 0 \quad \text{or} \quad \delta V = 0 \qquad \textbf{(7/7)}$$

Equation 7/7 expresses the requirement that the equilibrium configuration of a mechanical system is one for which the total potential energy V of the system has a stationary value. For a system of one degree of freedom where the potential energy and its derivatives are continuous functions of the single variable, say, x, which describes the configuration, the equilibrium condition $\delta V = 0$ is equivalent mathematically to the requirement

$$\frac{dV}{dx} = 0 \qquad \textbf{(7/8)}$$

Equation 7/8 states that a mechanical system is in equilibrium when the derivative of its total potential energy is zero. For systems with several degrees of freedom the partial derivative of V with respect to each coordinate in turn must be zero for equilibrium.*

There are three conditions under which Eq. 7/8 applies, namely, when the total potential energy is a minimum (*stable equilibrium*), a maximum (*unstable equilibrium*), or a constant (*neutral equilibrium*). Figure 7/15 shows a simple example of these three conditions. The potential energy of the roller is clearly a minimum in the stable position, a maximum in the unstable position, and a constant in the neutral position.

We may also characterize the stability of a mechanical system by noting that a small displacement away from the stable position results in an increase in potential energy and a tendency to return to the position of lower energy. On the other hand, a small displacement away from the unstable position results in a decrease in potential energy and

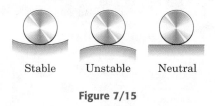

     Stable   Unstable   Neutral

Figure 7/15

---

*For examples of two-degree-of-freedom systems, see Art. 43, Chapter 7, of the first author's *Statics, 2nd Edition, SI Version*, 1975.

a tendency to move farther away from the equilibrium position to one of still lower energy. For the neutral position a small displacement one way or the other results in no change in potential energy and no tendency to move either way.

When a function and its derivatives are continuous, the second derivative is positive at a point of minimum value of the function and negative at a point of maximum value of the function. Thus, the mathematical conditions for equilibrium and stability of a system with a single degree of freedom x are:

$$
\begin{array}{ll}
\text{Equilibrium} & \dfrac{dV}{dx} = 0 \\[2ex]
\text{Stable} & \dfrac{d^2V}{dx^2} > 0 \\[2ex]
\text{Unstable} & \dfrac{d^2V}{dx^2} < 0
\end{array}
\qquad \textbf{(7/9)}
$$

The second derivative of V may also be zero at the equilibrium position, in which case we must examine the sign of a higher derivative to ascertain the type of equilibrium. When the order of the lowest remaining nonzero derivative is even, the equilibrium will be stable or unstable according to whether the sign of this derivative is positive or negative. If the order of the derivative is odd, the equilibrium is classified as unstable, and the plot of V versus x for this case appears as an inflection point in the curve with zero slope at the equilibrium value.

Stability criteria for multiple degrees of freedom require more advanced treatment. For two degrees of freedom, for example, we use a Taylor-series expansion for two variables.

© image100/Media Bakery

This lifting platform is an example of the type of structure that can be most easily analyzed with a virtual-work approach.

Sample Problem 7/4

The 10-kg cylinder is suspended by the spring, which has a stiffness of 2 kN/m. Plot the potential energy V of the system and show that it is minimum at the equilibrium position.

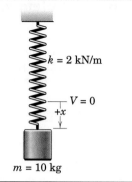

$k = 2$ kN/m

$V = 0$

$+x$

$m = 10$ kg

Solution. (Although the equilibrium position in this simple problem is clearly where the force in the spring equals the weight mg, we will proceed as though this fact were unknown in order to illustrate the energy relationships in the simplest way.) We choose the datum plane for zero potential energy at the position ① where the spring is unextended.

The elastic potential energy for an arbitrary position x is $V_e = \frac{1}{2}kx^2$ and the gravitational potential energy is $-mgx$, so that the total potential energy is

$$[V = V_e + V_g] \qquad\qquad V = \tfrac{1}{2}kx^2 - mgx$$

Equilibrium occurs where

$$\left[\frac{dV}{dx} = 0\right] \qquad \frac{dV}{dx} = kx - mg = 0 \qquad x = mg/k$$

Although we know in this simple case that the equilibrium is stable, we prove it by evaluating the sign of the second derivative of V at the equilibrium position. Thus, $d^2V/dx^2 = k$, which is positive, proving that the equilibrium is stable.

Substituting numerical values gives

$$V = \tfrac{1}{2}(2000)x^2 - 10(9.81)x$$

expressed in joules, and the equilibrium value of x is

$$x = 10(9.81)/2000 = 0.0490 \text{ m} \qquad \text{or} \qquad 49.0 \text{ mm} \qquad\qquad \textit{Ans.}$$

We calculate V for various values of x and plot V versus x as shown. The ② minimum value of V occurs at $x = 0.0490$ m where $dV/dx = 0$ and d^2V/dx^2 is positive.

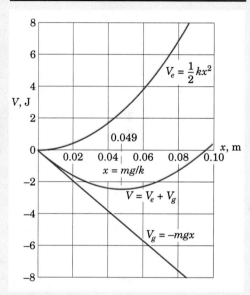

$V_e = \frac{1}{2}kx^2$

V, J

0.049

$x = mg/k$

x, m

$V = V_e + V_g$

$V_g = -mgx$

Helpful Hints

① The choice is arbitrary but simplifies the algebra.

② We could have chosen different datum planes for V_e and V_g without affecting our conclusions. Such a change would merely shift the separate curves for V_e and V_g up or down but would not affect the position of the minimum value of V.

Sample Problem 7/5

The two uniform links, each of mass m, are in the vertical plane and are connected and constrained as shown. As the angle θ between the links increases with the application of the horizontal force P, the light rod, which is connected at A and passes through a pivoted collar at B, compresses the spring of stiffness k. If the spring is uncompressed in the position where $\theta = 0$, determine the force P which will produce equilibrium at the angle θ.

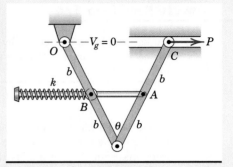

Solution. The given sketch serves as the active-force diagram of the system. The compression x of the spring is the distance which A has moved away from B, which is $x = 2b \sin \theta/2$. Thus, the elastic potential energy of the spring is

$$[V_e = \tfrac{1}{2}kx^2] \qquad V_e = \tfrac{1}{2}k\left(2b \sin \frac{\theta}{2}\right)^2 = 2kb^2 \sin^2 \frac{\theta}{2}$$

With the datum for zero gravitational potential energy taken through the support at O for convenience, the expression for V_g becomes

$$[V_g = mgh] \qquad V_g = 2mg\left(-b \cos \frac{\theta}{2}\right)$$

The distance between O and C is $4b \sin \theta/2$, so that the virtual work done by P is

$$\delta U' = P \, \delta \left(4b \sin \frac{\theta}{2}\right) = 2Pb \cos \frac{\theta}{2} \, \delta\theta$$

The virtual-work equation now gives

$$[\delta U' = \delta V_e + \delta V_g]$$

$$2Pb \cos \frac{\theta}{2} \, \delta\theta = \delta \left(2kb^2 \sin^2 \frac{\theta}{2}\right) + \delta \left(-2mgb \cos \frac{\theta}{2}\right)$$

$$= 2kb^2 \sin \frac{\theta}{2} \cos \frac{\theta}{2} \, \delta\theta + mgb \sin \frac{\theta}{2} \, \delta\theta$$

Simplifying gives finally

$$P = kb \sin \frac{\theta}{2} + \tfrac{1}{2}mg \tan \frac{\theta}{2} \qquad\qquad \textit{Ans.}$$

If we had been asked to express the equilibrium value of θ corresponding to a given force P, we would have difficulty solving explicitly for θ in this particular case. But for a numerical problem we could resort to a computer solution and graphical plot of numerical values of the sum of the two functions of θ to determine the value of θ for which the sum equals P.

Sample Problem 7/6

The ends of the uniform bar of mass m slide freely in the horizontal and vertical guides. Examine the stability conditions for the positions of equilibrium. The spring of stiffness k is undeformed when $x = 0$.

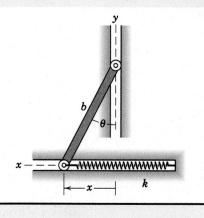

Solution.   The system consists of the spring and the bar. Since there are no external active forces, the given sketch serves as the active-force diagram. We will ① take the x-axis as the datum for zero gravitational potential energy. In the displaced position the elastic and gravitational potential energies are

$$V_e = \tfrac{1}{2}kx^2 = \tfrac{1}{2}kb^2 \sin^2 \theta \qquad \text{and} \qquad V_g = mg\,\frac{b}{2}\cos \theta$$

The total potential energy is then

$$V = V_e + V_g = \tfrac{1}{2}kb^2 \sin^2 \theta + \tfrac{1}{2}mgb \cos \theta$$

Equilibrium occurs for $dV/d\theta = 0$ so that

$$\frac{dV}{d\theta} = kb^2 \sin \theta \cos \theta - \tfrac{1}{2}mgb \sin \theta = (kb^2 \cos \theta - \tfrac{1}{2}mgb) \sin \theta = 0$$

The two solutions to this equation are given by

$$\sin \theta = 0 \qquad \text{and} \qquad \cos \theta = \frac{mg}{2kb}$$

We now determine the stability by examining the sign of the second derivative of V for each of the two equilibrium positions. The second derivative is

$$\frac{d^2V}{d\theta^2} = kb^2(\cos^2 \theta - \sin^2 \theta) - \tfrac{1}{2}mgb \cos \theta$$

$$= kb^2(2 \cos^2 \theta - 1) - \tfrac{1}{2}mgb \cos \theta$$

Solution I. $\sin \theta = 0, \theta = 0$

$$\frac{d^2V}{d\theta^2} = kb^2(2 - 1) - \tfrac{1}{2}mgb = kb^2\!\left(1 - \frac{mg}{2kb}\right)$$

$$= \text{positive (stable)} \qquad \text{if } k > mg/2b$$

$$= \text{negative (unstable) if } k < mg/2b \qquad\qquad \textbf{\textit{Ans.}}$$

Thus, if the spring is sufficiently stiff, the bar will return to the vertical position ③ even though there is no force in the spring at that position.

Solution II. $\cos \theta = \dfrac{mg}{2kb},\ \theta = \cos^{-1}\dfrac{mg}{2kb}$

$$\frac{d^2V}{d\theta^2} = kb^2\!\left[2\!\left(\frac{mg}{2kb}\right)^2 - 1\right] - \tfrac{1}{2}mgb\!\left(\frac{mg}{2kb}\right) = kb^2\!\left[\left(\frac{mg}{2kb}\right)^2 - 1\right] \qquad \textbf{\textit{Ans.}}$$

Since the cosine must be less than unity, we see that this solution is limited to the case where $k > mg/2b$, which makes the second derivative of V negative. ④ Thus, equilibrium for Solution II is never stable. If $k < mg/2b$, we no longer have Solution II since the spring will be too weak to maintain equilibrium at a value of θ between 0 and 90°.

Helpful Hints

① With no external active forces there is no $\delta U'$ term, and $\delta V = 0$ is equivalent to $dV/d\theta = 0$.

② Be careful not to overlook the solution $\theta = 0$ given by $\sin \theta = 0$.

③ We might not have anticipated this result without the mathematical analysis of the stability.

④ Again, without the benefit of the mathematical analysis of the stability we might have supposed erroneously that the bar could come to rest in a stable equilibrium position for some value of θ between 0 and 90°.

PROBLEMS

(Assume that the negative work of friction is negligible in the following problems.)

Introductory Problems

7/35 The potential energy of a mechanical system is given by $V = 6x^3 - 9x^2 + 7$ where x is the position coordinate associated with its single degree of freedom. Determine the position or positions of equilibrium of the system and the stability condition of the system at each equilibrium position.

Ans. $x = 0$ unstable, $x = 1$ stable

7/36 The uniform bar of mass m and length L is supported in the vertical plane by two identical springs each of stiffness k and compressed a distance δ in the vertical position $\theta = 0$. Determine the minimum stiffness k which will ensure a stable equilibrium position with $\theta = 0$. The springs may be assumed to act in the horizontal direction during small angular motion of the bar.

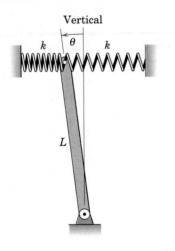

Problem 7/36

7/37 The bar of mass m with center of mass at G is pivoted about a horizontal axis through O. Prove the stability conditions for the two positions of equilibrium.

Ans. $\theta = 0$, unstable; $\theta = 180°$, stable

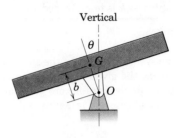

Problem 7/37

7/38 The small cylinder of mass m and radius r is confined to roll on the circular surface of radius R. By the methods of this article, prove that the cylinder is unstable in case (a) and stable in case (b).

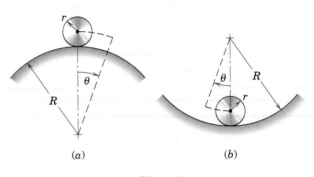

Problem 7/38

7/39 The uniform bar of mass m and length l is hinged about a horizontal axis through its end O and is attached to a torsional spring which exerts a torque $M = K\theta$ on the rod, where K is the torsional stiffness of the spring in units of torque per radian and θ is the angular deflection from the vertical in radians. Determine the maximum value of l for which equilibrium at the position $\theta = 0$ is stable.

$$Ans.\ l = \frac{2K}{mg}$$

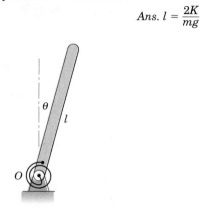

Problem 7/39

7/40 Determine the force P required to maintain equilibrium of the spring-loaded mechanism for a given angle θ. The spring has a stiffness k and is uncompressed at $\theta = 0$. The mass of the parts may be neglected.

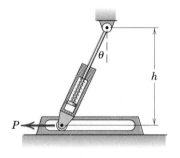

Problem 7/40

7/41 Two identical bars are welded at a 120° angle and pivoted at O as shown. The mass of the support tab is small compared with the mass of the bars. Investigate the stability of the equilibrium position shown.

Ans. Stable

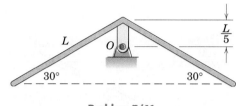

Problem 7/41

7/42 The figure shows the cross section of a uniform 60-kg ventilator door hinged along its upper horizontal edge at O. The door is controlled by the spring-loaded cable which passes over the small pulley at A. The spring has a stiffness of 160 N per meter of stretch and is undeformed when $\theta = 0$. Determine the angle θ for equilibrium.

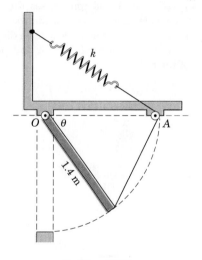

Problem 7/42

7/43 For the device shown the spring would be un-stretched in the position $\theta = 0$. Specify the stiffness k of the spring which will establish an equilibrium position θ in the vertical plane. The mass of the links is negligible compared with m.

$$\text{Ans. } k = \frac{mg}{2b} \frac{\cot \theta}{1 - \cos \theta}$$

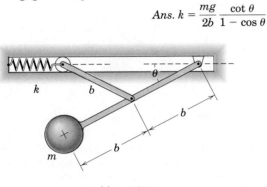

Problem 7/43

7/44 When $u = 0$, the spring of stiffness k is uncompressed. As u increases, the rod slides through the pivoted collar at A and compresses the spring between the collar and the end of the rod. Determine the force P required to produce a given displacement u. Assume the absence of friction and neglect the mass of the rod.

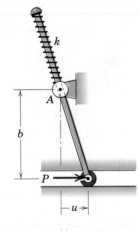

Problem 7/44

7/45 One of the critical requirements in the design of an artificial leg for an amputee is to prevent the knee joint from buckling under load when the leg is straight. As a first approximation, simulate the artificial leg by the two light links with a torsion spring at their common joint. The spring develops a torque $M = K\beta$, which is proportional to the angle of bend β at the joint. Determine the minimum value of K which will ensure stability of the knee joint for $\beta = 0$.

$$\text{Ans. } K_{\min} = \tfrac{1}{2}mgl$$

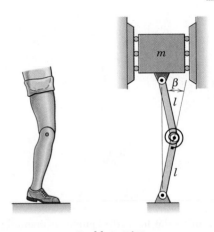

Problem 7/45

Representative Problems

7/46 The handle is fastened to one of the spring-connected gears, which are mounted in fixed bearings. The spring of stiffness k connects two pins mounted in the faces of the gears. When the handle is in the vertical position, $\theta = 0$ and the spring force is zero. Determine the force P required to maintain equilibrium at an angle θ.

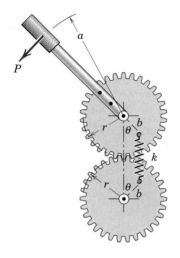

Problem 7/46

7/47 The cross section of a spring-loaded ventilator door hinged about a horizontal axis at O and having a mass m with mass center at G is shown in the figure. As the door opens, the spring is compressed by the hinged rod, which slides through the swivel collar at A. The spring is uncompressed when the door is closed. Show that the door will remain in equilibrium for any angle θ if the spring has the proper stiffness k.

Ans. $k = mg/a$

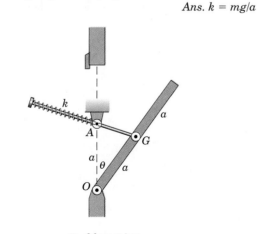

Problem 7/47

7/48 Determine the maximum height h of the mass m for which the inverted pendulum will be stable in the vertical position shown. Each of the springs has a stiffness k, and they have equal precompressions in this position. Neglect the mass of the remainder of the mechanism.

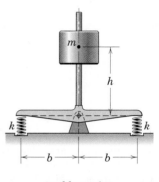

Problem 7/48

7/49 In the figure is shown a small industrial lift with a foot release. There are four identical springs, two on each side of the central shaft. The stiffness of each pair of springs is $2k$. In designing the lift specify the value of k which will ensure stable equilibrium when the lift supports a load (weight) L in the position where $\theta = 0$ with no force P on the pedal. The springs are equally precompressed initially and may be assumed to act in the horizontal direction at all times.

Ans. $k > \dfrac{L}{2l}$

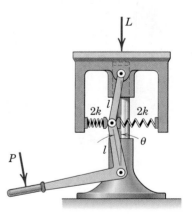

Problem 7/49

7/50 The system of freely pivoted light bar, end mass m, and spring of stiffness k is shown in its equilibrium configuration. The parameter n is a fraction which lies in the range $0 < n < 1$. Show that the system is stable for all values of n in the permitted range.

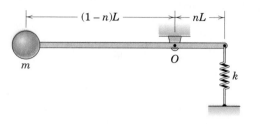

Problem 7/50

7/51 One end of the torsion spring is secured to the ground at A, and the other end is fastened to the shaft at B. The torsional stiffness K of the elastic spring is the torque required to twist the spring through an angle of one radian. The spring resists the moment about the shaft axis caused by the tension mg in the cable wrapped around the drum of radius r. Determine the equilibrium value of h measured from the dashed position, where the spring is untwisted.

$$Ans.\ h = \frac{mgr^2}{K}$$

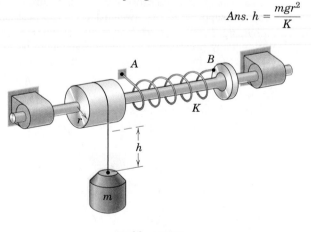

Problem 7/51

7/52 The 3-lb pendulum swings about axis O-O and has a mass center at G. When $\theta = 0$, each spring has an initial stretch of 4 in. Calculate the maximum stiffness k of each of the parallel springs which will allow the pendulum to be in stable equilibrium at the bottom position $\theta = 0$.

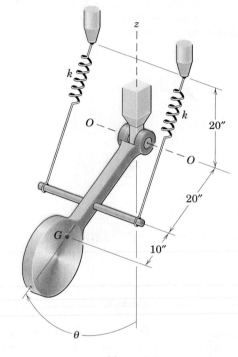

Problem 7/52

7/53 The cylinder of mass m and radius r rolls without slipping on the circular surface of radius ρ. Attached to the cylinder is a small body of mass $m/4$. Determine the maximum value of ρ for which the body is stable in the equilibrium position shown.

$$Ans.\ \rho < 6r$$

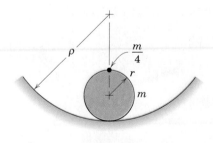

Problem 7/53

7/54 The solid hemisphere of diameter $2b$ and concentric cylindrical knob of diameter b are resting on a horizontal surface. Determine the maximum height h which the knob may have without causing the unit to be unstable in the upright position shown. Both parts are made from the same material.

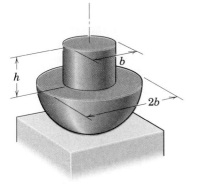

Problem 7/54

7/55 The uniform link AB has a mass m, and its left end A travels freely in the fixed horizontal slot. End B is attached to the vertical plunger, which compresses the spring as B falls. The spring would be uncompressed at the position $\theta = 0$. Determine the angle θ for equilibrium (other than the impossible position corresponding to $\theta = 90°$) and designate the condition which will ensure stability.

$$Ans. \ \sin \theta = \frac{mg}{2kl}, \ k > \frac{mg}{2l}$$

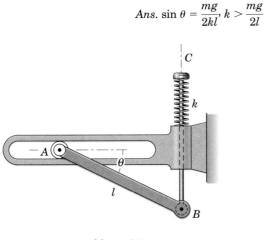

Problem 7/55

7/56 Predict through calculation whether the homogeneous semicylinder and the half-cylindrical shell will remain in the positions shown or whether they will roll off the lower cylinders.

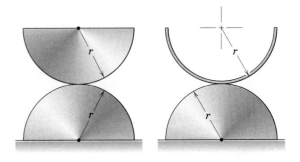

Problem 7/56

7/57 The figure shows a tilting desk chair together with the design detail of the spring-loaded tilting mechanism. The frame of the seat is pivoted about the fixed point O on the base. The increase in distance between A and B as the chair tilts back about O is the increase in compression of the spring. The spring, which has a stiffness of 96 kN/m, is uncompressed when $\theta = 0$. For small angles of tilt it may be assumed with negligible error that the axis of the spring remains parallel to the seat. The center of mass of an 80-kg person who sits in the chair is at G on a line through O perpendicular to the seat. Determine the angle of tilt θ for equilibrium. (*Hint:* The deformation of the spring may be visualized by allowing the base to tilt through the required angle θ about O while the seat is held in a fixed position.)

$$Ans. \ \theta = 11.19°$$

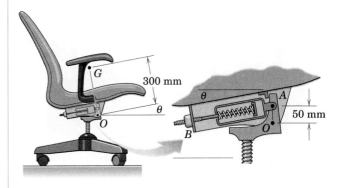

Problem 7/57

7/58 A proposed parallelogram linkage for an adjustable-position lamp is shown. If the unstretched length of the spring is $b/2$, determine the necessary spring stiffness k for equilibrium at a given angle θ with the vertical. The mass of the lamp and triangular fixture is m. Check the stability within the working range from $\theta = 2 \sin^{-1} \frac{1}{4} \cong 29°$ to $\theta = 180°$.

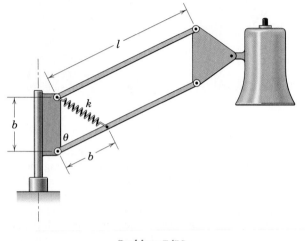

Problem 7/58

▶7/59 The double-axle front suspension is used on small trucks. In a test of the designed action, the frame F must be jacked up so that $h = 350$ mm in order to relieve the compression in the coil springs. Determine the value of h when the jack is removed. Each spring has a stiffness of 120 kN/m. The load L is 12 kN, and the central frame F has a mass of 40 kg. Each wheel and attached link has a mass of 35 kg with a center of mass 680 mm from the vertical centerline.

Ans. $h = 265$ mm

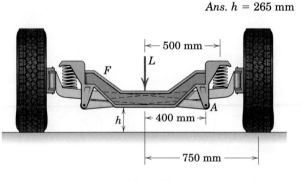

Problem 7/59

▶7/60 The uniform garage door AB shown in section has a mass m and is equipped with two of the spring-loaded mechanisms shown, one on each side of the door. The arm OB has negligible mass, and the upper corner A of the door is free to move horizontally on a roller. The unstretched length of the spring is $r - a$, so that in the top position with $\theta = \pi$ the spring force is zero. To ensure smooth action of the door as it reaches the vertical closed position $\theta = 0$, it is desirable that the door be insensitive to movement in this position. Determine the spring stiffness k required for this design.

Ans. $k = \dfrac{mg(r + a)}{8a^2}$

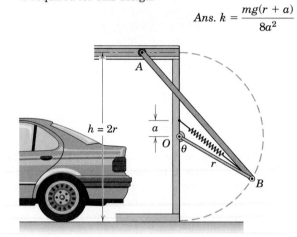

Problem 7/60

7/5 CHAPTER REVIEW

In this chapter we have developed the principle of virtual work. This principle is useful for determining the possible equilibrium configurations of a body or a system of interconnected bodies where the external forces are known. To apply the method successfully, you must understand the concepts of virtual displacement, degrees of freedom, and potential energy.

Method of Virtual work

When various configurations are possible for a body or a system of interconnected bodies under the action of applied forces, we can find the equilibrium position by applying the principle of virtual work. When using this method, keep the following in mind.

1. The only forces which need to be considered when determining the equilibrium position are those which do work (active forces) during the assumed differential movement of the body or system away from its equilibrium position.

2. Those external forces which do no work (reactive forces) need not be involved.

3. For this reason the active-force diagram of the body or system (rather than the free-body diagram) is useful to focus attention on only those external forces which do work during the virtual displacements.

Virtual Displacements

A virtual displacement is a first-order differential change in a linear or angular position. This change is fictitious in that it is an assumed movement which need not take place in reality. Mathematically, a virtual displacement is treated the same as a differential change in an actual movement. We use the symbol δ for the differential virtual change and the usual symbol d for the differential change in a real movement.

Relating the linear and angular virtual displacements of the parts of a mechanical system during a virtual movement consistent with the constraints is often the most difficult part of the analysis. To do this,

1. Write the geometric relationships which describe the configuration of the system.

2. Establish the differential changes in the positions of parts of the system by differentiating the geometric relationship to obtain expressions for the differential virtual movements.

Degrees of Freedom

In Chapter 7 we have restricted our attention to mechanical systems for which the positions of the members can be specified by a single variable (single-degree-of-freedom systems). For two or more degrees of freedom, we would apply the virtual-work equation as many times as there are degrees of freedom, allowing one variable to change at a time while holding the remaining ones constant.

Potential Energy Method

The concept of potential energy, both gravitational (V_g) and elastic (V_e), is useful in solving equilibrium problems where virtual displacements cause changes in the vertical positions of the mass centers of the bodies and changes in the lengths of elastic members (springs). To apply this method,

1. Obtain an expression for the total potential energy V of the system in terms of the variable which specifies the possible position of the system.

2. Examine the first and second derivatives of V to establish, respectively, the position of equilibrium and the corresponding stability condition.

REVIEW PROBLEMS

7/61 A "black box" contains a series of interconnected racks and pinions, gears, and other internal mechanical elements which transform the linear motion of the pushrod A to produce the linear motion of the pushrod B. For every unit of inward movement of A under the action of force P_1, rod B moves outward from the box one-fourth of a unit against the action of force P_2. If $P_1 = 100$ N, calculate P_2 for equilibrium. Neglect all friction and assume all mechanical components are ideally connected light rigid bodies.

Ans. $P_2 = 400$ N

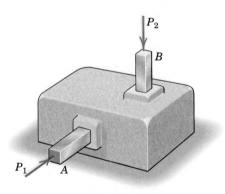

Problem 7/61

7/62 The semicylindrical shell of radius r is pivoted about a shaft through points O as shown. The mass of the two support tabs is small compared with the mass of the shell. Determine the maximum value of h for which equilibrium in the position shown is stable.

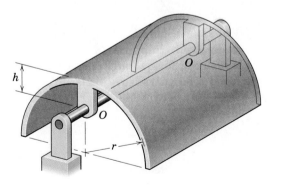

Problem 7/62

7/63 A uniform rectangular block of height h and mass m is centered in a horizontal position on the fixed circular surface of radius r. Determine the limiting value of h for stability.

Ans. $h < 2r$

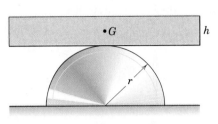

Problem 7/63

7/64 Identify which of the problems (a) through (f) are best solved (A) by the force and moment equilibrium equations and (B) by virtual work. Outline briefly the procedure for each solution.

Plane of each figure is vertical. Size and mass of each member and applied force are known.

Find θ for equilibrium
(a)

Find reactions at A and B
(b)

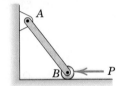

Find x for equilibrium
(c)

Find forces at A, B, and C
(d)

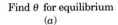

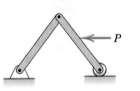

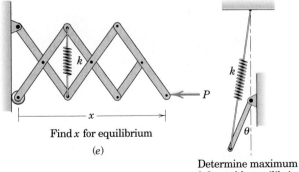

Find x for equilibrium
(e)

Determine maximum k for stable equilibrium at $\theta = 0$
(f)

Problem 7/64

7/65 The figure shows the cross section of a container composed of a hemispherical shell of radius r and a cylindrical shell of height h, both made from the same material. Specify the limitation of h for stability in the upright position when the container is placed on the horizontal surface.

Ans. $h < r$

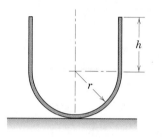

Problem 7/65

7/66 Determine the equilibrium values of θ and the stability of equilibrium at each position for the unbalanced wheel on the 10° incline. Static friction is sufficient to prevent slipping. The mass center is at G.

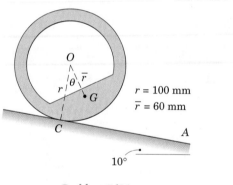

$r = 100$ mm
$\bar{r} = 60$ mm

Problem 7/66

7/67 The light bar OC is pivoted at O and swings in the vertical plane. When $\theta = 0$, the spring of stiffness k is unstretched. Determine the equilibrium angle corresponding to a given vertical force P applied to the end of the bar. Neglect the mass of the bar and the diameter of the small pulleys.

Ans. $\theta = \tan^{-1} \dfrac{2P}{ka}$

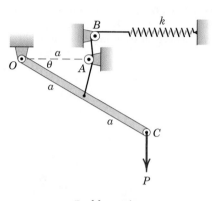

Problem 7/67

7/68 The figure shows the edge view of a uniform skylight door of mass m with center of mass midway between A and O. The door is counterbalanced by the action of the spring, which has a stiffness k and which is unstretched when the door is in the vertical plane with $\theta = 0$. Determine the necessary stiffness k to balance the door in an equilibrium position for $\theta > 0$.

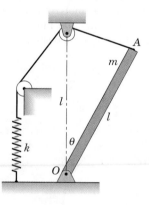

Problem 7/68

7/69 The figure shows the design of the cross section of a 35-kg overhead industrial door with hinged segments which pass over the rollers on the cylindrical guide. When the door is closed, the end A is at position B. On each side of the door there is a control cable fastened to the bottom of the door and wound around a drum C. Each of the two drums is connected to a torsion spring, which offers a torsional resistance which increases by 10 N·m for each revolution of the drum, starting from an unwound position of the spring where $x = 0$. Determine the designed value of x for equilibrium and prove that equilibrium in this position is stable.

Ans. $x = 2.05$ m

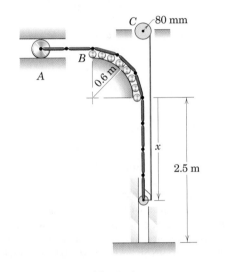

Problem 7/69

7/70 Two semicylindrical shells with equal projecting rectangles are formed from sheet metal, one with configuration (a) and the other with configuration (b). Both shells rest on a horizontal surface. For case (a) determine the maximum value of h for which the shell will remain stable in the position shown. For case (b) prove that stability in the position shown is not affected by the dimension h.

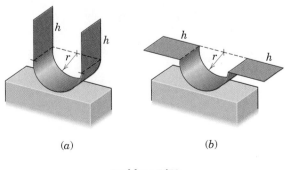

(a) (b)

Problem 7/70

7/71 The uniform rectangular panel of mass m with mass center at G is guided by its rollers—the upper pair in the horizontal tracks and the lower pair in the vertical tracks. Determine the force P, applied to the lower edge normal to the panel, required to maintain equilibrium at a given angle θ. (*Hint:* To evaluate the work done by P, replace it by its horizontal and vertical components.)

Ans. $P = \dfrac{mg \cos \theta}{1 + \cos^2 \theta}$

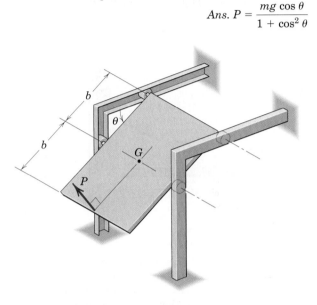

Problem 7/71

7/72 The uniform aluminum disk of radius R and mass m rolls without slipping on the fixed circular surface of radius $2R$. Fastened to the disk is a lead cylinder also of mass m with its center located a distance b from the center O of the disk. Determine the minimum value of b for which the disk will remain in stable equilibrium on the cylindrical surface in the top position shown.

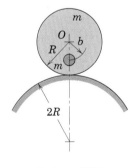

Problem 7/72

Computer-Oriented Problems*

*7/73 The end rollers of the uniform 20-lb bar AB move freely in the horizontal and vertical guides as shown. Determine the equilibrium position θ for the bar under the action of the 480-lb-in. couple M and the spring of stiffness $k = 1.25$ lb/in. The spring is uncompressed when $\theta = 0$.

Ans. $\theta = 23.0°$

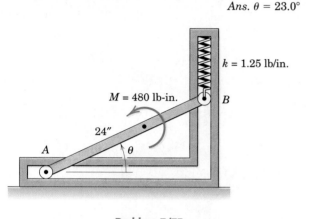

$k = 1.25$ lb/in.

$M = 480$ lb-in.

B

24″

A

θ

Problem 7/73

*7/74 Compute the total potential energy $V = V_e + V_g$ for the industrial door of Prob. 7/69 in terms of x and verify the results cited for that problem. Take $V_g = 0$ through A. Note that the potential energy of the portion of the door in contact with the cylindrical guide is a constant, which may be taken as zero for convenience.

*7/75 The bar OA, which weighs 50 lb with center of gravity at G, is pivoted about its end O and swings in the vertical plane under the constraint of the 20-lb counterweight. Write the expression for the total potential energy of the system, taking $V_g = 0$ when $\theta = 0$, and compute V_g as a function of θ from $\theta = 0$ to $\theta = 360°$. From your plot of the results, determine the position or positions of equilibrium and the stability of equilibrium at each position.

Ans. $\theta = 78.0°$, stable; $\theta = 260°$, unstable

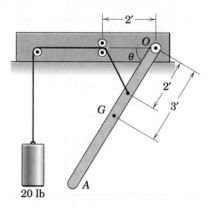

2′

O

θ

2′

G

3′

A

20 lb

Problem 7/75

*7/76 Determine the angle θ for equilibrium of the uniform 10-kg link with mass center at G as its position in the vertical plane is influenced by the 80-N horizontal force applied to its end. The spring has a constant of 1.5 kN/m and is unstretched when $\theta = 0$.

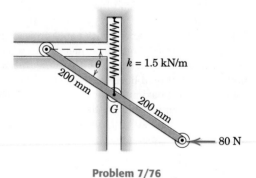

θ

200 mm

$k = 1.5$ kN/m

G

200 mm

80 N

Problem 7/76

*7/77 The figure shows the edge view of a 10-kg ventilator door with mass center at G and hinged about a horizontal axis through its edge at O. When $\theta = 90°$, the restraining spring of stiffness $k = 1250$ N/m is unstretched. Compute and plot the total potential energy of the door and spring as a function of θ from $\theta = 0$ to $\theta = 90°$. Take $V_g = 0$ at $\theta = 0$. From the plot, specify the equilibrium position, other than $\theta = \pm 90°$, and its stability condition. Can you justify the fact that the equilibrium at $\theta = +90°$ is stable with the spring attached and unstable without the spring?

Ans. $\theta = 49.9°$, unstable

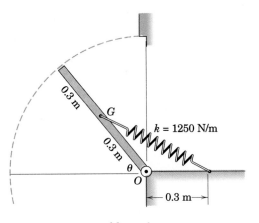

Problem 7/77

*7/78 The toggle mechanism is used to lift the 80-kg mass to a locked position when OB moves to OB' in the 3° position. To evaluate the design action of the toggle, plot the value of P required to operate the toggle as a function of θ from $\theta = 20°$ to $\theta = -3°$.

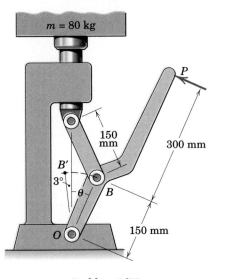

Problem 7/78

*7/79 The uniform link OA has a mass of 20 kg and is supported in the vertical plane by the spring AB whose unstretched length is 400 mm. Plot the total potential energy V and its derivative $dV/d\theta$ as functions of θ from $\theta = 0$ to $\theta = 120°$. From the plots identify the equilibrium values of θ and the corresponding stability of equilibrium. Take $V_g = 0$ on a level through O.

Ans. $\theta = 0$, unstable
$\theta = 51.1°$, stable

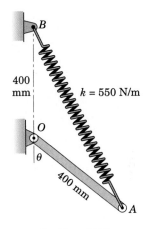

Problem 7/79

*7/80 Determine the equilibrium angle θ for the mechanism shown. The spring of stiffness $k = 12$ lb/in. has an unstretched length of 8 in. Each of the uniform links AB and CD has a weight of 10 lb, and member BD with its load weighs 100 lb. Motion is in the vertical plane.

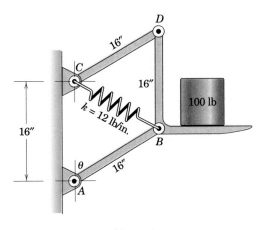

Problem 7/80

A

Area Moments of Inertia

A/1 Introduction

When forces are distributed continuously over an area on which they act, it is often necessary to calculate the moment of these forces about some axis either in or perpendicular to the plane of the area. Frequently the intensity of the force (pressure or stress) is proportional to the distance of the line of action of the force from the moment axis. The elemental force acting on an element of area, then, is proportional to distance times differential area, and the elemental moment is proportional to distance squared times differential area. We see, therefore, that the total moment involves an integral of form $\int$ (distance)^2 d (area). This integral is called the *moment of inertia* or the *second moment* of the area. The integral is a function of the geometry of the area and occurs frequently in the applications of mechanics. Thus it is useful to develop its properties in some detail and to have these properties available for ready use when the integral arises.

Figure A/1 illustrates the physical origin of these integrals. In part a of the figure, the surface area $ABCD$ is subjected to a distributed pressure p whose intensity is proportional to the distance y from the axis AB. This situation was treated in Art. 5/9 of Chapter 5, where we described the action of liquid pressure on a plane surface. The moment about AB due to the pressure on the element of area dA is $py\,dA = ky^2\,dA$. Thus, the integral in question appears when the total moment $M = k \int y^2\,dA$ is evaluated.

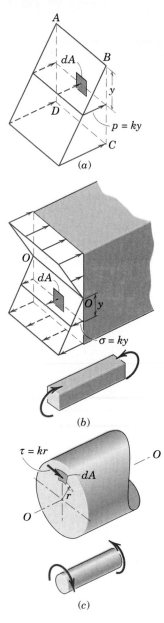

Figure A/1

In Fig. A/1b we show the distribution of stress acting on a transverse section of a simple elastic beam bent by equal and opposite couples applied to its ends. At any section of the beam, a linear distribution of force intensity or stress σ, given by $\sigma = ky$, is present. The stress is positive (tensile) below the axis O–O and negative (compressive) above the axis. We see that the elemental moment about the axis O–O is $dM = y(\sigma\,dA) = ky^2\,dA$. Thus, the same integral appears when the total moment $M = k \int y^2\,dA$ is evaluated.

A third example is given in Fig. A/1c, which shows a circular shaft subjected to a twist or torsional moment. Within the elastic limit of the material, this moment is resisted at each cross section of the shaft by a distribution of tangential or shear stress τ, which is proportional to the radial distance r from the center. Thus, $\tau = kr$, and the total moment about the central axis is $M = \int r(\tau\,dA) = k \int r^2\,dA$. Here the integral differs from that in the preceding two examples in that the area is normal instead of parallel to the moment axis and in that r is a radial coordinate instead of a rectangular one.

Although the integral illustrated in the preceding examples is generally called the *moment of inertia* of the area about the axis in question, a more fitting term is the *second moment of area*, since the first moment $y\,dA$ is multiplied by the moment arm y to obtain the second moment for the element dA. The word *inertia* appears in the terminology by reason of the similarity between the mathematical form of the integrals for second moments of areas and those for the resultant moments of the so-called inertia forces in the case of rotating bodies. The moment of inertia of an area is a purely mathematical property of the area and in itself has no physical significance.

A/2   DEFINITIONS

The following definitions form the basis for the analysis of area moments of inertia.

Rectangular and Polar Moments of Inertia

Consider the area A in the x-y plane, Fig. A/2. The moments of inertia of the element dA about the x- and y-axes are, by definition, $dI_x = y^2\,dA$ and $dI_y = x^2\,dA$, respectively. The moments of inertia of A about the same axes are therefore

$$I_x = \int y^2\,dA$$

$$I_y = \int x^2\,dA$$

(A/1)

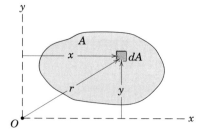

Figure A/2

where we carry out the integration over the entire area.

The moment of inertia of dA about the pole O (z-axis) is, by similar definition, $dI_z = r^2\, dA$. The moment of inertia of the entire area about O is

$$I_z = \int r^2\, dA \qquad \text{(A/2)}$$

The expressions defined by Eqs. A/1 are called *rectangular* moments of inertia, whereas the expression of Eq. A/2 is called the *polar* moment of inertia.* Because $x^2 + y^2 = r^2$, it is clear that

$$I_z = I_x + I_y \qquad \text{(A/3)}$$

For an area whose boundaries are more simply described in rectangular coordinates than in polar coordinates, its polar moment of inertia is easily calculated with the aid of Eq. A/3.

The moment of inertia of an element involves the square of the distance from the inertia axis to the element. Thus an element whose coordinate is negative contributes as much to the moment of inertia as does an equal element with a positive coordinate of the same magnitude. Consequently the area moment of inertia about any axis is always a positive quantity. In contrast, the first moment of the area, which was involved in the computations of centroids, could be either positive, negative, or zero.

The dimensions of moments of inertia of areas are clearly L^4, where L stands for the dimension of length. Thus, the SI units for area moments of inertia are expressed as quartic meters (m^4) or quartic millimeters (mm^4). The U.S. customary units for area moments of inertia are quartic feet (ft^4) or quartic inches (in.^4).

The choice of the coordinates to use for the calculation of moments of inertia is important. Rectangular coordinates should be used for shapes whose boundaries are most easily expressed in these coordinates. Polar coordinates will usually simplify problems involving boundaries which are easily described in r and θ. The choice of an element of area which simplifies the integration as much as possible is also important. These considerations are quite analogous to those we discussed and illustrated in Chapter 5 for the calculation of centroids.

Radius of Gyration

Consider an area A, Fig. A/3a, which has rectangular moments of inertia I_x and I_y and a polar moment of inertia I_z about O. We now visualize this area as concentrated into a long narrow strip of area A a distance k_x from the x-axis, Fig. A/3b. By definition the moment of inertia of the strip about the x-axis will be the same as that of the original area if $k_x{}^2 A = I_x$. The distance k_x is called the *radius of gyration* of the area

---

*The polar moment of inertia of an area is sometimes denoted in mechanics literature by the symbol J.

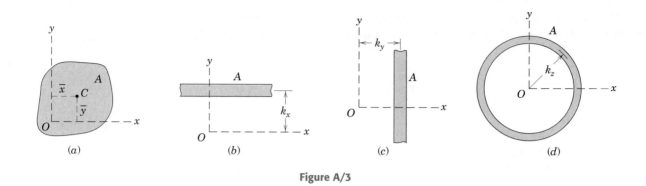

Figure A/3

about the x-axis. A similar relation for the y-axis is written by considering the area as concentrated into a narrow strip parallel to the y-axis as shown in Fig. A/3c. Also, if we visualize the area as concentrated into a narrow ring of radius k_z as shown in Fig. A/3d, we may express the polar moment of inertia as $k_z^2 A = I_z$. In summary we write

$$
\begin{aligned}
I_x &= k_x^2 A \\
I_y &= k_y^2 A \qquad \text{or} \qquad
\begin{aligned}
k_x &= \sqrt{I_x/A} \\
k_y &= \sqrt{I_y/A} \\
k_z &= \sqrt{I_z/A}
\end{aligned}
\\
I_z &= k_z^2 A
\end{aligned}
\qquad \textbf{(A/4)}
$$

The radius of gyration, then, is a measure of the distribution of the area from the axis in question. A rectangular or polar moment of inertia may be expressed by specifying the radius of gyration and the area.

When we substitute Eqs. A/4 into Eq. A/3, we have

$$
k_z^2 = k_x^2 + k_y^2 \qquad \textbf{(A/5)}
$$

Thus, the square of the radius of gyration about a polar axis equals the sum of the squares of the radii of gyration about the two corresponding rectangular axes.

Do not confuse the coordinate to the centroid C of an area with the radius of gyration. In Fig. A/3a the square of the centroidal distance from the x-axis, for example, is $\bar{y}^2$, which is the square of the mean value of the distances from the elements of the area to the x-axis. The quantity k_x^2, on the other hand, is the mean of the squares of these distances. The moment of inertia is *not* equal to $A\bar{y}^2$, since the square of the mean is less than the mean of the squares.

Transfer of Axes

The moment of inertia of an area about a noncentroidal axis may be easily expressed in terms of the moment of inertia about a parallel centroidal axis. In Fig. A/4 the x_0-y_0 axes pass through the centroid C of the area. Let us now determine the moments of inertia of the area about the

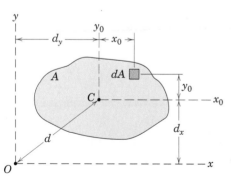

Figure A/4

parallel x-y axes. By definition, the moment of inertia of the element dA about the x-axis is

$$dI_x = (y_0 + d_x)^2 \, dA$$

Expanding and integrating give us

$$I_x = \int y_0{}^2 \, dA + 2d_x \int y_0 \, dA + d_x{}^2 \int dA$$

We see that the first integral is by definition the moment of inertia $\bar{I}_x$ about the centroidal x_0-axis. The second integral is zero, since $\int y_0 \, dA = A\bar{y}_0$ and $\bar{y}_0$ is automatically zero with the centroid on the x_0-axis. The third term is simply $Ad_x{}^2$. Thus, the expression for I_x and the similar expression for I_y become

$$\boxed{\begin{aligned} I_x &= \bar{I}_x + Ad_x{}^2 \\ I_y &= \bar{I}_y + Ad_y{}^2 \end{aligned}} \qquad \textbf{(A/6)}$$

By Eq. A/3 the sum of these two equations gives

$$I_z = \bar{I}_z + Ad^2 \qquad \textbf{(A/6a)}$$

Equations A/6 and A/6*a* are the so-called *parallel-axis theorems*. Two points in particular should be noted. First, the axes between which the transfer is made *must be parallel*, and second, one of the axes *must pass through the centroid of the area*.

If a transfer is desired between two parallel axes neither of which passes through the centroid, it is first necessary to transfer from one axis to the parallel centroidal axis and then to transfer from the centroidal axis to the second axis.

The parallel-axis theorems also hold for radii of gyration. With substitution of the definition of k into Eqs. A/6, the transfer relation becomes

$$k^2 = \bar{k}^2 + d^2 \qquad \textbf{(A/6b)}$$

where $\bar{k}$ is the radius of gyration about a centroidal axis parallel to the axis about which k applies and d is the distance between the two axes. The axes may be either in the plane or normal to the plane of the area.

A summary of the moment-of-inertia relations for some common plane figures is given in Table D/3, Appendix D.

Sample Problem A/1

Determine the moments of inertia of the rectangular area about the centroidal x_0- and y_0-axes, the centroidal polar axis z_0 through C, the x-axis, and the polar axis z through O.

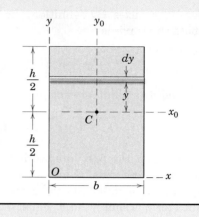

Solution. ① For the calculation of the moment of inertia $\bar{I}_x$ about the x_0-axis, a horizontal strip of area $b\ dy$ is chosen so that all elements of the strip have the same y-coordinate. Thus,

$$[I_x = \int y^2\, dA] \qquad \bar{I}_x = \int_{-h/2}^{h/2} y^2 b\, dy = \tfrac{1}{12}bh^3 \qquad Ans.$$

By interchange of symbols, the moment of inertia about the centroidal y_0-axis is

$$\bar{I}_y = \tfrac{1}{12}hb^3 \qquad Ans.$$

The centroidal polar moment of inertia is

$$[\bar{I}_z = \bar{I}_x + \bar{I}_y] \qquad \bar{I}_z = \tfrac{1}{12}(bh^3 + hb^3) = \tfrac{1}{12}A(b^2 + h^2) \qquad Ans.$$

By the parallel-axis theorem the moment of inertia about the x-axis is

$$[I_x = \bar{I}_x + Ad_x^2] \qquad I_x = \tfrac{1}{12}bh^3 + bh\left(\frac{h}{2}\right)^2 = \tfrac{1}{3}bh^3 = \tfrac{1}{3}Ah^2 \qquad Ans.$$

We also obtain the polar moment of inertia about O by the parallel-axis theorem, which gives us

$$[I_z = \bar{I}_z + Ad^2] \qquad I_z = \tfrac{1}{12}A(b^2 + h^2) + A\left[\left(\frac{b}{2}\right)^2 + \left(\frac{h}{2}\right)^2\right]$$

$$I_z = \tfrac{1}{3}A(b^2 + h^2) \qquad Ans.$$

Helpful Hint

① If we had started with the second-order element $dA = dx\, dy$, integration with respect to x holding y constant amounts simply to multiplication by b and gives us the expression $y^2 b\, dy$, which we chose at the outset.

Sample Problem A/2

Determine the moments of inertia of the triangular area about its base and about parallel axes through its centroid and vertex.

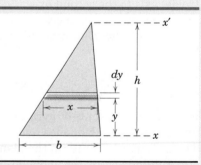

Solution. ① ② A strip of area parallel to the base is selected as shown in the figure, and it has the area $dA = x\, dy = [(h - y)b/h]\, dy$. By definition

$$[I_x = \int y^2\, dA] \qquad I_x = \int_0^h y^2 \frac{h - y}{h} b\, dy = b\left[\frac{y^3}{3} - \frac{y^4}{4h}\right]_0^h = \frac{bh^3}{12} \qquad Ans.$$

By the parallel-axis theorem the moment of inertia $\bar{I}$ about an axis through the centroid, a distance $h/3$ above the x-axis, is

$$[\bar{I} = I - Ad^2] \qquad \bar{I} = \frac{bh^3}{12} - \left(\frac{bh}{2}\right)\left(\frac{h}{3}\right)^2 = \frac{bh^3}{36} \qquad Ans.$$

A transfer from the centroidal axis to the x'-axis through the vertex gives

$$[I = \bar{I} + Ad^2] \qquad I_{x'} = \frac{bh^3}{36} + \left(\frac{bh}{2}\right)\left(\frac{2h}{3}\right)^2 = \frac{bh^3}{4} \qquad Ans.$$

Helpful Hints

① Here again we choose the simplest possible element. If we had chosen $dA = dx\, dy$, we would have to integrate $y^2\, dx\, dy$ with respect to x first. This gives us $y^2 x\, dy$, which is the expression we chose at the outset.

② Expressing x in terms of y should cause no difficulty if we observe the proportional relationship between the similar triangles.

Sample Problem A/3

Calculate the moments of inertia of the area of a circle about a diametral axis and about the polar axis through the center. Specify the radii of gyration.

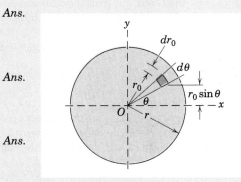

Solution.   A differential element of area in the form of a circular ring may be
① used for the calculation of the moment of inertia about the polar z-axis through O since all elements of the ring are equidistant from O. The elemental area is $dA = 2\pi r_0\, dr_0$, and thus,

$$\left[I_z = \int r^2\, dA\right] \qquad I_z = \int_0^r r_0{}^2(2\pi r_0\, dr_0) = \frac{\pi r^4}{2} = \tfrac{1}{2}Ar^2 \qquad\qquad Ans.$$

The polar radius of gyration is

$$\left[k = \sqrt{\frac{I}{A}}\right] \qquad\qquad k_z = \frac{r}{\sqrt{2}} \qquad\qquad\qquad Ans.$$

By symmetry $I_x = I_y$, so that from Eq. A/3

$$[I_z = I_x + I_y] \qquad\qquad I_x = \tfrac{1}{2}I_z = \frac{\pi r^4}{4} = \tfrac{1}{4}Ar^2 \qquad\qquad Ans.$$

The radius of gyration about the diametral axis is

$$\left[k = \sqrt{\frac{I}{A}}\right] \qquad\qquad k_x = \frac{r}{2} \qquad\qquad\qquad Ans.$$

The foregoing determination of I_x is the simplest possible. The result may also be obtained by direct integration, using the element of area $dA = r_0\, dr_0\, d\theta$ shown in the lower figure. By definition

$$[I_x = \int y^2\, dA] \qquad\qquad I_x = \int_0^{2\pi}\int_0^r (r_0\sin\theta)^2 r_0\, dr_0\, d\theta$$

$$= \int_0^{2\pi} \frac{r^4\sin^2\theta}{4}\, d\theta$$

② $$= \frac{r^4}{4}\,\frac{1}{2}\left[\theta - \frac{\sin 2\theta}{2}\right]_0^{2\pi} = \frac{\pi r^4}{4} \qquad\qquad Ans.$$

Helpful Hints

① Polar coordinates are certainly indicated here. Also, as before, we choose the simplest and lowest-order element possible, which is the differential ring. It should be evident immediately from the definition that the polar moment of inertia of the ring is its area $2\pi r_0\, dr_0$ times $r_0{}^2$.

② This integration is straightforward, but the use of Eq. A/3 along with the result for I_z is certainly simpler.

Sample Problem A/4

Determine the moment of inertia of the area under the parabola about the x-axis. Solve by using (a) a horizontal strip of area and (b) a vertical strip of area.

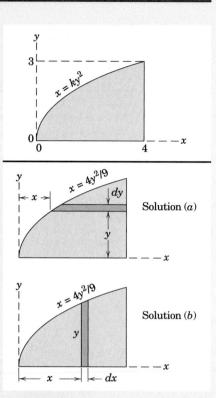

Solution. The constant $k = \frac{4}{9}$ is obtained first by substituting $x = 4$ and $y = 3$ into the equation for the parabola.

(a) Horizontal strip. Since all parts of the horizontal strip are the same distance from the x-axis, the moment of inertia of the strip about the x-axis is $y^2\,dA$ where $dA = (4 - x)\,dy = 4(1 - y^2/9)\,dy$. Integrating with respect to y gives us

$$[I_x = \int y^2\,dA] \qquad I_x = \int_0^3 4y^2\left(1 - \frac{y^2}{9}\right)dy = \frac{72}{5} = 14.40 \text{ (units)}^4 \qquad Ans.$$

(b) Vertical strip. Here all parts of the element are at different distances from the x-axis, so we must use the correct expressions for the moment of inertia of the elemental rectangle about its base, which, from Sample Problem A/1, is $bh^3/3$. For the width dx and the height y the expression becomes

$$dI_x = \tfrac{1}{3}(dx)y^3$$

To integrate with respect to x, we must express y in terms of x, which gives $y = 3\sqrt{x}/2$, and the integral becomes

① $$I_x = \frac{1}{3}\int_0^4 \left(\frac{3\sqrt{x}}{2}\right)^3 dx = \frac{72}{5} = 14.40 \text{ (units)}^4 \qquad Ans.$$

Helpful Hint

① There is little preference between Solutions (a) and (b). Solution (b) requires knowing the moment of inertia for a rectangular area about its base.

Sample Problem A/5

Find the moment of inertia about the x-axis of the semicircular area.

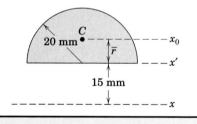

Solution. The moment of inertia of the semicircular area about the x'-axis is one-half of that for a complete circle about the same axis. Thus, from the results of Sample Problem A/3

$$I_{x'} = \frac{1}{2}\frac{\pi r^4}{4} = \frac{20^4\pi}{8} = 2\pi(10^4) \text{ mm}^4$$

We obtain the moment of inertia $\bar{I}$ about the parallel centroidal axis x_0 next. Transfer is made through the distance $\bar{r} = 4r/(3\pi) = (4)(20)/(3\pi) = 80/(3\pi)$ mm by the parallel-axis theorem. Hence,

$$[\bar{I} = I - Ad^2] \qquad \bar{I} = 2(10^4)\pi - \left(\frac{20^2\pi}{2}\right)\left(\frac{80}{3\pi}\right)^2 = 1.755(10^4) \text{ mm}^4$$

① Finally, we transfer from the centroidal x_0-axis to the x-axis. Thus,

$$[I = \bar{I} + Ad^2] \qquad I_x = 1.755(10^4) + \left(\frac{20^2\pi}{2}\right)\left(15 + \frac{80}{3\pi}\right)^2$$

$$= 1.755(10^4) + 34.7(10^4) = 36.4(10^4) \text{ mm}^4 \qquad Ans.$$

Helpful Hint

① This problem illustrates the caution we should observe in using a double transfer of axes since neither the x'- nor the x-axis passes through the centroid C of the area. If the circle were complete with the centroid on the x' axis, only one transfer would be needed.

Sample Problem A/6

Calculate the moment of inertia about the x-axis of the area enclosed between the y-axis and the circular arcs of radius a whose centers are at O and A.

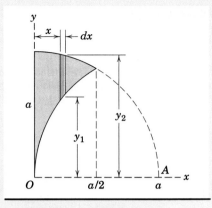

Solution. The choice of a vertical differential strip of area permits one integration to cover the entire area. A horizontal strip would require two integrations with respect to y by virtue of the discontinuity. The moment of inertia of the strip about the x-axis is that of a strip of height y_2 minus that of a strip of height y_1. Thus, from the results of Sample Problem A/1 we write

$$dI_x = \tfrac{1}{3}(y_2 \, dx)y_2{}^2 - \tfrac{1}{3}(y_1 \, dx)y_1{}^2 = \tfrac{1}{3}(y_2{}^3 - y_1{}^3) \, dx$$

The values of y_2 and y_1 are obtained from the equations of the two curves, which are $x^2 + y_2{}^2 = a^2$ and $(x - a)^2 + y_1{}^2 = a^2$, and which give $y_2 = \sqrt{a^2 - x^2}$

① and $y_1 = \sqrt{a^2 - (x - a)^2}$. Thus,

$$I_x = \tfrac{1}{3}\int_0^{a/2} \left\{ (a^2 - x^2)\sqrt{a^2 - x^2} - [a^2 - (x - a)^2]\sqrt{a^2 - (x - a)^2} \right\} dx$$

Simultaneous solution of the two equations which define the two circles gives the x-coordinate of the intersection of the two curves, which, by inspection, is $a/2$. Evaluation of the integrals gives

$$\int_0^{a/2} a^2 \sqrt{a^2 - x^2} \, dx = \frac{a^4}{4}\left(\frac{\sqrt{3}}{2} + \frac{\pi}{3}\right)$$

$$-\int_0^{a/2} x^2 \sqrt{a^2 - x^2} \, dx = \frac{a^4}{16}\left(\frac{\sqrt{3}}{4} + \frac{\pi}{3}\right)$$

$$-\int_0^{a/2} a^2 \sqrt{a^2 - (x - a)^2} \, dx = \frac{a^4}{4}\left(\frac{\sqrt{3}}{2} + \frac{2\pi}{3}\right)$$

$$\int_0^{a/2} (x - a)^2 \sqrt{a^2 - (x - a)^2} \, dx = \frac{a^4}{8}\left(\frac{\sqrt{3}}{8} + \frac{\pi}{3}\right)$$

Collection of the integrals with the factor of $\tfrac{1}{3}$ gives

$$I_x = \frac{a^4}{96}(9\sqrt{3} - 2\pi) = 0.0969a^4 \qquad\qquad Ans.$$

If we had started from a second-order element $dA = dx \, dy$, we would write $y^2 \, dx \, dy$ for the moment of inertia of the element about the x-axis. Integrating from y_1 to y_2 holding x constant produces for the vertical strip

$$dI_x = \left[\int_{y_1}^{y_2} y^2 \, dy\right] dx = \tfrac{1}{3}(y_2{}^3 - y_1{}^3) \, dx$$

which is the expression we started with by having the moment-of-inertia result for a rectangle in mind.

Helpful Hint

① We choose the positive signs for the radicals here since both y_1 and y_2 lie above the x-axis.

PROBLEMS

Introductory Problems

A/1 Use the differential element shown to determine the moment of inertia of the triangular area about the x-axis and about the y-axis.

$$Ans. \ I_x = \tfrac{1}{6}ab^3, \ I_y = \tfrac{1}{2}ba^3$$

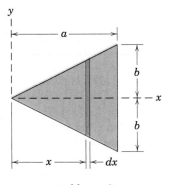

Problem A/1

A/2 Determine the moments of inertia of the rectangular area about the x- and y-axes and find the polar moment of inertia about point O.

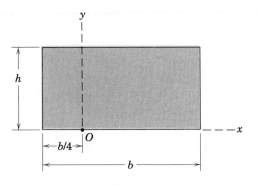

Problem A/2

A/3 The narrow rectangular strip has an area of 300 mm^2, and its moment of inertia about the y-axis is $35(10^3)$ mm^4. Obtain a close approximation to the polar radius of gyration about point O.

$$Ans. \ k_O = 18.48 \text{ mm}$$

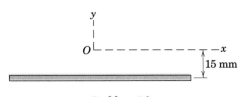

Problem A/3

A/4 Determine the ratio b/h such that $I_x = I_y$ for the area of the isosceles triangle.

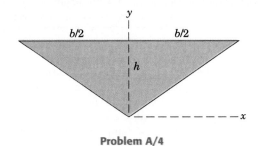

Problem A/4

A/5 Determine by direct integration the moment of inertia of the triangular area about the y-axis.

$$Ans. \ I_y = \frac{hb^3}{4}$$

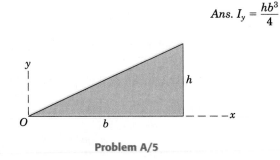

Problem A/5

A/6 The moments of inertia of the area A about the parallel p- and p'-axes differ by 50 in.^{4} Compute the area A, which has its centroid at C.

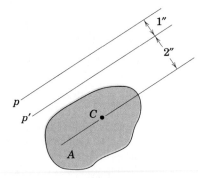

Problem A/6

A/7 The quarter-circular strip has a width b which is small compared with its radius r. Determine the polar moment of inertia of the area about its centroid.

Ans. $I_C = 0.298br^3$

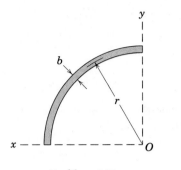

Problem A/7

A/8 Determine the polar moments of inertia of the semicircular area about points A and B.

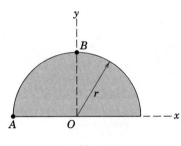

Problem A/8

A/9 Determine the polar radii of gyration of the right-triangular area about point A and about point O.

Ans. $k_A = \sqrt{(h^2 + b^2)/6}, k_O = \sqrt{(h^2 + b^2)/2}$

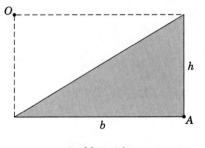

Problem A/9

A/10 Show that the moment of inertia of the rectangular area about the x-axis through one end may be used for its polar moment of inertia about point O when b is small compared with a. What is the percentage error n when $b/a = 1/10$?

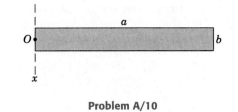

Problem A/10

Representative Problems

A/11 Determine a close approximation of the polar moment of inertia I_O of the thin circular strip of area about point O.

Ans. $I_O = 212(10^6)$ mm^4

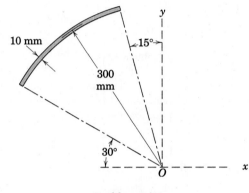

Problem A/11

A/12 Derive expressions for the moments of inertia I_x and I_y of the thin circular strip of area A about the x- and y-axes. Add your results and compare the sum with the answer to Prob. A/11 using the numerical values of that problem.

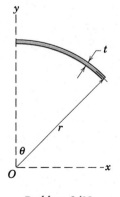

Problem A/12

A/13 Determine the moments of inertia I_x and I_y of the area of the thin semicircular ring about the x- and y-axes. Also find the polar moment of inertia I_C of the ring about its centroid C.

$$\text{Ans. } I_x = I_y = \tfrac{1}{2}\pi r^3 t$$
$$I_C = \pi r^3 t \left(1 - \frac{4}{\pi^2}\right)$$

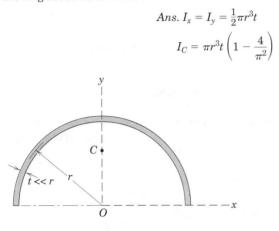

Problem A/13

A/14 Determine the moments of inertia of the quarter-circular area about the x- and y-axes, and find the polar radius of gyration about point O.

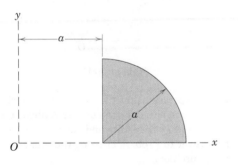

Problem A/14

A/15 Determine the rectangular and polar radii of gyration of the shaded area about the axes shown.

$$\text{Ans. } k_x = 0.754, \ k_y = 1.673, \ k_z = 1.835$$

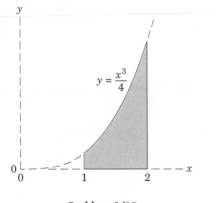

Problem A/15

A/16 Determine the polar radius of gyration of the area of the equilateral triangle of side b about its centroid C.

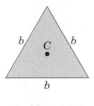

Problem A/16

A/17 Calculate the moment of inertia of the shaded area about the x-axis.

$$\text{Ans. } I_x = 9(10^4) \text{ mm}^4$$

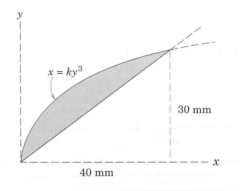

Problem A/17

A/18 Determine the moment of inertia of the elliptical area about the y-axis and find the polar radius of gyration about the origin O of the coordinates.

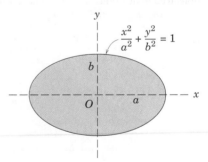

Problem A/18

A/19 Determine the moments of inertia of the shaded area about the x- and y-axes. Use the same differential element for both calculations.

Ans. $I_x = a^4/28$, $I_y = a^4/20$

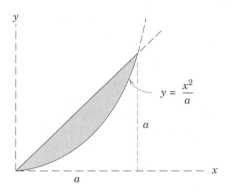

$$y = \frac{x^2}{a}$$

Problem A/19

A/20 Show that the moment of inertia of the area of the square about any axis x' through its center is the same as that about a central axis x parallel to a side.

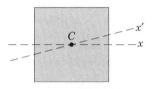

Problem A/20

A/21 Calculate by direct integration the moment of inertia of the shaded area about the x-axis. Solve, first, by using a horizontal strip having differential area and, second, by using a vertical strip of differential area.

Ans. $I_x = 51.2$ in.^{4}

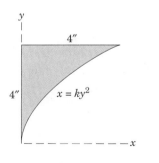

$x = ky^2$

Problem A/21

A/22 The plane figure is symmetrical with respect to the 45° line and has an area of 1600 mm^2. Its polar moment of inertia about its centroid C is $40(10^4)$ mm^4. Compute (*a*) the polar radius of gyration about O and (*b*) the radius of gyration about the x_0-axis.

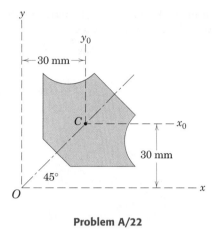

Problem A/22

A/23 Determine the moments of inertia of the shaded area about the y- and y'-axes.

Ans. $I_y = 30{,}000$ in.^{4}, $I_{y'} = 16{,}000$ in.^{4}

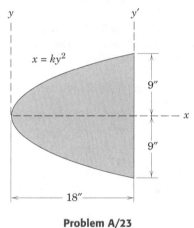

$x = ky^2$

Problem A/23

A/24 Determine the moments of inertia of the shaded area about the x- and y-axes.

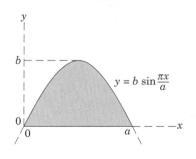

$$y = b \sin \frac{\pi x}{a}$$

Problem A/24

A/25 Determine the moments of inertia of the circular sector about the x- and y-axes.

$$Ans.\ I_x = \frac{R^4}{4}\left(\theta - \frac{1}{2}\sin 2\theta\right)$$

$$I_y = \frac{R^4}{4}\left(\theta + \frac{1}{2}\sin 2\theta\right)$$

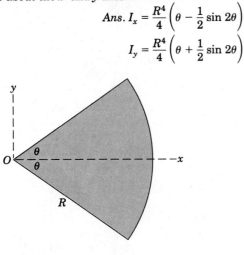

Problem A/25

A/26 Use the results cited for Prob. A/25 to determine the polar moment of inertia of the area of the circular sector about its centroid C.

A/27 Calculate the moment of inertia of the shaded area about the x-axis.

$$Ans.\ I_x = 140.8\ in.^4$$

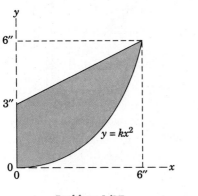

Problem A/27

A/28 Determine the radius of gyration about the y-axis of the shaded area shown.

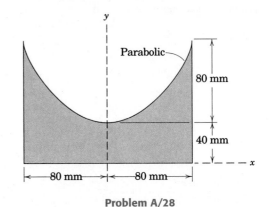

Problem A/28

A/29 Determine the moment of inertia of the shaded area about the y-axis.

$$Ans.\ I_y = 27.8\ (10^4)\ mm^4$$

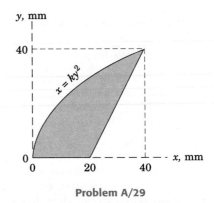

Problem A/29

A/30 From considerations of symmetry show that $I_{x'} = I_{y'} = I_x = I_y$ for the semicircular area regardless of the angle α.

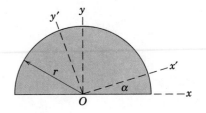

Problem A/30

A/31 Calculate the moments of inertia of the shaded area about the x- and y-axes, and find the polar moment of inertia about point O.

> *Ans.* $I_x = 25.6$ in.^{4}, $I_y = 30.5$ in.^{4}
> $I_O = 56.1$ in.^{4}

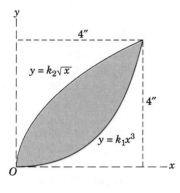

Problem A/31

A/32 The area shown is symmetrical about both the x-y and x'-y' axes. If its polar moment of inertia about O is $16(10^6)$ mm^4, determine the moments of inertia of the area about the x- and x'-axes. Can you also determine the moment of inertia about any axis through O?

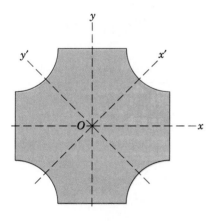

Problem A/32

A/33 By the methods of this article, determine the rectangular and polar radii of gyration of the shaded area about the axes shown.

> *Ans.* $k_x = k_y = \dfrac{\sqrt{5}}{4}\,a,\ k_z = \dfrac{\sqrt{10}}{4}\,a$

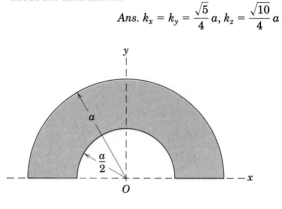

Problem A/33

▶**A/34** Calculate the moment of inertia of the shaded area of the two overlapping circles about the x-axis.

> *Ans.* $I_x = 0.1988r^4$

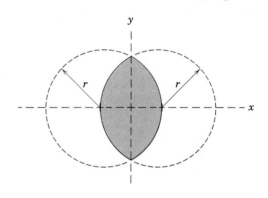

Problem A/34

A/3 COMPOSITE AREAS

It is frequently necessary to calculate the moment of inertia of an area composed of a number of distinct parts of simple and calculable geometric shape. Because a moment of inertia is the integral or sum of the products of distance squared times element of area, it follows that the moment of inertia of a positive area is always a positive quantity. The moment of inertia of a composite area about a particular axis is therefore simply the sum of the moments of inertia of its component parts about the same axis. It is often convenient to regard a composite area as being composed of positive and negative parts. We may then treat the moment of inertia of a negative area as a negative quantity.

When a composite area is composed of a large number of parts, it is convenient to tabulate the results for each of the parts in terms of its area A, its centroidal moment of inertia $\bar{I}$, the distance d from its centroidal axis to the axis about which the moment of inertia of the entire section is being computed, and the product Ad^2. For any one of the parts the moment of inertia about the desired axis by the transfer-of-axis theorem is $\bar{I} + Ad^2$. Thus, for the entire section the desired moment of inertia becomes $I = \Sigma\bar{I} + \Sigma Ad^2$.

For such an area in the x-y plane, for example, and with the notation of Fig. A/4, where $\bar{I}_x$ is the same as I_{x_0} and $\bar{I}_y$ is the same as I_{y_0} the tabulation would include

Part	Area, A	d_x	d_y	Ad_x^2	Ad_y^2	$\bar{I}_x$	$\bar{I}_y$
Sums	ΣA			ΣAd_x^2	ΣAd_y^2	$\Sigma\bar{I}_x$	$\Sigma\bar{I}_y$

From the sums of the four columns, then, the moments of inertia for the composite area about the x- and y-axes become

$$I_x = \Sigma\bar{I}_x + \Sigma Ad_x^2$$
$$I_y = \Sigma\bar{I}_y + \Sigma Ad_y^2$$

Although we may add the moments of inertia of the individual parts of a composite area about a given axis, we may not add their radii of gyration. The radius of gyration for the composite area about the axis in question is given by $k = \sqrt{I/A}$, where I is the total moment of inertia and A is the total area of the composite figure. Similarly, the radius of gyration k about a polar axis through some point equals $\sqrt{I_z/A}$, where $I_z = I_x + I_y$ for x-y axes through that point.

Sample Problem A/7

Calculate the moment of inertia and radius of gyration about the x-axis for the shaded area shown.

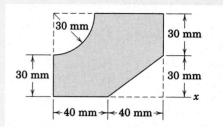

Solution. The composite area is composed of the positive area of the rectangle (1) and the negative areas of the quarter circle (2) and triangle (3). For the rectangle the moment of inertia about the x-axis, from Sample Problem A/1 (or Table D/3), is

$$I_x = \tfrac{1}{3}Ah^2 = \tfrac{1}{3}(80)(60)(60)^2 = 5.76(10^6) \text{ mm}^4$$

From Sample Problem A/3 (or Table D/3), the moment of inertia of the negative quarter-circular area about its base axis x' is

$$I_{x'} = -\frac{1}{4}\left(\frac{\pi r^4}{4}\right) = -\frac{\pi}{16}(30)^4 = -0.1590(10^6) \text{ mm}^4$$

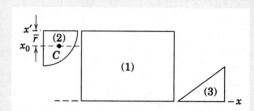

We now transfer this result through the distance $\bar{r} = 4r/(3\pi) = 4(30)/(3\pi) = 12.73$ mm by the transfer-of-axis theorem to get the centroidal moment of inertia of part (2) (or use Table D/3 directly).

① $[\bar{I} = I - Ad^2]$ $\quad \bar{I}_x = -0.1590(10^6) - \left[-\dfrac{\pi(30)^2}{4}(12.73)^2\right]$

$$= -0.0445(10^6) \text{ mm}^4$$

The moment of inertia of the quarter-circular part about the x-axis is now

② $[I = \bar{I} + Ad^2]$ $\quad I_x = -0.0445(10^6) + \left[-\dfrac{\pi(30)^2}{4}\right](60 - 12.73)^2$

$$= -1.624(10^6) \text{ mm}^4$$

Finally, the moment of inertia of the negative triangular area (3) about its base, from Sample Problem A/2 (or Table D/3), is

$$I_x = -\tfrac{1}{12}bh^3 = -\tfrac{1}{12}(40)(30)^3 = -0.90(10^6) \text{ mm}^4$$

The total moment of inertia about the x-axis of the composite area is, consequently,

③ $\quad I_x = 5.76(10^6) - 1.624(10^6) - 0.09(10^6) = 4.05(10^6) \text{ mm}^4$ **Ans.**

The net area of the figure is $A = 60(80) - \tfrac{1}{4}\pi(30)^2 - \tfrac{1}{2}(40)(30) = 3490 \text{ mm}^2$ so that the radius of gyration about the x-axis is

$$k_x = \sqrt{I_x/A} = \sqrt{4.05(10^6)/3490} = 34.0 \text{ mm} \qquad \textbf{Ans.}$$

Helpful Hints

① Note that we must transfer the moment of inertia for the quarter-circular area to its centroidal axis x_0 before we can transfer it to the x-axis, as was done in Sample Problem A/5.

② We watch our signs carefully here. Since the area is negative, both $\bar{I}$ and A carry negative signs.

③ If there had been more than the three parts to the composite area, we would have arranged a tabulation of the $\bar{I}$ terms and the Ad^2 terms so as to keep a systematic account of the terms and obtain $I = \Sigma\bar{I} + \Sigma Ad^2$.

PROBLEMS

Introductory Problems

A/35 Determine the percent reduction n in the polar moment of inertia of the rectangular plate due to the introduction of the rectangular hole. The x-y axes are centered on the plate.

Ans. $n = 6.25\%$

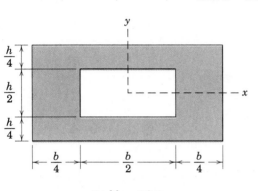

Problem A/35

A/36 Determine the moment of inertia about the x-axis of the square area without and with the central circular hole.

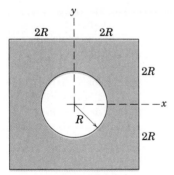

Problem A/36

A/37 Determine the polar moment of inertia of the circular area without and with the central square hole.

Ans. $I_z = 1.571R^4$, without hole
$I_z = 1.404R^4$, with hole

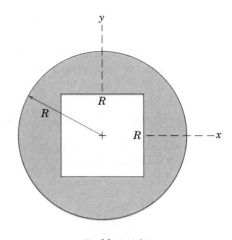

Problem A/37

A/38 Calculate the polar radius of gyration of the shaded area about the center O of the larger circle.

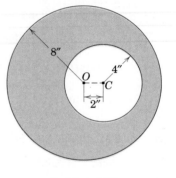

Problem A/38

A/39 The cross-sectional area of a wide-flange I-beam has the dimensions shown. Obtain a close approximation to the handbook value of $\bar{I}_x = 657$ in.^{4} by treating the section as being composed of three rectangles.

Ans. $\bar{I}_x = 649$ in.^{4}

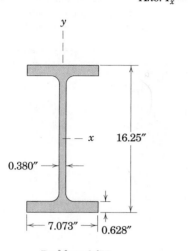

Problem A/39

A/40 Determine the moment of inertia of the shaded area about the x-axis in two ways. The wall thickness is 20 mm on all four sides of the rectangle.

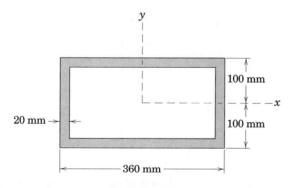

Problem A/40

A/41 Calculate the moment of inertia of the shaded area about the x-axis.

Ans. $I_x = 4.53(10^6)$ mm^4

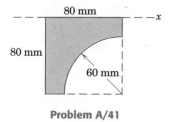

Problem A/41

A/42 Calculate the moment of inertia of the cross section of the beam about its centroidal x_0-axis.

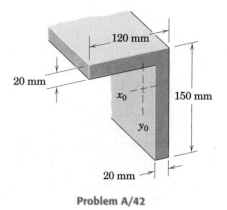

Problem A/42

Representative Problems

A/43 Determine the moments of inertia of the Z-section about its centroidal x_0- and y_0-axes.

Ans. $\bar{I}_x = 22.6(10^6)$ mm^4
$\bar{I}_y = 9.81(10^6)$ mm^4

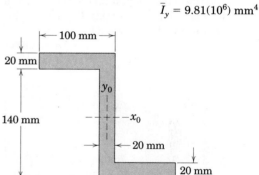

Problem A/43

A/44 Calculate the rectangular and polar moments of inertia of the shaded area about the axes shown.

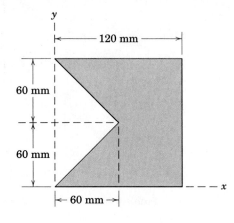

Problem A/44

A/45 A floor joist which measures a full 2 in. by 8 in. has a 1-in. hole drilled through it for a water-pipe installation. Determine the percent reduction n in the moment of inertia of the cross-sectional area about the x-axis (compared with that of the undrilled joist) for hole locations in the range $0 \le y \le 3.5$ in. Evaluate your expression for $y = 2$ in.

$$Ans.\ n = 0.1953 + 2.34y^2$$
$$n = 9.57\%$$

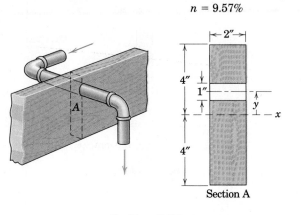

Problem A/45

A/46 The rectangular area shown in part a of the figure is split into three equal areas which are then arranged as shown in part b of the figure. Determine an expression for the moment of inertia of the area in part b about the centroidal x-axis. What percent increase n over the moment of inertia for area a does this represent if $h = 200$ mm and $b = 60$ mm?

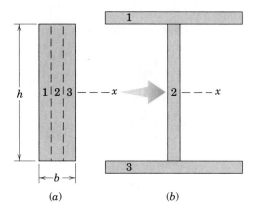

Problem A/46

A/47 Calculate the moment of inertia of the shaded area about the x-axis.

$$Ans.\ I_x = 15.64\ \text{in.}^4$$

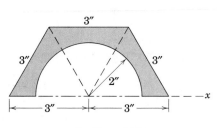

Problem A/47

A/48 Develop a formula for the moment of inertia of the regular hexagonal area of side b about its central x-axis.

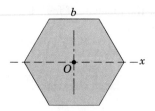

Problem A/48

A/49 Make use of the results of Prob. A/34 and determine the moment of inertia I_y of the shaded area shown.

Ans. $I_y = 0.388r^4$

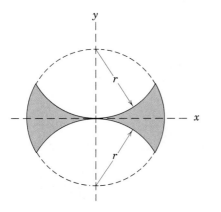

Problem A/49

A/50 Approximate the moment of inertia about the x-axis of the semicircular area by dividing it into five horizontal strips of equal width. Treat the moment of inertia of each strip as its area (width times length of its horizontal midline) times the square of the distance from its midline to the x-axis. Compare your result with the exact value.

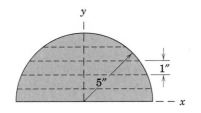

Problem A/50

A/51 Calculate the polar radius of gyration of the shaded area about its centroid C.

Ans. $k_C = 261$ mm

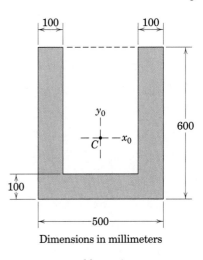

Dimensions in millimeters

Problem A/51

A/52 A hollow mast of circular section as shown is to be stiffened by bonding two strips of the same material and of rectangular section to the mast throughout its length. Determine the proper dimension h of each near-rectangle which will exactly double the stiffness of the mast to bending in the y-z plane. (Stiffness in the y-z plane is proportional to the area moment of inertia about the x-axis.) Take the inner boundary of each strip to be a straight line.

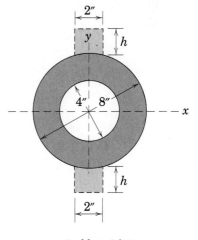

Problem A/52

A/53 The cross section of a bearing block is shown in the figure by the shaded area. Calculate the moment of inertia of the section about its base a-a.

Ans. $I_{a\text{-}a} = 886$ in.^{4}

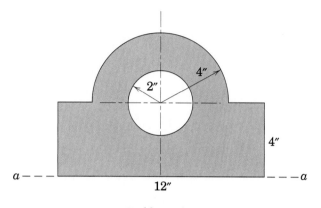

Problem A/53

A/54 Calculate the moments of inertia of the shaded area about the x- and y-axes.

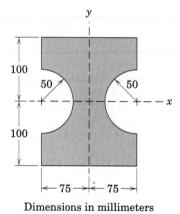

Dimensions in millimeters

Problem A/54

A/55 Calculate the moment of inertia of the standard 12 × 4-in. channel section about the centroidal x_0-axis. Neglect the fillets and rounds and compare with the handbook value of $\bar{I}_x = 16.0$ in.^{4}

Ans. $\bar{I}_x = 16.00$ in.^{4}

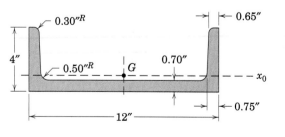

Problem A/55

A/56 Determine the polar moment of inertia about point O of the shaded area shown.

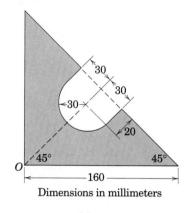

Dimensions in millimeters

Problem A/56

▶**A/57** For the 2 in. by 8 in. floor joist with the circular hole of Prob. A/45, determine an expression for the percent n by which the moment of inertia of the shaded area about its centroidal x'-axis (parallel to x) is reduced from the moment of inertia of the complete undrilled section about the x-axis. Express n in terms of y for the range $0 \leq y \leq 3.5$ in. Evaluate your expression for $y = 2$ in.

Ans. $n = 0.1953 + 2.68y^2$, $n = 10.91\%$

▶**A/58** For the H-beam section, determine the flange width b which will make the moments of inertia about the central x- and y-axes equal. (*Hint:* The solution of a cubic equation is required here. Refer to Art. C/4 or C/11 of Appendix C for solving a cubic equation.)

Ans. $b = 161.1$ mm

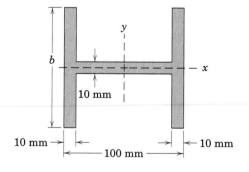

Problem A/58

A/4 PRODUCTS OF INERTIA AND ROTATION OF AXES

In this article, we define the product of inertia with respect to rectangular axes and develop the parallel-axis theorem for centroidal and noncentroidal axes. In addition, we discuss the effects of rotation of axes on moments and products of inertia.

Definition

In certain problems involving unsymmetrical cross sections and in the calculation of moments of inertia about rotated axes, an expression $dI_{xy} = xy \, dA$ occurs, which has the integrated form

$$I_{xy} = \int xy \, dA \qquad (A/7)$$

where x and y are the coordinates of the element of area $dA = dx \, dy$. The quantity I_{xy} is called the *product of inertia* of the area A with respect to the x-y axes. Unlike moments of inertia, which are always positive for positive areas, the product of inertia may be positive, negative, or zero.

The product of inertia is zero whenever either of the reference axes is an axis of symmetry, such as the x-axis for the area in Fig. A/5. Here we see that the sum of the terms $x(-y) \, dA$ and $x(+y) \, dA$ due to symmetrically placed elements vanishes. Because the entire area may be considered as composed of pairs of such elements, it follows that the product of inertia I_{xy} for the entire area is zero.

Transfer of Axes

By definition the product of inertia of the area A in Fig. A/4 with respect to the x- and y-axes in terms of the coordinates x_0, y_0 to the centroidal axes is

$$I_{xy} = \int (x_0 + d_y)(y_0 + d_x) \, dA$$

$$= \int x_0 y_0 \, dA + d_x \int x_0 \, dA + d_y \int y_0 \, dA + d_x d_y \int dA$$

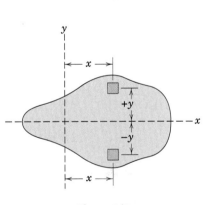

Figure A/5

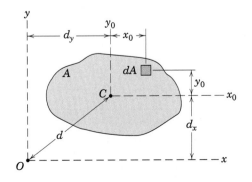

Figure A/4, Repeated

The first integral is by definition the product of inertia about the centroidal axes, which we write as $\bar{I}_{xy}$. The middle two integrals are both zero because the first moment of the area about its own centroid is necessarily zero. The fourth term is merely $d_x d_y A$. Thus, the transfer-of-axis theorem for products of inertia becomes

$$I_{xy} = \bar{I}_{xy} + d_x d_y A \qquad\qquad \textbf{(A/8)}$$

Rotation of Axes

The product of inertia is useful when we need to calculate the moment of inertia of an area about inclined axes. This consideration leads directly to the important problem of determining the axes about which the moment of inertia is a maximum and a minimum.

In Fig. A/6 the moments of inertia of the area about the x'- and y'-axes are

$$I_{x'} = \int y'^2 \, dA = \int (y \cos \theta - x \sin \theta)^2 \, dA$$

$$I_{y'} = \int x'^2 \, dA = \int (y \sin \theta + x \cos \theta)^2 \, dA$$

where x' and y' have been replaced by their equivalent expressions as seen from the geometry of the figure.

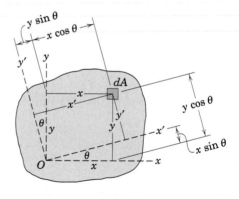

Figure A/6

Expanding and substituting the trigonometric identities

$$\sin^2 \theta = \frac{1 - \cos 2\theta}{2} \qquad \cos^2 \theta = \frac{1 + \cos 2\theta}{2}$$

and the defining relations for I_x, I_y, I_{xy} give us

$$I_{x'} = \frac{I_x + I_y}{2} + \frac{I_x - I_y}{2} \cos 2\theta - I_{xy} \sin 2\theta$$

$$I_{y'} = \frac{I_x + I_y}{2} - \frac{I_x - I_y}{2} \cos 2\theta + I_{xy} \sin 2\theta \qquad\qquad \textbf{(A/9)}$$

In a similar manner we write the product of inertia about the inclined axes as

$$I_{x'y'} = \int x'y' \, dA = \int (y \sin \theta + x \cos \theta)(y \cos \theta - x \sin \theta) \, dA$$

Expanding and substituting the trigonometric identities

$$\sin \theta \cos \theta = \tfrac{1}{2} \sin 2\theta \qquad \cos^2 \theta - \sin^2 \theta = \cos 2\theta$$

and the defining relations for I_x, I_y, I_{xy} give us

$$I_{x'y'} = \frac{I_x - I_y}{2} \sin 2\theta + I_{xy} \cos 2\theta \qquad \textbf{(A/9a)}$$

Adding Eqs. A/9 gives $I_{x'} + I_{y'} = I_x + I_y = I_z$, the polar moment of inertia about O, which checks the results of Eq. A/3.

The angle which makes $I_{x'}$ and $I_{y'}$ either maximum or minimum may be determined by setting the derivative of either $I_{x'}$ or $I_{y'}$ with respect to θ equal to zero. Thus,

$$\frac{dI_{x'}}{d\theta} = (I_y - I_x) \sin 2\theta - 2I_{xy} \cos 2\theta = 0$$

Denoting this critical angle by α gives

$$\tan 2\alpha = \frac{2I_{xy}}{I_y - I_x} \qquad \textbf{(A/10)}$$

Equation A/10 gives two values for 2α which differ by π, since $\tan 2\alpha = \tan (2\alpha + \pi)$. Consequently the two solutions for α will differ by $\pi/2$. One value defines the axis of maximum moment of inertia, and the other value defines the axis of minimum moment of inertia. These two rectangular axes are called the *principal axes of inertia*.

When we substitute Eq. A/10 for the critical value of 2θ in Eq. A/9a, we see that the product of inertia is zero for the principal axes of inertia. Substitution of $\sin 2\alpha$ and $\cos 2\alpha$, obtained from Eq. A/10, for $\sin 2\theta$ and $\cos 2\theta$ in Eqs. A/9 gives the expressions for the principal moments of inertia as

$$I_{max} = \frac{I_x + I_y}{2} + \frac{1}{2}\sqrt{(I_x - I_y)^2 + 4I_{xy}^2}$$
$$I_{min} = \frac{I_x + I_y}{2} - \frac{1}{2}\sqrt{(I_x - I_y)^2 + 4I_{xy}^2} \qquad \textbf{(A/11)}$$

Mohr's Circle of Inertia

We may represent the relations in Eqs. A/9, A/9a, A/10, and A/11 graphically by a diagram called *Mohr's circle*. For given values of I_x, I_y,

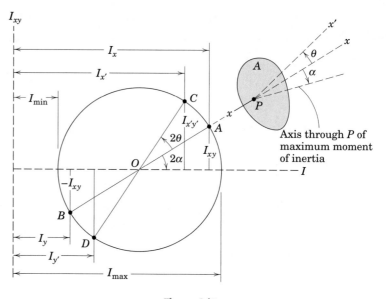

Figure A/7

and I_{xy} the corresponding values of $I_{x'}$, $I_{y'}$, and $I_{x'y'}$ may be determined from the diagram for any desired angle θ. A horizontal axis for the measurement of moments of inertia and a vertical axis for the measurement of products of inertia are first selected, Fig. A/7. Next, point A, which has the coordinates (I_x, I_{xy}), and point B, which has the coordinates $(I_y, -I_{xy})$, are located.

We now draw a circle with these two points as the extremities of a diameter. The angle from the radius OA to the horizontal axis is 2α or twice the angle from the x-axis of the area in question to the axis of maximum moment of inertia. The angle on the diagram and the angle on the area are both measured in the same sense as shown. The coordinates of any point C are $(I_{x'}, I_{x'y'})$, and those of the corresponding point D are $(I_{y'}, -I_{x'y'})$. Also the angle between OA and OC is 2θ or twice the angle from the x-axis to the x'-axis. Again we measure both angles in the same sense as shown. We may verify from the trigonometry of the circle that Eqs. A/9, A/9a, and A/10 agree with the statements made.

Sample Problem A/8

Determine the product of inertia of the rectangular area with centroid at C with respect to the x-y axes parallel to its sides.

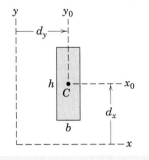

Solution. Since the product of inertia $\bar{I}_{xy}$ about the axes x_0-y_0 is zero by symmetry, the transfer-of-axis theorem gives us

$$[I_{xy} = \bar{I}_{xy} + d_x d_y A] \qquad\qquad I_{xy} = d_x d_y b h \qquad\qquad Ans.$$

In this example both d_x and d_y are shown positive. We must be careful to be consistent with the positive directions of d_x and d_y as defined, so that their proper signs are observed.

Sample Problem A/9

Determine the product of inertia about the x-y axes for the area under the parabola.

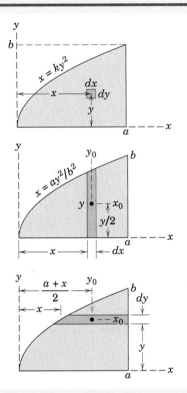

Solution. With the substitution of $x = a$ when $y = b$, the equation of the curve becomes $x = ay^2/b^2$.

Solution I. If we start with the second-order element $dA = dx\, dy$, we have $dI_{xy} = xy\, dx\, dy$. The integral over the entire area is

$$I_{xy} = \int_0^b \int_{ay^2/b^2}^{a} xy\, dx\, dy = \int_0^b \frac{1}{2}\left(a^2 - \frac{a^2 y^4}{b^4}\right) y\, dy = \frac{1}{6}a^2 b^2 \qquad\qquad Ans.$$

Solution II. Alternatively we can start with a first-order elemental strip and save one integration by using the results of Sample Problem A/8. Taking a vertical strip $dA = y\, dx$ gives $dI_{xy} = 0 + (\frac{1}{2}y)(x)(y\, dx)$, where the distances to the centroidal axes of the elemental rectangle are $d_x = y/2$ and $d_y = x$. Now we have

$$I_{xy} = \int_0^a \frac{y^2}{2} x\, dx = \int_0^a \frac{x b^2}{2a} x\, dx = \frac{b^2}{6a} x^3 \Big|_0^a = \frac{1}{6}a^2 b^2 \qquad\qquad Ans.$$

Helpful Hint

① If we had chosen a horizontal strip, our expression would have become $dI_{xy} = y\frac{1}{2}(a+x)[(a-x)\, dy]$, which when integrated, of course, gives us the same result as before.

Sample Problem A/10

Determine the product of inertia of the semicircular area with respect to the x-y axes.

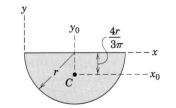

① **Solution.** We use the transfer-of-axis theorem, Eq. A/8, to write

$$[I_{xy} = \bar{I}_{xy} + d_x d_y A] \qquad I_{xy} = 0 + \left(-\frac{4r}{3\pi}\right)(r)\left(\frac{\pi r^2}{2}\right) = -\frac{2r^4}{3} \qquad\qquad Ans.$$

where the x- and y-coordinates of the centroid C are $d_y = +r$ and $d_x = -4r/(3\pi)$. Because y_0 is an axis of symmetry, $\bar{I}_{xy} = 0$.

Helpful Hint

① Proper use of the transfer-of-axis theorem saves a great deal of labor in computing products of inertia.

Sample Problem A/11

Determine the orientation of the principal axes of inertia through the centroid of the angle section and determine the corresponding maximum and minimum moments of inertia.

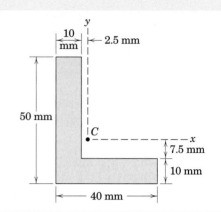

Solution. The location of the centroid C is easily calculated, and its position is shown on the diagram.

Products of Inertia. The product of inertia for each rectangle about its centroidal axes parallel to the x-y axes is zero by symmetry. Thus, the product of inertia about the x-y axes for part I is

$$[I_{xy} = \bar{I}_{xy} + d_x d_y A] \qquad I_{xy} = 0 + (-12.5)(+7.5)(400) = -3.75(10^4) \text{ mm}^4$$

where $\qquad d_x = -(7.5 + 5) = -12.5 \text{ mm}$

and $\qquad d_y = +(20 - 10 - 2.5) = 7.5 \text{ mm}$

Likewise for part II,

$$[I_{xy} = \bar{I}_{xy} + d_x d_y A] \qquad I_{xy} = 0 + (12.5)(-7.5)(400) = -3.75(10^4) \text{ mm}^4$$

where $\quad d_x = +(20 - 7.5) = 12.5 \text{ mm}, \qquad d_y = -(5 + 2.5) = -7.5 \text{ mm}$

For the complete angle

$$I_{xy} = -3.75(10^4) - 3.75(10^4) = -7.50(10^4) \text{ mm}^4$$

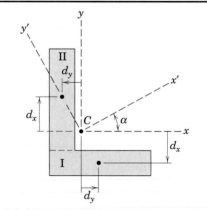

Moments of Inertia. The moments of inertia about the x- and y-axes for part I are

$$[I = \bar{I} + Ad^2] \qquad I_x = \tfrac{1}{12}(40)(10)^3 + (400)(12.5)^2 = 6.58(10^4) \text{ mm}^4$$
$$I_y = \tfrac{1}{12}(10)(40)^3 + (400)(7.5)^2 = 7.58(10^4) \text{ mm}^4$$

and the moments of inertia for part II about these same axes are

$$[I = \bar{I} + Ad^2] \qquad I_x = \tfrac{1}{12}(10)(40)^3 + (400)(12.5)^2 = 11.58(10^4) \text{ mm}^4$$
$$I_y = \tfrac{1}{12}(40)(10)^3 + (400)(7.5)^2 = 2.58(10^4) \text{ mm}^4$$

Thus, for the entire section we have

$$I_x = 6.58(10)^4 + 11.58(10)^4 = 18.17(10^4) \text{ mm}^4$$
$$I_y = 7.58(10^4) + 2.58(10^4) = 10.17(10^4) \text{ mm}^4$$

Principal Axes. The inclination of the principal axes of inertia is given by Eq. A/10, so we have

$$\left[\tan 2\alpha = \frac{2I_{xy}}{I_y - I_x} \right] \qquad \tan 2\alpha = \frac{2(-7.50)}{10.17 - 18.17} = 1.875$$

$$2\alpha = 61.9° \qquad \alpha = 31.0° \qquad\qquad Ans.$$

We now compute the principal moments of inertia from Eqs. A/9 using α for θ and get I_{max} from $I_{x'}$ and I_{min} from $I_{y'}$. Thus,

$$I_{max} = \left[\frac{18.17 + 10.17}{2} + \frac{18.17 - 10.17}{2}(0.471) + (7.50)(0.882) \right](10^4)$$

$$= 22.7(10^4) \text{ mm}^4 \qquad\qquad Ans.$$

$$I_{min} = \left[\frac{18.17 + 10.17}{2} - \frac{18.17 - 10.17}{2}(0.471) - (7.50)(0.882) \right](10^4)$$

$$= 5.67(10^4) \text{ mm}^4 \qquad\qquad Ans.$$

Helpful Hint

Mohr's circle. Alternatively we could use Eqs. A/11 to obtain the results for I_{max} and I_{min}, or we could construct the Mohr circle from the calculated values of I_x, I_y, and I_{xy}. These values are spotted on the diagram to locate points A and B, which are the extremities of the diameter of the circle. The angle 2α and I_{max} and I_{min} are obtained from the figure, as shown.

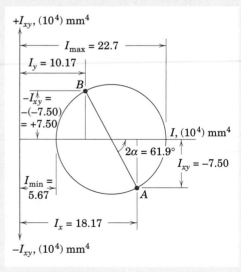

PROBLEMS

Introductory Problems

A/59 Determine the product of inertia of each of the three rectangular areas about the *x-y* axes.

Ans. (a) $I_{xy} = 360(10^4)$ mm^4, (b) $I_{xy} = -360(10^4)$ mm^4
(c) $I_{xy} = -90(10^4)$ mm^4

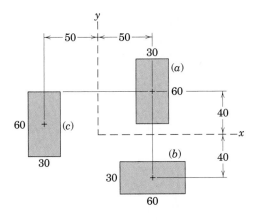

Dimensions in millimeters

Problem A/59

A/60 Determine the product of inertia about the *x-y* axes of the circular area with two equal square holes.

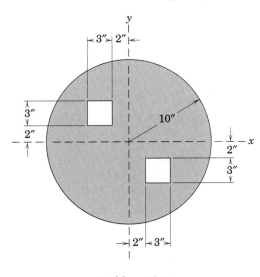

Problem A/60

A/61 Determine I_x, I_y, and I_{xy} for the rectangular plate with three equal circular holes.

Ans. $I_x = 2.44(10^8)$ mm^4, $I_y = 9.80(10^8)$ mm^4
$I_{xy} = -14.14(10^6)$ mm^4

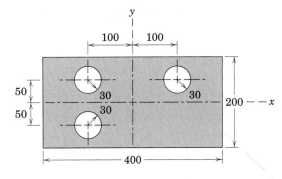

Dimensions in millimeters

Problem A/61

A/62 Determine the product of inertia of the rectangular area about the *x-y* axes. Treat the case where *b* is small compared with *L*.

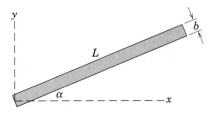

Problem A/62

A/63 Determine the product of inertia of each of the four areas about the *x-y* axes.

Ans. (a) and (c) $I_{xy} = 9.60(10^6)$ mm^4
(b) $I_{xy} = -4.71(10^6)$ mm^4
(d) $I_{xy} = -2.98(10^6)$ mm^4

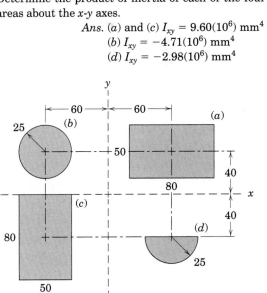

Dimensions in millimeters

Problem A/63

A/64 Determine the product of inertia of the shaded area about the *x-y* axes.

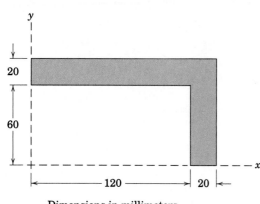

Dimensions in millimeters

Problem A/64

A/65 Determine the products of inertia of the area of the quarter-circular ring about the *x-y* axes and about the x_0-y_0 axes through the centroid C of the ring. Treat b as very small compared with r.

$$Ans.\ I_{xy} = \tfrac{1}{2}r^3b$$
$$I_{x_0y_0} = \left(\tfrac{1}{2} - \tfrac{2}{\pi}\right)r^3b$$

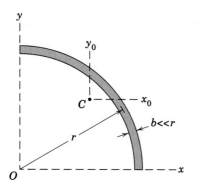

Problem A/65

A/66 Obtain the product of inertia of the quarter-circular area with respect to the *x-y* axes and use this result to obtain the product of inertia with respect to the parallel centroidal axes.

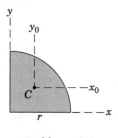

Problem A/66

A/67 Solve for the product of inertia of the semicircular area about the *x-y* axes in two different ways.

$$Ans.\ I_{xy} = \tfrac{2}{3}r^4$$

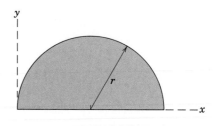

Problem A/67

A/68 Determine the product of inertia of the shaded area about the *x-y* axes.

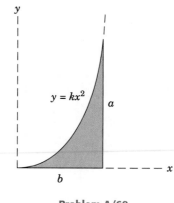

Problem A/68

Representative Problems

A/69 Determine the product of inertia of the trapezoidal area about the *x-y* axes.

$$Ans. \ I_{xy} = \frac{h^2}{24}(3a^2 + 2ab + b^2)$$

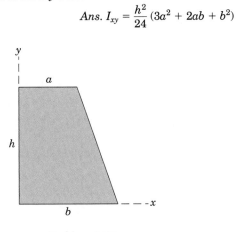

Problem A/69

A/70 Determine by direct integration the product of inertia I_{xy} for the shaded area shown. Indicate an alternative approach.

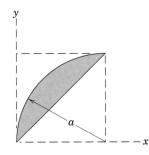

Problem A/70

A/71 Derive the expressions for the product of inertia of the right-triangular area about the *x-y* axes and about the centroidal x_0-y_0 axes.

$$Ans. \ I_{xy} = \frac{b^2h^2}{24}, \ I_{x_0y_0} = -\frac{b^2h^2}{72}$$

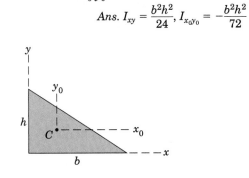

Problem A/71

A/72 Derive the expression for the product of inertia of the right-triangular area about the *x-y* axes. Solve, first, by double integration and, second, by single integration starting with a vertical strip as the element.

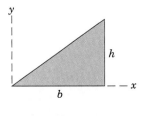

Problem A/72

A/73 By inspection indicate the axis through *O* of maximum moment of inertia. Find $I_{\max}$.

$$Ans. \ I_{\max} = 0.684r^4$$

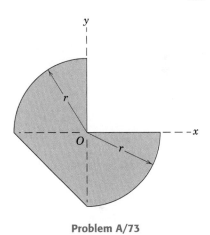

Problem A/73

A/74 Determine the moments and product of inertia of the area of the equilateral triangle with respect to the *x'-y'* axes.

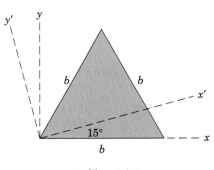

Problem A/74

A/75 Determine the moments and product of inertia of the quarter-circular area with respect to the x'-y' axes.

$$Ans. \ I_{x'} = \frac{r^4}{16}(\pi - \sqrt{3}), \ I_{y'} = \frac{r^4}{16}(\pi + \sqrt{3})$$

$$I_{x'y'} = \frac{r^4}{16}$$

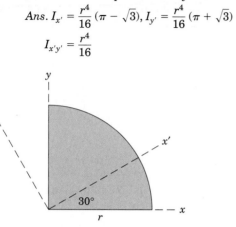

Problem A/75

A/76 Determine the moments and product of inertia of the area of the square with respect to the x'-y' axes.

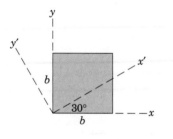

Problem A/76

A/77 Determine the maximum and minimum moments of inertia with respect to centroidal axes through C for the composite of the four square areas shown. Find the angle α measured from the x-axis to the axis of maximum moment of inertia.

$$Ans. \ I_{max} = 5.57a^4$$
$$I_{min} = 1.097a^4$$
$$\alpha = 103.3°$$

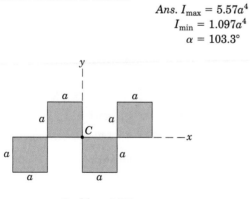

Problem A/77

A/78 Sketch the Mohr circle of inertia for each of the four rectangular areas with the proportions and positions shown. Indicate on each diagram point A which has coordinates (I_x, I_{xy}) and the angle 2α, where α is the angle from the x-axis to the axis of maximum moment of inertia.

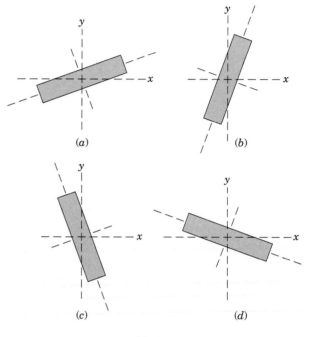

(a)

(b)

(c)

(d)

Problem A/78

A/79 Determine the proportions of the rectangular area for which the moment of inertia about an x'-axis through the midpoint C of the base is a constant value regardless of θ.

$$Ans. \ a = 2b$$

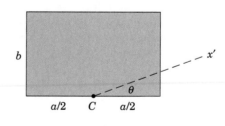

Problem A/79

A/80 Determine the maximum and minimum moments of inertia for the shaded area about axes through point O and identify the angle θ to the axis of minimum moment of inertia.

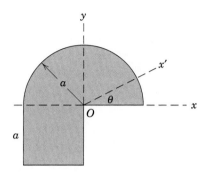

Problem A/80

A/81 Determine the minimum and maximum moments of inertia with respect to centroidal axes through C for the composite of the two rectangular areas shown. Find the angle α measured from the x-axis to the axis of maximum moment of inertia.

Ans. $I_{\min} = 0.505a^4$
$I_{\max} = 6.16a^4$, $\alpha = 112.5°$

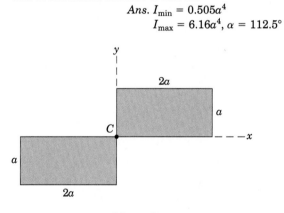

Problem A/81

A/82 The maximum and minimum moments of inertia of the shaded area are $12(10^6)$ mm^4 and $2(10^6)$ mm^4, respectively, about axes passing through the centroid C, and the product of inertia with respect to the x-y axes has a magnitude of $4(10^6)$ mm^4. Use the proper sign for the product of inertia and calculate I_x and the angle α measured from the x-axis to the axis of maximum moment of inertia.

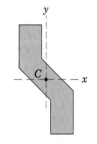

Problem A/82

A/83 The moments and product of inertia of an area with respect to the x-y axes are $I_x = 14$ in.^{4}, $I_y = 24$ in.^{4}, and $I_{xy} = 12$ in.^{4} Construct the Mohr circle of inertia and use it to determine the principal moments of inertia and the angle α from the x-axis to the axis of maximum moment of inertia.

Ans. $I_{\max} = 32$ in.^{4}, $I_{\min} = 6$ in.^{4}
$\alpha = 56.3°$ CW

A/84 Determine the angle α which locates the principal axes of inertia through point O for the rectangular area. Construct the Mohr's circle of inertia and specify the corresponding values of $I_{\max}$ and $I_{\min}$.

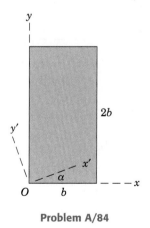

Problem A/84

A/85 Calculate the maximum and minimum moments of inertia of the structural angle about axes through its corner A and find the angle α measured counterclockwise from the x-axis to the axis of maximum moment of inertia. Neglect the small rounds and fillet.

Ans. $I_{max} = 1.782(10^6)$ mm^4, $I_{min} = 0.684(10^6)$ mm^4
$\alpha = -13.40°$

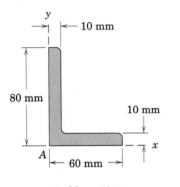

Problem A/85

 *Computer-Oriented Problems**

*A/86 Plot the moment of inertia of the shaded area about the x'-axis as a function of θ from $\theta = 0$ to $\theta = 90°$ and determine the minimum value of $I_{x'}$ and the corresponding value of θ.

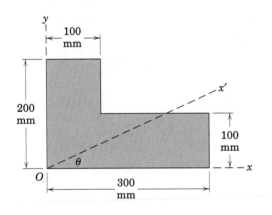

Problem A/86

*A/87 Plot the moment of inertia of the shaded area about the x'-axis for the range $0 \leq \theta \leq 90°$. Determine the minimum value of $I_{x'}$ and the corresponding value of θ.

Ans. $I_{min} = 3.03(10^6)$ mm^4 at $\theta = 64.1°$

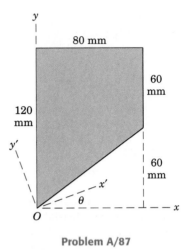

Problem A/87

*A/88 Plot the moment of inertia of the shaded area about the x'-axis as a function of θ from $\theta = 0$ to $\theta = 180°$. Determine the maximum and minimum values of $I_{x'}$ and the corresponding values of θ from the graph. Check your results by applying Eqs. A/10 and A/11.

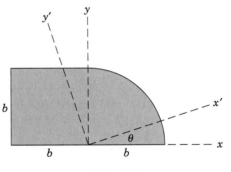

Problem A/88

*A/89 The figure shows the cross section of a structural concrete column. Determine and plot the product of inertia $I_{x'y'}$ of the sectional area about the x'-y' axes as a function of θ from $\theta = 0$ to $\theta = \pi/2$. Determine the angle θ for which $I_{x'y'} = 0$. This information is critical to the design of the column to determine the plane in which the column has a minimum resistance to bending. Make use of the results for Prob. A/71.

Ans. $I_{x'y'} = -203 \sin 2\theta - 192 \cos 2\theta$

$I_{x'y'} = 0$ at $\theta = 68.3°$

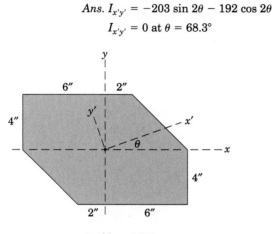

Problem A/89

*A/90 Plot the moment of inertia of the Z-section area about the x'-axis as a function of θ from $\theta = 0$ to $\theta = 90°$. Determine the maximum value of $I_{x'}$ and the corresponding value of θ from your plot, then verify these results by using Eqs. A/10 and A/11.

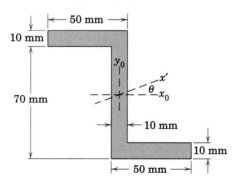

Problem A/90

*A/91 Plot the moment of inertia of the shaded area about the x'-axis as a function of θ from $\theta = 0$ to $\theta = 180°$. Determine the maximum and minimum values of $I_{x'}$ and the corresponding values of θ.

Ans. $I_{max} = 0.286b^4$ at $\theta = 131.1°$
$I_{min} = 0.0547b^4$ at $\theta = 41.1°$

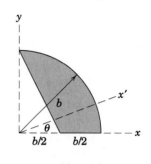

Problem A/91

*A/92 Determine the moment of inertia of the shaded area about the x'-axis through O in terms of θ and plot it for the range $\theta = 0$ to $\theta = 180°$. Find the maximum and minimum values and their corresponding angles θ.

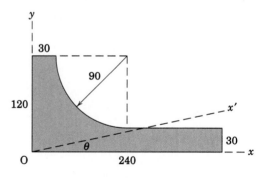

Dimensions in millimeters

Problem A/92

B MASS MOMENTS OF INERTIA

See *Vol. 2 Dynamics* for Appendix B, which fully treats the concept and calculation of mass moment of inertia. Because this quantity is an important element in the study of rigid-body dynamics and is not a factor in statics, we present only a brief definition in this *Statics* volume so that the student can appreciate the basic differences between area and mass moments of inertia.

Consider a three-dimensional body of mass m as shown in Fig. B/1. The mass moment of inertia I about the axis O–O is defined as

$$I = \int r^2 \, dm$$

where r is the perpendicular distance of the mass element dm from the axis O–O and where the integration is over the entire body. For a given rigid body the mass moment of inertia is a measure of the distribution of its mass relative to the axis in question, and for that axis is a constant property of the body. Note that the dimensions are (mass)(length)^2, which are kg·m^2 in SI units and lb-ft-sec^2 in U.S. customary units. Contrast these dimensions with those of area moment of inertia, which are (length)^4, m^4 in SI units and ft^4 in U.S. customary units.

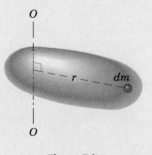

Figure B/1

C SELECTED TOPICS OF MATHEMATICS

C/1 INTRODUCTION

Appendix C contains an abbreviated summary and reminder of selected topics in basic mathematics which find frequent use in mechanics. The relationships are cited without proof. The student of mechanics will have frequent occasion to use many of these relations, and he or she will be handicapped if they are not well in hand. Other topics not listed will also be needed from time to time.

As the reader reviews and applies mathematics, he or she should bear in mind that mechanics is an applied science descriptive of real bodies and actual motions. Therefore, the geometric and physical interpretation of the applicable mathematics should be kept clearly in mind during the development of theory and the formulation and solution of problems.

C/2 PLANE GEOMETRY

1. When two intersecting lines are, respectively, perpendicular to two other lines, the angles formed by the two pairs are equal.

$$\theta_1 = \theta_2$$

2. Similar triangles

$$\frac{x}{b} = \frac{h - y}{h}$$

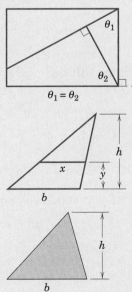

3. Any triangle

$$\text{Area} = \frac{1}{2}bh$$

4. Circle

Circumference $= 2\pi r$
Area $= \pi r^2$
Arc length $s = r\theta$

Sector area $= \frac{1}{2}r^2\theta$

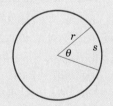

5. Every triangle inscribed within a semicircle is a right triangle.

$$\theta_1 + \theta_2 = \pi/2$$

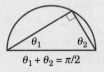

6. Angles of a triangle

$$\theta_1 + \theta_2 + \theta_3 = 180°$$
$$\theta_4 = \theta_1 + \theta_2$$

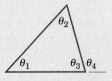

C/3   SOLID GEOMETRY

1. Sphere

Volume $= \frac{4}{3}\pi r^3$

Surface area $= 4\pi r^2$

2. Spherical wedge

Volume $= \frac{2}{3}r^3\theta$

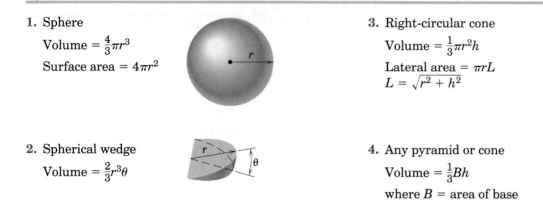

3. Right-circular cone

Volume $= \frac{1}{3}\pi r^2 h$

Lateral area $= \pi r L$
$L = \sqrt{r^2 + h^2}$

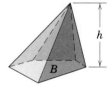

4. Any pyramid or cone

Volume $= \frac{1}{3}Bh$

where B = area of base

C/4   ALGEBRA

1. Quadratic equation

$ax^2 + bx + c = 0$

$x = \dfrac{-b \pm \sqrt{b^2 - 4ac}}{2a}$, $b^2 \geq 4ac$ for real roots

2. Logarithms

$b^x = y$, $x = \log_b y$

Natural logarithms

$b = e = 2.718\ 282$
$e^x = y$, $x = \log_e y = \ln y$

$\log (ab) = \log a + \log b$
$\log (a/b) = \log a - \log b$
$\log (1/n) = -\log n$
$\log a^n = n \log a$
$\log 1 = 0$
$\log_{10} x = 0.4343 \ln x$

3. Determinants

2nd order

$\begin{vmatrix} a_1 & b_1 \\ a_2 & b_2 \end{vmatrix} = a_1 b_2 - a_2 b_1$

3rd order

$\begin{vmatrix} a_1 & b_1 & c_1 \\ a_2 & b_2 & c_2 \\ a_3 & b_3 & c_3 \end{vmatrix} = \begin{matrix} +a_1 b_2 c_3 + a_2 b_3 c_1 + a_3 b_1 c_2 \\ -a_3 b_2 c_1 - a_2 b_1 c_3 - a_1 b_3 c_2 \end{matrix}$

4. Cubic equation

$x^3 = Ax + B$

Let $p = A/3$, $q = B/2$.

Case I: $q^2 - p^3$ negative (three roots real and distinct)

$\cos u = q/(p\sqrt{p})$, $0 < u < 180°$

$x_1 = 2\sqrt{p} \cos (u/3)$

$x_2 = 2\sqrt{p} \cos (u/3 + 120°)$

$x_3 = 2\sqrt{p} \cos (u/3 + 240°)$

Case II: $q^2 - p^3$ positive (one root real, two roots imaginary)

$x_1 = (q + \sqrt{q^2 - p^3})^{1/3} + (q - \sqrt{q^2 - p^3})^{1/3}$

Case III: $q^2 - p^3 = 0$ (three roots real, two roots equal)

$x_1 = 2q^{1/3}$, $x_2 = x_3 = -q^{1/3}$

For general cubic equation

$x^3 + ax^2 + bx + c = 0$

Substitute $x = x_0 - a/3$ and get $x_0{}^3 = Ax_0 + B$. Then proceed as above to find values of x_0 from which $x = x_0 - a/3$.

C/5   ANALYTIC GEOMETRY

1. Straight line

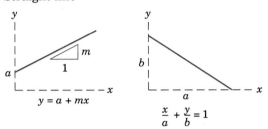

$$y = a + mx$$

$$\frac{x}{a} + \frac{y}{b} = 1$$

2. Circle

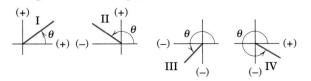

$$x^2 + y^2 = r^2$$

$$(x-a)^2 + (y-b)^2 = r^2$$

3. Parabola

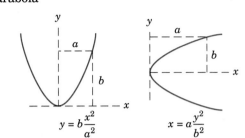

$$y = b\frac{x^2}{a^2}$$

$$x = a\frac{y^2}{b^2}$$

4. Ellipse

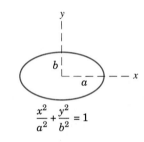

$$\frac{x^2}{a^2} + \frac{y^2}{b^2} = 1$$

5. Hyperbola

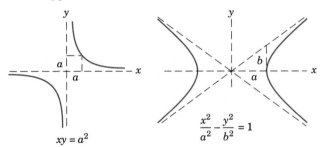

$$xy = a^2$$

$$\frac{x^2}{a^2} - \frac{y^2}{b^2} = 1$$

C/6   TRIGONOMETRY

1. Definitions

$\sin \theta = a/c \qquad \csc \theta = c/a$
$\cos \theta = b/c \qquad \sec \theta = c/b$
$\tan \theta = a/b \qquad \cot \theta = b/a$

2. Signs in the four quadrants

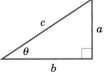

	I	II	III	IV
$\sin \theta$	+	+	−	−
$\cos \theta$	+	−	−	+
$\tan \theta$	+	−	+	−
$\csc \theta$	+	+	−	−
$\sec \theta$	+	−	−	+
$\cot \theta$	+	−	+	−

3. Miscellaneous relations

$$\sin^2 \theta + \cos^2 \theta = 1$$
$$1 + \tan^2 \theta = \sec^2 \theta$$
$$1 + \cot^2 \theta = \csc^2 \theta$$
$$\sin \frac{\theta}{2} = \sqrt{\frac{1}{2}(1 - \cos \theta)}$$
$$\cos \frac{\theta}{2} = \sqrt{\frac{1}{2}(1 + \cos \theta)}$$
$$\sin 2\theta = 2 \sin \theta \cos \theta$$
$$\cos 2\theta = \cos^2 \theta - \sin^2 \theta$$
$$\sin (a \pm b) = \sin a \cos b \pm \cos a \sin b$$
$$\cos (a \pm b) = \cos a \cos b \mp \sin a \sin b$$

4. Law of sines

$$\frac{a}{b} = \frac{\sin A}{\sin B}$$

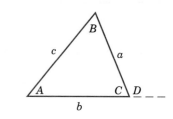

5. Law of cosines

$$c^2 = a^2 + b^2 - 2ab \cos C$$
$$c^2 = a^2 + b^2 + 2ab \cos D$$

C/7 VECTOR OPERATIONS

1. *Notation.* Vector quantities are printed in boldface type, and scalar quantities appear in lightface italic type. Thus, the vector quantity **V** has a scalar magnitude V. In longhand work vector quantities should always be consistently indicated by a symbol such as $\underline{V}$ or $\vec{V}$ to distinguish them from scalar quantities.

2. *Addition*
Triangle addition $\quad \mathbf{P} + \mathbf{Q} = \mathbf{R}$
Parallelogram addition $\quad \mathbf{P} + \mathbf{Q} = \mathbf{R}$
Commutative law $\quad \mathbf{P} + \mathbf{Q} = \mathbf{Q} + \mathbf{P}$
Associative law $\quad \mathbf{P} + (\mathbf{Q} + \mathbf{R}) = (\mathbf{P} + \mathbf{Q}) + \mathbf{R}$

3. *Subtraction*

$$\mathbf{P} - \mathbf{Q} = \mathbf{P} + (-\mathbf{Q})$$

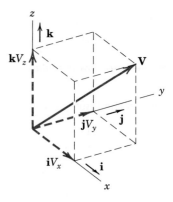

4. *Unit vectors* $\quad \mathbf{i, j, k}$

$$\mathbf{V} = V_x \mathbf{i} + V_y \mathbf{j} + V_z \mathbf{k}$$

where

$$|\mathbf{V}| = V = \sqrt{V_x^2 + V_y^2 + V_z^2}$$

5. *Direction cosines* $\quad l, m, n$ are the cosines of the angles between **V** and the x-, y-, z-axes. Thus,

$$l = V_x/V \qquad m = V_y/V \qquad n = V_z/V$$

so that

$$\mathbf{V} = V(l\mathbf{i} + m\mathbf{j} + n\mathbf{k})$$

and

$$l^2 + m^2 + n^2 = 1$$

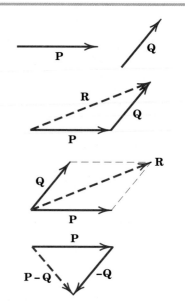

6. *Dot or scalar product*

$$\mathbf{P} \cdot \mathbf{Q} = PQ \cos \theta$$

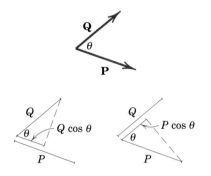

This product may be viewed as the magnitude of $\mathbf{P}$ multiplied by the component $Q \cos \theta$ of $\mathbf{Q}$ in the direction of $\mathbf{P}$, or as the magnitude of $\mathbf{Q}$ multiplied by the component $P \cos \theta$ of $\mathbf{P}$ in the direction of $\mathbf{Q}$.

Commutative law $\mathbf{P} \cdot \mathbf{Q} = \mathbf{Q} \cdot \mathbf{P}$

From the definition of the dot product

$$\mathbf{i} \cdot \mathbf{i} = \mathbf{j} \cdot \mathbf{j} = \mathbf{k} \cdot \mathbf{k} = 1$$

$$\mathbf{i} \cdot \mathbf{j} = \mathbf{j} \cdot \mathbf{i} = \mathbf{i} \cdot \mathbf{k} = \mathbf{k} \cdot \mathbf{i} = \mathbf{j} \cdot \mathbf{k} = \mathbf{k} \cdot \mathbf{j} = 0$$

$$\mathbf{P} \cdot \mathbf{Q} = (P_x\mathbf{i} + P_y\mathbf{j} + P_z\mathbf{k}) \cdot (Q_x\mathbf{i} + Q_y\mathbf{j} + Q_z\mathbf{k})$$
$$= P_xQ_x + P_yQ_y + P_zQ_z$$

$$\mathbf{P} \cdot \mathbf{P} = P_x{}^2 + P_y{}^2 + P_z{}^2$$

It follows from the definition of the dot product that two vectors $\mathbf{P}$ and $\mathbf{Q}$ are perpendicular when their dot product vanishes, $\mathbf{P} \cdot \mathbf{Q} = 0$.

The angle θ between two vectors $\mathbf{P}_1$ and $\mathbf{P}_2$ may be found from their dot product expression $\mathbf{P}_1 \cdot \mathbf{P}_2 = P_1 P_2 \cos \theta$, which gives

$$\cos \theta = \frac{\mathbf{P}_1 \cdot \mathbf{P}_2}{P_1 P_2} = \frac{P_{1_x}P_{2_x} + P_{1_y}P_{2_y} + P_{1_z}P_{2_z}}{P_1 P_2} = l_1 l_2 + m_1 m_2 + n_1 n_2$$

where l, m, n stand for the respective direction cosines of the vectors. It is also observed that two vectors are perpendicular to each other when their direction cosines obey the relation $l_1 l_2 + m_1 m_2 + n_1 n_2 = 0$.

Distributive law $\mathbf{P} \cdot (\mathbf{Q} + \mathbf{R}) = \mathbf{P} \cdot \mathbf{Q} + \mathbf{P} \cdot \mathbf{R}$

7. *Cross or vector product.*

The cross product $\mathbf{P} \times \mathbf{Q}$ of the two vectors $\mathbf{P}$ and $\mathbf{Q}$ is defined as a vector with a magnitude

$$|\mathbf{P} \times \mathbf{Q}| = PQ \sin \theta$$

and a direction specified by the right-hand rule as shown. Reversing the vector order and using the right-hand rule give $\mathbf{Q} \times \mathbf{P} = -\mathbf{P} \times \mathbf{Q}$.

Distributive law $\mathbf{P} \times (\mathbf{Q} + \mathbf{R}) = \mathbf{P} \times \mathbf{Q} + \mathbf{P} \times \mathbf{R}$

From the definition of the cross product, using a *right-handed coordinate system*, we get

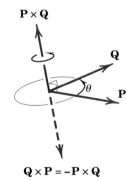

$$\mathbf{i} \times \mathbf{j} = \mathbf{k} \qquad \mathbf{j} \times \mathbf{k} = \mathbf{i} \qquad \mathbf{k} \times \mathbf{i} = \mathbf{j}$$

$$\mathbf{j} \times \mathbf{i} = -\mathbf{k} \qquad \mathbf{k} \times \mathbf{j} = -\mathbf{i} \qquad \mathbf{i} \times \mathbf{k} = -\mathbf{j}$$

$$\mathbf{i} \times \mathbf{i} = \mathbf{j} \times \mathbf{j} = \mathbf{k} \times \mathbf{k} = 0$$

With the aid of these identities and the distributive law, the vector product may be written

$$\mathbf{P} \times \mathbf{Q} = (P_x\mathbf{i} + P_y\mathbf{j} + P_z\mathbf{k}) \times (Q_x\mathbf{i} + Q_y\mathbf{j} + Q_z\mathbf{k})$$
$$= (P_yQ_z - P_zQ_y)\mathbf{i} + (P_zQ_x - P_xQ_z)\mathbf{j} + (P_xQ_y - P_yQ_x)\mathbf{k}$$

The cross product may also be expressed by the determinant

$$\mathbf{P} \times \mathbf{Q} = \begin{vmatrix} \mathbf{i} & \mathbf{j} & \mathbf{k} \\ P_x & P_y & P_z \\ Q_x & Q_y & Q_z \end{vmatrix}$$

8. *Additional relations*

Triple scalar product $(\mathbf{P} \times \mathbf{Q})\cdot\mathbf{R} = \mathbf{R}\cdot(\mathbf{P} \times \mathbf{Q})$. The dot and cross may be interchanged as long as the order of the vectors is maintained. Parentheses are unnecessary since $\mathbf{P} \times (\mathbf{Q}\cdot\mathbf{R})$ is meaningless because a vector $\mathbf{P}$ cannot be crossed into a scalar $\mathbf{Q}\cdot\mathbf{R}$. Thus, the expression may be written

$$\mathbf{P} \times \mathbf{Q}\cdot\mathbf{R} = \mathbf{P}\cdot\mathbf{Q} \times \mathbf{R}$$

The triple scalar product has the determinant expansion

$$\mathbf{P} \times \mathbf{Q}\cdot\mathbf{R} = \begin{vmatrix} P_x & P_y & P_z \\ Q_x & Q_y & Q_z \\ R_x & R_y & R_z \end{vmatrix}$$

Triple vector product $(\mathbf{P} \times \mathbf{Q}) \times \mathbf{R} = -\mathbf{R} \times (\mathbf{P} \times \mathbf{Q}) = \mathbf{R} \times (\mathbf{Q} \times \mathbf{P})$. Here we note that the parentheses must be used since an expression $\mathbf{P} \times \mathbf{Q} \times \mathbf{R}$ would be ambiguous because it would not identify the vector to be crossed. It may be shown that the triple vector product is equivalent to

$$(\mathbf{P} \times \mathbf{Q}) \times \mathbf{R} = \mathbf{R}\cdot\mathbf{P}\mathbf{Q} - \mathbf{R}\cdot\mathbf{Q}\mathbf{P}$$
or
$$\mathbf{P} \times (\mathbf{Q} \times \mathbf{R}) = \mathbf{P}\cdot\mathbf{R}\mathbf{Q} - \mathbf{P}\cdot\mathbf{Q}\mathbf{R}$$

The first term in the first expression, for example, is the dot product $\mathbf{R}\cdot\mathbf{P}$, a scalar, multiplied by the vector $\mathbf{Q}$.

9. *Derivatives of vectors* obey the same rules as they do for scalars.

$$\frac{d\mathbf{P}}{dt} = \dot{\mathbf{P}} = \dot{P}_x\mathbf{i} + \dot{P}_y\mathbf{j} + \dot{P}_z\mathbf{k}$$

$$\frac{d(\mathbf{P}u)}{dt} = \mathbf{P}\dot{u} + \dot{\mathbf{P}}u$$

$$\frac{d(\mathbf{P}\cdot\mathbf{Q})}{dt} = \mathbf{P}\cdot\dot{\mathbf{Q}} + \dot{\mathbf{P}}\cdot\mathbf{Q}$$

$$\frac{d(\mathbf{P} \times \mathbf{Q})}{dt} = \mathbf{P} \times \dot{\mathbf{Q}} + \dot{\mathbf{P}} \times \mathbf{Q}$$

10. *Integration of vectors.*   If **V** is a function of x, y, and z and an element of volume is $d\tau = dx\,dy\,dz$, the integral of **V** over the volume may be written as the vector sum of the three integrals of its components. Thus,

$$\int \mathbf{V}\,d\tau = \mathbf{i}\int V_x\,d\tau + \mathbf{j}\int V_y\,d\tau + \mathbf{k}\int V_z\,d\tau$$

C/8   SERIES

(Expression in brackets following series indicates range of convergence.)

$$(1 \pm x)^n = 1 \pm nx + \frac{n(n-1)}{2!}x^2 \pm \frac{n(n-1)(n-2)}{3!}x^3 + \cdots \quad [x^2 < 1]$$

$$\sin x = x - \frac{x^3}{3!} + \frac{x^5}{5!} - \frac{x^7}{7!} + \cdots \qquad\qquad [x^2 < \infty]$$

$$\cos x = 1 - \frac{x^2}{2!} + \frac{x^4}{4!} - \frac{x^6}{6!} + \cdots \qquad\qquad [x^2 < \infty]$$

$$\sinh x = \frac{e^x - e^{-x}}{2} = x + \frac{x^3}{3!} + \frac{x^5}{5!} + \frac{x^7}{7!} + \cdots \qquad [x^2 < \infty]$$

$$\cosh x = \frac{e^x + e^{-x}}{2} = 1 + \frac{x^2}{2!} + \frac{x^4}{4!} + \frac{x^6}{6!} + \cdots \qquad [x^2 < \infty]$$

$$f(x) = \frac{a_0}{2} + \sum_{n=1}^{\infty} a_n \cos\frac{n\pi x}{l} + \sum_{n=1}^{\infty} b_n \sin\frac{n\pi x}{l}$$

$$\text{where } a_n = \frac{1}{l}\int_{-l}^{l} f(x)\cos\frac{n\pi x}{l}\,dx, \quad b_n = \frac{1}{l}\int_{-l}^{l} f(x)\sin\frac{n\pi x}{l}\,dx$$

$$[\text{Fourier expansion for } -l < x < l]$$

C/9   DERIVATIVES

$$\frac{dx^n}{dx} = nx^{n-1}, \frac{d(uv)}{dx} = u\frac{dv}{dx} + v\frac{du}{dx}, \frac{d\left(\dfrac{u}{v}\right)}{dx} = \frac{v\dfrac{du}{dx} - u\dfrac{dv}{dx}}{v^2}$$

$$\lim_{\Delta x \to 0} \sin \Delta x = \sin dx = \tan dx = dx$$

$$\lim_{\Delta x \to 0} \cos \Delta x = \cos dx = 1$$

$$\frac{d\sin x}{dx} = \cos x, \qquad \frac{d\cos x}{dx} = -\sin x, \qquad \frac{d\tan x}{dx} = \sec^2 x$$

$$\frac{d\sinh x}{dx} = \cosh x, \qquad \frac{d\cosh x}{dx} = \sinh x, \qquad \frac{d\tanh x}{dx} = \operatorname{sech}^2 x$$

C/10   INTEGRALS

$$\int x^n \, dx = \frac{x^{n+1}}{n+1}$$

$$\int \frac{dx}{x} = \ln x$$

$$\int \sqrt{a+bx} \, dx = \frac{2}{3b}\sqrt{(a+bx)^3}$$

$$\int x\sqrt{a+bx} \, dx = \frac{2}{15b^2}(3bx-2a)\sqrt{(a+bx)^3}$$

$$\int x^2\sqrt{a+bx} \, dx = \frac{2}{105b^3}(8a^2-12abx+15b^2x^2)\sqrt{(a+bx)^3}$$

$$\int \frac{dx}{\sqrt{a+bx}} = \frac{2\sqrt{a+bx}}{b}$$

$$\int \frac{\sqrt{a+x}}{\sqrt{b-x}} \, dx = -\sqrt{a+x}\,\sqrt{b-x} + (a+b)\sin^{-1}\sqrt{\frac{a+x}{a+b}}$$

$$\int \frac{x \, dx}{a+bx} = \frac{1}{b^2}[a+bx-a\ln(a+bx)]$$

$$\int \frac{x \, dx}{(a+bx)^n} = \frac{(a+bx)^{1-n}}{b^2}\left(\frac{a+bx}{2-n} - \frac{a}{1-n}\right)$$

$$\int \frac{dx}{a+bx^2} = \frac{1}{\sqrt{ab}}\tan^{-1}\frac{x\sqrt{ab}}{a} \qquad \text{or} \qquad \frac{1}{\sqrt{-ab}}\tanh^{-1}\frac{x\sqrt{-ab}}{a}$$

$$\int \frac{x \, dx}{a+bx^2} = \frac{1}{2b}\ln(a+bx^2)$$

$$\int \sqrt{x^2 \pm a^2} \, dx = \tfrac{1}{2}[x\sqrt{x^2 \pm a^2} \pm a^2 \ln(x+\sqrt{x^2 \pm a^2})]$$

$$\int \sqrt{a^2-x^2} \, dx = \frac{1}{2}\left(x\,\sqrt{a^2-x^2} + a^2\sin^{-1}\frac{x}{a}\right)$$

$$\int x\sqrt{a^2-x^2} \, dx = -\frac{1}{3}\sqrt{(a^2-x^2)^3}$$

$$\int x^2\sqrt{a^2-x^2} \, dx = -\frac{x}{4}\sqrt{(a^2-x^2)^3} + \frac{a^2}{8}\left(x\sqrt{a^2-x^2} + a^2\sin^{-1}\frac{x}{a}\right)$$

$$\int x^3\sqrt{a^2-x^2} \, dx = -\frac{1}{5}(x^2+\tfrac{2}{3}a^2)\sqrt{(a^2-x^2)^3}$$

$$\int \frac{dx}{\sqrt{a + bx + cx^2}} = \frac{1}{\sqrt{c}} \ln \left(\sqrt{a + bx + cx^2} + x\sqrt{c} + \frac{b}{2\sqrt{c}} \right) \quad \text{or} \quad \frac{-1}{\sqrt{-c}} \sin^{-1} \left(\frac{b + 2cx}{\sqrt{b^2 - 4ac}} \right)$$

$$\int \frac{dx}{\sqrt{x^2 \pm a^2}} = \ln (x + \sqrt{x^2 \pm a^2})$$

$$\int \frac{dx}{\sqrt{a^2 - x^2}} = \sin^{-1} \frac{x}{a}$$

$$\int \frac{x \, dx}{\sqrt{x^2 - a^2}} = \sqrt{x^2 - a^2}$$

$$\int \frac{x \, dx}{\sqrt{a^2 \pm x^2}} = \pm \sqrt{a^2 \pm x^2}$$

$$\int x\sqrt{x^2 \pm a^2} \, dx = \frac{1}{3}\sqrt{(x^2 \pm a^2)^3}$$

$$\int x^2\sqrt{x^2 \pm a^2} \, dx = \frac{x}{4}\sqrt{(x^2 \pm a^2)^3} \mp \frac{a^2}{8} x\sqrt{x^2 \pm a^2} - \frac{a^4}{8} \ln (x + \sqrt{x^2 \pm a^2})$$

$$\int \sin x \, dx = -\cos x$$

$$\int \cos x \, dx = \sin x$$

$$\int \sec x \, dx = \frac{1}{2} \ln \frac{1 + \sin x}{1 - \sin x}$$

$$\int \sin^2 x \, dx = \frac{x}{2} - \frac{\sin 2x}{4}$$

$$\int \cos^2 x \, dx = \frac{x}{2} + \frac{\sin 2x}{4}$$

$$\int \sin x \cos x \, dx = \frac{\sin^2 x}{2}$$

$$\int \sinh x \, dx = \cosh x$$

$$\int \cosh x \, dx = \sinh x$$

$$\int \tanh x \, dx = \ln \cosh x$$

$$\int \ln x \, dx = x \ln x - x$$

$$\int e^{ax}\, dx = \frac{e^{ax}}{a}$$

$$\int xe^{ax}\, dx = \frac{e^{ax}}{a^2}\,(ax - 1)$$

$$\int e^{ax} \sin px\, dx = \frac{e^{ax}\,(a \sin px - p \cos px)}{a^2 + p^2}$$

$$\int e^{ax} \cos px\, dx = \frac{e^{ax}\,(a \cos px + p \sin px)}{a^2 + p^2}$$

$$\int e^{ax} \sin^2 x\, dx = \frac{e^{ax}}{4 + a^2}\left(a \sin^2 x - \sin 2x + \frac{2}{a}\right)$$

$$\int e^{ax} \cos^2 x\, dx = \frac{e^{ax}}{4 + a^2}\left(a \cos^2 x + \sin 2x + \frac{2}{a}\right)$$

$$\int e^{ax} \sin x \cos x\, dx = \frac{e^{ax}}{4 + a^2}\left(\frac{a}{2} \sin 2x - \cos 2x\right)$$

$$\int \sin^3 x\, dx = -\frac{\cos x}{3}\,(2 + \sin^2 x)$$

$$\int \cos^3 x\, dx = \frac{\sin x}{3}\,(2 + \cos^2 x)$$

$$\int \cos^5 x\, dx = \sin x - \frac{2}{3} \sin^3 x + \frac{1}{5} \sin^5 x$$

$$\int x \sin x\, dx = \sin x - x \cos x$$

$$\int x \cos x\, dx = \cos x + x \sin x$$

$$\int x^2 \sin x\, dx = 2x \sin x - (x^2 - 2) \cos x$$

$$\int x^2 \cos x\, dx = 2x \cos x + (x^2 - 2) \sin x$$

Radius of curvature

$$\rho_{xy} = \frac{\left[1 + \left(\dfrac{dy}{dx}\right)^2\right]^{3/2}}{\dfrac{d^2 y}{dx^2}}$$

$$\rho_{r\theta} = \frac{\left[r^2 + \left(\dfrac{dr}{d\theta}\right)^2\right]^{3/2}}{r^2 + 2\left(\dfrac{dr}{d\theta}\right)^2 - r\dfrac{d^2 r}{d\theta^2}}$$

C/11   Newton's Method for Solving Intractable Equations

Frequently, the application of the fundamental principles of mechanics leads to an algebraic or transcendental equation which is not solvable (or easily solvable) in closed form. In such cases, an iterative technique, such as Newton's method, can be a powerful tool for obtaining a good estimate to the root or roots of the equation.

Let us place the equation to be solved in the form $f(x) = 0$. Part a of the accompanying figure depicts an arbitrary function $f(x)$ for values of x in the vicinity of the desired root x_r. Note that x_r is merely the value

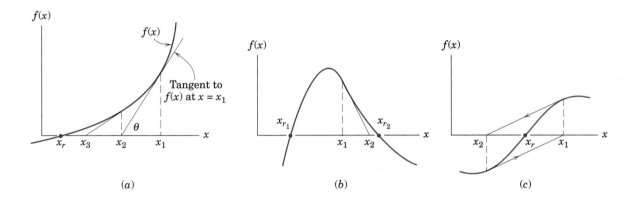

of x at which the function crosses the x-axis. Suppose that we have available (perhaps via a hand-drawn plot) a rough estimate x_1 of this root. Provided that x_1 does not closely correspond to a maximum or minimum value of the function $f(x)$, we may obtain a better estimate of the root x_r by extending the tangent to $f(x)$ at x_1 so that it intersects the x-axis at x_2. From the geometry of the figure, we may write

$$\tan \theta = f'(x_1) = \frac{f(x_1)}{x_1 - x_2}$$

where $f'(x_1)$ denotes the derivative of $f(x)$ with respect to x evaluated at $x = x_1$. Solving the above equation for x_2 results in

$$x_2 = x_1 - \frac{f(x_1)}{f'(x_1)}$$

The term $-f(x_1)/f'(x_1)$ is the correction to the initial root estimate x_1. Once x_2 is calculated, we may repeat the process to obtain x_3, and so forth.

Thus, we generalize the above equation to

$$x_{k+1} = x_k - \frac{f(x_k)}{f'(x_k)}$$

where

$$x_{k+1} = \text{the } (k+1)\text{th estimate of the desired root } x_r$$
$$x_k = \text{the } k\text{th estimate of the desired root } x_r$$
$$f(x_k) = \text{the function } f(x) \text{ evaluated at } x = x_k$$
$$f'(x_k) = \text{the function derivative evaluated at } x = x_k$$

This equation is repeatedly applied until $f(x_{k+1})$ is sufficiently close to zero and $x_{k+1} \cong x_k$. The student should verify that the equation is valid for all possible sign combinations of x_k, $f(x_k)$, and $f'(x_k)$.

Several cautionary notes are in order:

1. Clearly, $f'(x_k)$ must not be zero or close to zero. This would mean, as restricted above, that x_k exactly or approximately corresponds to a minimum or maximum of $f(x)$. If the slope $f'(x_k)$ is zero, then the tangent to the curve never intersects the x-axis. If the slope $f'(x_k)$ is small, then the correction to x_k may be so large that x_{k+1} is a worse root estimate than x_k. For this reason, experienced engineers usually limit the size of the correction term; that is, if the absolute value of $f(x_k)/f'(x_k)$ is larger than a preselected maximum value, that maximum value is used.

2. If there are several roots of the equation $f(x) = 0$, we must be in the vicinity of the desired root x_r in order that the algorithm actually converges to that root. Part b of the figure depicts the condition when the initial estimate x_1 will result in convergence to x_{r_2} rather than x_{r_1}.

3. Oscillation from one side of the root to the other can occur if, for example, the function is antisymmetric about a root which is an inflection point. The use of one-half of the correction will usually prevent this behavior, which is depicted in part c of the accompanying figure.

Example: Beginning with an initial estimate of $x_1 = 5$, estimate the single root of the equation $e^x - 10 \cos x - 100 = 0$.

The table below summarizes the application of Newton's method to the given equation. The iterative process was terminated when the absolute value of the correction $-f(x_k)/f'(x_k)$ became less than 10^{-6}.

k	x_k	$f(x_k)$	$f'(x_k)$	$x_{k+1} - x_k = -\dfrac{f(x_k)}{f'(x_k)}$
1	5.000 000	45.576 537	138.823 916	−0.328 305
2	4.671 695	7.285 610	96.887 065	−0.075 197
3	4.596 498	0.292 886	89.203 650	−0.003 283
4	4.593 215	0.000 527	88.882 536	−0.000 006
5	4.593 209	$-2(10^{-8})$	88.881 956	$2.25(10^{-10})$

C/12　Selected Techniques for Numerical Integration

1. Area determination.　Consider the problem of determining the shaded area under the curve $y = f(x)$ from $x = a$ to $x = b$, as depicted in part a of the figure, and suppose that analytical integration is not feasible. The function may be known in tabular form from experimental measurements or it may be known in analytical form. The function is taken to be continuous within the interval $a < x < b$. We may divide the area into n vertical strips, each of width $\Delta x = (b - a)/n$, and then add the areas of all strips to obtain $A = \int y\, dx$. A representative strip of area A_i is shown with darker shading in the figure. Three useful numerical approximations are cited. In each case the greater the number of strips, the more accurate becomes the approximation geometrically. As a general rule, one can begin with a relatively small number of strips and increase the number until the resulting changes in the area approximation no longer improve the accuracy obtained.

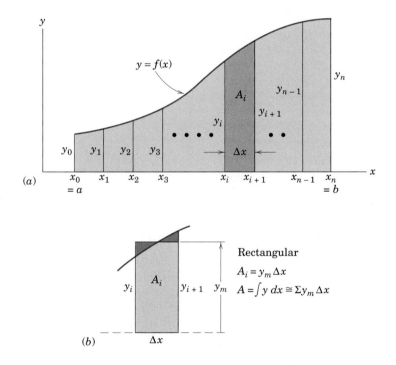

I. *Rectangular* [Figure (b)] The areas of the strips are taken to be rectangles, as shown by the representative strip whose height y_m is chosen visually so that the small cross-hatched areas are as nearly equal as possible. Thus, we form the sum Σy_m of the effective heights and multiply by Δx. For a function known in analytical form, a value for y_m equal to that of the function at the midpoint $x_i + \Delta x/2$ may be calculated and used in the summation.

II. *Trapezoidal* [Figure (c)] The areas of the strips are taken to be trapezoids, as shown by the representative strip. The area A_i is the average height $(y_i + y_{i+1})/2$ times Δx. Adding the areas gives the area approximation as tabulated. For the example with the curvature shown, clearly the approximation will be on the low side. For the reverse curvature, the approximation will be on the high side.

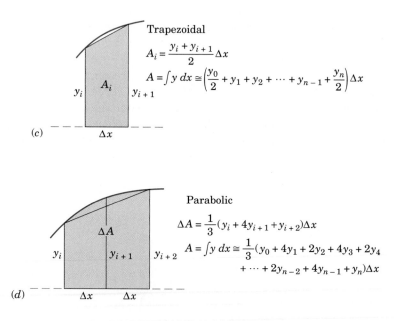

Trapezoidal

$$A_i = \frac{y_i + y_{i+1}}{2}\Delta x$$

$$A = \int y\,dx \cong \left(\frac{y_0}{2} + y_1 + y_2 + \cdots + y_{n-1} + \frac{y_n}{2}\right)\Delta x$$

(c)

Parabolic

$$\Delta A = \frac{1}{3}(y_i + 4y_{i+1} + y_{i+2})\Delta x$$

$$A = \int y\,dx \cong \frac{1}{3}(y_0 + 4y_1 + 2y_2 + 4y_3 + 2y_4$$
$$+ \cdots + 2y_{n-2} + 4y_{n-1} + y_n)\Delta x$$

(d)

III. *Parabolic* [Figure (d)] The area between the chord and the curve (neglected in the trapezoidal solution) may be accounted for by approximating the function by a parabola passing through the points defined by three successive values of y. This area may be calculated from the geometry of the parabola and added to the trapezoidal area of the pair of strips to give the area ΔA of the pair as cited. Adding all of the ΔA's produces the tabulation shown, which is known as Simpson's rule. To use Simpson's rule, the number n of strips must be even.

Example: Determine the area under the curve $y = x\sqrt{1 + x^2}$ from $x = 0$ to $x = 2$. (An integrable function is chosen here so that the three approximations can be compared with the exact value, which is $A = \int_0^2 x\sqrt{1 + x^2}\,dx = \frac{1}{3}(1 + x^2)^{3/2}\big|_0^2 = \frac{1}{3}(5\sqrt{5} - 1) = 3.393\,447$).

NUMBER OF SUBINTERVALS	AREA APPROXIMATIONS		
	RECTANGULAR	TRAPEZOIDAL	PARABOLIC
4	3.361 704	3.456 731	3.392 214
10	3.388 399	3.403 536	3.393 420
50	3.393 245	3.393 850	3.393 447
100	3.393 396	3.393 547	3.393 447
1000	3.393 446	3.393 448	3.393 447
2500	3.393 447	3.393 447	3.393 447

Note that the worst approximation error is less than 2 percent, even with only four strips.

2. Integration of first-order ordinary differential equations.
The application of the fundamental principles of mechanics frequently results in differential relationships. Let us consider the first-order form $dy/dt = f(t)$, where the function $f(t)$ may not be readily integrable or may be known only in tabular form. We may numerically integrate by means of a simple slope-projection technique, known as Euler integration, which is illustrated in the figure.

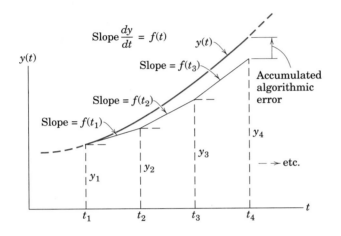

Beginning at t_1, at which the value y_1 is known, we project the slope over a horizontal subinterval or step $(t_2 - t_1)$ and see that $y_2 = y_1 + f(t_1)(t_2 - t_1)$. At t_2, the process may be repeated beginning at y_2, and so forth until the desired value of t is reached. Hence, the general expression is

$$y_{k+1} = y_k + f(t_k)(t_{k+1} - t_k)$$

If y versus t were linear, i.e., if $f(t)$ were constant, the method would be exact, and there would be no need for a numerical approach in that case. Changes in the slope over the subinterval introduce error. For the case shown in the figure, the estimate y_2 is clearly less than the true value of the function $y(t)$ at t_2. More accurate integration techniques (such as Runge-Kutta methods) take into account changes in the slope over the subinterval and thus provide better results.

As with the area-determination techniques, experience is helpful in the selection of a subinterval or step size when dealing with analytical functions. As a rough rule, one begins with a relatively large step size and then steadily decreases the step size until the corresponding changes in the integrated result are much smaller than the desired accuracy. A step size which is too small, however, can result in increased error due to a very large number of computer operations. This type of error is generally known as "round-off error," while the error which results from a large step size is known as algorithm error.

Example: For the differential equation $dy/dt = 5t$ with the initial condition $y = 2$ when $t = 0$, determine the value of y for $t = 4$.

Application of the Euler integration technique yields the following results:

NUMBER OF SUBINTERVALS	STEP SIZE	y at $t = 4$	PERCENT ERROR
10	0.4	38	9.5
100	0.04	41.6	0.95
500	0.008	41.92	0.19
1000	0.004	41.96	0.10

This simple example may be integrated analytically. The result is $y = 42$ (exactly).

USEFUL TABLES

TABLE D/1 PHYSICAL PROPERTIES

Density (kg/m^3) and *specific weight* (lb/ft^3)

	kg/m^3	lb/ft^3		kg/m^3	lb/ft^3
Air*	1.2062	0.07530	Lead	11 370	710
Aluminum	2 690	168	Mercury	13 570	847
Concrete (av.)	2 400	150	Oil (av.)	900	56
Copper	8 910	556	Steel	7 830	489
Earth (wet, av.)	1 760	110	Titanium	3 080	192
(dry, av.)	1 280	80	Water (fresh)	1 000	62.4
Glass	2 590	162	(salt)	1 030	64
Gold	19 300	1205	Wood (soft pine)	480	30
Ice	900	56	(hard oak)	800	50
Iron (cast)	7 210	450			

* At 20°C (68°F) and atmospheric pressure

Coefficients of friction

(The coefficients in the following table represent typical values under normal working conditions. Actual coefficients for a given situation will depend on the exact nature of the contacting surfaces. A variation of 25 to 100 percent or more from these values could be expected in an actual application, depending on prevailing conditions of cleanliness, surface finish, pressure, lubrication, and velocity.)

CONTACTING SURFACE	TYPICAL VALUES OF COEFFICIENT OF FRICTION	
	STATIC, μ_s	KINETIC, μ_k
Steel on steel (dry)	0.6	0.4
Steel on steel (greasy)	0.1	0.05
Teflon on steel	0.04	0.04
Steel on babbitt (dry)	0.4	0.3
Steel on babbitt (greasy)	0.1	0.07
Brass on steel (dry)	0.5	0.4
Brake lining on cast iron	0.4	0.3
Rubber tires on smooth pavement (dry)	0.9	0.8
Wire rope on iron pulley (dry)	0.2	0.15
Hemp rope on metal	0.3	0.2
Metal on ice		0.02

TABLE D/2   SOLAR SYSTEM CONSTANTS

Universal gravitational constant	G = 6.673(10^{-11}) m^3/(kg·s^2)
	= 3.439(10^{-8}) ft^4/(lbf-s^4)
Mass of Earth	m_e = 5.976(10^{24}) kg
	= 4.095(10^{23}) lbf-s^2/ft
Period of Earth's rotation (1 sidereal day)	= 23 h 56 min 4 s
	= 23.9344 h
Angular velocity of Earth	ω = 0.7292(10^{-4}) rad/s
Mean angular velocity of Earth–Sun line	ω' = 0.1991(10^{-6}) rad/s
Mean velocity of Earth's center about Sun	= 107 200 km/h
	= 66,610 mi/h

BODY	MEAN DISTANCE TO SUN km (mi)	ECCENTRICITY OF ORBIT e	PERIOD OF ORBIT solar days	MEAN DIAMETER km (mi)	MASS RELATIVE TO EARTH	SURFACE GRAVITATIONAL ACCELERATION m/s^2 (ft/s^2)	ESCAPE VELOCITY km/s (mi/s)
Sun	—	—	—	1 392 000 (865 000)	333 000	274 (898)	616 (383)
Moon	384 398* (238 854)*	0.055	27.32	3 476 (2 160)	0.0123	1.62 (5.32)	2.37 (1.47)
Mercury	57.3 × 10^6 (35.6 × 10^6)	0.206	87.97	5 000 (3 100)	0.054	3.47 (11.4)	4.17 (2.59)
Venus	108 × 10^6 (67.2 × 10^6)	0.0068	224.70	12 400 (7 700)	0.815	8.44 (27.7)	10.24 (6.36)
Earth	149.6 × 10^6 (92.96 × 10^6)	0.0167	365.26	12 742† (7 918)†	1.000	9.821‡ (32.22)‡	11.18 (6.95)
Mars	227.9 × 10^6 (141.6 × 10^6)	0.093	686.98	6 788 (4 218)	0.107	3.73 (12.3)	5.03 (3.13)

* Mean distance to Earth (center-to-center)

† Diameter of sphere of equal volume, based on a spheroidal Earth with a polar diameter of 12 714 km (7900 mi) and an equatorial diameter of 12 756 km (7926 mi)

‡ For nonrotating spherical Earth, equivalent to absolute value at sea level and latitude 37.5°

TABLE D/3   PROPERTIES OF PLANE FIGURES

FIGURE	CENTROID	AREA MOMENTS OF INERTIA
Arc Segment	$\bar{r} = \dfrac{r \sin \alpha}{\alpha}$	—
Quarter and Semicircular Arcs	$\bar{y} = \dfrac{2r}{\pi}$	—
Circular Area	—	$I_x = I_y = \dfrac{\pi r^4}{4}$ $I_z = \dfrac{\pi r^4}{2}$
Semicircular Area	$\bar{y} = \dfrac{4r}{3\pi}$	$I_x = I_y = \dfrac{\pi r^4}{8}$ $\bar{I}_x = \left(\dfrac{\pi}{8} - \dfrac{8}{9\pi}\right) r^4$ $I_z = \dfrac{\pi r^4}{4}$
Quarter-Circular Area	$\bar{x} = \bar{y} = \dfrac{4r}{3\pi}$	$I_x = I_y = \dfrac{\pi r^4}{16}$ $\bar{I}_x = \bar{I}_y = \left(\dfrac{\pi}{16} - \dfrac{4}{9\pi}\right) r^4$ $I_z = \dfrac{\pi r^4}{8}$
Area of Circular Sector	$\bar{x} = \dfrac{2}{3} \dfrac{r \sin \alpha}{\alpha}$	$I_x = \dfrac{r^4}{4}\left(\alpha - \dfrac{1}{2}\sin 2\alpha\right)$ $I_y = \dfrac{r^4}{4}\left(\alpha + \dfrac{1}{2}\sin 2\alpha\right)$ $I_z = \dfrac{1}{2} r^4 \alpha$

TABLE D/3 PROPERTIES OF PLANE FIGURES *Continued*

FIGURE	CENTROID	AREA MOMENTS OF INERTIA
Rectangular Area	—	$I_x = \dfrac{bh^3}{3}$ $\bar{I}_x = \dfrac{bh^3}{12}$ $\bar{I}_z = \dfrac{bh}{12}(b^2 + h^2)$
Triangular Area	$\bar{x} = \dfrac{a+b}{3}$ $\bar{y} = \dfrac{h}{3}$	$I_x = \dfrac{bh^3}{12}$ $\bar{I}_x = \dfrac{bh^3}{36}$ $I_{x_1} = \dfrac{bh^3}{4}$
Area of Elliptical Quadrant	$\bar{x} = \dfrac{4a}{3\pi}$ $\bar{y} = \dfrac{4b}{3\pi}$	$I_x = \dfrac{\pi ab^3}{16}, \quad \bar{I}_x = \left(\dfrac{\pi}{16} - \dfrac{4}{9\pi}\right)ab^3$ $I_y = \dfrac{\pi a^3 b}{16}, \quad \bar{I}_y = \left(\dfrac{\pi}{16} - \dfrac{4}{9\pi}\right)a^3 b$ $I_z = \dfrac{\pi ab}{16}(a^2 + b^2)$
Subparabolic Area $y = kx^2 = \dfrac{b}{a^2}x^2$ Area $A = \dfrac{ab}{3}$	$\bar{x} = \dfrac{3a}{4}$ $\bar{y} = \dfrac{3b}{10}$	$I_x = \dfrac{ab^3}{21}$ $I_y = \dfrac{a^3 b}{5}$ $I_z = ab\left(\dfrac{a^3}{5} + \dfrac{b^2}{21}\right)$
Parabolic Area $y = kx^2 = \dfrac{b}{a^2}x^2$ Area $A = \dfrac{2ab}{3}$	$\bar{x} = \dfrac{3a}{8}$ $\bar{y} = \dfrac{3b}{5}$	$I_x = \dfrac{2ab^3}{7}$ $I_y = \dfrac{2a^3 b}{15}$ $I_z = 2ab\left(\dfrac{a^2}{15} + \dfrac{b^2}{7}\right)$

TABLE D/4   PROPERTIES OF HOMOGENEOUS SOLIDS

(m = mass of body shown)

BODY	MASS CENTER	MASS MOMENTS OF INERTIA
Circular Cylindrical Shell	—	$I_{xx} = \frac{1}{2}mr^2 + \frac{1}{12}ml^2$ $I_{x_1x_1} = \frac{1}{2}mr^2 + \frac{1}{3}ml^2$ $I_{zz} = mr^2$
Half Cylindrical Shell	$\bar{x} = \frac{2r}{\pi}$	$I_{xx} = I_{yy}$ $\quad = \frac{1}{2}mr^2 + \frac{1}{12}ml^2$ $I_{x_1x_1} = I_{y_1y_1}$ $\quad = \frac{1}{2}mr^2 + \frac{1}{3}ml^2$ $I_{zz} = mr^2$ $\bar{I}_{zz} = \left(1 - \frac{4}{\pi^2}\right)mr^2$
Circular Cylinder	—	$I_{xx} = \frac{1}{4}mr^2 + \frac{1}{12}ml^2$ $I_{x_1x_1} = \frac{1}{4}mr^2 + \frac{1}{3}ml^2$ $I_{zz} = \frac{1}{2}mr^2$
Semicylinder	$\bar{x} = \frac{4r}{3\pi}$	$I_{xx} = I_{yy}$ $\quad = \frac{1}{4}mr^2 + \frac{1}{12}ml^2$ $I_{x_1x_1} = I_{y_1y_1}$ $\quad = \frac{1}{4}mr^2 + \frac{1}{3}ml^2$ $I_{zz} = \frac{1}{2}mr^2$ $\bar{I}_{zz} = \left(\frac{1}{2} - \frac{16}{9\pi^2}\right)mr^2$
Rectangular Parallelepiped	—	$I_{xx} = \frac{1}{12}m(a^2 + l^2)$ $I_{yy} = \frac{1}{12}m(b^2 + l^2)$ $I_{zz} = \frac{1}{12}m(a^2 + b^2)$ $I_{y_1y_1} = \frac{1}{12}mb^2 + \frac{1}{3}ml^2$ $I_{y_2y_2} = \frac{1}{3}m(b^2 + l^2)$

TABLE D/4 PROPERTIES OF HOMOGENEOUS SOLIDS *Continued*

(m = mass of body shown)

BODY	MASS CENTER	MASS MOMENTS OF INERTIA
Spherical Shell	—	$I_{zz} = \frac{2}{3}mr^2$
Hemispherical Shell	$\bar{x} = \frac{r}{2}$	$I_{xx} = I_{yy} = I_{zz} = \frac{2}{3}mr^2$ $\bar{I}_{yy} = \bar{I}_{zz} = \frac{5}{12}mr^2$
Sphere	—	$I_{zz} = \frac{2}{5}mr^2$
Hemisphere	$\bar{x} = \frac{3r}{8}$	$I_{xx} = I_{yy} = I_{zz} = \frac{2}{5}mr^2$ $\bar{I}_{yy} = \bar{I}_{zz} = \frac{83}{320}mr^2$
Uniform Slender Rod	—	$I_{yy} = \frac{1}{12}ml^2$ $I_{y_1y_1} = \frac{1}{3}ml^2$

TABLE D/4 PROPERTIES OF HOMOGENEOUS SOLIDS *Continued*

(m = mass of body shown)

BODY	MASS CENTER	MASS MOMENTS OF INERTIA
Quarter-Circular Rod	$\bar{x} = \bar{y}$ $= \dfrac{2r}{\pi}$	$I_{xx} = I_{yy} = \frac{1}{2}mr^2$ $I_{zz} = mr^2$
Elliptical Cylinder	—	$I_{xx} = \frac{1}{4}ma^2 + \frac{1}{12}ml^2$ $I_{yy} = \frac{1}{4}mb^2 + \frac{1}{12}ml^2$ $I_{zz} = \frac{1}{4}m(a^2 + b^2)$ $I_{y_1y_1} = \frac{1}{4}mb^2 + \frac{1}{3}ml^2$
Conical Shell	$\bar{z} = \dfrac{2h}{3}$	$I_{yy} = \frac{1}{4}mr^2 + \frac{1}{2}mh^2$ $I_{y_1y_1} = \frac{1}{4}mr^2 + \frac{1}{6}mh^2$ $I_{zz} = \frac{1}{2}mr^2$ $\bar{I}_{yy} = \frac{1}{4}mr^2 + \frac{1}{18}mh^2$
Half Conical Shell	$\bar{x} = \dfrac{4r}{3\pi}$ $\bar{z} = \dfrac{2h}{3}$	$I_{xx} = I_{yy}$ $\quad = \frac{1}{4}mr^2 + \frac{1}{2}mh^2$ $I_{x_1x_1} = I_{y_1y_1}$ $\quad = \frac{1}{4}mr^2 + \frac{1}{6}mh^2$ $I_{zz} = \frac{1}{2}mr^2$ $\bar{I}_{zz} = \left(\frac{1}{2} - \frac{16}{9\pi^2}\right)mr^2$
Right Circular Cone	$\bar{z} = \dfrac{3h}{4}$	$I_{yy} = \frac{3}{20}mr^2 + \frac{3}{5}mh^2$ $I_{y_1y_1} = \frac{3}{20}mr^2 + \frac{1}{10}mh^2$ $I_{zz} = \frac{3}{10}mr^2$ $\bar{I}_{yy} = \frac{3}{20}mr^2 + \frac{3}{80}mh^2$

TABLE D/4 **PROPERTIES OF HOMOGENEOUS SOLIDS** *Continued*
(m = mass of body shown)

BODY	MASS CENTER	MASS MOMENTS OF INERTIA
Half Cone	$\bar{x} = \dfrac{r}{\pi}$ $\bar{z} = \dfrac{3h}{4}$	$I_{xx} = I_{yy}$ $\qquad = \frac{3}{20}mr^2 + \frac{3}{5}mh^2$ $I_{x_1x_1} = I_{y_1y_1}$ $\qquad = \frac{3}{20}mr^2 + \frac{1}{10}mh^2$ $I_{zz} = \frac{3}{10}mr^2$ $\bar{I}_{zz} = \left(\frac{3}{10} - \frac{1}{\pi^2}\right)mr^2$
Semiellipsoid	$\bar{z} = \dfrac{3c}{8}$	$I_{xx} = \frac{1}{5}m(b^2 + c^2)$ $I_{yy} = \frac{1}{5}m(a^2 + c^2)$ $I_{zz} = \frac{1}{5}m(a^2 + b^2)$ $\bar{I}_{xx} = \frac{1}{5}m(b^2 + \frac{19}{64}c^2)$ $\bar{I}_{yy} = \frac{1}{5}m(a^2 + \frac{19}{64}c^2)$
Elliptic Paraboloid	$\bar{z} = \dfrac{2c}{3}$	$I_{xx} = \frac{1}{6}mb^2 + \frac{1}{2}mc^2$ $I_{yy} = \frac{1}{6}ma^2 + \frac{1}{2}mc^2$ $I_{zz} = \frac{1}{6}m(a^2 + b^2)$ $\bar{I}_{xx} = \frac{1}{6}m(b^2 + \frac{1}{3}c^2)$ $\bar{I}_{yy} = \frac{1}{6}m(a^2 + \frac{1}{3}c^2)$
Rectangular Tetrahedron	$\bar{x} = \dfrac{a}{4}$ $\bar{y} = \dfrac{b}{4}$ $\bar{z} = \dfrac{c}{4}$	$I_{xx} = \frac{1}{10}m(b^2 + c^2)$ $I_{yy} = \frac{1}{10}m(a^2 + c^2)$ $I_{zz} = \frac{1}{10}m(a^2 + b^2)$ $\bar{I}_{xx} = \frac{3}{80}m(b^2 + c^2)$ $\bar{I}_{yy} = \frac{3}{80}m(a^2 + c^2)$ $\bar{I}_{zz} = \frac{3}{80}m(a^2 + b^2)$
Half Torus	$\bar{x} = \dfrac{a^2 + 4R^2}{2\pi R}$	$I_{xx} = I_{yy} = \frac{1}{2}mR^2 + \frac{5}{8}ma^2$ $I_{zz} = mR^2 + \frac{3}{4}ma^2$

INDEX

507